PRIVATE EMPIRE

EXXONMOBIL **AND** AMERICAN POWER

Steve Coll

THE PENGUIN PRESS

NEW YORK

2012

THE PENGUIN PRESS
Published by the Penguin Group
Penguin Group (USA) Inc., 375 Hudson Street, New York, New York 10014,
U.S.A. • Penguin Group (Canada), 90 Eglinton Avenue East, Suite 700, Toronto,
Ontario, Canada M4P 2Y3 (a division of Pearson Penguin Canada Inc.) • Penguin
Books Ltd, 80 Strand, London WC2R 0RL, England • Penguin Ireland,
25 St. Stephen's Green, Dublin 2, Ireland (a division of Penguin Books Ltd) •
Penguin Books Australia Ltd, 250 Camberwell Road, Camberwell, Victoria 3124,
Australia (a division of Pearson Australia Group Pty Ltd)• Penguin Books
India Pvt Ltd, 11 Community Centre, Panchsheel Park, New Delhi – 110 017,
India • Penguin Group (NZ), 67 Apollo Drive, Rosedale, Auckland 0632,
New Zealand (a division of Pearson New Zealand Ltd) • Penguin Books
(South Africa) (Pty) Ltd, 24 Sturdee Avenue, Rosebank,
Johannesburg 2196, South Africa

Penguin Books Ltd, Registered Offices:
80 Strand, London WC2R 0RL, England

First published in 2012 by The Penguin Press,
a member of Penguin Group (USA) Inc.

Map illustrations by Gene Thorpe

LIBRARY OF CONGRESS CATALOGING IN PUBLICATION DATA

Coll, Steve.
Private empire : ExxonMobil and American power / by Steve Coll.
p.   cm.
Includes bibliographical references and index.
ISBN 978-1-59420-335-0
1. Exxon Corporation.   2. Exxon Mobil Corporation.   3. Petroleum industry and
trade—Political aspects—United States.   4. Corporate power—United States.
5. Big business—United States. I. Title.
HD9569.E95C65   2012
338.7'6223380973—dc23
2011044722

Printed in the United States of America
1   3   5   7   9   10   8   6   4   2

DESIGNED BY AMANDA DEWEY

Contents

PART ONE:
THE END OF EASY OIL

PART TWO:
THE RISK CYCLE

List of Maps

Author's Note

Four journalists made important contributions to this book while working as researchers during the four-year life of the project. Ben Van Huevelen, who is now the managing editor of the *Iraq Oil Report*, worked on that subject and ExxonMobil's litigation with Venezuela, as well as on corporate responsibility issues in Africa and Indonesia. Megha Rajagopalan, a 2008 graduate of the University of Maryland who is now studying in China under the Fulbright Scholar Program, worked on global warming, the *Exxon Valdez* spill, and phthalate regulation; chapters five and twenty-two benefited greatly from her research. Ann O'Hanlon, a former *Washington Post* reporter who now works at the Justice Department, reported on many subjects, but especially on campaign finance and lobbying; her work particularly supported chapters three, seventeen, twenty-two, and twenty-three. Haley Cohen, a 2011 graduate of Yale University who is now on a university fellowship in Latin America, recontacted many interview subjects, checked facts and interpretations, and added fresh reporting throughout. The book benefited from other supporters and collaborators; the acknowledgments provide an accounting. I am grateful and deeply indebted to all.

Selected Cast of Characters

AT EXXONMOBIL

Russell Bowen, Maryland territory manager

John Paul Chaplin, lead country manager, Nigeria, circa 2005–2009

Ken Cohen, vice president of public affairs

Tim Cutt, lead country manager, Venezuela, 2005–2007

Steven K. Davidson, outside lawyer in Venezuelan litigation

Theresa Fariello, director of the Washington office, 2009 to present

Brian Flannery, astrophysicist, climate policy adviser

Rosemarie Forsythe, Russia adviser, planner for international political strategy

Edward G. Galante, senior executive, contender to succeed Raymond, retired 2006

Otto Harrison, lead executive on *Exxon Valdez* cleanup

Ralph Daniel Nelson, lead country manager, Saudi Arabia, 2001–2004, director of the Washington office, 2005–2009

Lee R. Raymond, chairman and chief executive, 1993–2005

James Rouse, director of the Washington office, late 1990s–2005

Ron Royal, lead country manager, Chad, circa 2006

James F. Sanders, lead outside lawyer, *Alban v. ExxonMobil*

Frank Sprow, vice president, Safety, Health, and Environment, 2000–2005

Sherri Stuewer, senior executive, environmental policy, 2006 to present

Rex Tillerson, upstream executive with responsibility for Russia, later chairman and chief executive, 2006 to present

Glenn Waller, lead country manager, Russia, circa 2003

Martin J. Weinstein, lead outside lawyer, *John Doe v. ExxonMobil*

Ronald I. Wilson, lead country manager, Indonesia

IN THE U.S. GOVERNMENT

Representative Joe Barton, R-Texas, 1984 to present

George W. Bush, president, 2001–2009

Richard B. Cheney, vice president, 2001–2009

Representative John Dingell, D-Mich., 1955 to present

Don Evans, secretary of commerce, 2001–2005

Douglas Feith, undersecretary of defense for policy, 2001–2005

Robert Gelbard, ambassador to Indonesia, 1999–2001

Christopher Goldthwait, ambassador to Chad, 1999–2004

Barack Obama, president, 2009–

Judge Louis F. Oberdorfer, United States District Court, Washington, D.C.

Colin Powell, secretary of state, 2001–2005

Anton Smith, United States chargé d'affaires, Equatorial Guinea, 2008–2009

Alexander Vershbow, ambassador to Russia, 2001–2005

Marc Wall, ambassador to Chad, 2004–2007

Paul Wolfowitz, deputy secretary of defense, 2001–2005; president, World Bank, 2005–2007

IN AFRICA

Victor Attah, governor of Akwa Ibom State, Nigeria, 1999–2007

Idris Déby, president of Chad, 1990 to present

Simon Mann, former British Army officer, led coup attempt in Equatorial Guinea

Teodoro Obiang Nguema, president of Equatorial Guinea, 1979 to present

IN ALASKA

Joseph Hazelwood, Jr., captain, *Exxon Valdez*
Mandy Lindeberg, biologist, N.O.A.A.
Jeffrey Short, chemist, N.O.A.A.

IN INDONESIA

Abu Jack, guerrilla commander, Free Aceh Movement
Hasan di Tiro, leader of Free Aceh Movement

IN MARYLAND

Andrea Loiero, manager, Jacksonville Exxon
Stephen Snyder, plaintiffs' lawyer

IN THE MIDDLE EAST

Abdullah Bin Abdul Aziz, crown prince, later king of Saudi Arabia
Thamir Ghadhban, senior official, Iraq Ministry of Oil
Ali Al-Naimi, oil minister of Saudi Arabia
Prince Saud Al-Faisal, foreign minister of Saudi Arabia
Hussain Al-Shahristani, deputy prime minister, Iraq

IN RUSSIA

Mikhail Khodorkovsky, president, Yukos Oil
Bruce Misamore, chief financial officer, Yukos Oil
Vladimir Putin, president, Russian Federation, 2000–2008

IN VENEZUELA

Hugo Chavez, president, Venezuela, 1999 to present
Joseph Pizzurro, Venezuela's outside lawyer in litigation with ExxonMobil

PRIVATE EMPIRE

Prologue

"I'm Going to the
White House on This"

A s the *Exxon Valdez* churned through chalky turquoise port waters
toward the Gulf of Alaska, Captain Joseph Hazelwood descended to
his quarters. It was shortly after 9:30 p.m. on the evening of March
23, 1989, and he had some paperwork to complete, he told his subordinates. He was a taciturn man, forty-two years old, balding, about six feet
and 180 pounds. He dangled the Marlboro cigarette he smoked on the
corner of his lips. His father had flown torpedo bombers for the United
States Marine Corps in the Western Pacific and then served as an international pilot for Pan American World Airways. Joseph Jr. won admission to
the elite State University of New York Maritime College; the carefree
notation in his college yearbook read, "It Will Never Happen to Me." He
scored 138 on an I.Q. test. While at sea he read widely; in conversation,
he quoted Stonewall Jackson and Oscar Wilde. He had by now sailed for
the Exxon Corporation for twenty-one years, ten of those as an oil tanker
captain.[1]

He was attempting to recover that spring from what he would later
call a "midlife crisis." It had taken hold of him several years before. Long
stretches at sea had caused him to miss much of his daughter's childhood,

and this weighed on him. His wife, he recalled later, "detected that I was moodier than I had been before." He drank heavily—four or five doubles before dinner, wine with the meal, then several doubles afterward—but he did not feel immobilized by alcohol. Even after such a drinking regimen, although he could "detect a little clumsiness on my part," he "didn't trip over any furniture" and he "wasn't blotto," as he put it. While ashore, he periodically drove while intoxicated, attracted the attention of police several times, and lost his driver's license. He sensed that he might be in some sort of descent: "I didn't know what I was suffering from, if I was suffering from something." An Exxon supervisor told him, "If you've got a problem, take care of it." In 1985, he had checked himself into a New York hospital and underwent treatment for mild depression and alcohol abuse. Afterward he attended Alcoholics Anonymous meetings but continued periodically to drink. Exxon executives said that they had started to monitor his alcohol intake, prompted by incidents such as one in which the captain was overheard ordering beer over an Exxon ship's radio, but Hazelwood remained in service as a tanker captain and said he had no indication that he was being monitored by anyone. In the port town of Valdez on the afternoon of March 23 he drank what he would recall as two or three vodkas at the Pipeline Club before passing unexamined through the oil terminal gate and boarding his ship.[2]

The livelihood Hazelwood put at risk by his drinking was a privileged one; his salary at Exxon was about $180,000 a year, including benefits. He was one among many thousands of Americans whose incomes that spring could be traced in part to the work of a British Petroleum geologic field party that had surveyed Alaska's Brooks Range, north of the Arctic Circle, in the summer of 1958. The Suez Crisis and turmoil in Iran (the latter partially engineered by the Central Intelligence Agency) had made plain to B.P.'s executives that their oil holdings in the Middle East were politically insecure. Alaska's storm-swept seas and icy glaciers might look forbidding, but at least they were situated in a nation that welcomed private capital. American government surveys had suggested for years that Alaska's north was rich with oil and natural gas; B.P. was among the first of the major international oil companies to bear the uncertainties of the harsh climate and invest. By the early 1970s, it had established a large position as a leaseholder on Alaska's North Slope. Transport was the major obstacle to maximized profits in a region iced over for months at a time. Oilmen had talked for years about wild-eyed schemes to build an over-land pipeline from the Arctic to the south, but the project would require money and political alignments that seemed preposterously ambitious. Only another Middle Eastern crisis—the oil embargo of 1973, directed by Arab producers at the United States over its support of Israel—at last spurred construction of the Trans-Alaska Pipeline System to carry crude eight hundred miles from Prudhoe Bay across permafrost to the ice-free port of Valdez. To finance and operate the pipeline, B.P. formed a consor-tium with Exxon's precursor Humble Oil and with Atlantic Richfield. The first oil flowed in June and reached the Valdez Marine Terminal on July 28, 1977.[3]

A dozen years later, it poured through Valdez at a rate of about 2 million barrels per day—an amount equal to more than a quarter of all of America's domestic crude oil production. The Valdez terminal had grown into a labyrinth of pipes, oval storage tanks, and strings of festive-looking white safety lights—an improbable man-made installation tucked on a rise amid snow-draped mountain crags and majestic glaciers. Tankers as long as several football fields passed to and from the docks one after another. The Coast Guard funneled them through a ten-mile-wide ship-ping lane in Prince William Sound, an inland sea teeming with salmon, halibut, whales, seals, sea lions, porpoises, and sea otters. Inbound traffic traveled in a corridor to the east, outbound traffic to the west. That Thurs-day, the *Brooklyn* and the *ARCO Juneau* had departed Valdez for the Pacific only hours before Hazelwood embarked with a load of 1,264,155 barrels of crude. It was a misty night, but the winds were light, the seas were calm, and visibility extended eight miles. Hazelwood had navigated this passage at least a hundred times before.[4]

He returned to the bridge shortly after 11 p.m. His Valdez port pilot had disembarked, and a tugboat escort had also peeled away. Bligh Island now lay ahead to the southeast, shaped in the water like a sleeping croco-dile with a curling snout. To the island's west a red light pulsed every four seconds to mark Bligh Reef, which spread beneath the surface at between one quarter and nine fathoms. (The draft of a loaded oil tanker could be ten fathoms or more.) White icebergs from the Columbia Glacier bobbed ahead as well, visible to eye and radar, and several of them now appeared to block the outbound shipping lane. Hazelwood decided on a common maneuver, one taken earlier without incident by the two ships ahead of him. The *Exxon Valdez* would turn south across the inbound shipping lane toward Busby Island, near Bligh, evade the ice, and then turn back to the outbound corridor toward the Hinchinbrook Entrance and the open sea. The captain radioed the Coast Guard's Vessel Traffic Service to secure permission. "Judging by our radar, I will probably divert . . . and end up in the inbound lane if there's no conflicting traffic—over," he announced.[5]

The Coast Guard's monitors did not question him. One of the men on duty was on a coffee break. In any event, it was not as if he or anyone else in the Coast Guard could easily track the *Exxon Valdez*'s movements once it reached the vicinity of Bligh. Budget cuts during the 1980s had left the Vessel Traffic Service without a radar system that could follow ships reliably once they moved thirty or more miles south of Valdez. Even if the Coast Guard had possessed such radar, it might not have mattered; blood tests administered later to the two men on duty showed traces of marijuana and alcohol.[6] Regulators and the regulated had fallen into a slothful embrace, reflecting a national political atmosphere that emphasized the benefits of light government oversight. There was intense pressure to reduce costs within the oil industry. A decade of operations around Valdez had passed free of major accidents.

On the bridge, Hazelwood told Gregory Cousins, his third mate, "Bring it down to abeam of Busby and then cut back to the lanes."

Cousins was an experienced sailor who had made the passage through Prince William more than two dozen times, but he was not legally qualified to take control of the ship in these waters.

"Do you feel comfortable with what we are going to do?" Hazelwood asked him.

"Yes."

"Do you feel good enough that I can go below and get some paperwork out of the way?"

"I feel quite comfortable."[7]

At 11:50 p.m., Hazelwood left the bridge again for his quarters. Like the drinks he had downed less than four hours before boarding the ship, this decision was a violation of Exxon Shipping Company policy. He said later that he left "because there wasn't a compelling reason to stay."[8]

What occurred on the bridge in the minutes that followed would never be fully explained. Cousins, helmsman Robert Kagan, and other crew members became confused, attempted to turn the ship as their charts instructed, made technical mistakes, and soon lost track of their position altogether. At last Cousins telephoned Hazelwood: "I think we're in serious trouble."[9]

A terrible shock and sound engulfed them around ten minutes after midnight. Cousins felt a series of sharp jolts, heard some of the ship's relief valves open, and smelled oil. The ship's chief mate, James Kunkel, banged on a crew member's door to wake him: "Vessel aground. We're fucked." Bligh Reef had cut open the ship's belly across its length. Oil pools surfaced on the dark sea.

Hazelwood raced upstairs. He saw two officers peering overboard at the gushing oil, its roiling blackness illuminated by a spotlight. He retreated to a toilet and vomited. He found that he had trouble catching his breath; he felt that he had "been hit in the breadbasket with a ten-pound maul." He knew that "the world as I'd known it had come to an end."

About eighteen minutes after the first sounds of steel on rock echoed through his ship, Hazelwood radioed the Coast Guard. "We've—should be on your radar there," he said. "We fetched up hard aground north of Goose Island off Bligh Reef. Evidently we're leaking some oil and we're going to be here for a while."[10]

The watch stander in Valdez telephoned the Coast Guard's commander, Steve McCall. "I've got the *Exxon Valdez* hard aground Bligh Reef."

"Are you serious?"

"I'm serious as a heart attack."

Just over thirty minutes later McCall pulled off the Valdez dock in a fast boat with two other Coast Guard officers, a representative of the state Department of Environmental Conservation, and two local pilots. The night had turned crisp and clear, and when they reached the *Exxon Valdez*, "you could see oil bubbling out from underneath," recalled Mark Delozier, one of the Coast Guard officers aboard. As the oil surfaced it "made a gurgling noise and big bloops would leap right out of the water two to four feet. Then it would settle down to the surface." The slick around the ship was already twelve to eighteen inches thick.

They boarded and approached Hazelwood. "How did this happen?"

"You're looking at it."

Delozier stepped back and pulled a colleague into a huddle. "Did you smell what I smelled?"

"Yeah."

"We need to get someone out here to do an alcohol test on the captain as well as the crew."[11]

In the days ahead Hazelwood's intoxication would simplify perceptions of the accident. It provided Exxon a means to narrow its responsibility. The corporation soon dispatched a telegram to its tanker captain informing him that he had been fired. (Hazelwood later testified that he learned of his dismissal only through media reports.) For late-night comedians the drunk-driving imagery proved irresistible. Hazelwood's number-one excuse, according to a David Letterman Top Ten list: "I was just trying to scrape some ice off the reef for my margarita." It did not take long for government investigators to discover, however, that the grounding had been caused by a more complex chain of human error, abetted by inadequate regulations and corporate safety systems.

When Steve Cowper, the governor of Alaska, arrived on the scene the next afternoon, one of his colleagues told him that Hazelwood "may have been drinking but I'm not sure that had anything to do with it." A Coast Guard study would soon conclude that the service's own multiple deficiencies contributed significantly, along with a pattern of expediency within the oil and shipping industries: "The game rules now are for a professional investor to move freely within the marketplace spending as little as is necessary. Today's adage is to do more with less, make two tankers do the work done by three previously."

That had certainly been the adage at Exxon in the years before the wreck. In 1982, the corporation employed 182,000 people. Unexpectedly, oil prices dropped. In response, chief executive Lawrence G. Rawl advanced a slashing campaign begun by his predecessor, Clifton C. Garvin Jr. The campaign eliminated about 80,000 jobs by 1989—more than 40 percent of the workforce in just seven years. At Exxon's headquarters in a white skyscraper on Sixth Avenue in New York City, employment fell from 1,362 to 330. The corporation's top environmental officer at headquarters was demoted; his staff was reorganized and absorbed by a research group. The experts in oil spill response that wrote Exxon's manual for disaster management also lost their jobs. The cuts buoyed Exxon's financial performance at a time when competitors struggled. In 1987, Exxon reported more annual profit per employee than any other major American corporation.

The National Transportation Safety Board concluded that third mate Cousins and his shipmates were overworked and that cutbacks of the number of crew assigned to Exxon tankers had compromised the ship's ability to detect potential hazards. The N.T.S.B. cast blame not only on Exxon Shipping, but also on the state of Alaska, the Coast Guard, and the performance of the individuals aboard, including the captain. Hazelwood's decision to leave the bridge was a factor in the accident, and his drinking may have contributed to that decision, but the failings that led the *Exxon Valdez* onto the rocks ran deeper than his own.[12]

B y Saturday, Exxon executives estimated that 240,000 barrels of crude had poured into Prince William Sound, more than had ever been dumped into American waters at one time. Television camera crews arrived in Valdez by the scores and beamed out images of saturated birds and blackened sea otters. The immediate damage to wildlife caused by a spill arises from direct contact with the oil while it is exposed on the water or the shore. A marine mammal such as an otter will lose vital insulation on the contaminated section of its fur, preen itself for relief, ingest the petroleum, and soon die. Nearly 1,000 sea otter carcasses appeared in Prince William Sound after the spill. Federal scientists never established an exact count of the total dead, but they estimated that about 2,800 sea otters perished. With their furry faces and pleading black eyes the otters became the symbols of a broader wildlife massacre. Scientists later estimated that about 300 harbor seals, 250 bald eagles, 22 killer whales, 250,000 seabirds, and billions of salmon and herring eggs were destroyed by the initial exposure to *Exxon Valdez* oil.

On Saturday afternoon, March 25, Don Cornett, Exxon's director of its office in Anchorage, a silver-haired veteran of the corporation, telephoned George Nelson, one of his counterparts at the pipeline consortium. Cornett would play a leading role in representing Exxon before the media and the public in the days and years ahead. His past assignments had included a stint as an environmental manager in the corporation's marine department; he had experience with oil spills. He flew by chartered helicopter to Valdez and hit the telephones. He would later reflect that because of the chaotic flow of information, "it was hard to always make the right calls. . . . The emotions surrounding the damage to the mammals and the birds were totally understandable and totally unmanageable." That Saturday, because he was calling from an emergency response center, his telephone line was recorded.

"Are you getting a lot of press contacts into your office?" Nelson asked him.

"Jesus Christ. I can't get off the telephone."

"I thought you probably were, but many, many hundreds of them—how many of them are down in Valdez? A whole bunch, aren't there?"

"'Yeah, yeah. Well, they're taking some in Houston, some in New York, and I'm taking some here. . . . They're getting prank calls."

"They're getting what?"

"Prank calls."

"Like what?"

"'You dirty bastard.' . . . We're getting those over here too," Cornett said.

"Well, this is going to be a public relations nightmare."

"Nightmare," Cornett agreed.

"Nightmare, to say the least. Yes, to say the very least."

"Do you know how I feel?"

"How?"

"Do you remember when Patton looked out over the battlefield and said, 'God help me, I do love it so.' . . . When they were going to invade

Europe, he said, 'God wouldn't let this happen and not make me be in on it.' That's the way I feel."[13]

Lee R. Raymond, the president of Exxon Corporation and then its number-two executive, heard about the grounding of the *Exxon Valdez* while on company business in Jacksonville, Florida. Raymond had helped to design the ambitious reorganization plan that had eliminated more than 40 percent of the corporation's employees in the years before the wreck. He was in Jacksonville because Lawrence Rawl had sent him there to scout real estate. As part of their campaign to remake Exxon, Rawl and Raymond had decided to move the company's head office out of Manhattan; Jacksonville was a possible destination. As he absorbed the news from Alaska, Raymond said later, he was "chagrined . . . horrified and to an extent devastated." His wife, Charlene, told him, "It's the first time I have ever been embarrassed that we work for Exxon."[14]

Raymond was ill with a severe spring cold and could not fly for a few days, on doctor's orders. Pent up in Jacksonville, restless, he began to assess the cause of the accident and to coordinate Exxon's response. Raymond had grown up in Watertown, South Dakota. His father was a railroad engineer who drove trains between Watertown and Aberdeen. In high school the younger Raymond decided to study chemistry, particularly its mathematical aspects; he attended the University of Wisconsin as an undergraduate and ultimately earned a doctoral degree in chemical engineering at the University of Minnesota as a National Science Foundation fellow and on other scholarships. He took his first job at Exxon and rose through the ranks. He could mist up when speaking about his father or the people who worked for him at Exxon. Normally, however, he did not come across as a sentimental man and could be a blunt and demanding manager. At the time of the spill he was fifty years old and had worked at Exxon for twenty-five years.

In time Raymond would draw a number of conclusions about the *Exxon Valdez*. One of his earliest assessments was that environmentalists and confused politicians in Alaska—particularly Alaska's governor,

Cowper—had prevented Exxon from reducing the effects of the disaster by refusing to allow the company to spray chemical dispersants on the oil slick during the first days, as provided for in a spill response plan previously filed by Exxon with the state. Chemical dispersants do not eliminate oil, but if applied correctly in favorable conditions, they can break up concentrations and drive oil droplets underwater. That can reduce the impact on birds and marine mammals that feed or travel at the surface. Chemically dispersed oil may also be less likely to wash up on beaches in dangerous concentrations. Oil driven beneath the surface might harm fish or other subsurface life, however. Fisheries were the most important source of income and employment in Prince William Sound. Also, several factors can limit the effectiveness of dispersants. The chemicals are less effective in cold waters than in warmer waters. They are typically released by aerial spraying. If the seas below are calm, as they were in Prince William Sound during the first days after the *Exxon Valdez* gashed itself on Bligh Reef, the chemicals may not churn and mix adequately. If the particular composition of spilled oil or the chemistry of the water is unknown, then the impact of dispersants may also be uncertain.

Chaos reigned around the decision makers on-site in Valdez on Friday and Saturday. Local fishermen arrived at hastily arranged press conferences at the town's civic center, shouted questions, made speeches, and threatened to take the cleanup into their own hands. On the stage at the press conferences stood Exxon employees "wearing three-piece suits," recalled Dennis Kelso, then in charge of Alaska's Department of Environmental Conservation. Meanwhile, in the crowd, "there was so much fear and anger you could hear it crackling through the audience."[15] Valdez was a small and isolated town; local oil and government representatives struggled to make decisions while consulting with their superiors over long-distance telephone lines. There were multiple sources of overlapping authority: the Coast Guard, the state of Alaska, the pipeline consortium, and Exxon. British Petroleum, the lead owner of the pipeline consortium, Alyeska, was supposed to have ensured that preparations for response to an oil spill in the sound would be adequate, but the consortium had failed to equip tiny Valdez with adequate boats, vehicles, booms, leased aircraft, and other vital materials.

The Coast Guard had emergency procedures to respond to oil spills and had supervised cleanups in the Gulf of Mexico and elsewhere, but it, too, lacked the equipment to take full charge in Prince William Sound. Exxon lacked the means as well, because it had been relying on the B.P.-led pipeline consortium. Exxon "had more experience," as a senior Coast Guard officer put it, so the Coast Guard yielded to the corporation. Raymond later said that Exxon in fact had access to "a lot of cleanup equipment on the ground," but he blamed Alaskan officials for not granting permission to use it. He also deflected any suggestion that his and Rawl's decision to lay off Exxon's oil spill specialists during their cost-cutting spree had hindered the corporation's response: "We have people all over the world trained to handle oil spills, even if they don't have the exact title of oil spill specialist."

Fishermen in Valdez believed that spraying chemical dispersants would do more harm than good to salmon and other fish populations on which they depended. Hundreds of thousands of young salmon were about to be released into the sound at the start of their annual migration—if they swam through toxic oil driven beneath the surface by dispersants, they might be destroyed before they reached maturity. Locals voiced these fears to Steve Cowper and his advisers. Cowper authorized a few tests, which did not turn out promisingly—the chemicals were dumped accidentally onto cleanup crews. Coast Guard officers were not enthusiastic about using dispersants; neither Alaskan state officials nor Environmental Protection Agency specialists recommended going forward aggressively.

Lee Raymond fumed. Running limited experiments in these circumstances was "like testing the fire hose after the house is on fire," he thought. He accepted that in cases where the chemistry of oil and water was unknown, a dispersant plan might best be implemented cautiously—but that was not the case here. "There is only one kind of crude on a vessel leaving Valdez," he said later. He was referring to the fact that only well-known Alaska crude blends came down the pipeline and their chemistry had been well studied. "It is one of the most susceptible of all crude oils" to dispersants. "Therefore, that information didn't need to be established."

Early in his Exxon career Raymond had worked at the lab that developed COREXIT, the dispersant available to use in the sound; he felt he knew the issues from the molecular level on up. But Exxon executives on the scene kept telling him that "the state and special interest groups trying to influence the state" were opposed to using Exxon's previously approved dispersants, and Dennis Kelso, in particular, was "flat opposed."[16] Alaskan officials and federal scientists later concluded that there had been neither enough chemicals nor delivery systems to make a decisive impact in the time available, but Raymond held just as firmly to the opposite view. The deeper this conviction took hold of him, the more it seemed to harden his belief that once the oil began to pour into Prince William Sound, the corporation acted blamelessly, while environmentalists did not.

"I asked you a moment ago . . . what, if anything, you felt Exxon did wrong, and I think your answer began by saying, well, you didn't really think it was a matter of right and wrong," Jim Sherman, a lawyer for the State of Alaska, asked Raymond later at a deposition.

"Well, I don't mean to be argumentative, but assigning blame isn't the same as being right or wrong," Raymond said.

"Well, do you think the State of Alaska's actions in the first seventy-two hours after the spill in regard to dispersant use were wrong?"

"My own view is that dispersants should have been applied. If you are suggesting that the state didn't think they should be applied, then I guess we would have a difference of view. And since I'm right, I guess by your supposition you are wrong."

"By those same terms, did Exxon do anything in the course of the weeks that followed the spill that was wrong?"

"The state may have a view on that and I have a different view."

"I'm asking for your view, sir."

"I think— I'm never going to say that we are always doing everything exactly right. I would be naive to do that; but if you are asking me, are there any major decision points that we faced in how to respond to that spill, that in hindsight we would go back and say we think we were wrong, and I don't think there are any."[17]

As the oil spread, Samuel Skinner, the secretary of transportation, summoned Admiral Paul Yost, commandant of the United States Coast Guard, to a meeting at the White House. They waited in the West Wing for President George H. W. Bush, a former oil wildcatter who earned his fortune in West Texas before embarking on his career in politics, intelligence, and diplomacy. That spring President Bush was preoccupied by events abroad—spreading dissent in Eastern and Central Europe, pro-democracy students camped out in Beijing's Tiananmen Square, and the rising radicalism of Mikhail Gorbachev's perestroika.

Admiral Yost had made his professional reputation as a patrol boat commander in Vietnam. As Coast Guard commandant since 1986, he had pulled the service toward military discipline. He banned beards, earning the enmity of a generation of officers, and he moved to install naval weapons systems aboard Coast Guard vessels.

In the Oval Office, Yost briefed President Bush on the militarylike dimensions of the *Exxon Valdez* crisis, "trying to explain what was needed to mobilize in a major oil spill, and what Valdez looked like, with one or two motels, and one or two little restaurants," as Yost recalled it.

Bush looked at his watch. "I've had the German ambassador waiting for ten minutes," he said. "I've got to go see him."

The president turned to his chief of staff, John Sununu, a former governor of New Hampshire. "You take this into your office and get this thing moving," he said.

Yost and Skinner trailed Sununu through the West Wing. When they sat down, Sununu told him, "Admiral, you're going to Alaska and you're going to supervise this oil spill."

"Mr. Sununu, I'm not going to Valdez," Yost answered. "I can't run the Coast Guard from Valdez. It's a worldwide operation."

Sam Skinner laid his hand on Yost's shoulder. "Paul, you're going to Alaska."[18]

By the time he arrived, the debate about chemical dispersants was no longer relevant.

On the night of Sunday, March 26, about seventy-two hours after the initial grounding, a fierce spring storm raged through Prince William Sound. Southwesterly gales up to seventy miles per hour blew and scattered the oil from the sea surfaces around Bligh across to rocky island beaches dozens of miles away, on the far side of the sound: Knight Island, Eleanor Island, Ingot Island, Disk Island, Naked Island. It was as if someone had blown very hard on an ashtray and scattered its ashes. The swirling winds were so strong that crude even appeared in treetops on the distant islands. The gales rendered the earlier debates about chemical dispersants academic; if they had been applied all out during the first three days, they might have reduced by a little the amount of oil that the winds blew, but not enough to forestall catastrophe.

The storm transformed the cleanup. Now the challenge became to remove the contamination from dozens of beaches during the summer months before the snow and harsh weather of late autumn returned. Otto Harrison, a bespectacled veteran of Exxon's international operations and offshore oil production, led the effort. Harrison was to work alongside Commandant Yost. Exxon reported that it spent $2.1 billion on cleanup operations in 1989, and even some critics of the company credited the vigor of its efforts once the operation became organized. Disagreements persisted, however, about whether Exxon was doing all that it could.

In public, Exxon president Lee Raymond suggested that the corporation would take its orders about its cleanup decisions from the Coast Guard's on-scene commander. "That is the man we look to," he said. "That is the man who approves our plans."[19]

In private, Exxon and the Coast Guard found themselves in conflict. The commandant "didn't get along with [Otto] Harrison at all," Yost recalled. His Exxon counterpart "was a big man. He made decisions very quickly. He stood by his guns and he wouldn't be pushed around. . . . I told him what to do and he sometimes did what he wanted. It was that kind of a relationship, but he was good. He was plenty good."

To build political support for Harrison's decisions, Raymond and Lawrence Rawl flew regularly to Washington to meet with Sununu, Skinner,

Interior secretary Manuel Lujan Jr., and Environmental Protection Agency director William K. Reilly. In Juneau they pressed Governor Cowper and his aides to back Exxon's cleanup plan.

Commandant Yost argued about the number of workers they should deploy on the beaches. Yost told Harrison that he wanted five thousand people hired for the summer crews.

"Admiral, I can't support five thousand people on the beaches," Harrison replied.

"Then get the support up there—that's your problem. I want five thousand on the beaches."

"I'm going to the White House on this," the Exxon executive said.

"Go ahead," Yost replied.[20]

Soon the commandant received a call to return to Washington. He met with Skinner again, and the Transportation secretary accompanied him to the White House to see John Sununu.

"Admiral," Sununu announced, "I'm not going to require five thousand people on the beach."

"In that case," the Coast Guard commandant answered, "I can't guarantee the president that this is going to be cleaned up this summer."

He added, "Let the record show that you've got a very unhappy commandant."[21]

At a congressional hearing that spring, Senator Slade Gorton of Washington State pointed out to Exxon's chairman that Japanese executives routinely accepted responsibility for serious corporate failings, no matter the cause, by resigning from their positions. Gorton asked Lawrence Rawl whether he had considered doing the same.

Rawl was the Irish American son of a New Jersey truck driver who had enlisted in the United States Marines, made sergeant, and then became a petroleum engineer at the University of Oklahoma using the G.I. bill. He had spent his entire professional life at Exxon. "A lot of Japanese kill themselves as well," Rawl answered Gorton, "and I refuse to do that."[22]

Lee Raymond never surrendered his conviction that irrational environmentalists had exacerbated Exxon's problems in Alaska by their opposition to dispersant use, but he did scrutinize the catastrophe for other lessons. One of these was that "no matter what you decide is the right thing to do in terms of trying to deal with the spill, you have to get after it very quickly. The lesson learned here was to try and make sure that there were procedures both in the company and in the respective governments that they knew and we knew that if an incident were to happen, exactly what to do and how to do it."

Raymond conceded that the *Exxon Valdez* episode suggested the need for "perhaps a rebalancing of risk-reward in many of our operations." The risks of accidents of the sort that poisoned Prince William Sound that spring and summer of 1989, he said three years later, "apparently are much higher than anybody in either our company or the industry had envisioned."[23]

John Browne—later Lord Browne of Madingley—joined British Petroleum, as it was then known, in 1966, as a university apprentice. In comparison with Lee Raymond of South Dakota, Browne was an international and cosmopolitan figure. He was half Hungarian, half British; he was born in Germany and spent parts of his childhood in Singapore and Iran. As a young oil executive assigned to New York, he lived in Greenwich Village, taught himself to cook, and spent his spare time at the opera and in Soho art galleries. He was a charismatic young man with floppy ears and a mop of dark hair. As he rose through B.P.'s leadership ranks, Browne began to think that corporations "must behave consistently with the will of society," as he put it. He puzzled over what that insight might imply for the practices of a large oil corporation.

The oil in the holding tanks of the *Exxon Valdez* had been pumped from Arctic Alaskan fields partially owned by B.P. On the morning of the ship's grounding, Browne happened to be asleep at a company base camp on the North Slope, where he had come to say good-bye to colleagues as he departed for a new assignment. At 5:00 a.m., B.P.'s Alaska general manager woke him up. "We've got a message," he reported. "There's some oil seeping around Valdez. It's from a tanker and they say it's Exxon's. But no one seems to be doing anything."

Browne would recall that he "knew right away that something terrible had happened." He boarded a plane and flew over Prince William Sound, "home to precious wildlife" where "whales would be returning from the warm water in the south soon." As he peered down from above, he could see that white ice floes already were tinged with black. It seemed to him that too little was happening by way of response and cleanup. Where were the response boats and the booms to keep oil off the beaches? In fact, that was a question that British Petroleum's senior executives should have been able to answer; the inadequate response was their failure, too, but it would soon be overshadowed by Exxon's culpability.

Browne sensed that the spill's "repercussions for the industry would be huge. It was the start of a new chapter."

The *Exxon Valdez* had "damaged not just a fragile environment but also the flimsy trust in oil companies." Environmental groups would "have a field day," he expected. Unfortunately, "it was no use" saying to them "that B.P. was better than its competitors. The industry was now measured by its weakest member, the one with the worst reputation. That oil company was now Exxon."[24]

A few days before the *Exxon Valdez* ran onto Bligh Reef, tens of thousands of Hungarians marched through Budapest. The demonstrators turned the commemoration of an 1848 uprising against Austrian rule into a revolt against Soviet-backed communism. "Resign!" they shouted outside downtown buildings housing Communist Party bureaucrats. "Freedom! . . . No more shall we be slaves!" They carried flags from Hungary's pre-Communist era and demanded the withdrawal of Soviet military forces. "Ivan, Aren't You Homesick?" and "Legal State, Not a Police State" declared their protest signs.

The defiant march added to the cracks spreading that spring through the structures of global politics. The Berlin Wall fell a few months later, in November. The Soviet Union fissured and then disappeared. Democratic and free-market revolutions and revivals swept through Central Europe, Africa, Asia, and Latin America. Ethnic, religious, and territorial conflicts, long subdued by the cold war, erupted one after another. The world was remade, tossed, liberated—and reopened for international business.

The *Valdez* wreck stunned Exxon and its rising leader, Lee Raymond. The disaster would change the corporation profoundly. Internal reforms imposed by Raymond in response to the accident would turn one of America's oldest, most rigid corporations into an even harder, leaner place of rule books and fear-inspiring management techniques. At the same time, Raymond and the rest of Exxon's leaders would gradually pass through the introspection triggered by the *Valdez* spill and seek out the oil and gas plays that opened so unexpectedly after 1989. An age of empire beckoned America and Exxon alike.

In a bracingly short time, Anglo-American optimism and idealism about free markets, foreign investment, and the rule of law found adherents in the most unlikely world capitals. Brand-new nations brimming with oil and gas and others previously closed to Western corporations hung out FOR LEASE signs to lure geologists from Houston and London: Russia, Kazakhstan, Azerbaijan, Angola, Qatar, and tiny Equatorial Guinea, on the West African coast, soon to market itself through its Washington lobbyists as the "Kuwait of Africa." These post–cold war opportunities for American, British, French, and Italian oil companies could be ambiguous, risky, and sometimes fleeting. Resentful nationalism and suspicion of the United States and Europe persisted in many capitals of the new oil powers. State-owned petroleum companies from China, India, Brazil, and elsewhere were rising quickly as competitors. Exxon might be America's largest and most powerful oil corporation, but it would require all the political influence, financial resources, dazzling technology, speed, and stamina that its leaders could muster to seize the lucrative oil deals made possible by communism's fall and global capitalism's revival.

The United States now stood unchallenged as a worldwide military power. Exxon's empire would increasingly overlap with America's, but the two were hardly contiguous. Pentagon policy, after the Soviet Union's demise, sought to keep international sea-lanes free; to reduce the global danger of nuclear war, terrorism, and transnational crime; to manage or contain Russia and China; to secure Israel; and to foster, against long odds, a stable Middle East from which oil supplies vital for global economic growth could flow freely. Exxon benefited from the new markets and global commerce that American military hegemony now protected. Yet the corporation's activity also complicated American foreign policy; Exxon's far-flung interests were at times distinct from Washington's. Lee Raymond would manage Exxon's global position after 1989 as a confident sovereign, a peer of the White House's rotating occupants. Raymond aligned Exxon with America, but he was not always in sync; he was more akin to the president of France or the chancellor of Germany. He did not manage the corporation as a subordinate instrument of American foreign policy; his was a private empire.

Exxon's power within the United States derived from an independent, even rebellious lineage. The corporation had been hived off from John D. Rockefeller's Standard Oil monopoly in 1911, after a bruising antitrust campaign led by economic reformers and populist politicians. The visceral hostility toward Washington sometimes eschewed by Exxon executives eight decades later suggested some of them had still not gotten over it.

Exxon's size and the nature of its business model meant that it functioned as a corporate state within the American state. Like its forebearer, Standard, Exxon proved across decades that it was one of the most powerful businesses ever produced by American capitalism. From the 1950s through the end of the cold war, Exxon ranked year after year as one of the country's very largest and most profitable corporations, always in the top five of the annual Fortune 500 lists. Its profit performance proved far more consistent and durable than that of other great corporate behemoths of America's postwar boom, such as General Motors, United States Steel, and I.B.M. In 1959, Exxon ranked as the second-largest American corporation by revenue and profit; four decades later it was third. And more than any of its corporate peers, Exxon's trajectory now pointed straight up. The corporation's revenues would grow fourfold during the two decades after the fall of the Berlin Wall, and its profits would smash all American records.

As it expanded, Exxon refined its own foreign, security, and economic policies. In some of the faraway countries where it did business, because of the scale of its investments, Exxon's sway over local politics and security was greater than that of the United States embassy. In impoverished African countries increasingly important to Exxon's strategy, such as Chad, the weight of the corporation's investments and the cash flow it shared with local governments overwhelmed the economy and became the central prize in violent local contests for power. In Moscow and Beijing, Exxon's independent power and negotiating agenda competed with and sometimes attracted more attention than the démarches issued by American secretaries of state. Yet the corporation could also be insular and even passive in the faraway places where it acquired and produced oil and gas. It fenced off local operations and separated its workforce from upheaval outside its gates. If its oil flowed and its contract terms remained intact, then Exxon often followed a directive of minimal interference in local politics, especially if those politics were controversial, as in the case of the African dictatorships with which the corporation partnered, or the countries, such as Indonesia and Venezuela, where civil conflict swirled around Exxon properties. In Washington, Exxon was a more confident and explicit political actor. The corporation's lobbyists bent and shaped American foreign policy, as well as economic, climate, chemical, and environmental regulation. Exxon maintained all-weather alliances with sympathetic American politicians while calling as little attention to its influence as possible.

The cold war's end signaled a coming era when nongovernmental actors—corporations, philanthropies, terrorist cells, and media networks—all gained relative power. Exxon's size, insularity, and ideology made its position distinct. Unlike Walmart or Google (to name two other multinational corporations that would rise after 1989 to global influence), the object of Exxon's business model lay buried beneath the earth. Exxon drilled holes in the ground and then operated its oil and gas wells for many years, and so its business imperatives were linked to the control of physical territory. Increasingly, the oil and gas Exxon produced was located in poor or unstable countries. Its treasure was subject to capture or political theft by coup makers or guerrilla movements, and so the corporation became involved in small wars and kidnapping rackets that many other international companies could gratefully avoid.

The time horizons for Exxon's investments stretched out longer than those of almost any government it lobbied. "We see governments come and go," Lee Raymond once remarked, an observation that was particularly true of Washington, with its constitutionally term-limited presidency.[25] Exxon's investments in a particular oil and gas field could be premised on a production life span of forty or more years. During that time, the United States might change its president and its foreign and energy policies at least half a dozen times. Overseas, a project's host country might pass through multiple coups and political upheavals during the same four decades. It behooved Exxon to develop influence and lobbying strategies to manage or evade political volatility.

American spies and diplomats who occasionally migrated to work at Exxon discovered a corporate system of secrecy, nondisclosure agreements, and internal security that matched some of the most compartmented black boxes of the world's intelligence agencies. The corporation's information control systems guarded proprietary industrial data but also sought to protect its long-term strategic position by minimizing its visibility. Exxon's executives deflected press coverage; they withheld cooperation from congressional investigators, if the letter of the law allowed; and they typically spoke in public by reading out sanitized, carefully edited speeches or PowerPoint slides. Their strategy worked: Exxon made a fetish of rules, but it rarely had to justify or explain publicly how it operated when the rules were gray.

As the *Valdez* wreck made obvious, Exxon's massive daily operations— soon to produce 1.5 billion barrels of oil and gas pumped from the ground each year, and 50 billion gallons of gasoline sold worldwide—posed huge environmental risks. After the *Valdez*, Exxon would become again, as it had been in the first decades of Standard Oil's existence, the most hated oil company in America.

When gasoline prices soared, American commuters felt powerless before its influence. In effect, Exxon *was* America's energy policy. Certainly there was no governmental policy of comparable coherence. After fitful, failed efforts to wean itself from imported oil during the 1970s, the United States had evolved no effective government-led energy strategy. Its de facto policy was the operation of free markets amid a jumble of patchwork subsidies, contradictory rules, and weak regulatory agencies. The very weakness of policy favored Exxon. As the public's frustration grew over rising pump prices and dependence on oil imports that transferred billions of dollars to hostile regimes overseas, Exxon became a natural lightning rod. The corporation managed this criticism with the same coolheaded patience and indifference that it employed to endure political risk in tinpot African dictatorships. Compromise was not the Exxon way.

PART ONE

THE END OF EASY OIL

One

"One Right Answer"

Sidney J. Reso was typical of the men who rose into Exxon's senior leadership ranks: an engineer by academic training; an Exxon employee for life; married for thirty-seven years to his wife, Patricia; and quietly appreciative of his privileges as his wealth grew. He maintained a membership at the Spring Brook Country Club near his office in New Jersey and owned a vacation condominium by the shore in Florida. He was not a man given to radical decisions or departures.

It did not augur well, then, when a neighbor discovered his car idling with the driver-side door open at the end of his 250-foot driveway on a wooded cul-de-sac in Morris Township at 8:00 a.m. on the morning of April 29, 1992. Reso had passed through the front door of his large brick-and-clapboard home as usual at 7:30 a.m. that Thursday morning to make the fifteen-minute drive to his office in Florham Park. There he served as the president of a large international Exxon division responsible for oil and gas exploration and production outside of North America. Police quickly circulated fliers seeking information about a missing white man, five feet ten inches tall, 180 pounds, with blue eyes and gray hair showing a reddish tint.[1]

Lawrence G. Rawl, Exxon's soon-to-retire chief executive, and his successor, Lee Raymond, were together at an Exxon board of directors meeting in Dallas. The annual shareholder meeting would soon begin; each year, the board held a meeting beforehand. Resolutions and board member election voting passed in a ritualized, scripted session. A senior executive in Exxon's security department entered and leaned over Lee Raymond's shoulder as he read out to the room from prepared materials. "I've got to talk to you," he said. "Right now."

Raymond excused himself and returned a few minutes later to report, "Sid's been kidnapped."

The board sat in silence. Kidnappings were a periodic threat; attempts against executives came and went in waves. In 1974, Exxon had paid $14 million to free one of its executives, Victor Samuelson, from the Marxist People's Revolutionary Army in Argentina. The feeling in the room was, "Not another one." Rawl was upset; Reso had worked directly for him for years.

A telephone caller had already issued a ransom demand to the corporation, security reported. Rawl called in the Federal Bureau of Investigation. Its director, William Sessions, began to call Raymond each morning to deliver updates. Running their investigation out of Newark, F.B.I. agents required several days to conclude that the initial caller seemed to be authentic. A ransom note demanded that Exxon gather $18.5 million in old one-hundred-dollar bills, load them into laundry bags, and prepare for a drop. The demand came from the Fernando Pereira Brigade, Warriors of the Rainbow. The name referred to the freelance photographer who drowned in the Pacific Ocean in 1985 when French intelligence agents sank the *Rainbow Warrior*, a vessel belonging to Greenpeace, the environmental crusaders, as it led a seaborne protest against nuclear weapons testing in French Polynesia. Since the *Exxon Valdez* spill, Greenpeace had made Exxon a prominent target of its anti-oil campaigning, but the group propounded nonviolence and civil disobedience, not kidnapping. ("This tragic allegation does a real disservice to legitimate environmental organizations working to protect our global environment," its executive director, Steve D'Esposito, told a reporter while denying any involvement.)[2]

The kidnappers communicated sporadically after their initial ransom demand. The F.B.I.'s agents spread out across New Jersey to conduct a massive investigation. Patricia Reso twice appeared on television to issue appeals on behalf of her family. ("Wherever he is, I wonder if he's cold," she said of her husband, "because his overcoat was in the car.") As the weeks passed, the kidnappers threatened a wider war against Exxon. "If you choose not to pay, Reso will die in 24 hours," a letter delivered in early June declared. "If you interfere in any way with the [money] delivery prior to Reso's release we will strike at our selected targets. These people will not be seized [that is, kidnapped] but will be treated as soldiers in war."[3]

Rawl and Raymond visited the F.B.I.'s task force in Newark and were impressed by the scale of the effort. In this age before cell phones, the bureau's investigators hypothesized that the kidnapper would use a public pay phone to communicate. They also figured he was probably still in the area. The task force rounded up enough agents to stake out every pay phone within a twenty-mile radius of the kidnapping site. On the night of June 18, one of these F.B.I. surveillance teams watched a blond man wearing gloves make a telephone call at a pay phone at a New Jersey shopping mall. The agents followed the caller's Oldsmobile and arrested the man shortly after midnight. In the car they found laundry bags and a briefcase containing a 1985 directory of the home addresses of Exxon executives.[4] Sessions called Raymond: "I think we got him."[5]

Arthur Seale grew up as the son of a policeman in Hillside, New Jersey, a middle-class town of about twenty thousand. At twenty-one he married a wealthy town girl, Jackie Szarko, whose parents owned properties, a liquor store, and a delicatessen. Arthur followed his father onto the Hillside police force but was suspended twice and fined three times in six years for defying orders and drawing his gun inappropriately. He later resigned with a $10,000 annual injury pension and took a job in Exxon's security department in the New York area. He worked initially as a chauffeur. Whether Exxon knew of his trouble in Hillside before it hired him is not clear, but Seale performed well enough to move up to a corporate security position in Florham Park, where he earned as much as $60,000 in salary. He became angry, however, when former F.B.I. agents were promoted ahead of him, and in 1987, Exxon dismissed him. He nurtured a grudge against the oil company and the F.B.I.

Arthur and Jackie Seale moved to Hilton Head, South Carolina, where they purchased a furniture store, bought a marsh-front house in an exclusive neighborhood, enrolled their children in private schools, and seemed to have reestablished themselves. After a little more than a year, however, they hastily moved away from South Carolina, evading $715,000 in debts and court claims. Before long they had returned to New Jersey to live with Arthur's parents.[6]

According to court records, in December 1991, Arthur and Jackie had started to covertly survey Sidney Reso's cul-de-sac and to plot a kidnapping scheme. The couple constructed a wooden box six feet four inches in length and three feet six inches in width and placed it in a rented storage locker. Arthur consulted bankers in the Bahamas about how he might avoid taxes if he came into a large pile of cash.

On that Thursday spring morning at the base of his driveway, when Sidney Reso stopped as usual to pick up his newspaper, Arthur Seale grabbed him by the collar, wrestled him toward a white van, and in the scuffle, accidentally fired a .45-caliber pistol, wounding Reso in the forearm. Seale said later that his wife treated the wound with hydrogen peroxide and that the victim called him "sir" as he bound and gagged him with duct tape and placed him in the prefabricated wooden box. In Seale's estimation the container was "much larger than a coffin . . . more like a closet."[7]

On the afternoon of May 2, the kidnapper and his wife inspected their container and discovered that Reso had died. They hauled his body to a state forest in southern New Jersey and buried him. Afterward they continued to demand ransom from Exxon until the F.B.I. arrested them.

The *Exxon Valdez* accident had been preventable. It exposed the risks that arise when industrial systems of enormous scale and consequence are entrusted to imperfect human beings without adequate safeguards. Sidney Reso's death three years later was of a different character and perhaps not preventable at all. Like the spill in Prince William Sound, however, it shocked Exxon's leaders and employees. Lawrence Rawl was crushed by the loss of his colleague. The kidnapping reinforced a broader sense within the corporation, reeling from criticism and lawsuits over the *Valdez*, that it was under siege. It reinforced, too, a sense that Exxon's leaders might need to find new ways to exert greater control over the world in which they operated—to seek a "rebalancing," in Lee Raymond's phrase, of the management of risk. The changes that Raymond would soon impose on Exxon would alter the experience of every employee and manager who worked at the corporation in the years to come.

During the early 1990s, Exxon increasingly became Lee Raymond's company. Lawrence Rawl retired as chairman in 1993 at age sixty-five, but the practical transfer of power had begun earlier. After the *Exxon Valdez* grounding, at the corporation's monthly board meetings, it was Raymond who reported to the board about the results of his investigations into the accident's causes and about his assessment of what corporate policies should be changed in response. His updates and recommendations for reform, linked to the *Valdez* investigations, were part of "every board meeting for probably two and a half years," he recalled.[8]

The *Valdez* wreck and soft global oil prices, which argued for cost cutting beyond the steep reductions of the 1980s, offered an opportunity to push through sweeping management reforms within Exxon at a pace that would have been difficult to achieve without the rationale of crisis. Even before the accident, Rawl had been trying to shake up Exxon's bureaucratic ways. The 1980s had been a painful decade for oil corporations. Two Arab embargoes during the previous decade (designed to punish the United States for its support of Israel), followed by the 1979 Iranian Revolution, had driven crude prices to unprecedented highs, but by late 1985, prices had collapsed steeply. The unexpected drop squeezed cash flow so badly that for a short period Exxon borrowed money to pay its shareholder dividend. According to Raymond, Rawl, the former marine sergeant, believed that Exxon was "top heavy" and "didn't have the accountability it needed. . . . We had committees on committees." Advancing his mentor's drive to tear up the old Exxon organization charts and march forward in double time, Raymond drove home an operating

philosophy in which managers would be measured more directly for
their performances, and safety systems would be driven relentlessly
toward zero defects.[9]

Raymond set up a program for every Exxon division and affiliate worldwide to "reappraise risk." Many large industrial corporations sought to emphasize worker safety, but after the *Valdez* and Reso episodes, Exxon's
system became deeper and more pervasive than that of any of its peers.
To encourage internal whistle-blowing about safety, fraud, or discipline
problems, Raymond established a "hotline or anonymous post office box"
where employees could report violations. He oversaw changes in Exxon's
drug and alcohol policy: He scrutinized every job category and named
about 13 percent of them as "designated safety positions" that would henceforth be subject to special rules—these positions included not only oil
tanker captains, but also gasoline delivery truck drivers and equipment
operators in refineries and chemical plants. In the future, if an employee
voluntarily entered drug or alcohol rehabilitation, he or she would not lose
employment at Exxon but would be prohibited from ever working again
in one of the designated safety jobs. A new drug- and alcohol-testing program took hold, affecting both those who had sought treatment and all
of those who worked in the designated safety jobs, irrespective of their
personal histories. Raymond decided that the latter category should include the corporation's top three hundred executives, even if all they did
was push paper; this edict became known as the Raymond Rule. The rules
seemed more likely to drive alcoholism among senior executives into the
shadows than to ensure sobriety at the top, but the emphasis now was on
universal, mechanical systems.

To revamp its internal Global Security organization and prevent
any recurrence of a crime like the Reso kidnapping, Exxon hired Joseph
R. Carlon, a former assistant director for investigations and intelligence at
the United States Secret Service. Soon Exxon's senior executives enjoyed
personal protection regimes similar to those of American presidential candidates or holders of high national office. If a board member participated
in a confidential discussion by telephone in his or her home, corporate
security vetted the caller's houseguests and surveyed the number of telephone extensions, to prevent anyone from sneaking onto the line.[10]

The corporation's revitalized safety and risk management drive increasingly took on the trappings of a cult. Exxon departments worldwide organized regular safety meetings and competitions. Groups of employees that had no reportable accidents or safety incidents might win gift cards to Walmart or blue safety jackets with the names of the winning employees stitched onto the breast pockets. The prize-chasing worker collectives ensured that office clerks did not leave their file drawers open, lest someone bump against them. Failing to turn off a coffeepot might draw a written reprimand. Cars had to be backed in to parking spaces, so that in case of an emergency, the driver could see clearly while speeding away and would not inadvertently injure colleagues. "You would not believe the number of hours we listened to them talk about driving slowly in the parking garage," a former manager recalled. To discourage speeding down long plant driveways, the corporation installed electric signs linked to radar guns.

Every meeting at every Exxon office, no matter the agenda and no matter the personnel assembled, had to begin with a "safety minute," akin to a blessing before a meal, in which a randomly chosen employee would speak briefly about one safety issue or another. "Please take note of the Exit sign in the hallway," the briefer might say, "and note that the stairway to the outdoor plaza lies to the left of the meeting room door." If a group of employees worked together for years in the same office and held a lot of meetings, it could be very difficult to come up with a fresh safety minute, and so the briefings could become as repetitive as the routines of commercial flight attendants before takeoff. Safety minutes gradually became commonplace at many corporations engaged in dangerous industrial operations, but few companies enforced them like Exxon. (Chevron Corporation and British Petroleum later adopted the safety minute idea, and a scientist at one of the competitors reported to a friend at Exxon, "They've been assimilated into the Exxon Borg.") Reportable injuries tracked in statistical reports would soon include food poisoning, bee stings, stapler pricks, and paper cuts. As one of the corporation's senior safety managers would later explain: "If we have a whole lot of paper cuts going on, we have to ask ourselves, 'Well, what do we do to avoid paper cuts? Do we ask people to use gloves when they use the copy machine?'"[11]

The group safety confessionals at Exxon offices and plants covered conduct beyond the workplace: The correct use of a ladder while cleaning gutters at home might be discussed, or the imperative of wearing seat belts during the daily commute, or the danger of getting too much sun on a beach vacation. At these meetings employees stood and shared with their colleagues stories of "near-misses," as in a 12-step recovery program. One twenty-eight-year manager recalled listening to a colleague confess that an object had flown out of his lawn mower while he was cutting the grass at home and had struck him in the leg.

On Exxon billboards, office walls, and corporate vehicles worldwide the company would ubiquitously post a motto adopted from its oil drilling division: "Nobody Gets Hurt." In Africa, workers were required to submit to blood tests to prove that they had taken their antimalaria medication, Malarone; if they failed the test, the workers could be fired and sent home on a plane ticket they paid for themselves. Particularly in poorer countries without traffic enforcement, if accidents became a chronic problem, Exxon would install electronic monitoring systems in its vehicles to track drivers' whereabouts remotely, to ensure they did not exceed the company's own imposed speed limit. Managers purchased radar guns and dispatched oil workers onto rudimentary clay African roads to monitor their colleagues' speeds. Drivers might be fired for a single violation.[12]

Raymond integrated the new corporate safety rules into an intensified top-down culture of command management emanating from Exxon's headquarters. At his yearly meeting with Wall Street analysts, he conspicuously announced Exxon's safety record before enumerating the corporation's profit performance. He described his safety drive as a proxy for more far-reaching changes that would ultimately manifest themselves on the bottom line: "The only way you can be successful in the area of safety is through disciplined commitment and day-to-day management of the business."[13]

In 1992, the year of Sidney Reso's death, Exxon unveiled to its employees and executives a universal new management regime, the Operations Integrity Management System, or O.I.M.S., "more vinyl binders than you can possibly imagine, every single goddamn aspect of how we operate," as a former executive put it. "So there could be no excuses." O.I.M.S.

involved "Framework Expectations" about eleven "Elements." These included the basic challenge of risk assessment and management. O.I.M.S. section 2.1 declared, "Risk is managed by identifying hazards, assessing consequences, and probabilities." Five subsections of the rule outlined how to achieve this goal through the use of data, documentation, and outside evaluators.

The system also addressed human frailty in the workplace. Section 5.5 prescribed that Exxon employees should "routinely identify and eliminate their at-risk behaviors and those of their co-workers" while ensuring that "Human Factors, workforce engagement, and leadership behaviors are addressed."[14]

The legacy of catastrophic failure in Prince William Sound proved nonetheless to be persistent. *Fortune* had ranked the corporation as America's sixth most admired before the accident; afterward, it fell to one hundred and tenth. Telephone operators in the Exxon credit card department heard so much abuse from angry customers who used the *Valdez* accident to vent their spleens that the corporation made counselors available to console its employees. Many years after the grounding, the corporation's public affairs department organized focus groups with North American opinion leaders. When the moderator pronounced the word "Exxon" and asked for a free-association response, more than half of the participants blurted out, *"Valdez."*[15]

Initially Raymond sought to address the claims of Alaskan fishermen, cannery workers, and small business owners affected by the *Valdez* spill by handing out $300 million in compensation without asking for legal releases. Soon he chose to defend Exxon's position by fighting lawsuits filed by the state of Alaska, the federal government, Alaskan businesses, and individuals. Raymond rejected all efforts to extract punitive damages from Exxon. He accepted in principle that his corporation was liable for actual damages in Alaska where such claims could be proven—he settled virtually all of those claims by 1994. But punitive judgments levied as a deterrent or as a source of emotional satisfaction Raymond would fight as long as it took. "It was a very tough time for them," but increasingly Exxon's leadership group concluded that the anti-Exxon campaigning after the *Valdez* spill "was unfair," recalled Kathleen Cooper, who joined Exxon as its chief economist in 1990. "They paid compensation immediately—sooner than some companies might have. . . . At some point we said, 'We've spent a lot of money, we have done it on a proactive basis, and we just can't keep going.' We need to say, 'This is it.' That is what Raymond was saying and I think the whole company was behind him."[16]

The seeming virulence of Exxon's permanent opposition—Greenpeace and other environmentalists, dissident shareholders, Manhattan and Hollywood liberals, and assorted magical thinkers about wind and solar power (as Exxon executives tended to view those who believed renewable sources could meet America's energy requirements anytime soon)—strengthened the solidarity among Exxon's besieged executives. Gradually they returned to the operation of their oil and gas business for the profit of their shareholders. And they found a setting more compatible with their Alamo attitudes: They moved to Texas.

A *click, click, click* of heels on marble echoed through the vast lobby at intervals as women in charcoal pantsuits and men in dark suits and white shirts slipped through electronically controlled glass security chambers and crossed before a reception desk. The passing Exxon executives politely acknowledged the uniformed guards, who replied in turn with a formal "Mr." or "Ms." The corporation's new campus in the featureless exurban city of Irving resembled a high-end condominium community or a prosperous modern college set amid pine trees, wind-bent mesquite trees, and green lawns. Breezes rippled a small man-made lake. The main building was of modest height and sleekly constructed from granite, smoked glass, and polished marble. On one side of the lobby rose a tall, photo-realistic oil painting of a pristine alpine village on a lake with snow-capped mountains in the distance; opposite was an equally large canvass depicting a desert canyon. Visitors killed time in square-backed leather club chairs beneath the paintings. The aesthetic suggested a Four Seasons hotel without many guests. A second wall of security-controlled glass doors awaited visitors entering the top floor. There, it was necessary to wait for the doors behind to close before access to the inner executive suite—known to employees as the God Pod—would be granted. The God

Pod contained about twenty thousand square feet of office space housing just four or five executive suites, including Lee Raymond's, as well as conference rooms. Inside, it sometimes felt as if a neutron bomb had recently detonated, killing off the local population but leaving the elegant physical facilities intact. The persistent quiet and formality of the headquarters building had an ominous quality; some employees referred to Irving as the "Death Star."[17]

Until its retreat to Texas in 1993, Exxon had been rooted in Manhattan since 1885, when John D. Rockefeller and his founding partners at Standard Oil of Ohio had moved their headquarters from the city of Cleveland to 26 Broadway. The son of a traveling elixir salesman, Rockefeller had rebelled against his father's example by following his frugal mother's advice and growing up to become disciplined, orderly, circumspect, earnest, and religiously devout. As American oil consumption boomed in the late nineteenth century, he and his partners methodically seized control of the industry, destroyed their competitors, innovated with technology, and built the first "integrated" oil company, meaning that they controlled the profitable exploitation of oil from the wellhead through the refining process to the retail sale of gasoline. At its peak Standard Oil controlled 90 percent of the American oil market. From its early days it attracted the same kind of opposition that would shadow Exxon a century later—muckrakers, journalists, trustbusters, and other American factions suspicious of concentrated industrial power. The muckraker Ida Tarbell's nineteen-part *McClure's Magazine* series, published in 1904 as the book *The History of the Standard Oil Company*, attacked the corporation's power but acknowledged the strengths of its scientific culture: "From the beginning the Standard Oil Company has studied thoroughly everything connected with the oil business. It has known, not guessed at, conditions. It has had a keen authoritative sight. It has applied itself to its tasks with indefatigable zeal."[18] "Bringing order to chaos" was the way Rockefeller had once described his monopoly. That ambition had not ebbed within Exxon almost a century later.

Tarbell's investigation accelerated a movement to break up Standard Oil on antitrust grounds. By the time the United States Supreme Court ordered the company's dismantlement in 1911, John D. Rockefeller had retired and taken up philanthropy. The largest of the "baby Standards" born from the breakup was Standard Oil of New Jersey. It later marketed itself and its products under the Esso, Enco, and Humble Oil labels before modern branding specialists settled on Exxon in 1973. At the time of the *Exxon Valdez* spill the corporation remained by far the biggest oil company in the United States—twice the size of the next largest, Mobil Oil, another baby Standard, the successor to Standard Oil of New York; larger still than Chevron, the successor to Standard Oil of California; and ten times the size of the Atlantic Richfield Company, initially born of the Standard monopoly as a refiner.[19]

Exxon hewed most closely to the Rockefeller inheritance of discipline, rigor, technological research, and unsentimental competition. By the 1990s, there were "lots of wrong ways of doing projects—and then there [was] the Exxon way," as Ed Chow, a longtime Chevron executive, put it. Exxon's managers and engineers were "very, very prickly as partners . . . and they don't like to be partners, unless they're the operator," a competing executive said. At industry meetings the Exxon participants could be easily identified: conservatively dressed, hairstyles that seemed influenced by military rules, cliquish, secretive, and businesslike. Senior executives who rose through Exxon's ranks reinforced with one another that they served a corporation whose "fundamentals" traced in important ways all the way back to Rockefeller, as Raymond put it.[20]

Executives at other oil companies tended to regard their Exxon cousins as ruthless, self-isolating, and inscrutable, but also as priggish Presbyterian deacons who proselytized the Sunday school creed Rockefeller had lived by: "We don't smoke; we don't chew; we don't hang with those who do." Ethics rooted in Judeo-Christian religious tradition were part of the fabric of Exxon. "They encourage you to get married," a former manager recalled. Such values were "not just a lot of lip service," said another longtime executive. "J. D. Rockefeller went to church every Sunday and his employees better by God go to church on Sunday or they were not good employees. It is kind of a legacy. When I went to work for the company in the 1970s, managers would have employees join hands around the table and pray for the success of Exxon."[21] Compared with executives at San Francisco–based Chevron or the international behemoths of British Petro-

leum and Royal Dutch Shell, a British-Dutch conglomerate, senior executives at Exxon sometimes lacked what bicoastal American or European
executives would call worldliness. Many of Exxon's U.S.-based executives
traveled extensively but remained insulated, introverted; when they mingled, it was to golf or hunt with others like themselves.

Manhattan no longer seemed a suitable base. Striving senior executives would typically arrive at Exxon's modern headquarters, a towering
white skyscraper, at around 7:30 a.m., only to find it vacant because there
were no early-morning go-getters. Long commutes from the suburbs
seemed to deter early birds; in any event, the sense among some executives was that a lethargy had set in. "You could have thrown a bowling ball
down the fifty-third floor," where top executives work, "and it wouldn't
have hit anybody," recalled one manager. Howard Kauffmann, the corporation's president at the time, advised the executives he met who were
anxious for change: "If you ever get this place in a van, make sure it drives
at least two days before it stops." When Rawl and Raymond decided to
move, around 1987, Rawl pulled a map of the United States out of his
desk and they quickly drew Xs through one section of the country after
another—the West Coast because its taxes were high, the North because
it was cold, the far Southwest because it seemed too out of the way. That
left them with the Confederacy, essentially. They scouted new headquarters sites in Atlanta, Jacksonville, Charlotte, Houston, Austin, and Dallas;
narrowed the choice to Texas; and bought some land in Austin. They
ultimately selected Dallas because it was easy to reach from around the
world and would keep the headquarters away from the oil provincialism
of Houston, where Exxon already had a large presence. They sold the
Sixth Avenue building in Manhattan and reaped $477 million.[22]

Exxon recruited heavily from the petroleum engineering departments
of the public universities of America's South, Southwest, and Midwest.
By locating its headquarters in Texas, the corporation placed itself in the
landscape to which many of its long-tenured American employees belonged. Exxon maintained "kind of a 1950s southern religious culture,"
said an executive who served on the corporation's board of directors during the Raymond era. "They're all engineers, mostly white males, mostly
from the South. . . . They shared a belief in the One Right Answer, that

you would solve the equation and that would be the answer, and it didn't need to be debated."

The executive was startled to discover at one point that the corporation's top five leaders, all white males, were the fathers, combined, of fourteen sons and zero daughters. The mathematical probability that such a quirk had no basis in the corporation's social mores was low. "What is there in the culture here that promotes people with sons?" he wondered. Sports? Hunting? He could not figure it out, if there was indeed something other than a fluke to discern.[23]

Lee Raymond, a son of the working-class Great Plains, considered himself unabashedly to be a "free-market capitalist" and resisted government intervention and regulation instinctively; Dallas suited him. When Raymond found himself in a public battle with gay rights organizations over his decision to deny corporate benefits to same-sex partners of his employees, a board member challenged him to at least make a public statement saying the corporation did not discriminate on the basis of sexual orientation. Raymond declined. He told colleagues that he did not pay much attention to matters such as sexual preference, but he was not going to make an announcement.

"Do you discriminate against people based on sexual preference?" the director asked.

"Of course not," Raymond answered.

"Then why don't you say it?"

"Well, it's not required by law."

"But it's a freebie," the director persisted, speaking later to one of Raymond's lieutenants.

The executive retorted: "What's next? Polygamy?"[24]

The Exxon way included an updated version of a decades-old employee ranking system in which each year all managers were required to assign a number rating to all personnel under their supervision, ranking those of similar pay grade from best to worst, and to recommend high performers for assignments that would groom them for later leadership. The evaluations covered judgment, creativity, leadership, competence, sensitivity, and other subjective qualities, but there was no grading on a curve; supervisors were required to distribute outstanding and inadequate grades in even proportions. Those initially ranked in the top tier were then promoted into a group of similarly ranked high performers to determine which of those would emerge as the best of the best. The winners could expect to be promoted quickly, but also to have to pick up and move every few years. The system, analogous to natural selection, hardened Exxon's culture and wrote the corporation's D.N.A.: It was driven by numbers, focused tightly on performance, and in many ways inflexible. "The forced ranking system was poisonous," said one manager who went through it successfully. "It created feelings of distrust with your coworkers because of the competition and the zero-sum consequences." Amid cost reductions, reassignments, demotions, salary reductions, and job cuts, the pressure only intensified. A former executive once described the ranking system as "dog-eat-dog competition under the patina of working together."

Even before the *Valdez*, Exxon had been a place that emphasized procedure and cultivated orthodoxy; with the inauguration of O.I.M.S., the i-dotters and t-crossers rose to predominant authority. Because of the nature of its business, Exxon's recruitment was biased toward engineers, scientists, accountants, and personalities who were comfortable with rules—people who were pleased, even eager, to work for one company all their lives, and to move from place to place in its service. Senior executives noticed that employees tracking for management tended to reach a point—somewhere around four to seven years after joining the corporation—when they either committed to Exxon or left. Those who stayed did not find O.I.M.S. ironic or extreme; they *liked* the culture of discipline and accountability. Restless free thinkers and habitual dissenters who accidentally got hired (often as scientists) tended to decide quickly that they would be happier elsewhere. The result was a corporation led in its upper management ranks by people who were not only supporters of the O.I.M.S. reforms, but true believers.[25]

Around an industry conference table, Exxon's delegation usually dominated. "You don't like them, but you respect them a lot because you know that they're really smart," said a competing executive. An Exxon delegate to an industry committee meeting typically arrived with a binder full of research, colorful PowerPoint slides, and carefully outlined remarks that reinforced the impression that he was "the smartest guy in the room."

Exxon's sheer size meant it enjoyed advantages of research and scale; if Mobil dispatched one lawyer to an industry conference, he might arrive to see two or three from Exxon across the table, shuffling through the papers they had spent many hours preparing as a team.[26]

The ultimate measure (and the chief purpose) of this management culture was Exxon's financial performance. Even during the early Lee Raymond era, a time when oil prices gyrated disruptively and at one point fell to historic lows, the corporation's performance was superior from quarter to quarter and year to year. Exxon earned $6.47 billion in profit in 1996 on $110 billion in revenue, more profit than any American corporation that year except General Electric and General Motors. Mobil, the next-largest American oil competitor, posted about half of Exxon's profit margin. Exxon made more profit on each dollar it invested than any of its American or international competitors. Its exceptional ability to complete massive, complex drilling and construction projects on time and under budget meant that, in comparison to industry peers, it remained exceptionally profitable in recessions and boom times alike, when oil prices were high and when prices were low.[27]

The Exxon way came across as arrogance to many outsiders. Raymond once stepped before a large conference of Wall Street analysts and announced, "What you're hearing today may seem boring. . . . You'll just have to live with outstanding, consistent financial and operating performance." As to the performance of Exxon's competitors at Chevron, Royal Dutch Shell, and British Petroleum, Raymond added, "I have to [say] I am surprised at the apparent lack of focus."[28]

"Exxon's attitude toward the other majors has always been, 'We are Oil—the rest of you are kids,'" said a long-tenured executive at a competing company. Exxon became such an oft-cited antagonist in this oilman's household that once, when the industry seemed troubled, his daughter asked him, "Dad, do you think things will get so bad that you'd go to work for Exxon?" ("No, I sure hope not," he answered.) Rockefeller Goodwin, a descendant of the founding family who became a critic of modern Exxon management, acknowledged that the company enjoyed a "strong corporate culture. . . . Unfortunately, it includes a lack of interest in listening to outsiders, an assumption that they know the answers." The shareholder activist Robert Monks, another persistent critic, found Exxon managers "self-referential" and "good operators [but] not good citizens." A senior civil servant who worked on international energy issues at the White House recalled, "It doesn't take you more than five minutes dealing with Exxon people to kind of get the full two-by-four-across-the-head sense of some of their culture," because of their blunt directness.[29]

Engineers and financial controllers influenced the corporation more than its global business strategists or brand marketers did. The latter tended toward habits of dreamy ambition and improvisation difficult to reconcile with O.I.M.S. By the mid-1990s, Exxon operated in almost two hundred countries with about eighty thousand regular employees; overseas, 98 percent of its employees were non-American. To operate such a business in proximity to the sorts of daily risks illuminated by the *Exxon Valdez* grounding did require discipline.

"We don't run this company on emotions," Lee Raymond liked to say. "We run it on science and principles." He sought "the relentless pursuit of efficiency," he once said.[30] As Standard Oil had discovered a century earlier, however, the larger, more profitable, and more powerful Exxon became, the more it attracted attention as a political actor. And in politics, discipline, performance-to-budget, and error-free design were not common qualities; instead there was a surfeit of "Human Factors," in the O.I.M.S. vernacular. As Exxon rose to greater global influence in the early twenty-first century, the corporation's leaders persistently struggled to find a supple human touch.

Two

"Iron Ass"

Lee Raymond lived and worked within a bubble of privilege. He traveled the world with round-the-clock support from the corporation's Aviation Services and Global Security departments. If his day began at his 8,642 square-foot, five-bedroom brick-façade home in Dallas, then his longtime chauffeur and bodyguard, a retired New York City police officer, would meet him there and usher him into a dark sedan. Raymond rarely drove himself anywhere. Nor did the indignities of commercial airline travel encroach on him. Citing kidnapping and other security threats, the corporation's board of directors had decided that its chief executive should not fly on commercial carriers. Raymond had use of Exxon's corporate planes for both personal and professional travel. Aviation Services managed about nine jets—around the turn of the decade, the inventory included several Gulfstream aircraft, a Bombardier Challenger, and two Bombardier Global Express jets. Lee Raymond's principal plane—a ten-passenger G-IV, and later an eleven-passenger Global Express, each with sleeping and mess facilities, satellite telephones, a defibrillator, and CPR-trained flight attendants—bore a tail number expressive of his position: N-100-A, or as it was referred to by corporate aviation personnel, "November One Hundred Alpha."

The crews catered to Raymond's onboard tastes: a glass of milk with popcorn in it, within arm's reach of his executive chair. Aviation Services's approximately two dozen pilots and several dozen additional support staff also tended to his wife, Charlene, who often traveled with the chairman and who favored bowls of wrapped chocolates. "When you take care of her, you take care of me," Raymond told them.[1]

Lee and Charlene Babette Raymond were inseparable. By the late 1990s, the couple had developed a typical annual migration: a late-January trip to Pebble Beach, California, where Lee sometimes played in the pro-am golf tournament with the likes of P.G.A. professional Ronnie Black; an April sojourn to Augusta, Georgia, to attend the Masters golf tournament, where Lee might catch up with his friend Phil Mickelson or have dinner with Tom Watson; and then Easter at their winter home in Palm City, Florida. Golf was Raymond's most discernible passion away from the office—he regularly joined in corporate tournaments. When Raymond had business in Asia, he and Charlene sometimes managed to fit in a vacation break in Hawaii on the way out or the way back. In the autumn he often spent Thanksgiving at the Augusta golf club again and might include a stag trip to Exxon's vast, corporate-owned bird-hunting ranch in southeastern Texas, near the town of Alice. At Christmas the Raymonds typically retreated again to Palm City, Florida.

Throughout the year they made weeks-long international trips on which the Exxon chief might negotiate for or ratify the final terms of new oil production contracts, attend a ribbon cutting at a new refinery, deliver a speech at an industry conference, or chair a board meeting. On a trip to London in 1997, the couple picked up an expensive painting; panicked Aviation Services and Global Security employees, fearing theft, guarded the artwork for nearly two weeks aboard November One Hundred Alpha as the Raymonds hopped on Exxon business from city to city. As their wealth grew, they collected not only art, but real estate. They added a $3.8 million house in the desert near Palm Springs, California, and a $7 million home in Scottsdale, Arizona.[2]

Raymond and Charlene had both grown up in modest circumstances in the American heartland. Both were devout Christians. Raymond's Plains-bred parents had raised him as a member of the Evangelical United Brethren in Watertown, South Dakota, a denomination that later became part of the United Methodist Church. Charlene came of age in a German Catholic family from Kohler, Wisconsin. They met at the University of Wisconsin and married when Raymond was twenty-three. Raymond converted to Catholicism and thereafter rarely missed a mass; in Saudi Arabia, which banned Christian churches, he attended services inside the U.S. embassy.

Although Charlene had earned a college degree in journalism, when she gave birth to triplet boys, she devoted herself to them and to her husband. Even at home, Raymond worried about discipline. After he rose within Exxon, he tried to control his family's use of corporate jets—he barred his triplet sons from flying on them, fearing that if he allowed them the privilege, it would encourage lax behavior by other Exxon executives. Charlene could be as demanding as her husband, and she could also be extremely frugal, as if clinging to lessons imparted during the Depression-influenced era of her youth. Deplaning in Berlin or Paris, she might fill a bag with snacks while complaining about the prices charged for breakfast in the luxury hotels where she and her husband stayed.[3]

Aviation Services staff talked among themselves about which Exxon-Mobil executives with jet privileges were the most arrogant or prone to temper over petty problems. The capacity of some of Exxon's multimillionaire leaders to become abusively angry over delays caused by bad weather, pilot changes, or mechanical problems never ceased to amaze their more modestly salaried crews.

Lee Raymond could be sharp-tongued, but he was not the worst offender in that regard. He tried to maintain a cordial formality with his travel crews and won respect, if not affection, from some of them. That was about the most that could be said of the reputation Raymond enjoyed among Exxon executives and employees more generally: He was respected. He was also feared.

Some managers who had worked in other corporations, even notably hierarchical and disciplined ones, found striking the atmosphere of terror and deference Raymond generated in the minds of many who worked for him. Although it was possible to locate people who would say that Raymond was not insulting or mean to them personally, even these exceptional people acknowledged that he was often unpleasant to large numbers of others. Some of those who knew Raymond well, and liked him overall, felt he badgered colleagues in part to keep people away from him. If this was his strategy, it worked. He won the nickname "Iron Ass" among some employees. Behind his desk in the God Pod hung a painting of a fierce tiger.

He calculated that in a corporation as large and diverse as Exxon, with tens of thousands of employees scattered in offices, refineries, and oil production compounds worldwide, the only way a chief executive could hope to extract disciplined results was to overdo it—that is, unless Raymond used his bully pulpit at Irving to pound hard and even intimidate his employees, the natural drift and compromising tendencies of such a large workforce would produce mediocre results.

In a small group or a social setting, Raymond could be relaxed and pleasant company. There was a South Dakota–bred reticence about him that could be confused with coldness. His manner masked a streak of sentimentality. He could be fiercely loyal to ExxonMobil colleagues and sometimes wept openly when subordinates faced illnesses or other personal struggles. At a retirement party for his longtime assistant Adrienne Hurtt, Raymond recounted that he had been on a business trip when his mother died, and that Adrienne had called and imparted the news with perfect grace. As he told the story, Raymond broke down and cried before his colleagues.[4]

He worked hard. When in Dallas, he typically left the Irving headquarters around 5:30 p.m. with a bulging, battered-looking, soft Hartmann briefcase and a pair of plastic legal binders full of memos and reports. At home, he and Charlene kept separate bedrooms, in part because Raymond snored, but mainly because he stayed up until about midnight to read and mark up his files. "His life was the company," said a former member of the board of directors. Beyond Charlene, Raymond's friendships were mainly drawn from a small clan of retired and serving chief executives of international oil companies. Traveling in Europe, Raymond would take Charlene to dinner with Lodewijk van Wachem, a retired chairman of Royal Dutch Shell, and his wife. Long dinners where the men could trade stories about the global industry were often Raymond's idea of evening entertainment. As to hobbies, "Golf was about it," the former director said.

Before larger audiences and workplace groups, Raymond often seemed to go looking for a fight. It seemed the worst thing an Exxon manager could be in Raymond's eyes was dishonest, but the second-worst thing was to be stupid. He could be withering with senior executives, Wall Street analysts, journalists, and dissident shareholders who asked what he considered to be a dumb question or who disappointed him with the quality of their analyses. "Stupid shits" was one of the direct phrases by which he conveyed his judgments.

Raymond "definitely had a sense of humor," a subordinate recalled, and he "didn't bother belittling people below a certain level. You had to be up to where you had significant responsibility before you could get both barrels." In those cases Raymond did not hold back. He had been a champion debater in high school in South Dakota and he took transparent pride in his ability to knock down an opposing speaker. During his rise, Raymond ran Esso Inter-America, the corporation's Latin American division. There he reshaped an Aruban refinery losing $10 million a month into a $25-million-a-month profit center. He did not fashion this turnaround timidly. In front of the subsidiary's senior managers and board of directors he once turned on a subordinate whose comment had underwhelmed him: "And what little birdie flew in the window and whispered that dumb-shit idea in your ear?" Later, when he reigned over all of Exxon, he would preside over company town hall meetings and question sessions. Sensitive employees in the amphitheater cringed when, as inevitably happened, some incautious manager stood to ask Raymond an impertinent question about when one or another employee benefit might be granted. Raymond "would look at the person who asked as if he could will death," another former manager recalled. Raymond believed he had never belittled a colleague in front of others, but belittlement is an experience usually defined by the victim. Raymond admitted, "I'm not known to suffer fools gladly."

His physical appearance did nothing to soften the impression he made. He wore square wire-framed glasses and kept his straight, side-parted light brown hair closely cropped. He had large ears. He had grown into a fleshy man and the jowls beneath his chin could billow like a bullfrog's neck. A childhood cleft palate had left him with a prominent harelip. Exxon employees who found themselves on the receiving end of Raymond's ridicule sometimes referred to him darkly as "the Lip." The amateur psychologists among them speculated that it might have required a certain learned toughness and even meanness, an ability to tune out taunts, to grow up in a small midwestern town with such a visible defacement: "I can envision [him] as a child being absolutely persecuted by other kids," recalled a former employee. Raymond obviously got through it, the in-house analysis went, but after he achieved success, perhaps it was not surprising that "he doesn't take any prisoners."[5]

Raymond's predecessor, Lawrence Rawl, had created an unforgiving climate within Exxon while tearing into the corporation's bloated cost structure and overseeing a campaign of staff reductions that federal investigators found had contributed to the *Exxon Valdez* fiasco, but which had also protected the corporation from financial distress. Rawl had also belittled colleagues at meetings, engendering an atmosphere in which his principal deputy, Raymond, seemed to believe as he ascended, one executive said, that he should be "out-Rawling Rawl" in toughness.[6]

Raymond saw himself as an oil and gas purist. He told colleagues that outside its headquarters the corporation should carve in stone the words, "crude oil." He felt that it was critical that Exxon "not get confused about what we are trying to do around here."[7] He and Rawl had employed their drill sergeant–inspired ethos to direct a sharp turn in corporate strategy away from an era, during the 1960s and 1970s, when Exxon had tried to adapt itself partially to environmentalism. (Rawl's predecessor as chairman, Clifton Garvin, a chemical engineer, had gone so far as to install solar panels to heat the swimming pool at his suburban New Jersey home.) Besides cost cutting, they sought to restore Exxon's focus on its core business.[8]

Some of his colleagues believed Raymond possessed a one-of-a-kind analytical mind and memory, and that his acuity contributed mightily to Exxon's superior business and Wall Street performance. At a time of wage stagnation and other rising cost pressures on working and middle-class American families, this success enriched many Exxon employees and increasingly set those located in the United States apart as an economically secure class. Exxon managers had jobs for life if they could hack the corporation's internal systems and were willing to move from place to place. They enjoyed secure defined-benefit pension plans and restricted stock that would make many middle and upper managers millionaires if they stayed long enough and managed their personal finances carefully. Exxon managers tolerated Raymond's tirades in part because they understood, as the years passed, that he was making them rich. This was not Silicon Valley: The corporation's scientists and division chiefs did not walk away with fortunes at thirty-five, but if they conformed and performed, they would rise gradually on a tide of oil profits into an economically privileged elite.

Raymond reserved a particular scorn for the Wall Street analysts who published commentary about Exxon's business strategy. After years of sporadic efforts to engage with stock commentators, the corporation began to stage an annual meeting with analysts. It was an unusual hearts-and-minds campaign because during the question-and-answer session, it was rare for Raymond to respond to any query without first challenging the analyst's assumptions or intelligence.

"This is going to be a tough meeting; you ought to take two patience pills," Peter Townsend, the vice president for investor relations and corporate secretary, would warn him before these sessions.

"No, three."[9]

If Raymond began his answer to an analyst's question with "Frankly" or "To be candid with you," it was a signal to duck. He started one meeting at the New York Stock Exchange by noting that executives from The Walt Disney Company were also present in the building that morning: "I don't think Mickey or his friend Goofy are going to join us, but I may have to hold my judgment on that until after the Q&A session."[10]

Raymond served in effect as the corporation's chief financial officer, in possession of all of the critical numbers. By the time he became chairman, he had also served for years as a director at J.P. Morgan, the Wall Street investment bank. During the mid-1980s, Exxon's financial performance looked respectable but undistinguished, in comparison with its oil industry peers. Rawl's cost cutting and reorganization campaign was intended to force improvements. In 1987, Rawl began to place heavy emphasis on a metric called "return on capital employed" or R.O.C.E. (often spoken of as "row-see").

This was a performance measure that sought to show how well a particular Exxon business unit—and overall, the corporation—used the cash it borrowed or recycled from earnings to reap returns from new projects. After he took charge, Raymond campaigned on Wall Street to have his particular measure of R.O.C.E. recognized as the premiere number by which oil corporations should be judged. He argued repeatedly to analysts that oil companies were very long-term businesses that consumed a great deal of capital, and that, ultimately, they should be judged not by quarterly profits or share-price fluctuations, but by how well they managed their investments—whether, for example, they regularly destroyed capital by leasing unproductive oil fields, going over budget on huge drilling projects, or by building unprofitable refineries.

The deep cost cutting continued by Raymond raised Exxon's rates of return on capital. So did the drive Raymond advanced to improve Exxon's relatively low-profit divisions, particularly gasoline refining. The "downstream" divisions of integrated oil companies like Exxon were generally much less profitable than the "upstream" units that found, pumped, and sold crude oil and gas. ("Downstream" was an industry term that referred to what took place after oil was pumped from the ground: the refining of oil into gasoline or aviation fuel, and retail sales to motorists at thousands of Exxon-branded gasoline stations across the United States.) Exxon had long tolerated low downstream profit margins and even occasional losses because having huge refineries worldwide gave the corporation a built-in market for its own oil sales. In effect, upstream profits subsidized the downstream. By maintaining a focus on R.O.C.E. inside Exxon and preaching about it on Wall Street, and by tying performance on that number to promotions and bonuses for Exxon managers, Raymond hoped to create change.

Exxon's rates of return on capital rose sharply during the 1990s, declined as oil prices fell late in the decade, and then recovered to a record level of about 20 percent by the decade's end, superior to any competitor. Raymond was only partly successful in persuading others to embrace his math—although no other major industry adopted his ideas about R.O.C.E.'s centrality as a metric, Exxon's major oil company rivals did start to report their own R.O.C.E. numbers, to Exxon's benefit. R.O.C.E. was, in any event, a somewhat arbitrary figure by which to compare oil giants. The measure favored ExxonMobil's relative strengths as an operator in low-margin downstream and chemical divisions; for investors, all companies highlighted the numbers that made them look best. Certainly R.O.C.E. was a long-standing and valid way to measure a corporation's ability to maintain profit discipline across many projects over many years. Still, "about two thirds" of an oil company's R.O.C.E. is typically explained by "commodity prices," as James J. Mulva, the chief executive officer of ConocoPhillips, once remarked. (His company's R.O.C.E. scores lagged.) That is, the most sparkling annual R.O.C.E. numbers, in comparison with returns seen in other industries, often reflected factors in the global oil market beyond any one company's control. That was apparent in Exxon-Mobil's own yo-yoing numbers. Yet within the oil industry, R.O.C.E. scores did provide a basis to compare operating and capital efficiency. Raymond understood the distortions caused by swings in oil prices, but he thought the number was as good a way as any available to judge management's long-term investment discipline. "Our competitors hated it," Raymond recalled. "The reason they hated it is that it's a report card, and while everyone can talk about individual projects and how attractive they may appear to be, ultimately, over time, you have to look at, 'Well, how do all of those individual projects add up?'"[11]

Raymond's relentless proselytizing about R.O.C.E. was part of a larger pattern of his leadership: He chose his own metrics; he declared that other metrics were wrong; he delivered profits; and he ignored criticism.

That worked well enough when the subject was Exxon's increasingly strong quarterly profit performance, in comparison to peers. It worked less well when the subject was Exxon's ability to find enough new oil and gas to replace the hundreds of millions of barrels the corporation pumped and sold every year. As profitable as Raymond was making Exxon by the late 1990s, he struggled increasingly with a challenge that had never shadowed John D. Rockefeller: how to keep the corporation's oil reserves—the foundation of its business—from shrinking.

The cold war's end initially promised new bounty for Western oil corporations. Vast reserves did initially open for bidding in the former Soviet Union, Africa, and elsewhere. But it did not take long for the opening to become constricted. Raymond concluded by the late 1990s that while there was plenty of oil in the ground worldwide, the amount that Exxon could access might make it difficult over time to replace the vast quantities the corporation pumped each year—Exxon produced about 585 million barrels of oil and gas liquids in 1997. The reason involved the persistence of "resource nationalism," or the inclination of governments that owned oil and natural gas to maintain control of their treasure.

The business models of the major international oil companies such as Exxon, Chevron, Royal Dutch Shell, and British Petroleum—and the prices that their shares commanded on Wall Street and on the London exchange—depended in part on the size of the underlying trove of oil and gas the corporations could claim to own. "Booked reserves" or "equity oil" referred to those proven reserves that a corporation controlled legally and could exploit for sale in future years.

In the world's wealthy free-market political economies—the United States, Norway, the United Kingdom, Canada, Australia—virtually all of the oil available was "equity oil" in the sense that any company that found it and acquired it could own it legally under contract, in a manner akin to property rights. Exxon and its peers could display such equity reserves to shareholders as "proved" or "booked" oil under regulatory and accounting rules enforced in the United States by the Securities and Exchange Commission. The size of these booked reserves allowed shareholders to estimate future profits with relatively high confidence; equity oil was fundamental to Exxon's stock market valuation, just as the number of shopping malls or office buildings owned by a real estate company would be fundamental to its market value.

Before the 1970s, Exxon, BP, and other large oil companies owned and operated large oil and gas fields in Saudi Arabia, Iraq, Iran, Venezuela, and elsewhere. Rising anticolonialism and nationalism stoked a period of upheaval that caused them to lose major properties. The twin anti-American oil embargoes of the 1970s signaled the arrival of the new era. The embargoes coincided with the rise of a price cartel, the Organization of the Petroleum Exporting Countries (O.P.E.C.), which served the interests of oil-producing governments. In 1979, the Iranian Revolution's philosophy of Islamic self-determination advanced the spread of anti-Western political attitudes in Middle Eastern capitals. Populist leaders of oil-producing governments competed to prove themselves as resource nationalists—proud owners of their own geological wealth and unwilling to allow foreign corporations to possess a single barrel. Saudia Arabia, Iraq, Iran, Venezuela, Algeria, Libya, and other governments all seized back oil and gas fields from Western corporations, including Exxon. The expropriations decimated Exxon's holdings and rates of daily oil production. By the late 1990s, Exxon and the other large private oil and gas companies based in the United States and Europe owned less than 20 percent of the world's oil reserves. In 1973, Exxon had produced just over 6.5 million barrels of oil and gas liquids per day from its worldwide properties. By the late 1990s, that figure had fallen by more than two thirds.[12]

The corporation spent much of this period "struggling with the issue," Lee Raymond recalled. "We lost our equity position in the Middle East," where 60 percent of the world's proved oil reserves were located. The most fundamental question facing the corporation was, "What's that mean for the company?"[13]

Even the most nationalistic governments might welcome Western companies as technology partners and hire them strictly as contractors to drill, produce, and refine oil and gas, as a homeowner might hire out a contractor to renovate a house. Such fee-for-service contracts could be the basis of a profitable business—Schlumberger and Halliburton were examples of companies that built lucrative franchises in this way. But deals of this kind stopped short of allowing the contractor to own any oil beneath the ground. Exxon's business had always been premised on owning oil, which required greater risk taking but promised much higher profit than a contractor could hope to earn.

By the 1990s, virtually all of the oil in the Middle East was off-limits to corporate ownership because of resource nationalism. The strategic problem facing Exxon was that apart from new frontiers in offshore ocean waters or above the Arctic Circle, there did not seem to be much new oil or gas to discover in territory controlled by wealthy, free-market countries. Elsewhere, in Africa, Asia, Latin America, and the former Soviet Union, oil geologists advised, there were still major discoveries to be made, but inconveniently, much of this new oil seemed likely to be found in places where governments would be skeptical about allowing Western corporate ownership. Notwithstanding communism's fall, many politicians in developing countries, responding to popular feeling, still held that it was neither just nor necessary to give away ownership of oil and gas reserves, least of all to Western capitalists like Lee Raymond.

Although Exxon jockeyed for position as the world's largest privately owned oil company, by the late 1990s it ranked only fourteenth or lower on a worldwide basis if the list included government-owned companies such as Saudi Aramco, Kuwait Petroleum, Gazprom of Russia, Petrobras of Brazil, or Sonangol of Angola. These state-owned giants not only showed large inventories of booked reserves, they also increasingly prowled outside their borders to compete with Exxon to capture next-generation oil reserves in Africa, Asia, and Latin America.

Exxon's strategy was to emphasize its superior record of project execution, budget management, and cutting-edge technology. Its executives tried to persuade oil-owning governments that Exxon's efficiencies could deliver an enormous cash windfall over the long life of a project, in comparison with what a less-efficient state-owned company could deliver. Computing power had started to remake oil exploration techniques. Three-dimensional imaging and algorithms that sorted reams of seismic data into patterns transformed the ability of Exxon's geologists to find oil and gas. Many state-owned oil companies lagged behind. Not all oil-endowed countries had the capacity or the political and economic stability to build and manage a state-owned oil company that was competent in all sectors of the business. These were the openings Exxon tried to seize. Yet Chinese and Russian competitors could make offers to African or Latin American or Central Asian host governments that Exxon couldn't touch—government-to-government loans, arms transfers, and political favors.

All this placed unprecedented pressure on Raymond and his peers to show Wall Street that they could find or buy new oil and gas reserves to replace what they produced annually. If an oil company failed to replace production for a sustained period, it would be on a path to liquidation. Under U.S. Securities and Exchange Commission–supervised accounting rules, the oil and gas reserves Exxon and its peers reported to shareholders were not carried on corporate balance sheets, but they were reported each year in S.E.C. filings. The reserves were among the most important assets oil companies described to investors because they suggested the scope of a particular company's potential future profits and its sustainability. The sheer size of a company like Exxon or Shell increasingly made the math of annual reserve replacement daunting—more than a billion barrels had to be found and booked as new equity reserves each year if the company did not want to appear to be shrinking.

The temptation for Wall Street to fudge the numbers was powerful, as events at Shell would soon bear out. Yet the management of annual reserves reporting—having in-house geologists count up "proved" holdings, field by field; reviewing those counts at higher levels of management annually; and applying objective standards that could hold up if the S.E.C. inspected them—was a relatively new priority. During the long history of Standard Oil and its successor companies, down to offspring Exxon, executives had not had to worry much about reserve counting or replacement. There was plenty of oil to drill worldwide, and because Exxon, in particular, had owned stakes in Saudi Aramco and Iraqi and Iranian companies with massive reserves, the issue seemed of little material importance. It was only after the nationalization waves of the 1970s that annual reserve replacement became precarious, and counting methods drew attention from regulators. Gradually, Wall Street analysts and investors focused down on the question of which oil companies were renewing their reserves healthily each year and which were struggling and even in danger of spiraling smaller. Raymond took up the challenge of reserve counting with characteristic aggression and disdain for Washington regulation.

He had cause for concern: Between December 31, 1996, and December 31, 1997, the proved reserves of oil and liquid natural gas that Exxon reported to the S.E.C. fell from 6.34 billion barrels to 6.17 billion barrels, a decline of more than 2 percent. The amount of that decline was the equivalent of 465,000 barrels of oil production per day, annualized, a sizable amount by industry standards. During the same year, the corporation's reported proved natural gas holdings increased slightly, but not enough to offset the fall in oil reserves. Judging by the rules enforced by the S.E.C., then, Exxon's total oil and gas reserves shrank during 1997.

The news that Exxon had retracted a bit, at least temporarily, was hardly shocking: All of the major oil and gas companies struggled during the 1990s to replace produced reserves. To the public and Wall Street, however, Lee Raymond made no such admission. On February 5, 1998, Exxon announced in a press release that it had replaced *121 percent* of its 1997 production by adding new proved reserves—in other words, its reserves had grown, not shrunk. "This year's strong performance is the fourth year in a row that we've exceeded 100 percent replacement," Raymond declared.[14]

How could Exxon tell the public and Wall Street one thing and the S.E.C. another? The answer involved a catalog of legalese known as S.E.C. Rule 4-10, the binder of regulations that governed what oil and gas companies could and could not report as "proved reserves." A purpose of Rule 4-10 was to prevent oil companies from puffing up their reserve numbers to lure share buyers and bolster their stock market prices.

One of the rule's provisions held that "oil and gas producing activities do not include . . . the extraction of hydrocarbons from shale, tar sands, or coal."[15] Tar sands refer to bitumen, a thick form of oil that usually has to be dug out of the ground by techniques that resemble mining, after which it is mixed with chemicals to create a liquid suitable for transport by pipeline to an oil refiner. Canada holds some of the world's largest tar sands reserves. Exxon, through local affiliates, had been buying into these holdings under Raymond's spur. Raymond believed it was wrong for the S.E.C. to exclude tar sands from proved oil reserve counting—and so he simply ignored the commission's rules when he issued press releases.

The S.E.C. had enacted the tar sands rule because it wanted investors to know when a company was engaged in mining and when it was engaged in oil production—for one thing, the expense of mining was typically higher than the expense of oil drilling, so the value of tar sands reserves over time might be less than an equivalent amount of purer oil. Raymond thought this was wrong, too: The purpose of S.E.C. regulation was to ensure accurate documentation of resources, not to force companies to dance around in outmoded categories devised by bureaucrats. There was no doubt that Exxon owned the Canadian tar sands resources it claimed— outright fraud was not the issue. In any event, if Exxon's Canadian tar holdings were included with other oil and gas holdings, the corporation's total proved reserves had indeed gone up during 1997, not down— and this is what Raymond chose to emphasize to Wall Street and the public.

Did his defiance of the S.E.C. rules matter? Raymond might disagree with the regulations on the books, but the purpose of S.E.C. regulation was to ensure that all investors had accurate information about the companies whose shares they owned. Through fraud-inflated bubbles, Wall Street reminded the world every ten or twenty years why such regulation was vital. Raymond declared publicly each winter in press releases and Wall Street presentations that Exxon enjoyed smooth, year-to-year reserve replacement when, in fact, no such picture existed, according to S.E.C. regulations. And the corporation's spinning did help to mask a strategic issue at Exxon, the challenge of resource nationalism and reserve replacement, which was of genuine and enduring importance.

On December 31, 1996, Exxon stock traded at $24.50 per share. On February 6, 1998, the day after the corporation issued a press release boasting of "the fourth year in a row that we've exceeded 100 percent replacement," Exxon shares closed at $31.00. Exxon's profitability was unchallengeable, but it could be questioned whether all of that 10 percent plus annual gain in stock price, which benefited Exxon executives and employees as well as ordinary shareholders, was honestly earned.

Raymond's defiance of the S.E.C. was not illegal—the corporation's annual 10-K filings to the commission appeared to be carefully parsed, and they broke out tar sands (or "oil sands") as a separate reserve figure, in fine print. Rule 4-10 enforced disclosures in official filings to the commission, not in press releases or oral statements to Wall Street analysts.

Exxon's practice of reserve spinning to the public reflected a broader mind-set of chutzpah toward Washington. The commission staff was by far the weaker party in this particular contest; for many years the S.E.C. did not have a single oil geologist on its payroll to assess the "proved reserve" claims made by the oil corporations it oversaw.

Exxon's argumentative press releases also signaled how heavily the reserve replacement conundrum weighed on Raymond and his colleagues. The underlying challenge of reserve booking did have far-reaching consequences: It would draw Exxon to far-flung corners of the earth in pursuit of reportable reserves the corporation might previously have ignored on the grounds that they involved too much political and economic risk and dragged the company into dictatorships and other violent settings for which it was not well prepared.

"What is the one item that concerns you most, that disturbs your sleep?" the Wall Street oil industry analyst Fadel Gheit recalled asking Raymond over a meal.

"Reserve replacement," Raymond had replied.[16]

John Browne ascended to become chief executive of British Petroleum two years after Lee Raymond took over Exxon. As individuals, their personal interests could hardly have been more disparate—Browne, the Soho-inspired gourmet; Raymond, the duck hunter and country club golfer. As business competitors, they faced a common challenge. As Browne put it later, "We believed governments of oil-producing nations would increasingly prefer to work with very big and influential oil companies. They did not think small was beautiful—that was clear when you spoke to them. They wanted to see a big balance sheet, global political clout, and technological prowess, and they wanted to be sure that you would be around for a long time."[17]

Moreover, the geographical distribution of Exxon and British Petroleum's oil holdings increasingly isolated those two companies from the regions with the most promising new opportunities. Exxon's reserves and those of British Petroleum were the most heavily weighted toward the politically safe, economically free but geologically mature oil regions of the West. About 80 percent of Exxon's reported proved oil reserves lay beneath the United States, Canada, and Europe. British Petroleum had struck out aggressively in the former Soviet Union after communism's collapse, but about three quarters of its assets remained in Britain and the United States. Chevron had acquired sizable reserves in Kazakhstan and Africa. Mobil owned large holdings in Africa and Asia; the majority of its reserves lay outside of North America and Europe.[18]

In the autumn of 1996, at the Four Seasons hotel in Berlin, John Browne presented a plan to British Petroleum's board of directors in which he argued that BP should seek a merger with another large international company in order to compete with state-owned oil companies and improve the geographical diversity of its oil holdings. Browne's first choice was Mobil.

The BP chief regarded his Mobil counterpart, Lou Noto, as a like-minded cosmopolitan. Noto had grown up in Bensonhurst, Brooklyn, as the son of a labor organizer; he was a stocky, lively, charismatic business strategist with a sizable ego. He lived in Manhattan and indulged his fondnesses for cigars, Porsches, and opera. Browne found him to be a "warm, friendly person and '*bon viveur*.'" They smoked cigars together from time to time and talked oil.

The gigantic investments and industrial operations required to produce and refine oil meant that international companies often found it financially prudent to partner on projects, much as syndicates of Wall Street investment banks shared the risks of selling stocks and bonds. This pattern of coinvestment and coexistence meant that global oil executives kept up steady contact with one another—it was a form of continuous diplomatic relations, involving both cooperation and dispute.

BP and Mobil had embarked on a joint venture that would combine their European refining businesses. Browne introduced the idea of a full merger that would create the world's largest privately owned oil company.[19]

Noto shared Browne's view of Big Oil's predicament: "We need to face some facts," he said later. "The world has changed. The easy things are behind us. The easy oil, the easy cost savings—they're done." Noto was "worried." He "expected the environment to become more volatile, and more competitive, and more difficult geographically and geologically."[20]

Mobil had inherited a large share of the downstream assets of Standard Oil. The corporation operated adeptly on the commercial side of the oil business—wheeling and dealing, negotiating customer contracts, maneuvering amid price volatility, and the like. It won major new upstream plays in newly independent Kazakhstan after the Soviet Union's demise. Yet Mobil was highly dependent on the profits generated by a single large natural gas field in Indonesia and offshore properties in Nigeria; both countries were politically unstable and wracked by violence.

Late in 1996 and early in 1997, Noto and Browne held a lengthy series of secret meetings to design a full merger of their corporations; their work climaxed at a long conference in the New York offices of the law firm Davis Polk & Wardwell. Yet as a decision point neared, Noto thought that Mobil might still be better off on its own. On March 28, 1997, he met Browne in Mobil's corporate jet hangar outside of Washington, D.C., and "made it clear that we could go no further," as Browne recalled it. Browne felt that he had "wasted a lot of time and effort." He flew back to London and announced to a colleague, "Well, we'd better think of something else."[21]

Plummeting oil prices compounded the pressures he faced. The causes of the price fall emanated from Saudi Arabia, the world's leading oil producer, at about 9 million barrels per day of capacity at the time, and the leading source of American oil imports. After Saddam Hussein's invasion of Kuwait, Venezuela's government decided to break from O.P.E.C. policy and produce as much oil as possible. It looked as if Venezuela might be trying to steal some of Saudi Arabia's American market share. The usurper attempted to almost double its oil production, from 3.2 million barrels a day to 5.5 million a day. For a while, the gambit worked; Venezuela gained more and more of the U.S. import market and replaced Saudi Arabia as America's number-one outside oil supplier. In 1997, however, Saudi Arabia retaliated by authorizing a surge in its own oil production, a program "explicitly designed to punish Venezuela" and to establish a "deterrent," as the industry consultant Edward L. Morse would describe it, to dissuade any other oil-rich country that might harbor similar ambitions. The Saudi production surge drove global oil prices to historic lows in 1998. By the end of that year, oil would fall to just ten dollars per barrel; adjusted for inflation, that was the lowest price the world had enjoyed since the 1960s.[22]

Disciplined Exxon could weather such a sudden price collapse. After the cost reduction binge of the 1980s, Raymond had reduced Exxon's operating expenses an additional $1.3 billion annually in the five years until 1997. Less-efficient companies such as Mobil struggled. Nobody knew how long prices might stay so low. The long-term challenge of resource nationalism compounded the anxiety. All this coaxed Lou Noto back to the possibility of a merger.

In June 1998, he attended a meeting with Lee Raymond organized by the American Petroleum Institute (A.P.I.), the Washington-headquartered oil industry trade group. Raymond raised the possibility of a minor deal to combine Exxon and Mobil refinery operations in Japan.

"Maybe we should talk about that," the Exxon chief said.

"That and other things," Noto replied.

Mobil's top Management Committee met in New York every Tuesday and Thursday. One morning that summer, Noto arrived and said, "Guess who I had dinner with last night? I had dinner with Lee Raymond."

The news shocked his colleagues. Exxon was more than twice Mobil's size by revenue. Layoffs would be the one inevitable by-product of such a combination, and the job losses would reach the highest ranks of the Mobil hierarchy. "There was a massive anxiety," an executive involved recalled. They worried as well about the culture shift if conservative Exxon took charge; by comparison, Mobil had been loosely governed.

That summer, John Browne advanced a fallback plan to merge with Amoco, the offspring of Standard Oil of Indiana, headquartered in Chicago. Browne and Laurance Fuller, Amoco's chief executive, held a series of private dinners in a back room of Le Pont de la Tour, the London restaurant, where Fuller "could smoke his cigarettes and we could all drink Puligny-Montrachet," as Browne recalled it. "Remarkably, no one noticed."

On August 11, 1998, they announced that their companies intended to merge, with Browne to be in charge of the successor corporation. The deal would create the largest corporation in Great Britain and one of the largest private oil companies in the world.

Browne's announcement galvanized his competitors. "It was as if the industry had been standing by waiting for someone to make the first move; it felt like we had broken a dam," as he put it.[23] Every North American and European leader of a large oil corporation seemed to conclude simultaneously that his company needed to merge to get bigger. Chevron and Texaco would soon combine, as would Conoco and Phillips, and Total with Petrofina and Elf.

Raymond believed that Exxon was primed for transformational change. As he had taken full control during the mid-1990s, he had concluded that the corporation had a management that could handle a lot more than it was being asked to do. The post-*Valdez* reformers were in place. They had restructured, streamlined, and reduced costs. They were down to "the fine grind," as he put it to his colleagues. Now what? How could they convert their emerging efficiency into a strategic leap, something that would have global scale?

Around this time, DuPont and Exxon discussed a swap of DuPont's Conoco oil division for Exxon's chemical division, but the idea did not ripen. Raymond's rationale for any proposed merger was not complicated. He ran a resource company. Replacement of resource stocks was fundamental; an acquisition at the right price was a common way for resource companies to replace reserves and grow. It was a part of Exxon's own history—Standard Oil of New Jersey had grown by acquiring Humble Oil and Refining and other reserve-rich firms. In this case, a big merger might provide a new source of leverage for Raymond to accelerate the drive for efficiency and accountability, the vanquishing of bureaucracy, that he had started after the *Valdez* debacle. It would be the "last brick in the wall of remaking Exxon," he declared.[24]

That summer of 1998, Lee Raymond and Lou Noto intensified discussions about a recombination of the baby Standards they each led. There would be antitrust issues in the United States if they proposed a deal, but their lawyers advised them that if they sold off some retail gas stations

and perhaps a few refinery properties, the deal should be approved. From
Exxon's perspective, the fit with Mobil was well tailored, particularly
because the ends-of-the-earth map of Mobil's oil reserves complemented
Exxon's more conservative profile, so heavily weighted in North America
and Europe. Mobil's holdings included substantial assets in West Africa,
Venezuela, Kazakhstan, and Abu Dhabi. It also held important natural
gas positions in Qatar and Indonesia. By purchasing Mobil, Exxon could
scale up to compete with state-owned oil giants and leapfrog onto new
geographical frontiers.

Exxon had the currency—its own stock—to make such a gargantuan
deal without incurring debt or financial risk. During the 1960s, Exxon had
handled its cash flow conventionally, paying out most of its earnings as
cash dividends. This practice rewarded small shareholders by providing
them reliable income. The corporation had about eight hundred thousand
such shareholders by the early 1980s. During inflation-menaced 1982,
Exxon's dividend was a hefty 10 percent. The next year, however, Clifton
Garvin embarked on a campaign of share "buybacks" as a substitute for
some of the spending on dividends. He was advised by Jack Bennett, a
finance wizard and mentor to Raymond who left Exxon during the mid-
1970s to work at the Treasury Department and then returned to the
corporation's board. He and Garvin concluded that after 1980 global oil
prices looked fundamentally unstable. Given the volatility that seemed
likely, it would be cheaper for at least a while for Exxon to buy oil and gas
reserves by purchasing its own stock than by investing in long-term projects at a high price point. "We had a tremendous amount of cash and no
debt, we were convinced that the price structure was unstable, and thus
we had no other option," Raymond recalled, but to buy back shares. Exxon
management had long raised cash dividends to beat inflation, but Garvin,
and later Rawl and Raymond, were reluctant to raise the dividend too
much higher, to match the 5 and 6 percent payouts offered by Shell and
British Petroleum, for fear that in a down cycle for oil prices "you can get
yourself in a real squeeze on cash," Raymond said.[25]

Each year, therefore, the corporation went into the stock market and
used some of the cash generated by its operations to buy its own shares.
Between 1983 and 1991, Exxon bought a net total of 518 million shares

worth $15.5 billion—a whopping 30 percent of the shares then outstand-ing.[26] As a result, each remaining share owned by an investor controlled a progressively higher percentage of Exxon's profits and oil reserves: In 1983, a single Exxon share owned 6.7 barrels of oil and gas equivalents, but at the end of 1989 it owned 8.4 barrels.

During the buybacks, Exxon's dividend yields fell in relative terms, leaving small shareholders with less cash in their pockets. Did this mat-ter? Arguably, dividend payments and buybacks were equivalent—in one case, a shareholder received cash and in the other, the value of shares rose proportionately. One question was who would make better use of ExxonMobil's cash, its executives or its dispersed shareholders. By the time he took charge of the corporation, Raymond answered emphatically, "We can."

Arguably, too, share buybacks could be justified only if the price of Exxon shares at the time of purchase was so low that buying them was a better use of cash than looking for new oil reserves. In the decades to come, however, Exxon would make buybacks continuously, in all price environments, joining an American corporate fashion. Academic studies showed that many corporate leaders had a poor record of buying back shares only when prices were low. "The implied returns . . . from buybacks by big companies would have been laughed out of the boardroom if they had been proposed for investment in bricks and mortar or other more conventional projects," wrote Richard Lambert, a British critic of the prac-tice.[27] Such programs also raised red flags with some corporate governance specialists because of the manipulations they might mask. Corporate man-agers might deliberately suppress earnings before a buyback campaign by front-loading expenses to temporarily drive down the price of shares they intended to buy. Repurchases might also smooth out publicly reported earnings per share, to sell Wall Street investors on a story of placid growth when the underlying business was more volatile. In Exxon's case, all of these concerns hovered, but the buybacks at times also seemed a way to dispose of a problem that other companies could only envy: too much cash.

The shares Exxon bought back did not simply vanish; they were parked in the form of "treasury shares" belonging to the corporation. Raymond and his management team chose to use the parked shares to purchase

Mobil tax free. If Exxon could not discover on its own great gobs of oil, it would buy what it could not find: This was an extraordinary payoff for two decades of cash flow discipline.

In late November, Raymond flew on an Exxon Gulfstream IV corporate jet to Augusta, Georgia. The corporation maintained a membership at the Augusta National Golf Club, the site of the annual Masters tournament, and the club hosted an annual Thanksgiving party for families. Raymond typically attended and played golf with his three sons, who had grown into better golfers than he was. This time, Raymond ensconced himself in one of the club's cabins and played as many rounds as possible while reviewing the final deal terms. Late one night a messenger had to find Raymond's bungalow to hand over documents.[28] Everyone involved in the deal negotiations, including Lou Noto, knew this would not be a merger of equals. There would be a weighted exchange of shares, but as a practical matter Exxon would take over Mobil. Noto volunteered to accept a subordinate position as vice chairman, reporting to Raymond; he would serve in that role for a transition period and then depart, to enjoy his life in New York.

John Browne later asked Noto if he would have preferred to merge with BP or Exxon. "BP, of course, but I couldn't make it work," Noto told him, as Browne recalled it. "When you bought Amoco it was inevitable that Exxon would buy us. It was only a matter of time." BP had merged its way to a size that Exxon had to match if it wanted to compete, and the acquisition of Mobil was the easiest way for Raymond to get there.[29]

On December 1, 1998, Raymond and Noto stood side by side in a J.P. Morgan conference facility in New York to announce their $81 billion deal. There was no mistaking the new company's hierarchy: Raymond opened the meeting and spoke for twenty straight minutes. He laid out the merger terms, described the prospective business advantages, and announced future plans in bland press-release prose, displaying all the charm of "the proverbial shoe salesman," as one newspaper reporter covering the announcement put it.

When at last his turn at the microphone arrived, Noto hastened to say that his decision to merge with Exxon was "not a combination based on desperation."

The Shape of an Empire:
Upstream Investments of Exxon and Mobil Before Their Merger

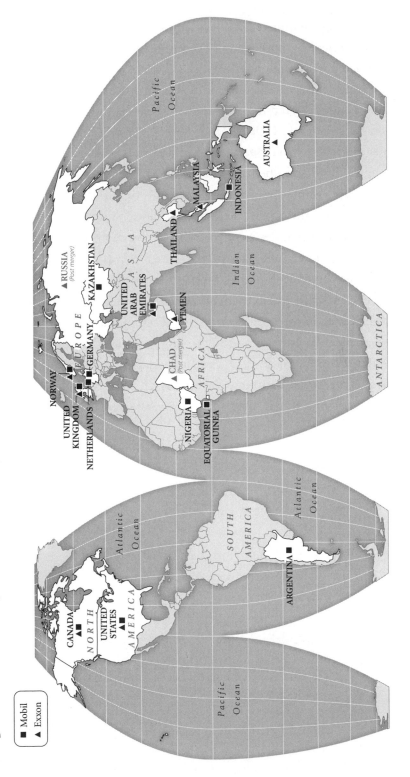

Note: Map shows countries where Exxon and Mobil were producing more than 40,000 barrels per day of oil or 1 million cubic feet of gas in 1998, according to materials published by the companies. Exxon's properties in Russia and Chad, not yet in production, are also included.

Map by Gene Thorp

It was, instead, a requirement of the times. "Competition has changed," he said. "We're here because we're trying to respond to these changes."[30]

ExxonMobil Corporation—the world's largest nongovernmental producer of oil and natural gas, and soon to become the largest corporation of any kind headquartered in the United States—formally came into existence on December 1, 1999. During its first year of combined operations, the corporation would earn $228 billion in revenue, more than the gross domestic product—the total of all economic activity—of Norway. If its revenue were counted strictly as gross domestic product, the corporation would rank as the twenty-first-largest nation-state in the world. A United Nations analysis, designed to calculate by more subtle measures the relative economic influence of particular companies and nations, concluded that ExxonMobil ranked forty-fifth on the list of the top one hundred economic entities in the world, including national governments, during its first year. Its net profit alone—$17.7 billion that inaugural year—was greater than the gross domestic product of more than one hundred nation-states, from Latvia to Kenya to Jordan. As Lee Raymond told his colleagues, "If we haven't gotten to 'economy of scale,' we're never going to find it." He was optimistic. Oil prices were rising again. "It's a great time to be ExxonMobil," he declared."[31]

Three

"Is the Earth Really Warming?"

O n February 8, 2001, nineteen days after George W. Bush's inauguration as president, ExxonMobil chairman Lee Raymond met with Vice President Dick Cheney in Cheney's West Wing office at the White House. They knew each other "very well," as Raymond put it later. Indeed, Raymond had known Cheney for more than two decades, dating back to the period when Cheney was a congressman from Wyoming. They had hunted quail and pheasants together. They were compatible personalities— both reticent, bred on the cold plains of the upper Midwest, and both educated at the University of Wisconsin. They were ardent in their free-market views, inclined to a certain tough bluntness, and not very much worried about what others thought about them, particularly bicoastal media elites and liberal intelligentsia.

During the 1990s, the Cheneys and the Raymonds had lived near one another in the old-line Preston Hollow and Highland Park neighborhoods of Dallas. Cheney served after 1995 as chief executive of Halliburton; his company contracted regularly to provide services to Exxon. (Halliburton did not seek to own and produce oil and gas directly, as ExxonMobil did, but made its money by providing construction and engineering services under contract to oil producers, whether they were government-owned companies or private corporations.) Socially, Raymond's wife, Charlene, and Cheney's wife, Lynne, saw each other not only in Dallas, but at retreats and meetings of the American Enterprise Institute for Public Policy Research, a free-market think tank where Raymond served on the board and Lynne served as a senior fellow. When Raymond sat down with Cheney after the latter's swearing in as vice president, the meeting was best understood as a discussion between like-minded peers and friends who were comparing notes at the cusp of a new project.

"Look," Raymond told Cheney, as he recalled it later, "my view is this country is going to be an importer [of oil] for as far as the eye can see. If you believe that to be the case, what things can we do?" Raymond answered his own question: "The first thing anybody will tell you is you have to have diversification of supply—you can't get yourself in the position where you only have one supplier."

That situation had developed between the United States and the Middle East. To overcome this bottleneck of geopolitical dependency, Raymond said, the United States had to be "engaged every place, every place you can go. . . . Even in the North Slope of Alaska." Raymond felt that "one of the key issues in foreign policy" was how to manage the challenge of increasing access to new oil supplies. As he recalled it, he told Cheney, "We should be trying to encourage, as a matter of foreign policy, having countries develop their natural resources," so that the United States could have "as diverse a portfolio of supplies as you can have, so that you don't get yourself beholden to any one person."[1]

ExxonMobil had enjoyed easy access to high-ranking government officials during the Clinton administration; when the corporation's Washington representatives needed a meeting, they almost always got one. Raymond told colleagues that ExxonMobil enjoyed access to the administration that was comparable to the halcyon years of the Reagan presidency. Clinton appointees approved the Exxon and Mobil merger with a minimum of fuss. Al Gore's candidacy for the White House, however, had attracted considerable resistance from the oil industry, in part because of Gore's record of environmental activism. Oil and gas companies had donated $34 million to political candidates during the 2000 cycle—more than three fourths of that funding had gone to Republicans.[2]

George W. Bush's father had made his fortune in the oil patch, and the new president had run a wildcatting firm earlier in his career, less successfully. There was little mystery about how the Bush administration would proceed on energy policy. A few days before his meeting with Raymond, Bush had assigned Cheney to head a cabinet task force to make rapid recommendations. The speed with which the panel planned to issue its findings—just a few months would pass from the panel's formation to the issuance of a finished report—made the initiative recognizable to Washingtonians as one of those precooked packages where the authors know much of the outcome in advance. The purpose of such a task force is typically not to deliberate over difficult problems, but to create a process whereby Congress, industry, and the incoming cabinet all become invested in a set of recommendations formally endorsed by a new president. These can then be quickly translated into executive orders, new regulations, and proposed laws.

Cheney ordered the task force to work in secret, to protect the prerogatives of the White House, and by doing so he made the group's work a lightning rod for criticism and conspiracy fears. It took years of litigation to discover through partial document releases what intuition might have suggested at the time: Cheney favored energy deregulation and an aggressive push to open up oil and gas production in the United States, and he identified himself with the priorities of Lee Raymond, among other industry executives.

James J. Rouse, a United States Army veteran educated in Mississippi who ran ExxonMobil's Washington, D.C., office, attended one of the Cheney panel meetings next door to the White House and presented some of the corporation's forecasts about future global energy demand. It was rising, he noted, and so more oil and gas production would be required. But ExxonMobil did not require access to a midlevel panel staffed by civil servants to make its points to the Bush White House. Lee Raymond flew to Washington about every other month and met privately with Cheney on some of his visits, perhaps two or three times a year. If he needed to make a request or share an observation more urgently, all he had to do was pick up the telephone.

Cheney had seen for himself at Halliburton how geopolitics, resource nationalism, and the emergence of large state-owned oil companies had pinched Western oil companies as they attempted to replace the equity reserves they pumped and sold each year. One purpose of the preconceived energy task force Cheney led was to prepare for legislation designed to open up Alaska's Arctic National Wildlife Refuge to oil drilling. Raymond supported Cheney's plan and railed regularly in public against what he regarded as the self-defeating energy policies of the United States, which restricted access to oil and gas on free-market American territory while exacerbating the country's dependency on imports from unstable and nationalistic regimes. Compared with other oil majors, however, ExxonMobil was no longer a dominant player inside the United States. Chevron had inherited some of the longest-lived of Standard Oil's American oil properties, in California, and Chevron and British Petroleum had moved more boldly than Exxon into the Gulf of Mexico when leasing opened during the Clinton administration. Exxon had opportunities to exploit oil and natural gas in Alaska, but held back from some expensive deals because Raymond had learned after the *Valdez* that the political risks posed by Alaska's frontier-minded political culture and populist governors were comparable to those in West Africa.

Buoyed by the trust embedded in their long-lived friendship, Raymond and Cheney typically talked less about domestic oil policy issues than about developments overseas in the world's best-endowed regions. Raymond might report on conversations he had recently had with the president of Kazakhstan or the foreign minister of Saudi Arabia. He might relay to Cheney insights about the Saudi royal court or requests about some problem Saudi leaders were having with the Bush administration. Cheney might share similar notes from his own conversations with foreign leaders. In protocol, power, and habit of mind, Raymond and Cheney were each, in effect, deputy heads of state—when they traveled, they met with kings and presidents, and perhaps ministers or chiefs of national oil companies, but rarely anyone less powerful. When the friends gathered in Washington, their meetings were strictly business. If Raymond needed only ten minutes to make his request about a Kazakh trade issue, for example, and Cheney was quickly satisfied with the ExxonMobil chief's arguments, Raymond would not waste the vice president's time any further. In the White House, a one-hour meeting was a lengthy one, and Lee Raymond knew what it was like to have a daily schedule that was oversubscribed.

ExxonMobil's interests were global, not national. Once, at an industry meeting in Washington, an executive present asked Raymond whether Exxon might build more refineries inside the United States, to help protect the country against potential gasoline shortages.

"Why would I want to do that?" Raymond asked, as the executive recalled it.

"Because the United States needs it . . . for security," the executive replied.

"I'm not a U.S. company and I don't make decisions based on what's good for the U.S.," Raymond said.

ExxonMobil executives managed the interests of the corporation's shareholders, employees, and worldwide affiliates that paid taxes in scores of countries. The corporation operated and licensed more gas stations overseas than it did in the United States. It was growing overseas faster than at home. Even so, it seemed stunning that a man in Raymond's position at the helm of an iconic, century-old American oil company, a man who was a political conservative friendly with many ardently patriotic office-holders, could "be so bold, so brazen." Raymond saw no contradiction; he did indeed regard himself as a very patriotic American and a political conservative, but he also was fully prepared to state publicly that he had fiduciary responsibilities. Raymond found it frustrating that so many people—particularly politicians in Washington—could not grasp or would not take the time to think through ExxonMobil's multinational dimensions, and what the corporation's global sprawl implied about its relationship with the United States government of the day.[3]

After the merger, ExxonMobil moved its Washington, D.C., office from Pennsylvania Avenue to K Street, in the heart of the capital's lobbying district. The office occupied one high floor of a new complex constructed from pink granite and concrete. A turret, topped by an American flag, distinguished the building from the LEGOLAND of downtown Washington. Landscape paintings lined the walls of ExxonMobil's suite; antique Mobil oilcans and Esso signs decorated the shelves of its conference rooms. The furniture and dark cherry wood paneling created a formal and professional atmosphere, but the office eschewed the garish luxuries of the capital's big law firms and the D.C. outposts of Wall Street investment banks.

By the time of the merger with Mobil, the office's director, James Rouse, had been with the company for thirty-seven years. The résumés of the lobbyists he hired spoke of technical competence and stolidity. ExxonMobil's strategy was not so much to dazzle or manipulate Washington as to manage and outlast it.

The Washington office functioned not only as a liaison to the White House and Congress, and to industry associations such as the American Petroleum Institute, but also as a kind of embassy, to deepen ExxonMobil's influence and connections in the foreign countries where it did the most business. International issues managers developed contacts with ambassadors and commercial representatives of foreign embassies. As this international group in the K Street office expanded, Robert Haines, a West Point graduate and Vietnam veteran, handled Asian embassies; Sim Moats, a former State Department diplomat, handled Africa; and Oliver Zandona, a Mobil career manager, watched the Middle East. They also worked the American bureaucracy, particularly the State Department and the National Security Council, to influence and understand U.S. policy. They maintained some contacts at the Central Intelligence Agency and other federal intelligence departments as well. The C.I.A. ran a station in Houston, where oil industry executives who traveled globally occasionally stopped by for informal exchanges about leaders and events abroad, but in Washington the corporation's lobbyists tried to keep their distance from Langley, site of the C.I.A.'s headquarters, for fear that host governments where the corporation produced oil might mistake ExxonMobil as some sort of wing of the spy agency.

ExxonMobil relied mainly on lifelong career employees for much of its Washington lobbying. Lee Raymond kept the number of full-time lobbying staff relatively modest, partly to avoid attracting unfavorable attention from journalists or opponents in the environmental lobbies. Where possible, the corporation lobbied as part of an industry coalition, principally through the American Petroleum Institute. Yet Raymond maintained a strong core of ExxonMobil staff to work Capitol Hill, the White House, and regulatory agencies. Many of the corporation's employee-lobbyists rotated into Washington from other corporate disciplines, with little or no prior experience of government affairs. "I think we did it in-house so we could get it done right," said Joseph A. Gillan, who worked on environmental issues in the office around the time of the Mobil merger. "Exxon's culture was, 'Let's do it ourselves.'"[4]

The corporation did retain Washington law firms and outside lobbying shops as consultants, although relatively few during the first Bush term. But these hired guns dispensed advice to ExxonMobil on contracts and were not typically relied upon to deliver access—some of the outside retainers were even forbidden from attending meetings with public officials. Some of the key outsiders on retainer—former oil-friendly senators such as Don Nickles of Oklahoma and J. Bennett Johnston of Louisiana—attended a weekly meeting at ExxonMobil's office to exchange intelligence and analysis about legislation, electoral politics, and energy issues. The meetings sometimes had an air of jockeying and showboating as the lobbyists competed to show how much inside knowledge they possessed (and therefore how much value their retainer contracts generated for the oil corporation). The corporation spent heavily on its Washington operation, but the great majority of its costs involved salaries and benefits for career employees. In 2001, ExxonMobil spent $6 million on its lobbying operation, but only $300,000 on outsiders. The next year, it spent $8.3 million in total, but only another $300,000 on contractors.[5]

Rouse oversaw only one lobbyist to watch the entire House of Representatives: Jeanne O. Mitchell, a University of Florida graduate, a cheerful woman who briefed ExxonMobil's policy and tax PowerPoint slides with genuine enthusiasm. Her counterpart for lobbying the Senate, William "Buford" Lewis, was a balding specialist in gasoline and other transport fueling systems.

Lobbyists and consultants newly hired at the office were instructed that ExxonMobil sought to avoid asking for specific favors, such as earmarks, on Capitol Hill. "We don't need the government's help" was the prevailing instruction. "We just want to know the rules of the road." The line used to indoctrinate new arrivals to the Washington office was that ExxonMobil did not want anything *from* the American government, but it did not want the government to do anything *to* the company, either. Lee Raymond saw his Washington operation as being 180 degrees opposite from, say, General Electric, which ExxonMobil executives regarded as a Washington rent seeker, always trying to bend Congress and the administration to policies that would subsidize or enhance G.E. business divisions. When Raymond saw G.E.'s celebrity chief executive, Jack Welch, he did not hesitate to give him a hard time about his corporation's favor seeking in Washington. In fact, Raymond was kidding himself. There were many favors, executive orders, lobbying meetings, and laws ExxonMobil sought and obtained from the American government. Yet the above-it-all slogans imparted to newcomers did reflect the corporation's aversion to backroom deal making on legislation in Congress. ExxonMobil had long had a policy of refusing to give rides to members of Congress on its fleet of corporate jets. The advantage of advertising an official, declared policy of asking for no favors from members of Congress, Raymond explained, was that "when they come and ask us for favors, we can say no." The corporation's typical lobbying meeting on the Hill involved briefing and then leaving behind preprinted PowerPoint slides vetted in Irving. When ExxonMobil lobbyists looked for serious help, they more often turned to the executive branch, discreetly.[6]

A sardonic line among ExxonMobil lobbyists in the Washington office held that the corporation's number-one issue of concern was taxation; its number-two issue was tax; its number-three issue was tax; and its number-four issue varied from year to year. With gross global revenues well north of $200 billion, even small changes in the U.S. corporate tax code could cost ExxonMobil dearly. Raymond instructed the K Street office not to ask for specific tax earmarks, but to concentrate on preventing unfavorable changes in the code, such as oil industry–specific windfall taxes or changes to depreciation rules that might raise ExxonMobil's effective tax rates—this lobbying work was often defensive and involved trying to talk industry-friendly coalitions in Congress into blocking unfavorable changes.[7]

ExxonMobil's lobbyists operated within the same disciplined, hierarchical corporate system that refinery managers and offshore drilling platform operators did. Their talking points could be mechanical sounding: "This is what the corporation believes is the best course of action." In the scrum of final conference talks over a particular bill, if Jeanne Mitchell or Buford Lewis was asked for an opinion about an alternative option to the one she or he had briefed, the ExxonMobil lobbyist would simply repeat the talking points and conclude again: "This is what the corporation believes is the best course of action." A lobbyist for another oil corporation recalled, "You'd think, 'Why do I feel like I'm talking to a wall?' Because you are: They can't move off a position" without permission from headquarters. They were "honest as the day is long," recalled a former Republican staffer, speaking of the corporation's Washington staff. "Sometimes blunt."

ExxonMobil commanded the in-house expertise "to run to ground all the possible technical arguments around a particular policy area," said an industry advocate. "They have people there who can talk about epidemiological studies down to the parts per billion." Yet the Washington office's influence was also limited by its rigidity and perceived arrogance. "They have a terrible reputation and they deserve criticism for having allowed that to develop," reflected a Republican lobbyist who worked with them. "They very much have the opinion that 'We are ExxonMobil; we're right. And we will prevail.' In Washington, it doesn't always work that way. Most people don't understand the economics of energy and ExxonMobil hasn't explained that very well. They have the money to educate people but they don't do it very well."[8]

ExxonMobil's Washington strategists divided the capital's political population into four broad tiers, in descending order of sympathy for Irving's agenda. There were those who represented or otherwise had emerged from the oil patch, where many thousands of jobs were at stake—senators and congressmen from Texas, Louisiana, Oklahoma, and Wyoming, some of them industry veterans. This group included, after 2000, President George W. Bush and Vice President Dick Cheney. The second tier consisted of free-market Republicans who didn't particularly understand the oil and gas industry, but who usually would be supportive of the industry's positions. The third tier consisted of Democrats or liberal Republicans who tended not to trust ExxonMobil and its ilk, and who regularly voted against the corporation's interests, but who had been around Washington long enough to become pragmatic about oil industry issues; they were at least open to constructive discussion and might occasionally vote the industry's way. ExxonMobil's lobbyists and executives cultivated ties and made generous campaign contributions to all three of these Washington subspecies. Lee Raymond had friendships with industry-leaning Democrats—Representative John Dingell, the longtime automotive industry champion from Michigan, for example.

Then there was tier four: the enemy, as some of the military veterans who manned ExxonMobil's Washington office did not mind putting it from time to time. These were Democrats and environmental activists who, it seemed to the corporation's executives, wanted to disenfranchise ExxonMobil and to use the corporation's unpopularity to galvanize liberal constituents and funders. These activists did not believe in the legitimacy of the profit motive, in the estimation of some ExxonMobil executives. Senators Charles Schumer and Dick Durbin fell into this category; Senator Hillary Clinton, on the other hand, did not. Lee Raymond accepted that there was not much he could do about the company's permanent opposition in Washington; these people were not going to change their views. Nor should ExxonMobil bend to them. Clifton Garvin's flirtation with solar investments during the Carter administration was regarded within the corporation as an object lesson in what not to do. Even Raymond accepted that it had made sense for Garvin to explore alternative energy businesses during the oil supply upheavals of the 1970s, but the lesson, he believed, was that alternatives to oil were not economically competitive and would not be for the foreseeable future. The corporation should stick to its core expertise and not chase after fleeting political or policy fashions. "In hindsight it appeared that we were abdicating who we were," Raymond recalled. "Presidents come and go; Exxon doesn't come and go."[9]

Some of ExxonMobil's Washington lobbyists also believed that the most extreme anti-oil activists could be contained only by direct counter-attack and pressure. There was no sense in pretending otherwise, they argued. The industry's uncompromising opponents had to be taken on uncompromisingly.

As the Bush administration took office, one issue was rising in Washington that was of far-reaching, even existential importance to the oil industry, an issue that would test ExxonMobil's lobbying prowess: climate change. As policy debates about global warming intensified, two men, one in government and the other in industry, would increasingly distinguish themselves by their ardent skepticism toward climate scientists and their opposition to all government regulation: Dick Cheney and Lee Raymond.

In a large color ad taken out in *Life* magazine in 1962 by ExxonMobil's precursor Humble Oil, a small, smartly dressed cartoon character saluted a photograph of a majestic glacier. "Each Day Humble Supplies Enough Energy to Melt Seven Million Tons of Glacier!" the ad's headline declared.

Such was the John F. Kennedy era of scientific optimism, as marketed by Madison Avenue. Four decades later, of course, many scientists regarded the retreat of glaciers and mankind's unembarrassed capacity to melt them as signs of a slowly unfolding global catastrophe. (Of the 144 glaciers monitored by researchers between 1900 and 1980, 2 advanced and 142 retreated, an indicator of the earth's warming surface temperature during the twentieth century.)[10] Climate change became a galvanizing priority of science and public policy in a remarkably short time. It took less than two decades from the time of Humble's ice-melting ad campaign for Exxon's executives to recognize that climate change would arrive as a public policy challenge in Washington and other global capitals, and that it might undermine the corporation's business model. Characteristically, Exxon began to prepare itself well before the phrase "global warming" saturated public consciousness.

The "greenhouse effect" is a natural process in which sunlight is trapped by the planet's atmosphere; without it, the earth would be very cold. The question that scientists gradually examined during the twentieth century was whether additional gases released by the clearing of forests and the burning of fossil fuels—coal, oil, and natural gas—accelerated the greenhouse effect. It was not until the 1950s and 1960s that a few scientists began to document credibly that human activity was releasing more and more carbon dioxide, a greenhouse gas, and that warming might be the result. Their findings did not immediately stick. The earth's climate had undulated across millennia. Orbital variations, the intensity of sunlight, and other natural factors had produced alternating ice ages and periods of boiling seas long before the rise of smokestacks and automobiles. Scientists remained divided as late as the 1970s about essential questions such as whether the earth was warming markedly, whether a warming or cooling trend posed the greatest future danger, and how human and industrial activity fit into the picture.

James Hansen, an astrophysicist at the National Aeronautics and Space Administration, was among the first scientists to call attention to the danger that greenhouse gas emissions could produce dramatic warming within a relatively short time. His and other work prompted the first National Academy of Sciences examination in 1979. The academy's study group found that if man-made CO_2 emissions—from coal-burning electricity plants, automobile exhaust, truck fumes, airplane exhaust, deforestation, and other sources—continued to grow, there was "no reason to doubt that climate changes will result and no reason to believe that these changes will be negligible," as the chairman of a study for the National Research Council's Climate Research Board put it. The findings might be "disturbing to policymakers" because "a wait-and-see policy may mean waiting until it is too late."[11]

The National Academy report attracted Exxon's attention. Any effort to tax, limit, or eliminate carbon dioxide emissions on environmental grounds would have obvious implications for Big Oil. Exxon emitted tens of millions of metric tons of carbon dioxide in the course of its own oil production, refining, chemical manufacturing, and electricity-generating operations. Not only did the corporation burn carbon-laden fuels, it then sold such fuels for profit to other users, who also burned them.

In 1980, just after the publication of the National Academy study, the corporation hired its own astrophysicist, Brian Flannery, who had taught at Harvard University. A few years later Flannery recruited a chemical engineer named Haroon Kheshgi, who had worked at the Lawrence Livermore National Laboratory. Flannery and Kheshgi started to produce, while salaried employees for Exxon, peer-reviewed research for the United Nations's Intergovernmental Panel on Climate Change, or I.P.C.C. This was a network of many dozens of mostly academic and government scientists established to create definitive assessments, at multiyear intervals, of the scientific evidence about global warming. Exxon's climate scientists also used corporate funds to support climate-modeling research at the Massachusetts Institute of Technology. They produced internal assessments of the scientific and policy questions for Exxon's Management Committee. In the early years of this Exxon climate work, constructing an accurate model of future Earth temperatures seemed daunting. When Flannery arranged contracts with the M.I.T. climate modelers, he told them, by his own account, "Embrace the uncertainty in all of this."[12]

Lee Raymond had no particular background in climate science, but as a chemical engineer whose doctoral thesis had concerned mathematical modeling, he considered himself adequately qualified to reach his own judgments on the underlying scientific questions. (He was the first Exxon chief executive to have a doctorate.) At the University of Minnesota, where he had earned his advanced degree on a scholarship, his academic mentor, Neal Amundson, a renowned figure in chemical engineering, had instructed him, "Science is science, and don't let these damn politicians ever screw you up."

During the 1980s, when global warming first emerged as a public policy matter, Raymond turned to the scientists in Exxon's oil exploration department, who by profession studied the history of the planet. The corporation's scientists told him that climate measurements on Earth were very recent relative to the planet's longevity, and that this was a reason to be skeptical about extrapolating data.

Raymond also entered the incipient climate debate with deep skepticism about nonpolluting alternatives to oil and gas such as solar and wind power, which were relatively costly but might seem more attractive if climate change was a concern. "I've been there and done that," Raymond would say of his history with green technologies.[13] He was referring to successive management assignments he undertook during the late 1970s and early 1980s. In one, he served as second-in-command at New York–based Exxon Enterprises, which housed alternative energy initiatives, including solar power. The division had been conceived during the 1960s as a kind of in-house venture capital arm that might incubate new and profitable businesses outside of oil and gas. The embargoes and oil price shocks of the 1970s made this goal seem all the more appealing. Exxon went into the office equipment business, selling electronic typewriters, fax machines, ink-jet printers, flat-panel displays, voice recognition hardware, home computers, and computer chips. The corporation studied the possibility of merging with Bristol-Myers, the drug maker; Colgate-Palmolive, the consumer products giant; and Hewlett-Packard, the computer company. In retrospect, such diversification looked like folly to oil industry strategists, but at the time, it was a corporate fashion.

Raymond's contribution to Exxon's experimental thrust was to recommend that it be shut down. He dumped the corporation's solar investments. Any business that required government subsidies to be viable was not for Exxon, he declared.

In the summer of 1988, amid a record-breaking heat wave, James Hansen testified before Congress about the findings of a paper he had coauthored with six other N.A.S.A. scientists. Using three different forecasts of releases of CO_2 into the atmosphere during the century to come, Hansen and his colleagues predicted that even in the best case, future temperature changes would be "sufficiently large to have major impacts on people and other parts of the biosphere."[14]

His work fortified the first attempt by governments to regulate greenhouse gases. Delegates to the 1992 Rio Earth Summit negotiated a treaty, the United Nations Framework Convention on Climate Change. President George H. W. Bush signed the agreement, and the United States Senate ratified it. The treaty divided the world's governments into categories, distinguishing between wealthy industrialized countries and poorer, industrializing ones. It embraced the principle that wealthy countries should pay the greenhouse gas reduction costs of poorer countries, on the grounds that the privileged nations had created much of the problem in the first place and could afford to fix it, whereas it would be unfair to penalize or restrain the industrial growth of poor countries as they tried to lift their citizens out of poverty. The convention exacted no binding commitments from any of its parties. However, the governments and leaders of industrialized countries, including President George H. W. Bush, pledged to adopt national policies that would "aim" to reduce their overall greenhouse gas emissions to 1990 levels by the year 2000.

Three years later, the United Nations's assessment group, the I.P.C.C., reported that most of the observed warming on Earth's surface since 1950 was likely to have been caused by human and industrial activity. "The balance of evidence . . . suggests a discernible human influence on global climate," its summary report stated.[15]

Lee Raymond publicly rejected even the qualified formulations of the 1995 assessment. In October 1997 (which would prove to be the fifth-warmest year on the planet, to that point, since the mid-nineteenth century), he flew to Beijing to deliver a speech to the Fifteenth World Petroleum Congress, an event hosted by the People's Republic of China. At the time, the Clinton administration was in the last round of international negotiations that would produce the Kyoto Protocol, an enhancement of the 1992 Framework Convention, with commitments that would require rich governments to reduce their emissions. Raymond's purpose in Beijing was to denounce the Clinton administration's negotiating position. He devoted thirty-three paragraphs of his seventy-eight-paragraph speech to the argument that evidence about man-made climate change was an illusion and that a binding agreement to reduce greenhouse gas emissions was therefore unnecessary:

Is the Earth really warming? Does burning fossil fuels cause global warming? And do we now have a reasonable scientific basis for predicting future temperature?

In answer to the first question, we know that natural fluctuations in the Earth's temperature have occurred throughout history—with wide temperature swings. The ice ages are a good example.

In fact, one period of cooling occurred from 1940 to 1975. In the 1970s, some of today's prophets of doom from global warming were predicting the coming of a new ice age. . . . The Earth is cooler today than it was twenty years ago.

We also have to keep in mind that most of the greenhouse effect comes from natural sources. . . . Only four percent of the carbon dioxide entering the atmosphere is due to human activities—96 percent comes from nature.

Leaping to radically cut this tiny sliver of the greenhouse pie on the premise that it will affect climate defies common sense and lacks foundation in our current understanding of the climate system. . . . It is highly unlikely that the temperature in the middle of the next century will be affected whether policies are enacted now or 20 years from now.

He went further: He urged poor, rapidly industrializing countries such as China to defy the United States and Europe by blocking any agreement in Kyoto that would result in "slower economic growth, lost jobs, and a profound and unpleasant impact on the way we live." China and other developing nations might be exempted from the treaty's direct economic costs, but this "will not prevent them from being hurt. Their exports will suffer as the economies of industrialized nations slow. So all of us would suffer from these proposals." Moreover, China and other poorer countries had an obligation, on behalf of their impoverished citizens, to ignore the fears of environmentalists comfortably ensconced in the wealthy West, Raymond argued:

The most pressing environmental problems of the developing nations are related to poverty, not global climate change. Addressing these problems will require economic growth, and that will necessitate increasing, not curtailing, the use of fossil fuels.[16]

It was extraordinary for the chief executive of a U.S.-headquartered multinational to lobby against a treaty he disliked by appealing to a Chinese Communist government, among others, to adopt a negotiating position opposed to a sitting American president.

Raymond believed, however, that his obligation as Exxon's chief executive was not primarily to support American diplomacy—and certainly not when he disagreed with its assumptions so profoundly. The Beijing address was "seminal," recalled Frank Sprow, a senior Exxon executive who worked closely with Raymond on the climate issue.

Exxon's message was that governments should avoid steps that would curtail economic growth. Raymond adamantly believed that Kyoto was both an impractical and an unjust economic agreement—impractical because it would require the United States to make sacrifices in its national way of life that its people would never undertake, and unfair because it laid too much of the climate policy burden on developing economies whose governments had an urgent moral duty to lift their people out of poverty, which required, in his estimation, burning fossil fuels.

Exxon might withstand the financial and business burdens that would likely follow from treaty-imposed limits on greenhouse gas emissions, but Raymond feared that the global economy would slow markedly and the knock-on effects of reduced growth would hurt the oil industry and the country. He also viscerally resented what he regarded as the fear-mongering of the environmentalist movement. Only by hyping the threat could they justify immediate, even drastic policy intervention: "Just give me a break!" Raymond told his colleagues.

"They had come to the conclusion that the whole debate around global warming was kind of a hoax," said an executive who had direct access to Raymond. "Nobody inside Exxon dared question that."

China and scores of other poor countries ignored Raymond's pleadings and signed the Kyoto Protocol, along with the United States, in December 1997. Thirty-seven industrialized nations, including America, accepted binding targets (although without any enforcement mechanism)

that between 2008 and 2012, they would reduce their emissions 5 percent below 1990 levels. For the first time—almost two decades after the
first National Academy of Sciences study had suggested that climate
change might be "disturbing to policymakers"—a regime to control greenhouse gas emissions threatened to impose real costs on industrial corporations like Exxon.

Arthur G. "Randy" Randol III, who served as ExxonMobil's senior environmental adviser in Washington at the time, led climate lobbying for
the corporation's K Street team. Randol had earned his doctoral degree
in nuclear engineering at the University of Florida in Gainesville. During
the 1990s, he had immersed himself in the issues around climate change.
A large man, he could be blunt in argument. He was "brilliant," an admir-
ing colleague said, but he had "a reputation for being pretty aggressive.
Lots of people in Washington are very polite in meetings, and Randy is a
bull in a china shop." He could "talk about climate studies and carbon
technology projects the way other people I know talk about the 1986 Red
Sox outfield," another colleague recalled.

For political cover, Exxon increasingly worked the climate account
through the American Petroleum Institute, the industry trade and advocacy group. Randol provided technical expertise, while Raymond offered
authority and funds. During the late 1990s, with emphatic support from
Exxon, climate became the "eight-hundred-pound gorilla" within the institute, a "really, really big issue—bigger than anything else," a former
executive recalled. The oil industry did not want "to risk a reduced reliance on petroleum based upon provisional science, emerging science, or
based upon harmful public policies," as Philip Cooney, an A.P.I. attorney
who worked on climate policy at the time, put it. Lee Raymond took the
lead within A.P.I., strengthened by the expertise of Exxon's in-house astrophysicist, Brian Flannery.[17]

They recognized that if they oiled the opposition to Kyoto in
Washington—if they allowed environmental groups to frame the issue as
one pitting greedy oil corporations against planet Earth—they would un-

dermine their own interests. To evade direct assaults by environmentalists, Exxon and other A.P.I. members joined a newly invented and more broadly based group, the Global Climate Coalition, with influential members from every part of the country and many different industries. Exxon and Royal Dutch Shell joined, but so did the Aluminum Association, General Motors, Ford, and DaimlerChrysler. They won endorsements from autoworkers concerned that Kyoto would lead to American job losses. During the last years of the Clinton administration, the coalition became "the most effective industry association I've ever seen at working to block progress on climate change," Kert Davies, the research director for Greenpeace, said later. Under Raymond's spur, A.P.I. also poured money into independent think tanks and advocacy groups that were predisposed to attack Kyoto, or were invented for the purpose by individual anti-Kyoto campaigners aligned with industry. Their strategies emphasized "the promotion of free-market principles," as the institute's lawyer, Phil Cooney, later put it.[18]

Greenpeace launched its own well-funded campaign to strengthen Kyoto. In the United Kingdom it attacked British Petroleum. In the United States it focused its efforts on the Global Climate Coalition's most unpopular member: Exxon.

From Greenpeace's Washington, D.C. office, the group's highly committed activist cadres scoured the capital for evidence that might discredit their oil-funded opponents. An allied group, the National Environmental Trust, dug up an A.P.I. document suggesting that the oil industry association had decided to rerun the tactics of the tobacco industry. Between the 1960s and 1980s, that industry had spent millions of dollars to fund dissident scientists and think tanks willing to challenge scientific evidence about smoking's dangers.

The document was an "Action Plan" drafted by the American Petroleum Institute's Global Climate Science Team that called for up to $7.9 million in spending to influence public opinion about Kyoto. It declared that "victory" would be achieved when:

Average citizens "understand" (recognize) uncertainties in climate science;

Recognition of uncertainties becomes part of the "conventional wisdom";

Media "understands" (recognizes) uncertainties in climate science;

Media coverage reflects balance on climate science and recognition of the validity of viewpoints challenging the current "conventional wisdom"; . . .

Those promoting the Kyoto treaty on the basis of extant science appear to be out of touch with reality.

The document also recommended that A.P.I. "identify, recruit, and train" a team of scientists "who do not have a long history of visibility and/or participation in the climate change debate" and fund them to "add their voices to those recognized scientists who already are vocal."[19]

This, increasingly, was the underlying structure of Washington policy debates: a kaleidoscope of overlapping and competing influence campaigns, some open, some conducted by front organizations, and some entirely clandestine. Strategists created layers of disguise, subtlety, and subterfuge—corporate-funded "grassroots" programs and purpose-built think tanks, as fingerprint-free as possible. In such an opaque and untrustworthy atmosphere, the ultimate advantage lay with any lobbyist whose goal was to manufacture confusion and perpetual controversy. On climate, this happened to be the oil industry's position.

Raymond's public affairs chief, Kenneth P. Cohen, directed a network of allies and grantees in Washington who created havoc in the climate science debate. Walt Buchholtz, like Cohen a veteran of Exxon's Chemical Company, served as a policy adviser to The Heartland Institute, a Chicago-based free-market group that frequently published tracts challenging the scientific basis for global warming fears. The Competitive Enterprise Institute, on L Street, received hundreds of thousands of dollars from Cohen's department; its free-market advocates filed lawsuits challenging the implementation of climate reviews by the Clinton administration, on the grounds that the scientific data relied upon was unreliable. Exxon provided $373,500 in 1998 and 1999 to the Annapolis Center for Science-Based Public Policy, a nonprofit that backed some of the most prominent scientists skeptical of mainstream science on climate; the center would eventually honor Oklahoma senator James Inhofe, the Congress's most ardent doubter of global warming, for his work in promoting "science-based public policy."[20] The individuals writing and lobbying in the network Exxon funded described themselves as honest, libertarian skeptics who had the courage to challenge conventional scientific wisdom. They did not feel polluted by the receipt of Exxon money any more than liberal-minded campaigners might feel polluted by the receipt of grant funding from, say, the George Soros–backed, left-leaning philanthropy, the Open Society Institute. Relatively few of the thinkers in the network aligned with Exxon's views were climate scientists, however. They typically concentrated on economics and public policy matters. The books authored by members of this movement included titles such as *Red Hot Lies: How Global Warming Alarmists Use Threats, Fraud, and Deception to Keep You Misinformed* and *The Global-Warming Deception: How a Secret Elite Plans to Bankrupt America and Steal Your Freedom*. Inside Exxon-Mobil's K Street office, the sense among some of the lobbying staff was that a lot of this provocative activity was being stoked by the public affairs department in Irving with the idea that it would please the boss, Raymond, whose views on climate policy were well known; a few worried that the fringe campaigners might ultimately endanger shareholders by creating litigation or regulatory risk for the corporation.

The A.P.I. internal documents rooted out by investigators for environmental groups did not contain the kind of smoking-gun evidence about climate science that was earlier unearthed from the tobacco companies. The tobacco industry's documents made clear that corporate scientists knew that smoking was harmful, but nonetheless buried the facts and published misleading studies. In the case of the emerging controversies over climate, there was no evidence that A.P.I. or Exxon maliciously distorted in-house scientific research. The corporation's advocacy campaigners were now inching toward dangerous legal territory, but in the main, the "Action Plan" documented a subtle strategy involving the use of money to advance corporate interests by exploiting the uncertainties and argumentation that can be innate to science.

On May 31, 2000, in Dallas, six months after the Mobil merger, Lee Raymond stood before the first annual meeting of ExxonMobil shareholders—an unruly gathering of religious leaders, environmentalists, and other dissidents who regularly used the meeting, which was required by law, to pressure Raymond over his corporation's public policies, particularly on the environment, alternative energy, and climate. One such activist had just accused Raymond of ridiculing those in the audience who disagreed with him.

"I'm not ridiculing anybody," Raymond answered. "And I resent the assertion that I am. We have a difference of view. This is a democracy. . . . And frankly, I'm not interested in being ridiculed. . . ."

Another speaker demanded "a long-term solution to global warming"; applause erupted.

Raymond possessed no impulse to restrain himself on this subject. "If the data were compelling, I would change my view," he once said. "Ninety percent of the people thought the world was flat. No?"

Now, Raymond went further than he had ever gone in locating his corporation's place in the global warming debate.

"Mark, would you provide me a slide on the seventeen thousand scientists?" Raymond asked an aide.

A slide duly flashed on a wide screen. It depicted a petition organized by anti-Kyoto campaigners and signed by thousands of scientists. The idea was to demonstrate that many respectable scientists doubted key aspects of the I.P.C.C. consensus about the likelihood of human contributions to global warming. The petition's credibility had already been undermined by testimony presented to Congress demonstrating that its signatures included those of pop musicians such as the Spice Girls and James Brown. If Raymond knew about these problems, he did not care.

"This is a petition signed by seventeen thousand scientists. . . . 'There is no convincing scientific evidence that any release of carbon dioxide, methane, or other greenhouse gases is causing or will in the foreseeable future cause catastrophic heating of the earth's atmosphere and disruption of the earth's climate.' So, contrary to the assertion that has just been

made that everybody agrees, it looks like at least seventeen thousand scientists don't agree. My point is not that these seventeen thousand are right
and you're wrong. Your point is you're right and I'm wrong. I'm not saying
you're wrong. What I am saying is there is a substantial difference of view
in the scientific community as to what exactly is going on. . . . We're not
going to follow what is politically correct. . . ."

He went on. "Mark, would you first give me the three-thousand-
year slide?"

Another image flashed on the screen. It showed lines undulating on
a graph.

"That's the earth's temperature as best these scientists are able to
estimate what it was for the past three thousand years," Raymond continued. "It's been a long time since I went to graduate school. But if you just
eyeball that, you could make a case statistically that, in fact, the temperature is going down.

"I'm not asserting that. Similarly, I reject the assertion that it's
going up."[21]

The 2000 presidential campaign was a dead heat to the finish. Al Gore,
concerned about winning coal states, muted his views about the dangers of global warming. The handful of quotations and policy statements
George W. Bush offered on climate were rife with contradictions. Asked
about global warming during a debate with Gore, he said that "it's an issue
that needs to be taken very seriously," but he also suggested that some
climate scientists were "changing their opinion a little bit," without ex-
plaining himself further. Bush denounced the Kyoto Protocol as too harmful to industrialized countries like the United States, but his campaign also
issued a policy document urging mandatory reductions of four major pollutants, including carbon dioxide. Bush's decision to name CO_2 as a pollutant suggested that he might accept Kyoto's broad goals.[22]

After his inauguration, in addition to Vice President Cheney's energy
policy task force, the president named a less-publicized cabinet-level
working group to review climate change science and policy. The members
included Secretary of State Colin Powell, Treasury Secretary Paul O'Neill,

Commerce Secretary Don Evans, Energy Secretary Spencer Abraham, and Christine Todd Whitman, head of the Environmental Protection Agency. Cheney also took part. John Bridgeland, director of the White House's Domestic Policy Council, and Gary Edson, a deputy national security adviser, organized the work. They recruited a half dozen career climate scientists working in federal departments to move temporarily to the Eisenhower Executive Office Building, next to the White House. There they organized climate science and policy briefings for the new cabinet members.

"It was a heady time," recalled one of the participating scientists, Aristides Patrinos. "The potential was so great." He and other career scientists summoned to the White House had concluded, on the basis of the evidence from the campaign and the transition, that Bush shared their sense of urgency about the need to control greenhouse gases. Because the president was a Republican with a background in the oil industry, Patrinos thought, "This was like when [Richard] Nixon went to China—Bush could really be the one who would do something with respect to climate change."[23]

Patrinos and his colleagues delivered science lectures to the cabinet group at rotating sites—one time at State, the next at Agriculture, and so on. James Hansen of N.A.S.A. delivered one private lecture; James Edmonds of the Pacific Northwest National Laboratory gave another concerning the mix of policies and technologies that might be required to stabilize greenhouse gas concentrations. During these sessions, which were unpublicized and closed to all but senior staff, Patrinos was impressed at how open-minded some of the cabinet members, such as Colin Powell and Don Evans, seemed to be. As the lectures went on, however, he also became concerned about the demeanor of Vice President Cheney.

The scientists laid out vivid, illustrated accounts of the damage global warming could bring in the future: melting glaciers, rising sea levels, droughts, and severe storms. They offered specific forecasts about the impact global warming could have on public health and on the economy. One of the lecturing government scientists described the possibility of rising sea levels in "lowland areas in Miami, south Florida," as Patrinos recalled it.

Hearing this, Cheney shifted uncomfortably, Patrinos remembered. He looked like a "raging bull. . . . He got up, paced back and forth, then stood next to me, and I could sense that he was not a happy camper." Cheney remained silent.[24]

The vice president soon preempted the climate task force's work. Haley Barbour, a former chairman of the Republican National Committee who had become a lobbyist for a utility firm that stood to lose if greenhouse gases were regulated, urged Cheney in a March 1 memo to persuade Bush not to align with the "eco-extremism" of those who saw carbon dioxide as a pollutant. Two weeks after Barbour's memo landed, Cheney arranged for Bush to sign a letter to Congress repudiating his campaign position about CO_2—without so much as informing Christine Whitman, the new Environmental Protection Agency (E.P.A.) chief, in advance.

Whitman called Treasury Secretary O'Neill. "Energy production is all that matters," she said. "[Cheney] couldn't have been clearer."

"We just gave away the environment," O'Neill replied.[25]

A few weeks later, ExxonMobil's climate policy specialist Randy Randol sought a meeting with Under Secretary of State Paula Dobriansky, the administration's lead diplomat on global warming issues. One of Dobriansky's senior aides, career foreign service officer Ken Brill, prepared a briefing memo. It noted: "Mr. Randol has asked for this meeting at the suggestion of our Ambassador-designate to Sweden, Charles Heimbold, who served on the board of ExxonMobil." Heimbold, the former chief executive of the drug maker Bristol-Myers Squibb, felt that "we should hear from Exxon/Mobil scientists who have perspectives on the climate change debate that are not consistent with the science that has supported our climate policy until now." Brill suggested some talking points for the under secretary that might assuage the corporation's lobbyist:

Understand Exxon/Mobil's position that there should be no precipitous policy decisions if scientific uncertainties remain. . . .
Administration will continue to oppose the Protocol, but must move
forward on improving our scientific understanding. . . . We will,
however, continue to rely on input from industry and other friends
as to what constitutes a realistic market-based approach.[26]

Four

"Do You Really Want Us as an Enemy?"

E arly in March 2001, an Acehnese rebel commander known as Abu Jack ("Father of Jack," in Arabic) telephoned Ron Wilson, a Texas A&M graduate who ran ExxonMobil's operations in Indonesia. The time had arrived, the caller said, for the corporation to make payments to his separatist guerrilla force. Other oil and gas companies paid for the right to operate in the disputed province of Aceh, Abu Jack claimed. "So must ExxonMobil" was the essence of his message.

Wilson told Abu Jack—whose given name was Zackaria Ahmad—that he would take the demand to his supervisors. He hung up and soon called the United States embassy. He and other ExxonMobil officials disclosed that they had evidence that rebels were stockpiling heavy weapons near their facilities. They also declared they would never pay extortion money. "We are very close to closing down," they reported.[1]

When Lee Raymond acquired Mobil Oil, he also acquired a small war. It was a conflict that Mobil had been struggling with for decades. In any merger, the acquiring party often finds that the target company has a few problems that are worse than expected. Mobil's role as a party in one of fractious Indonesia's most violent separatist insurgencies quickly emerged as such a case. The war was emblematic of ExxonMobil's dilemmas in the era of resource nationalism. The corporation's options to acquire "equity" oil and gas outside of the United States, Europe, and Australia were increasingly limited to poor and weak states prone to internal violence. And in a period of Internet-enabled corporate responsibility campaigns, oil drilling in such countries seemed to attract guerrillas and human rights researchers in equal measure. Exxon had largely avoided the problems that arose from extracting oil and gas in the midst of small wars. The acquisition of Mobil's far-flung properties—in Indonesia and West Africa, especially—would force Raymond and his management team to come to terms with issues they had little experience managing, including the conduct of security forces guarding ExxonMobil oil and gas fields and the geopolitics and diplomacy required to bring oil-related insurgencies to a negotiated end. Raymond's one-size-fits-all Operations Integrity Management System was not especially well suited for the murky violence, corruption, and shifting politics Exxon now confronted in Indonesia.

Mobil had been present in the country for decades. During the 1970s, it had acquired access to a lucrative natural gas field on the northern tip of Sumatra, in the province of Aceh (pronounced *Aah-chay*). The latest round of separatist conflict had been under way for almost twenty-five years in a poor but lush seaside region of rain forests, mountains, rice paddies, and palm oil plantations. Aceh had been an independent Muslim kingdom ruled by sultans for more than four centuries. A Dutch colonial army landed in 1873; the invading commander died within a week and so did many of his men. The first Acehnese resistance war lasted forty years. It calmed and then resumed after Indonesia gained independence from the Netherlands in 1949. From the 1970s, Aceh's struggle to control its own affairs revolved considerably around natural gas and the question of who should benefit from its sale. The gas lay buried in the Arun field, as it was known, beneath an expanse of fertile, palm-laden land along the northern mouth of the Strait of Malacca, near the town of Lhokseumawe.

A large share of the Arun field belonged to Mobil. It contained about

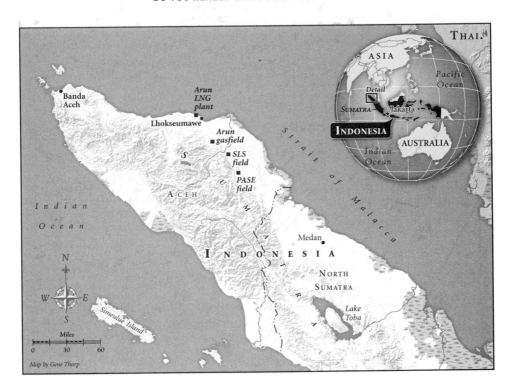

Map by Gene Thorp

17 trillion cubic feet of gas (the equivalent of just under 3 billion barrels of oil) and proved to be highly remunerative: In the decade leading to the Exxon merger, the Arun field accounted for about a fifth of Mobil's overseas revenue from oil and gas production. The subsidiary that extracted Aceh's gas and then liquefied it for transportation to Japan and other markets earned $295 million in profits in 1998, $311 million the next year, and $498 million the year after that. The earnings reflected lucrative contracts Mobil had negotiated during the panicked period of the Arab oil embargoes and the early Iranian Revolution, when many energy-importing nations in Asia feared they would not have access to supply at any cost and proved willing to pay relatively high prices for guaranteed long-term deliveries.

"My nightmare is to pick up the *New York Times* and read that both Nigeria and Indonesia are in flames," Lou Noto, Mobil's chairman, told industry colleagues in the late 1990s. Those two countries accounted for a lopsided share of Mobil's profits; both were wracked by internal rebellions.

Noto, therefore, had extra incentive to muddle through the Aceh war. ExxonMobil, under Lee Raymond, was not going to lightly set aside half a billion dollars in annual profit, either, but the merged corporation had more financial flexibility to tell the likes of Abu Jack to go away. The Indonesian government's position was more like Mobil's had been—it was dependent on keeping the gas profits flowing. Its take from Aceh was about $1.2 billion in 2000, more than a fifth of the government's total oil and gas receipts that year, and about 6 percent of its revenues from all sources, before international aid.[2]

To a great extent Aceh's war had evolved into a contest over who could bargain or shoot his way to control the Arun field's cash flow. One of the contenders was a former New Yorker named Hasan di Tiro, a charismatic Acehnese nationalist leader to some, and to others, "a quixotic, self-promoting political dabbler prone to hysterics and exaggeration," as one biographer put it. Di Tiro was a great-grandson of a heroic nineteenth-century anti-Dutch guerrilla fighter. He grew up in unassuming circumstances in Aceh, migrated to Indonesia's main island of Java to attend law school, and then won a scholarship to the United States in 1950, at the age of twenty-five. He attended Columbia University and later worked in the information department of the Indonesian mission to the United Nations. He made the acquaintance of Edward Lansdale, the Central Intelligence Agency's legendary Asia hand during the cold war. Di Tiro found himself "circulating in the eddies and backwaters of international diplomacy" in New York.[3]

He absorbed the radical ideas of Marxist-influenced, postcolonial liberation movements that spread worldwide during the 1960s, but he also tried to provide for his family through business ventures back in Indonesia. In 1974, one of Di Tiro's companies, Doral Inc., bid for a contract to build a pipeline connected to Mobil's Acehnese gas field; the job went instead to the San Francisco–based Bechtel Corporation. Di Tiro's opponents later emphasized this commercial setback as a cause of his final radicalization. Di Tiro told a different story: Soon after he lost the pipeline bid, he was flying aboard a private jet when its engines died. He promised himself that if he survived, he would lead a revolution for Acehnese independence. He had reached late middle age and believed that he had "lived long enough," he told an interviewer. His biographer felt that Di Tiro was describing a "midlife crisis of sorts."[4]

He founded the Gerakan Aceh Merdeka, or "Free Aceh Movement," known as G.A.M. His followers snuck into the province from Malaysia and opened their "armed struggle" in the rain forests and volcanic hills of the rugged Pidie region on October 30, 1976. Di Tiro issued a declaration of independence six weeks later, drawing on his American education: "We, the people of Aceh . . ." he began. His war strategy, he later wrote, was to shut down "foreign oil companies . . . to prevent them from further stealing our oil and gas." G.A.M. leaflets warned Mobil and Bechtel employees to "pack and leave this country immediately." Di Tiro organized about three hundred fighters and managed to make contact with Muammar Gaddafi, the Libyan dictator; he sought training for his men at Libyan camps. "They have Mobil Oil," Di Tiro reportedly told Gaddafi, "so you must support us."[5]

Because Mobil employed nearly three thousand Acehnese directly or by contract, it proved risky for G.A.M. to target the company; job losses would alienate the rebels' population base. Abu Jack's extortion demand reflected the murky war that had evolved in reaction to these constraints: Rather than throw Mobil out, G.A.M. sought to access the corporation's revenues, directly and indirectly. Racketeering had become commonplace on both sides of the conflict.

The violence was sporadic, but it was often most intense around the sprawling, fenced-in Mobil gas facilities. On the north side of Lhokseumawe (the town's name meant "Everything Deep," a reference to the swampy terrain in which it sat) stood the factory-size Arun liquefied natural gas (L.N.G.) plant. On the other side of town lay the gas fields themselves, spread out across tens of square miles, intermingled with inhabited villages. A modern gas well is relatively unobtrusive in comparison to an oil well: a chest-high, cylindrical metal structure with no moving parts. Mobil installed these robot-looking creatures in fenced areas with names such as Cluster I and Cluster II.

Point A was the main administrative and engineering headquarters for the gas fields. ExxonMobil compounds worldwide displayed a universal design: yellow security lights, high double fences at the entrance, and just inside (after the post-*Valdez* safety campaigns evolved) a large billboard declaring "Nobody Gets Hurt." About 220 expatriate employees—Americans, Australians, Japanese—lived within ExxonMobil's compounds in Aceh. Around them lay clayroads, rice paddies, grazing fields with a few stray cattle, and tin-roofed village homes.

G.A.M. fielded a few thousand guerrillas; its most sophisticated weapons were semiautomatic rifles and rocket-propelled grenade launchers. The guerrillas taxed and extorted villagers and erected roadblocks to take money and property from passing vehicles. A few miles away from a G.A.M. roadblock constructed from fallen tree trunks, Indonesian soldiers might man their own barrier, peering into cars in search of suspicious-looking young Acehnese men.

Mobil had adapted to the war without ever missing a gas delivery. Early in 2001, however, a stream of extortion letters and phone calls started to arrive at ExxonMobil's offices. The corporation's security department, which ran its own intelligence operations in the province, heard "widely divergent rumors," as a U.S. embassy cable put it, about what lay behind the letters and whether the commander known as Abu Jack was, in fact, responsible: "One source says he's now in jail and that imposters are making the threats; others say he is a double agent in the employ of the security forces."[6]

ExxonMobil's security department had also received reports that Indonesian executives at a partner company had recently paid a $100,000 ransom to win the release of a kidnapped Indonesian-born executive. The alleged payoff had "heightened concern" that rebels might now be encouraged to kidnap someone at ExxonMobil, perhaps an expatriate. To deter G.A.M., ExxonMobil suggested to the U.S. embassy in Jakarta that the corporation take out newspaper advertisements declaring its "refusal to make illicit payments"; the embassy judged, however, that "any effort to publicly defy the G.A.M. . . . is not advisable."[7]

Abu Jack, or whoever he was, telephoned ExxonMobil once more on the morning of March 9. A large rebel force had gathered to attack the

corporation's gas fields, the caller reported; G.A.M. had ordered villagers in the area to leave. ExxonMobil's local employees could see that
nearby residents were, in fact, leaving. Around the same time, a mortar
attack and a roadside pipe-bombing targeted a bus carrying corporate
personnel. The evidence suggested to ExxonMobil that G.A.M.—or some
faction of the undisciplined rebel movement—had changed its targeting
policy to go after the company directly, either to advance its extortion
campaign or because senior G.A.M. leaders had quietly decided that ExxonMobil was now an enemy of its rebellion, in a way it had not been seen
as before.[8]

Ron Wilson, who was the president of Mobil Oil Indonesia, the subsidiary that managed all of ExxonMobil's oil and gas operations in the
country, decided he could wait no longer. He was accustomed to managing
risk on behalf of expatriate and local employees, but he concluded that
G.A.M. had now crossed a line. Wilson reported through ExxonMobil's
chain of command ultimately to Harry Longwell, Raymond's executive
vice president for upstream operations on the Management Committee, the corporation's supreme governing council. Raymond's judgment
about the war he had inherited in Aceh, he recalled, was that "Mobil
wasn't shooting anybody, but obviously the military was going to protect" the gas field, "driven by orders from Jakarta, and Mobil was kind of
in the middle of it." Raymond entrusted the day-to-day decision making
to Longwell.

He decided to shut down operations in Aceh. The decision shocked
Indonesia—newspapers covered the story under front-page banners.
Until the Indonesian government created conditions in which Exxon-
Mobil's employees felt safe, Ron Wilson and other ExxonMobil spokes-
people declared, Jakarta's billion-dollar annual revenue flow would be
turned off.

Robert Gelbard served that winter as the United States ambassador
to Indonesia. He was a large, balding, and sometimes combative war-zone
diplomat who had served in the Balkans during the Kosovo conflict. He
told Ron Wilson that he supported ExxonMobil's decision. Gelbard's
responsibilities included the safety of American citizens in Indonesia, and

he felt the situation in Aceh was becoming "dramatically dangerous," he said later, and he was "really worried" that some of ExxonMobil's people "were going to get killed."[9]

Still, Gelbard wanted to intervene to help restart gas production as soon as possible. Indonesia had embarked only recently on a shaky, unstable democratic transition after decades of military rule. The country's president, Abdurrahman Wahid, could ill afford the loss of taxes and royalties from ExxonMobil's Aceh gas fields.

The United States formally rejected G.A.M.'s independence drive and supported Indonesia's claims of sovereignty over Aceh. G.A.M. leaders nonetheless considered the United States to be friendly to their aspirations because American diplomats advocated autonomy negotiations that would grant Acehnese leaders greater control over local affairs within a united Indonesia. With approval from Washington and the Indonesian government, Gelbard flew to Singapore for a secret meeting with one of G.A.M.'s most senior leaders. Gelbard recalled that he "wanted to be very clear with them: Yes, we like them, but no, we didn't support independence." At the same time, Gelbard believed that a military victory was not feasible for either side in the war—only successful autonomy negotiations between Jakarta and G.A.M. could end the violence. Wahid's democratic government was inclined toward such peace negotiations, but Wahid had attracted opposition from hard-liners in the Indonesian military who wanted to eradicate G.A.M. by force. If Gelbard could restart ExxonMobil's gas production, he might, among other things, deliver a victory to Jakarta's beleaguered civilian peace-promoting forces. Gelbard said later that his intervention in the Aceh war had nothing to do with ExxonMobil's business interests or the profits it produced from the Arun gas fields; he sought to reduce the Aceh conflict's violence so that Indonesia would have a better chance to move from dictatorship toward democracy.

ExxonMobil could not be precise about what improvements in security would be necessary to persuade the corporation to restart operations in Aceh. "We'll know it when we see it," an ExxonMobil executive told one of the ambassador's colleagues at the embassy.[10]

One way to reassure the corporation would be to persuade G.A.M. to publicly declare that ExxonMobil was off-limits in Aceh's war. Gelbard thought that the United States should "read the riot act" to G.A.M. about its decision to target ExxonMobil. The ambassador and other senior Bush administration officials decided that spring to embark on an extraordinary campaign to restore ExxonMobil's Aceh operations, and by doing so relieve pressure on Indonesia's wobbly elected president. It was unusual for an American administration to negotiate directly with a guerrilla force over its targeting strategies, and even more unusual for it to apply American pressure to remove from insurgent target lists a lucrative field operated by ExxonMobil.

Aceh's conflict was a dirty war characterized not only by kidnapping and extortion, but also by a brutal campaign carried out by the Indonesian military, a campaign that included torture and summary executions of suspected guerrillas. By aligning itself with ExxonMobil and Indonesia's government to pressure G.A.M., the Bush administration risked associating itself with the Indonesian military's tactics. Sections of the military were on ExxonMobil's payroll to provide security at the perimeter of the Arun fields. These payments to Indonesian soldiers by the corporation were mandated by ExxonMobil's contract. In return, the corporation's Indonesian partners agreed to "assist and expedite" ExxonMobil "by providing . . . security protection . . . as may be requested" by the oil company. As a practical matter this meant that the Indonesian government supplied troops from the Tentara Nasional Indonesia, or Indonesian National Army, known as the T.N.I., to protect the gas fields. Under the arrangement, ExxonMobil paid the Indonesian soldiers' salaries; by the time of the extortion campaign in early 2001, the going rate was about $294 per month for a typical enlisted man. The soldiers were by all accounts—including that of the Bush administration—engaged in appalling human rights violations.

As ExxonMobil prepared to shut down in Aceh, Ambassador Gelbard signed a confidential cable to Washington. He reported his embassy's judgment that G.A.M. was guilty of "atrocities." He also described, however, the ongoing crimes of Indonesian security forces that protected Exxon-

Mobil's gas fields: "The military/police offensive [in Aceh] is resulting in significantly growing human rights abuses. Many civilian corpses bear marks of torture and their hands are tied behind their backs. Neighbors of those later found dead often report that non-Acehnese men in plain-clothes kidnapped the victims."[11] ExxonMobil's daily operations were fixed in the middle of that dark violence.

The Indonesian military's brutality in Aceh traced to the authoritarian "New Order" government of Indonesian president Mohammed Suharto, a former general who took power during the 1960s after a violent purge of the Indonesian Communist Party. The United States saw Suharto as a vital link in its anti-Communist strategy in Southeast Asia. Indonesia is an unwieldy archipelago of about seventeen thousand islands spread out over three thousand square miles. Suharto consolidated his power by allowing the military to enrich itself during deployments around the country's resource-rich islands; he also constructed a tight-knit circle of family and ethnic Chinese business cronies in the capital of Jakarta. To shore up his security alliance with Washington, the president allowed American corporations to enjoy access to Indonesia's minerals, oil, and gas. Suharto offered mining concessions to Freeport-McMoRan Copper & Gold and he delivered to Mobil the large stake in Aceh's Arun field. (Mobil offered a share of the field to Exxon at the time, but the latter's upstream executives demurred, to their enduring regret.)

Mobil entered into a production-sharing contract with Indonesia's P. T. Pertamina, then a state-owned oil company. Under Indonesian law Suharto could name certain "Vital National Objects" that required military protection; in 1983, the Aceh property was so designated. Thousands of T.N.I. soldiers poured into North Sumatra to protect Mobil from the threats and sporadic attacks carried out by G.A.M.

Suharto had tried and failed to win the war in Aceh by force during the 1990s. He declared the province a special military zone. Torture and disappearances became commonplace. The T.N.I. rounded up thousands of young Acehnese men, interred them in camps, and forced them to sing the national anthem as part of their reeducation. According to human rights investigators, army officers set up schemes to profit from their deployments to Aceh—they ran logging operations, marijuana farms, and other rackets.

Security posts and unmarked interrogation houses became the settings for the blackest chapters of Aceh's conflict during this period. Some of the interrogations took place on Mobil property or very nearby. The T.N.I. units set up posts along the fenced perimeters of the gas fields; the posts were sometimes separated by just a few hundred yards. Two of the most notorious facilities around Mobil's fields were known as Post A13 and Rancong Camp. A post might consist of a two-story concrete building or just a barbed wire, sandbagged encampment with makeshift sleeping quarters.

One area with a particularly heavy security presence lay toward the south of Lhokseumawe, where pipes gathered gas from scattered wellheads and drew it into trunk lines for transport to the liquefied natural gas plant. A large trunk line ran down a straight, miles-long corridor known as the Pipeline Road. By early 2001, G.A.M. had taken to planting bombs and digging up pipes along the road. The T.N.I. erected security posts at intervals along the Pipeline Road and ran patrols in the area.

During the mid-1990s, Indonesian soldiers and intelligence officers arrested a number of G.A.M. leaders, including Sofyan Daoud, at the Lhokseumawe port, as they returned from exile in Malaysia. "They were taken to the Mobil facility for interrogation," according to Ifdhal Kasim, the chairman of Indonesia's National Human Rights Commission, which collected evidence about the case. There were more than twenty detainees and "they were tortured at that complex," according to Kasim. "There was all sorts of torture by the soldiers."[12] Over the years, hundreds of young men arrested in the vicinity of the Mobil gas fields disappeared, according to Acehnese separatist activists and independent human rights investigators. Acehnese villagers assumed that the missing men had been killed in custody and had probably been buried near the T.N.I.'s security posts.

Only after Suharto's regime cracked and collapsed under pressure from democracy campaigners in May 1998 did it prove possible to investigate past abuses. That summer human rights researchers interviewed villagers around the Mobil gas fields, documented the names of missing young men, and, guided by informants, dug in the ground for evidence. B. N. Marbun, a member of the National Human Rights Commission, estimated that at least two thousand Acehnese torture victims lay buried in secret graves. He and other investigators identified a dozen such locations and found remains in six of them; in one grave, in the village of Bukit Sentang, they dug up at least a dozen bodies.[13]

On October 10, 1998, a coalition of seventeen Indonesian human rights groups issued a statement alleging that Mobil Oil "provided crucial logistic support to the army, including earth-moving equipment that was used to dig mass graves" to bury Aceh's torture victims and missing young men. *BusinessWeek* published a cover story two months later under the headline, "What Did Mobil Know?"

The oil company's executives told the magazine that the answer was, essentially, "nothing." Their employees had occasionally loaned the Indonesian army heavy equipment such as excavators during the New Order years, but only for "peaceful purposes." The Mobil executives said they believed their equipment had been used to build roads.[14]

These human rights allegations had surfaced just as Lee Raymond and Lou Noto entered into their final talks about merging Exxon and Mobil. Noto flew hurriedly to Jakarta. He met with Gelbard's predecessor as U.S. ambassador to Indonesia, J. Stapleton Roy, who "expressed concern" about the issue. Noto said that Mobil was unaware of any abuses by T.N.I. soldiers guarding its facilities and did not know anything about its bulldozers being used to dig graves. He flew back and appeared with Raymond in New York on December 1 to announce their merger deal.

Along the Pipeline Road and around the gas fields Indonesian human rights researchers continued to dig for corpses.

The Clinton administration cut off aid to the Indonesian military and suspended training contacts because of the human rights abuses committed by the T.N.I. Many of the abuses that concerned the administration took place in East Timor, another disputed province of Indonesia, where the T.N.I. sought, unsuccessfully, in 1999, to prevent a separatist-minded population from voting for independence in a United Nations–sponsored referendum. East Timor's history and status under international law made it a special case; G.A.M.'s independence drive in Aceh enjoyed none of the same U.N.–sanctioned legitimacy. Isolated and embittered after losing East Timor, the T.N.I.'s commanders redoubled their focus on suppressing the rebellion in Aceh. Mobil still paid the salaries of T.N.I. soldiers and officers deployed to protect its fields, despite the official American sanctions over human rights abuses. Legally, Mobil was a subsidiary partner of the Indonesian state oil company in Aceh, and the security payments were one of its contractual commitments. Agus Widjojo, a serving Indonesian general at the time, recalled that his colleagues in the military's high command felt "confusion and ambivalence" about Aceh's rebellion as democracy took hold in their country. Indonesia seemed fragile and beset by centrifugal forces; the generals regarded themselves as the last guardians of national integrity.[15]

In 2000, Indonesian security forces "were responsible for numerous instances of, at times indiscriminate, shooting of civilians, torture, rape, beatings and other abuse, and arbitrary detention in Aceh" and elsewhere, the U.S. State Department reported. "Army forces, police, and G.A.M. members committed numerous extrajudicial killings." The U.S. embassy in Jakarta did not regard Mobil as culpable, however. Its diplomats accepted the corporation's account of itself: "The companies are unable to control military/police actions, including the use of equipment, that may result in human rights abuses," a cable to Washington reported. "Mobil faces this dilemma in Aceh."[16]

P rivate profit-making companies had been waging war independent of their home country governments since at least the days of the East India Company and the colonization of the Americas in the eighteenth century. The idea that such corporations had a legal or moral duty to refrain from facilitating organized violence in their areas of operations was more recent. During the nineteenth century, Quaker ethical movements and antislavery campaigners in the United States and Great Britain, among other places, presaged the ideas that were lumped, toward the end of the twentieth century, under the rubric of "corporate social responsibility." The 1970s brought an expansion of popular and political campaigns to codify corporate conduct for the sake of the public interest. By 2001, reports of human rights abuses carried out by military forces protecting oil and gas operations in Colombia and Nigeria—as well as the questions raised about Mobil's complicity in Aceh's violence—had given birth to a formal compact, the Voluntary Principles on Security and Human Rights, cosponsored by the Clinton administration and Tony Blair's Labor Party–led government in Great Britain.

The compact, as its title indicated, was not binding. It advocated that oil companies undertake human rights risk assessments when they worked in violence-prone regions; communicate their human rights values to host armies that protected their facilities; avoid working with "individuals credibly implicated in human rights abuses"; and permit the use of force "only when strictly necessary and to an extent proportional to the threat." The companies should also "to the extent reasonable, monitor the use of equipment provided by the Company and to investigate properly situations in which such equipment is used in an inappropriate manner." The language suggested a corporate version of the Rules of Engagement guidance typically issued by the White House to the Pentagon in wartime.[17]

Chevron, Shell, British Petroleum, Conoco, and a number of mining companies signed the agreement. Lee Raymond refused. "Exxon just didn't see the relevance to them," recalled Arvind Ganesan, a Human Rights Watch lawyer who participated in the negotiations. "They just disengaged." There was some skepticism among the corporation's decision makers about whether the initiative would outlast the expiring Clinton administration, and in any event, ExxonMobil did not habitually join political compacts initiated by outsiders; it wrote its own rules worldwide. ExxonMobil's place in the Aceh conflict created legal and reputational risks that adherence to the Voluntary Principles could help reduce, but the corporation was convinced that it could handle those risks. Particularly during the first Bush term, ExxonMobil displayed unilateralism in its foreign and security policies. "We don't sign on to other people's principles," an executive later explained. The corporation said it would monitor the accord, perhaps to reevaluate later.[18]

ExxonMobil's security team was aware of the T.N.I.'s human rights record, internal corporate documents show. An ExxonMobil e-mail acknowledged "the poor reputation of the Indonesian military, especially in the area of respecting human rights and in their predilection for 'rogue'/ clandestine operations." Another internal report found that the Indonesian soldiers around the Aceh gas fields "were undisciplined, lacked professional deportment and were not in any state of readiness." As a third internal assessment put it:

> Local security forces [are] ineffectual and often present as great a threat as the activists. The military presence is a double-edged sword, with some military personnel acting as information brokers, thieves, extortionists and intimidators.[19]

There was no evidence that ExxonMobil's security advisers encouraged or participated in the T.N.I.'s torture and extrajudicial killing in Aceh. Exactly how the corporation handled from day to day its knowledge that such human rights abuses were taking place is not clear. Because of the Indonesian military's political power and the sensitivities surrounding the conflict, neither Indonesia's government nor independent human rights investigators could interview or examine the records of the T.N.I. units that worked in partnership with ExxonMobil or, before it, Mobil Oil. Evidence trails faded as the years passed.

Within ExxonMobil, responsibility for assessing Aceh's violence and managing relations with the Indonesian military fell to the Global Security department. Global Security's roster of overseas employees and contractors conjured the lineup of a Hollywood action film: former K.G.B. officers, veterans of the British Special Air Service and French special forces, and retired officers of the Central Intelligence Agency and the United States military.

At the time of the Mobil merger, Lee Raymond elevated Mike Farmer, a career corporate security professional, to lead Global Security. Under

Farmer, the Aceh case fell to Tommy Chong, who had a background in Singapore law enforcement; he ran ExxonMobil's Southeast Asian security operations out of an office in his native country.

ExxonMobil's executives understood the reputational and other risks they bore. Aceh had witnessed a "complete breakdown of law and order," Robert Haines, an international relations manager in Washington, wrote in a memo to his superiors on December 13, 1999. Haines emerged as an important adviser to ExxonMobil on its Aceh problem after the merger. He was a West Point graduate who had commanded an armored cavalry troop in Vietnam, leading rural sweep operations near Da Nang. After that tour, he resigned his commission and entered law school; a long career in Mobil's office of general counsel had led him eventually to Fairfax, Virginia, where he headed up the international section of the corporation's public affairs office at the time of the merger. He was one of the few Mobil hands in Washington that Exxon kept on. His Vietnam experience had equipped him to assess Aceh: The presence of Indonesian troops around the gas fields "only serves to inflame the population and results in suspicions that [ExxonMobil] is linked to the military," he wrote.[20]

In the spring of 2000, after the merger closed, ExxonMobil Global Security concluded that it could use some fresh eyes on its Aceh problem. Mike Farmer assigned John Alan Connor, an Arabic-speaking retired U.S. Army lieutenant colonel who had served as a Special Forces officer with the Green Berets and worked extensively in the Middle East, to the Indonesia security team. Connor had joined Exxon after leaving the U.S. Army. (There was a sizable contingent of former military men at the corporation.) In Yemen he had successfully negotiated truces with tribal sheiks around Exxon's oil fields, and in Africa he had helped scope out security for oil field operations in areas prone to insurgency. Farmer asked Connor to undertake a "risk assessment" of ExxonMobil's position in Aceh.[21]

Connor looked at how Indonesian soldiers used and misused ExxonMobil equipment. He found that T.N.I. soldiers occasionally approached Indonesian-born employees to demand a bulldozer or dump truck, according to accounts of his study that circulated within ExxonMobil. If the employee refused, he might be beaten up or threatened. Connor's assessment found "nothing as dramatic as mass graves dug with ExxonMobil equipment," according to a person familiar with the internal reporting. The review did document cases of equipment being hijacked by Indonesian soldiers for unknown purposes in the midst of a conflict rife with abuses. This sort of strong-arming of ExxonMobil equipment by local security forces was a chronic problem for the company worldwide, particularly in Africa. Such "borrowing" by local security forces posed legal risks to the corporation. Under the Foreign Corrupt Practices Act, the American antibribery statute, there were limitations to what equipment or services ExxonMobil could provide to host governments and militaries without charging market prices. Each time the T.N.I. demanded a free ride on one of ExxonMobil's corporate airplanes or asked to "borrow" a truck, the request had to be reviewed by the corporation's lawyers—who often turned down the requests. In Indonesia, this had left T.N.I. officers frustrated and even more inclined than before to take what they wanted at gunpoint. ExxonMobil told its local employees not to sacrifice their "physical safety" if threatened, but if possible to resist demands by soldiers to take equipment and to call for help.[22]

High-level executives reviewed the assessments of the corporation's relationship with the Indonesian military. Farmer forwarded a report entitled "Indonesia Strategic Security Study" to ExxonMobil vice president Lance Johnson, noting that it "identifies a range of critical tasks that must be completed quickly . . . to respond to ongoing and potential security concerns." A second internal report concluded that it would be necessary to enforce "uncompromising controls across the board."[23]

Connor stayed on in Aceh to support the security mission. He and other Exxon security officers—some permanently stationed in the province, others rotating in and out—tried to develop close working relationships with the Indonesian army battalion and company commanders deployed around the gas fields. The corporation's security executives felt it was "ludicrous" to think that ExxonMobil should be held responsible for T.N.I. brutality or the use of excavation equipment outside of their control. Yet there could be little doubt that ExxonMobil exercised some authority over the T.N.I. soldiers assigned to its Acehnese fields. ExxonMobil's contract, for example, gave the corporation the right to influence

the Indonesian forces' "deployment logistics," and it "assisted in the management of security affairs" with the T.N.I.[24]

As Mike Farmer recalled it, ExxonMobil's corporate security officers
on-site would take "business requirements to the military and say, 'This
is what we'd like to do over the next week or over the next ten days—
can you take the appropriate steps to make sure that that's done.'" For
example, the corporation might be starting up gas wells in a certain field
or might be moving employees in a convoy, and it would ask the T.N.I. to
deploy support. A typical written instruction from Tommy Chong carried
the subject heading "Deployment of Military Resources" and began, "We
have revised the deployment logistics of the new military resources as
follows: POINT A: 40 soldiers inclusive of 15 to handle military escorts
for employee travels. . . ." Another internal document made explicit ExxonMobil's authority over the T.N.I. units it paid. It carried the heading
"Increase in Military Deployment" and instructed that the Indonesian
army be asked to confirm that ExxonMobil "has the right to influence the
security plan."[25]

The Global Security department sought to reduce the risk that Indonesian soldiers would engage in abuses by requesting that the soldiers
refrain from sweep or offensive operations. Yet the corporation endorsed
the Indonesian army's and police's plans to construct a layered defense
around the Mobil property, including a forward perimeter of security
posts, to catch G.A.M. guerrillas as they tried to approach. In effect, this
defensive system created an infrastructure of patrolling and interrogation
on and adjacent to Mobil's fields. ExxonMobil urged the T.N.I. units in
Aceh to be "defensive, not offensive," according to an individual involved,
as "nobody wanted to have any sort of cloud over our operations." The
Indonesian military units "were supposedly in static defensive positions
that would go out roughly five kilometers on each side to prevent direct
or indirect fire from coming at us."

Inevitably, even defensive patrolling would involve detentions and
interrogations of G.A.M. suspects. Published human rights reports—
including by the U.S. government—made clear that the questioning of
guerrilla suspects by Indonesian officers was not likely to be polite. Yet as
late as 2003, ExxonMobil had no written internal codes or guidelines for

the use of force that could be handed out to soldiers or police protecting the corporation's property, according to statements made by ExxonMobil executives to American officials in another country with mounting security problems, Nigeria. Even if ExxonMobil did not like the Voluntary Principles, there were other international standards for police conduct. These included the United Nations Convention on Human Rights, the United Nations Code of Conduct for Law Enforcement Officials, and the United Nations Guidelines on Use of Force. The International Committee of the Red Cross also published standards for the appropriate use of minimal force by police. Royal Dutch Shell, which had already confronted allegations arising from police and military excesses in defense of its oil properties in Africa, had developed "Rules for Guidance in the Use of Firearms by the Police," which it wrote down on two-sided laminated cards and handed out to personnel assigned to defend its properties. ExxonMobil resisted writing down any such rules. American lawyers advised their international oil clients that "such a formal move could expose the company to undue liabilities," according to a State Department account.

The constraints ExxonMobil sought to impose on the T.N.I. were therefore conveyed informally. In private meetings, ExxonMobil's security officers told their Indonesian counterparts, "We couldn't operate without you guys, we recognize the sacrifice you're making and we respect the professionalism—and no human rights issues."

These lectures on human rights reached "the point of being a cliché," recalled an individual involved. "The instruction we got was, 'Do not look like you're aiding or abetting the Army in any way,'" recalled a second individual involved. The Indonesian army officers sometimes resented the lectures they received. Some of the Indonesian officers battling G.A.M. made clear to their ExxonMobil liaisons that they thought the Americans were out of their depth.[26]

Gusty winds blew a cold rain across Washington on March 12, 2001. Alwi Shihab, Indonesia's foreign minister, an Islamic scholar with a doctoral degree in religious studies from Temple University in Philadelphia, arrived by limousine at the State Department. He entered under a canopy and ascended to the ornate seventh-floor office occupied by Colin Powell, his counterpart in the Bush administration.

After an exchange of pleasantries, Shihab raised the conundrums of Aceh's war. G.A.M.'s threats against ExxonMobil amounted to "blackmail," the foreign minister told Powell. He hoped investments by American corporations in Indonesia would lead the new administration to support his government. "With $38 billion at stake in Indonesia, the United States would not want to see the country disintegrate."

President Wahid's government would be grateful for American backing in the effort to calm Aceh's violence, he continued, "not in terms of public intervention," but through private messages to G.A.M. that might have "great weight on the other side." As in the Middle East, the United States had the leverage in Aceh to force the two sides to negotiate, Shihab believed. His government was willing to give the Acehnese "everything short of independence," he said.

Powell said that if the Indonesian military abused civilians in Aceh, it could cause "the greatest harm" to the country's relationship with the Bush administration. It was critical that the Indonesian military apply only that force that was "reasonable and necessary to the task," he said. The secretary made clear, however, that he had taken note of Shihab's message that there might be a role for the United States to "send signals to the other side" in Aceh's war.[27]

ExxonMobil's gas operations in Aceh had now become embedded in U.S. diplomatic and intelligence priorities in Indonesia. The Bush administration sought, overall, to support Indonesia's fragile democracy, improve civilian control over the military, stanch human rights violations, and suppress Islamist radicals—goals that sometimes competed with one another, because the Indonesian military was at once a potential source of stability and instability. ExxonMobil seemed to be both part of the problem and part of the solution in Aceh. On the one hand, its gas production seemed to provoke and exacerbate guerrilla violence, and that violence encouraged abuses by the military. Yet the revenue ExxonMobil's gas sales provided Jakarta was critical to the country's young democracy. The Bush administration found itself simultaneously under pressure from Exxon-Mobil to do something about the deteriorating Aceh war and from the Indonesian government to do something about ExxonMobil's unwilling-ness to operate amid guerrilla violence. The corporation's decision to shut down gas production that spring had provoked an outcry in Indonesia's parliament. Politicians threatened to nationalize the gas fields; they sum-moned Ron Wilson, the ExxonMobil country manager, to a parliamentary hearing to explain the corporation's decision to suspend production. Other politicians spoke darkly about American conspiracies to undermine Indonesia's fragile democratic government; some accused Gelbard of forc-ing ExxonMobil to close down.

That accusation so aggravated the ambassador that he shot off letters to local newspapers refuting the charge. The decision to cease production had been ExxonMobil's alone, he wrote, although the corporation enjoyed the support of the U.S. government. In Washington, Robert Haines met repeatedly with frontline Bush administration officials in charge of Indo-nesia policy at the National Security Council and the State Department—Karen Brooks at the N.S.C. and Ralph "Skip" Boyce at State. The corporation's message, crafted by an informal Indonesia crisis committee that included Haines and senior executives in Houston and Irving, was that only the United States could resolve the Aceh war by brokering some sort of agreement between Jakarta and G.A.M. Haines also made clear that G.A.M.'s decision to target ExxonMobil directly was a new factor in the corporation's experience of the war, that it placed Exxon-Mobil personnel at risk, and that this was the reason they had shut down their operations. "You really need to get in there and do some-thing," Haines told Bush administration officials. ExxonMobil did not have a specific blueprint or plan of action, but like Indonesia's foreign minister, the corporation felt that only the United States government had the necessary leverage on both sides of the war. "We are not diplomats, but we do know this is a problem and you are the guys that can do it," Haines said.

ExxonMobil refused to negotiate with G.A.M. Its Acehnese employ-ees included many G.A.M. sympathizers and probably a few formal members. Some of these local employees urged cooperation with G.A.M., but the corporation's executives in the United States concluded that their contractual and political position with the Indonesian government required them to be careful. ExxonMobil's position was that G.A.M. was an illegal armed group, and therefore the corporation would have no direct dealings with its leaders. What the Bush administration might do was another matter.

Ambassador Gelbard arranged a meeting for ExxonMobil at the Ministry of Industry and Trade, which was headed by Luhut Panjaitan, a retired four-star general. Ron Wilson arrived with Gelbard and other embassy officers at the ministry's headquarters, located on a riverbank beside one of congested Jakarta's major highways.

It had not been a single event, Wilson explained to the Indonesian officials, but an accumulation of threats and near misses that had led to ExxonMobil's decision to shut down Aceh's gas operations. Snipers had fired upon ExxonMobil airplanes and had wounded employees, Wilson said. Hijackers had stolen more than fifty company vehicles. Assailants had bombed four company convoys by remote control.

"ExxonMobil has had an Indonesian presence for one hundred years," Wilson said. The company had shipped more than five thousand liquefied natural gas cargoes without missing a single one until now. It was in Indonesia for the long run and had made its decision to suspend production reluctantly. The safety of its employees was paramount, however.

Panjaitan explained that Indonesia intended to restore security in Aceh by launching a "limited offensive" against G.A.M. New battalions of Indonesian forces were arriving around Lhokseumawe as they spoke.

Wilson chose his words carefully. "I understand how difficult it is to restore peace," he said. "I appreciate that the military is preparing to carry out operations in a careful, selected way. As a company, ExxonMobil cannot condone human rights abuses. The whole world is watching events in Aceh. Charges of human rights abuses could cripple efforts to resume operations."

Wilson emphasized that his corporation had never paid money to G.A.M., despite the demands of Abu Jack and other commanders. Payoffs would only aggravate the situation and lead to more extortion, he said.

ExxonMobil was not demanding that Indonesia's government reduce the risk faced by its employees in Aceh to zero, Wilson declared as the meeting concluded. But the corporation's employees had to "feel safe traveling by road and assured that the workplace was not likely to come under mortar attack, or that they might be kidnapped."[28]

After the meeting with Panjaitan, the Indonesian government continued to try to persuade Wilson that it could meet his standards. Purnomo Yusgiantoro, the energy minister, called Wilson and suggested they fly into Aceh on a government plane to tour the area, so that the ExxonMobil manager could see that order was being restored and that it was safe enough to resume production.

Wilson called Gelbard and asked if he should accept Yusgiantoro's invitation.

"Don't be insane," the ambassador advised. "Don't go."

The minister went anyway, alone, and G.A.M. rebels shot at his plane.[29]

During the first week of April 2001, Ambassador Gelbard flew to Banda Aceh, the seaside provincial capital, a flat and humid expanse of low-slung, water-streaked concrete buildings shaded by palm trees. A Swiss peacemaking organization, then known as the Henry Dunant Centre, maintained a local forum for on-again, off-again talks between Indonesian and G.A.M. representatives. Gelbard scheduled separate meetings with leaders on each side of the conflict. He raised the subject of human rights with Indonesia's government delegation: G.A.M. certainly committed abuses, Gelbard told them, but the international community holds democratically elected governments to higher standards than guerrilla groups.

ExxonMobil had no covert agenda in closing its Aceh operations, Gelbard said. The corporation had been entirely justified in its concerns about security; the United States supported ExxonMobil's decision but had not instigated it.

The ambassador became more forceful when the G.A.M. delegation arrived. "G.A.M. is clearly responsible for the attacks on ExxonMobil,"

Gelbard announced. "Some G.A.M. leaders are now even boasting about shutting down ExxonMobil." He said that Hasan di Tiro had promised in private meetings with Clinton administration officials that he would issue a public statement that ExxonMobil was not a target of the guerrilla campaign; he had never done so. G.A.M.'s attacks on the oil company now were a "major mistake," Gelbard declared.

The United States would not tolerate terrorism against U.S. citizens and economic interests. G.A.M. had been "very lucky" that no American citizens working for ExxonMobil had been killed thus far. Even so, he warned, there would be "severe consequences" if G.A.M. did not stop the attacks immediately. The Bush administration had so far refrained from naming G.A.M. a terrorist organization under American law. A terrorist designation would mean travel and banking restrictions for G.A.M. leaders. The administration might reconsider that decision, unless the assaults on ExxonMobil property and interests ended. Moreover, the United States received many requests from the Indonesian military and police for help in fighting against G.A.M.—intelligence, training, and equipment.

"Do you really want us as an enemy?" Gelbard asked.[30]

The G.A.M. representatives acknowledged responsibility for the attacks on ExxonMobil. They said that Indonesian troops guarding the gas fields were fair military targets. The troops used ExxonMobil property as a "sanctuary" from which to launch raids into nearby villages. Therefore, in their analysis, ExxonMobil facilitated the killing of Acehnese.

G.A.M. leaders said years later that they felt increasingly agitated at the time by ExxonMobil's possible complicity in extrajudicial killings of their cadres. The corpses unearthed along the Pipeline Road and elsewhere late in 1998 legitimized ExxonMobil as a target, they said. The corporation "seemed to support the Indonesian government," recalled Nordin Abdul Rahman, one of G.A.M.'s political leaders. "People concluded that ExxonMobil provided heavy equipment for the burials." Not only was "ExxonMobil land used for mass graves," said Munawar Zainal, a G.A.M. student leader and occasional representative of the movement in Washington, but "they gave the Indonesian security forces money. This to us was unacceptable."[31] Gelbard, for his part, felt that

ExxonMobil had "behaved very responsibly and very sensibly," as he put it later. He regarded the corporation's dilemma as a "textbook example" of "a dangerous situation when a U.S. energy company behaved very well."[32]

At the Banda Aceh meeting, Gelbard told G.A.M. that its guerrillas had mounted attacks on ExxonMobil's civilian housing, employee buses, and other targets clearly unconnected to the Indonesian military. This had to end.

The ambassador flew back to Jakarta, but the Bush administration's campaign to coerce G.A.M. to stop targeting ExxonMobil continued. On April 23, Skip Boyce arrived in Banda Aceh from Washington; Boyce was a career foreign service officer who now ran the East Asia and Pacific portfolio out of Foggy Bottom. The envoy met Indonesian officials and assured them that the United States opposed Acehnese independence, but he urged negotiations that would address the legitimate grievances of the Acehnese.

"We are deeply concerned by attacks on ExxonMobil facilities in Aceh," Boyce said. He warned against cracking down on G.A.M. now that the American oil corporation had withdrawn: "The closure of ExxonMobil should not be a pretext for launching a military offensive, which would only worsen the security situation."

He also took up G.A.M.'s concerns about the offensive operations waged by Indonesian forces from inside the corporation's property. Indonesian forces guarding ExxonMobil's fields "should not perform any other mission—specifically, they should not sweep or raid neighboring villages, which only exacerbates the violence," Boyce said.

When the envoy met with G.A.M.'s leaders, he reinforced Gelbard's earlier warning: Attacks on ExxonMobil "risked turning the U.S. into G.A.M.'s enemy." The separatist guerrillas would want to "consider carefully before making an enemy of a superpower like the U.S."[33]

Hasan di Tiro and several of his top political aides had found asylum in Sweden. The Bush administration pressed its warnings further, through the Swedish foreign ministry, two weeks later. A Swedish official met with two senior aides to Di Tiro, Zaini Abdullah and Malik Mahmud, and told them that attacks on ExxonMobil had become "self-defeating" and should be stopped. The G.A.M. men stated that it was not their policy to attack foreign property. As to the wider war in Aceh, they believed it was the T.N.I. that was defeating itself: Human rights abuses by the Indonesian military against Acehnese civilians would soon produce international sympathy for G.A.M. and its cause.[34]

Indonesian security forces killed Abu Jack in an operation in Aceh on June 4. By then G.A.M.'s leaders seemed to be wavering about ExxonMobil. Boyce sought an audience with Di Tiro and Mahmud on June 15 and repeated his warnings.

The telephone rang in ExxonMobil's office in Aceh in late June. The caller claimed to be a lieutenant of G.A.M.'s senior military commander on the ground in Aceh, Abdul Syafie. The guerrilla movement had received orders "from Sweden," the caller reported, not to attack ExxonMobil facilities anymore.[35] The corporation could return to gas production without fear.

Ron Wilson conveyed to the U.S. embassy that production would resume soon—probably in July. The disruption to Mobil's operations had come to an end, due in part to the Bush administration's quiet threats to designate G.A.M. leaders as terrorists. The loss of revenue had lasted about five months.

Robert Gelbard and his colleagues could not in the end protect President Wahid and those around him who favored peace talks in Aceh. A political crisis, stirred by hard-liners in the Indonesian army, gathered in the parliament. In July, as ExxonMobil moved in expatriate engineers to check valves on the Aceh wells and restart gas production, Wahid fell from power. Megawati Sukarnoputri, the third president of Indonesia since Suharto's fall, succeeded him. She was close to the T.N.I. That month she declared martial law and ordered thousands more soldiers into Aceh to defeat G.A.M. once and for all by military force.

As the new troops arrived that August, ExxonMobil officials met with the U.S. embassy to provide an update on their security regime. Gas production was ramping up again; revenues were flowing. The executives "expressed satisfaction with current levels of security," the embassy's reporting officer informed Washington. "The military had changed its operations from one of passively occupying [ExxonMobil's] facilities to providing a secure perimeter. About 3,000–5,000 soldiers, a large increase from last year, were patrolling an area out to five kilometers. . . . The military had also more than tripled the stationary posts along the Pipeline Road. The improved security had netted individuals attempting to infiltrate bombs."[36] Yet even under renewed military pressure, G.A.M. for the most part refrained from turning its guns back on ExxonMobil. The Bush administration had made clear that the consequences of such targeting could be grave.

G.A.M.'s international lobbying activities were, at best, ad hoc. Acehnese students scattered around the world, inflamed by the violence in their homeland, organized chapters and agitated for attention. From Sweden, Hasan di Tiro and his aides ran a makeshift political and communications campaign. In Gelbard's judgment, they showed "no realistic attitude or skillful diplomatic strategy" and apparently preferred "the morally repugnant and totally flawed position that 'losing is winning,' i.e., less dialogue and more . . . violence and atrocities might win international sympathy."[37]

G.A.M.-aligned students won visas to study in the United States or were resettled there as refugees; one cluster of younger refugees lived in Harrisburg, Pennsylvania, about two hours' drive northwest from Washington. That group took advantage of its proximity to the capital to try to win appointments with anyone who would listen to them. They had few allies.

In 2001, an Acehnese student activist named Faisal knocked "out of the blue" on the Dupont Circle door of Terry Collingsworth's office. Collingsworth was then general counsel of the International Labor Rights Forum, a nonprofit that campaigned against child labor and sweatshops in developing countries. Collingsworth belonged to a network of American human rights lawyers who employed novel legal arguments and a previously obscure eighteenth-century law, the Alien Tort Claims Act, to

sue corporations, individuals, and governments for civil damages arising
from human rights atrocities overseas. In 1997, he had supported a lawsuit, *Doe v. Unocal*, in which thirteen Burmese villagers asserted that they
had been forced at gunpoint by the Burmese military to build a pipeline
for Union Oil Company of California.

One of Collingsworth's assistants, who happened to speak the Indonesian language of Bahasa, took the meeting. Faisal, it turned out, had
heard of the Unocal lawsuit and explained that "he had a case just like it,
involving Exxon," Collingsworth recalled being told. The lawyer flew to
Aceh within two weeks. Traveling secretly with local activists, Collingsworth
snuck into the villages on the edges of the Indonesian military's defensive
perimeter around Lhokseumawe and took notes during interviews with
victims and witnesses.[38]

That June, just as Robert Gelbard succeeded in his unpublicized
campaign to persuade G.A.M. not to target ExxonMobil any longer,
Collingsworth and his colleagues filed *John Doe I et al. v. ExxonMobil
Corporation et al.* in United States District Court in Washington, D.C. The
lawsuit drew upon the allegations of eleven Acehnese villagers, whose
names were withheld to protect them from T.N.I. reprisals. The Acehnese
plaintiffs lived in the vicinity of the ExxonMobil gas fields. Plaintiff John
Doe I alleged that in January 2001, "while riding his bicycle cart to the
local market to sell his vegetables, he was accosted by soldiers who were
assigned to ExxonMobil's T.N.I. Unit 113. The soldiers shot him in the
wrist, threw a hand grenade at him and then left him for dead." John Doe
II alleged that soldiers from the same unit beat him, took him to Rancong
Camp near the gas fields, and "detained and tortured him there for a period of three months, all the while keeping him blindfolded." Later the
soldiers removed his blindfold, took him outside, and showed him "a large
pit where there was a large pile of human heads. The soldiers threatened
to kill him and add his head to the pile."[39]

Lee Raymond still owned Aceh's deteriorating war—and after
Collingsworth's lawsuit, its potential legal liabilities as well. Notwithstanding its posture of independence and self-sufficiency in Washington,
ExxonMobil had required the Bush administration to sort out G.A.M.,
and it would soon lobby the administration vigorously to quash Collings-

worth's case. In a pinch, the corporation did not hesitate to seek and accept direct help from the United States. Managing civil violence in remote, complex countries would not prove to be one of ExxonMobil's notable competencies. Yet beyond Aceh, ExxonMobil's portfolio of risk-producing small wars would only grow.

Five

"Unknown Injury"

On most mornings during the summer of 2001, Mandy Lindeberg tried to rise early, to beat ExxonMobil's biologists onto the beaches. She slept aboard the *Kittywake II*, a seventy-two-foot converted wooden tug that she, and a small team of researchers, had chartered on behalf of her employer, the United States government, and in particular the National Oceanographic and Atmospheric Administration (N.O.A.A.), which monitored the world's oceans and weather. Her goal that summer was to survey ninety-one beach segments in Prince William Sound. Under the design of Lindeberg's study, she and her team would dig at least seven thousand holes. Each day, they shoveled away rocks and sediment on the beaches to a depth of fifty centimeters and then examined the pits for evidence of oil—perhaps left over from the *Exxon Valdez* spill twelve years earlier, or perhaps from some other source. When they found some, they scooped samples into jars.

As she moved from place to place, Lindeberg could often see scientists contracted by ExxonMobil following her in the *Spirit of Glacier Bay*, a 178-foot cruise ship with thirty staterooms. She and her fellow government scientists consulted a Web site about the cruise ship's luxury features, which they mocked among themselves. An air of rivalry tinted with class and cultural warfare took hold as the summer progressed. Lindeberg could not tell exactly what the ExxonMobil scientists were doing, but they seemed to be monitoring the extent of her pit digging on the beaches. They also dug some of their own holes on the same stretches where she worked. David Janka, a long-haired banjo player and charter captain who worked on related oil research projects with Lindeberg's team, would peer from the bridge of his motor vessel at the trailing corporate scientists. "The bio-stitutes," he called them. At least once the ExxonMobil team hired a helicopter to track the movements of the government scientists, Lindeberg recalled.

At times the ExxonMobil scientists would complain that Lindeberg had used up all of the good sampling spots on a particular beach. Lindeberg thought to herself, "You were having eggs Benedict; we were having our gruel and going to the beach first—that's not my problem." But she tried to be diplomatic: "My crew arrived here early this morning," she told them. "You're welcome to sample here as soon as we are done."[1]

She was an informal, stout, brown-haired woman in her late thirties who had grown up in the Puget Sound area of the state of Washington. She had studied marine biology in college and then moved to Alaska to work on the marine and wildlife injury assessments after the *Exxon Valdez* spill; in 1996, she took a position at the National Marine Fisheries Service of N.O.A.A. Most of the year, Lindeberg worked in the state capital of Juneau at the agency's Auke Bay Laboratories, which included a dilapidated campus of docks, labs, warehouses, and trailers located just off the Glacier Highway. The lab stood on a slope that afforded a spectacular view of Lynn Canal, a part of Alaska's Inside Passage, which contains fjords teeming with whales, sea lions, and bald eagles. The dress code at Auke Bay was casual. On those rare summer days when the sun shined, the scientists might turn up in shorts and Hawaiian shirts and leave their dogs tied up outside their trailer doors. Almost all of the biologists, chemists, and toxicologists at Auke Bay were, like Lindeberg, long-settled refugees from the Lower 48. Alaska attracted them because of its abundance of understudied natural life. The state also seemed to appeal to personalities with an ornery or independent streak, and the Auke Bay group was no exception.

After the *Exxon Valdez* spill, the laboratory had become a center for research about the effects of spilled oil on the natural environment. The Auke Bay team increasingly had to cope with the bands of academic scientists ("from back East") who turned up in Alaska with lucrative contracts from the oil corporation. Initially, ExxonMobil funded forty or fifty researchers to travel to Alaska each summer to work on the subjects that N.O.A.A.'s smaller network of government-funded scientists also explored; by the summer of 2001, the corporate-funded researchers numbered about a dozen. By processes that remained mysterious to the Auke Bay team, but which they chalked up to the ways of a world fueled by money, the studies published with oil corporation funding never seemed to damage ExxonMobil's legal position that Prince William Sound had fully recovered from the *Exxon Valdez* spill. The corporation's studies sometimes produced similar data to those from the government teams, but the ExxonMobil scientists usually reached different conclusions about what the data implied. Still, the Auke Bay team had never experienced anything quite like the shadowing and monitoring that unfolded after Mandy Lindeberg started digging her seven thousand holes.[2]

Twelve years after the accident, Prince William Sound's rocky beaches looked unsoiled. The initial cleanup undertaken by Exxon in the summers of 1989 and 1990 was almost universally judged a success. But was the oil really gone? Had the fish and wildlife in the area fully recovered? The answers could have legal and financial implications. The original approximately $1 billion settlement among Exxon, the federal government, and the state of Alaska, reached in 1991, contained a Reopener for Unknown Injury clause that allowed the two government parties to seek up to an additional $100 million from ExxonMobil if they could prove environmental damage that was unforeseeable at the time of the original settlement.[3]

There had been signs that oil remained in pockets underneath some of the beaches. Lindeberg's hole digging might provide evidence to support such a reopener claim. Her summer study was the latest in a series of attempts by N.O.A.A.'s Auke Bay team of biologists and toxicologists to document spilled oil's lingering and less visible impacts. That research involved fundamental questions about the sources of oil's harmful effects on natural environments. In the long run, ExxonMobil and the entire oil industry had an economic interest in those findings, too.

The battle between ExxonMobil and N.O.A.A. over Mandy Lindeberg's work illuminated a larger, recurring aspect of the corporation's influence over American public life. Whether the subject was the damage caused by oil and gasoline spills, climate change, the safety of chemicals ExxonMobil manufactured, or other critical matters involving public health and the environment, the corporation joined directly in scientific controversies to protect its interests. It contracted with academic scientists, and it brought staff scientists out of ExxonMobil laboratories to lobby Congress and regulatory agencies. ExxonMobil's science bore all the hallmarks of the corporation's worldwide strategy: It was well funded, carried out by highly competent individuals, unrelenting in its focus on core business issues, and influenced by the litigation strategies of aggressive lawyers. Even the corporation's most ardent opponents conceded that the individual ExxonMobil staff scientists they encountered were typically ethical and professional. The question that nagged those on the receiving end of ExxonMobil's blended campaigns of research, lawsuits, and political lobbying was whether the corporation's science could be judged honest.

Jeffrey Short, the chemist who served as the lead scientist for Mandy Lindeberg's hole-digging enterprise, first came north to take a job excavating ditches for N.O.A.A.'s fisheries division on the Alaskan peninsula. He had grown up during the Sputnik era around Edwards Air Force Base, in Lancaster, California, where his father was a rocket engineer. Once, playing outside on a summer evening, Short saw a bright light on the horizon, in the direction of the base; when he came home, his father explained that one of the Atlas rockets he worked on had exploded. Perhaps not surprisingly, the younger Short grew into "one of those nerd kids that was blowing stuff up." He once forced an evacuation of his family's house when an experimental vacuum chamber he had made from an old refrigerator compressor spewed sulfur dioxide gas. At the University of California he studied philosophy and biochemistry. He moved into physical chemistry in graduate school and then earned a doctoral degree in fishery biology at the University of Alaska. He grew into a wiry man with thinning brown hair and a face that seemed to radiate bemused curiosity.[4]

Short's training in both biology and quantitative chemistry drew him toward the chemical mysteries of oil as far back as the 1970s, when the Trans-Alaska Pipeline System first began to pump crude to Valdez. At that time, the U.S. government had not conducted much study about what effects spilled or seeping oil might have on a marine environment such as Prince William Sound. Federal government and oil company research programs provided funding for Short and other scientists to examine the subject.

As of the mid-1970s, most of the research into oil's poisonous effects on fish and mammals had been derived from the methods used to assess chemical compounds for the insecticide industry. Those methods focused on short-term, or "acute," toxicity—how much of a particular compound was required to kill half of exposed animals after ninety-six hours of continuous exposure. Such assessments could make clear to manufacturers and regulators which compounds were the most immediately poisonous and required special handling. But ninety-six-hour bioassays, as research chemists refer to them, constitute a narrow way to consider the full toxic potential of a chemical compound. As he began to think about oil, Jeffrey considered that there might be other, longer-term effects on an animal after an initial oil exposure.

Petroleum is referred to as a fossil fuel because it was formed from the remains of ancient algae and zooplankton. (Early in the twentieth century, scientists believed oil came from the remains of dinosaurs; the more recent theory that the source was mainly microscopic plant life is widely accepted, but still relies on some speculation.) The plant residues were gradually transformed into oil across eons by heat and pressure beneath the earth's surface. Because oil originated in biomass, it is chemically complex; each batch of petroleum presents a distinct blend of hundreds of thousands of chemical compounds. Researchers have characterized only a small percentage of oil's full chemical makeup, but they have divided the most abundant and easily separable compounds into several classes of hydrocarbons—that is, combinations with distinct arrangements of the elements hydrogen and carbon. One class, known as aliphatic hydrocarbons, is essentially safe for living creatures. Another class, the asphaltenes, is often what is left over after oil is refined by industrial processes; these compounds are used to glue rocks together as asphalt. A third class, called aromatic compounds, has the potential to damage living tissue and biological systems.

About a week after the *Exxon Valdez* ran aground on Bligh Reef, Jeffrey Short found himself on a boat headed into Prince William Sound to participate in the first round of environmental damage assessments. He was interested in which compounds from the spilled oil were dissolving into seawater, at what concentrations, and at what levels of depth. At first he collected seawater directly, but soon he began to use bay mussels as his measuring instruments. A single mussel will pump a liter of water through itself in an hour as it scavenges for nutritious particles. In the process it will gather and concentrate pollutants with unusual efficiency. Short dropped cages full of mussels into Prince William Sound and lowered them to varying depths—at one, five, and twenty-five meters. "We weren't really sure how big the impacts were going to be below the surface," he recalled. "The predominant thinking at the time was that there would not be much in the way of effects." His mussels provided an initial baseline measurement of oil dissolved in the sound's seawater.[5]

The traditional studies suggested there should not be large fish kills because the dissolved concentrations of aromatic compounds would not be high enough. Yet scientists never really had had the chance to study this assumption in the field or to explore the possible "sublethal" or subtler, long-term effects of spilled oil, which might damage fish or animals without killing them outright. Short knew it sounded coldhearted, but he regarded the *Exxon Valdez* accident as a historic opportunity to see how a big oil spill might affect marine life outside a lab.

Of Prince William's marine inhabitants, salmon and herring were the two species that mattered most, economically. Five large commercial hatcheries dotted the sound's shores. Together, they formed one of the

largest pink salmon hatchery systems in the world. Pink salmon particularly suited Jeffrey Short's research agenda because the fish's life cycle and
migration patterns are strictly predictable. Whether it is wild or commercially hatched, a pink salmon born in Prince William Sound will swim out
from its birthplace to the Gulf of Alaska and return two years later to its
exact place of origin.

After their initial water measurements using mussels, Short and his
colleagues, along with other government-funded scientists at the Auke
Bay Laboratories and elsewhere, studied the mortality rates of salmon
hatched from streams along beaches that had been heavily oiled by the
Valdez spill. They compared these rates of mortality to those of fish
hatched along beaches that had not been oiled. A mystery soon presented
itself. Several years after the initial spill, when the surface oil had been
cleaned up and the beaches seemed restored, the scientists observed lower
survival rates among fish reared downstream from beaches that had earlier
been oiled. In the places with the lower survival rates, fish embryos and
young fish had likely been exposed to dissolved oil, but in very low concentrations—not enough to harm them, according to traditional bioassay
studies.

One possibility was that dissolved aromatic compounds from oil
might have harmful effects on fish embryos or fish development at much
lower levels of concentration than previously believed. If so, the toxic
compounds might create defects in young fish that could be difficult to
detect through clinical observation because the fish wouldn't necessarily
all die of the same cause; their weakened condition might play out in an
ocean environment, over the fish's lifetime, in varied and unpredictable
ways. As evidence emerged to support this hypothesis, one of the scientists
working with Short, Ron Heintz, had an inspiration: Auke Bay could set
up its own pink salmon hatchery, expose embryos and young fish to varying levels of oil, send the fish out to sea, and count their mortality rates
two years later, when they reliably returned to their birthplace. "Here's an
idea!" Short exclaimed when he heard the proposal. "We should do that!"
It was an expensive and risky experiment by government standards—
more than a half million dollars. But they won approval in 1996.

Over the next several springs, Auke Bay's scientists and their collaborators tagged tens of thousands of pink salmon and then counted and examined the fish as they returned. This work produced a significant scientific discovery: Dissolved or exposed oil *did* have a sublethal toxic effect at levels of concentration many hundreds of times lower than previous research had suggested was dangerous. (The scientists could not initially explain why oil caused elevated mortality rates, only that it did. Later research by other scientists showed that oil exposure could damage a fish's heart as it developed, which in turn damaged the circulation system and sometimes produced early death. The Auke Bay scientists later found the same effect when they studied herring and cod embryos; other scientists would reproduce the results with zebra fish and mummichogs.) The damage caused by oil exposure did not seem to be passed down from one generation of fish to the next, however; at least, Jeffrey Short's team could not demonstrate such an intergenerational effect. Salmon populations steadily recovered in Prince William Sound after the initial disruptions. The single-generation effects of oil toxicity meant ExxonMobil was probably off the hook for further financial damages on that score.[6]

Still, as a result of the Auke Bay's post-*Valdez* work, the underlying science about the dangers of oil spills to marine environments had been revised, at least in the opinion of the N.O.A.A. team and other scientists who reviewed and duplicated their findings. This might influence the environmental liabilities of ExxonMobil and other oil corporations when other spills occurred. "It was a really unexpected and pretty profound change" in how scientists "viewed oil toxicity," Short said.

As the team's work was published, ExxonMobil began to fund competing studies using other methodologies and sample sizes; all of the studies the corporation supported challenged the premise that oil was dangerous in the way that the N.O.A.A. team suggested. ExxonMobil employed many chemists in its refinery and research divisions. Its scientists did not dispute the notion that certain aromatic compounds such as benzene, toluene, and xylene could be dangerous to living beings. However, ExxonMobil did not accept the finding by the N.O.A.A. scientists that dissolved oil present in a natural breeding ground might harm embryo development, even after the Auke Bay's original salmon study was replicated and extended to other species.

The corporation's resistance and argumentation did not particularly bother Short, once the confirming studies from other noncorporate scientists came in. "I think in the wider scientific community we've won that battle, because other people have independently replicated it and figured out the biochemical mechanisms underlying it, and in fact there's some really elegant work done by people who are not us. So they [ExxonMobil] can have that position if they like, but most people think it's flawed."[7]

In 1999, to memorialize the tenth anniversary of the *Exxon Valdez* spill, reporters and camera crews descended on Prince William Sound. ExxonMobil spokespeople emphasized that the sound's beaches were free of oil—as was true, at least on the surface—and that wildlife in the region had recovered, as was also true, generally speaking. And yet some area residents claimed that they had accidentally set a few oiled Prince William Sound beaches on fire while camping. Dave Janka had discovered beaches around Knight Island where he could easily dig beneath the rocks and find pockets of fresh oil. Where had the oil come from? Was it left over from the *Valdez*? During the tenth anniversary season, Janka ferried media crews to those beaches and helped the reporters dig out handfuls of oil to show their audiences. "There was a considerable range of opinion" about whether Prince William Sound remained burdened by submerged *Valdez* oil, Jeffrey Short recalled. So, around the time of the anniversary, he proposed a study "to actually measure how much oil was on the beach, which had never been done before and was widely viewed as impossible." Given the findings about oil's sublethal effects, if the oil was still around the Sound, hidden, it might pose persistent dangers.[8]

Two scientists at the United States Geological Survey, Jim Bodkin and Brenda Bellachey, were in the meantime intrigued by a second biological mystery in local wildlife populations. Scientists funded by a trust established with proceeds from the Exxon legal settlement had discovered elevated levels of an enzyme known as P450 in sea otters and harlequin ducks. Biologists sometimes track the enzyme because it increases in an animal's liver if the creature is exposed to oil or other pollutants. Where was the pollution coming from? All of the scientists studying Prince William Sound could see that residual oil on the surface of the beaches was declining almost to the vanishing point, and that tides and rain were chipping away at what little remained on the surfaces of rocks. Did significant quantities of oil that nobody could see persist around the sound and were they somehow getting into the sea otters' food chain or ecosystem?[9]

If there was persistent oil, it had to lie below the surface. Jeep Rice, a biologist at Auke Bay who specialized in toxicology, Mandy Lindeberg, and Jeffrey Short recruited a statistician to help them design a random sampling on the sound's beaches. On each beach segment, they would stretch out surveyor's tape and implant stakes to initially divide beaches into squares, at randomly selected locations, and then dig. Short feared that the whole project could prove to be an embarrassment; they would dig seven thousand holes, spend hundreds of thousands of taxpayer dollars, and find perhaps four or five pockets of persistent *Valdez* oil. As it turned out, Mandy Lindeberg's pit-digging teams almost immediately struck fresh oil—oil that had not been weatherized into relatively harmless tar balls, but which seemed to be preserved beneath the rocks, as fresh—and toxic—as the day it spilled from the *Exxon Valdez*. Her initial findings meant that fresh oil had survived in many more places inhabited by the sound's wildlife than had previously been contemplated, which might have implications for ExxonMobil's liability under the reopener clause. Accounts of her initial findings reached the Alaskan press in May 2001. "Within about a week," recalled her colleague Jeep Rice, "she's being followed" by the ExxonMobil cruise ship.

Letters arrived at Auke Bay from O'Melveny & Myers, a large corporate law firm based in Los Angeles. They contained Freedom of Information Act requests from ExxonMobil demanding all of the documents, plans, and preliminary research findings in the federal government's possession concerning not only the seven-thousand-hole study Lindeberg had started, but other studies the Auke Bay scientists had undertaken about the possible toxic effects of oil on the environment.[10]

Jeffrey Short published a newspaper essay that summer in which he described the findings of N.O.A.A.'s work examining the legacy of the *Exxon Valdez* spill: "Much more oil was found than anticipated—around 200 times more than claimed by Exxon's contractor." Sea otters and some

bird species that forage on beaches where oil remained beneath the surface "have biochemical markers that indicate they are still exposed to oil.
It appears that oil may still be a factor impeding their recovery, possibly
through ingestion of oiled prey."[11]

The Auke Bay scientists knew that their findings would be provocative, but the response they drew this time went beyond any line of argument they had heard before. David S. Page, a professor at Bowdoin College
in Maine and a scientist under contract with ExxonMobil, published a
rejoinder that came close to accusing the Auke Bay scientists of faking
their evidence.

It was Page, as it turned out, who had overseen the effort to shadow
Lindeberg around the sound during the summer. After inspecting the
beaches studied, he wrote, "We saw no evidence that Short dug 7,000
pits. . . . Had thousands been dug, we would have located many more."
The pit sites he could find "were chosen subjectively" by the N.O.A.A.
team, he argued; the government scientists had employed an approach
that "exaggerates the extent of remaining residues. . . . It indicates a strong
bias in Short's study and raises questions about the scientific validity of
its conclusions." Overall, Page wrote, "Prince William Sound today is as
healthy as it would have been if the spill hadn't happened."

Page's published accusations prompted an internal review at N.O.A.A.
to determine whether Short and his colleagues had indeed committed
fraud. "It's against the law for civil servants to take the public's money and
make stuff up," as Short put it later. Eventually, the investigators exoner-
ated the Auke Bay team. Short hired lawyers and fired off cease-and-desist
letters to Page and to the administration of Bowdoin College; he accused
his adversary of defamation. Neither Page nor the college took any action
in response.[12]

David Page was an academic scientist who had been working on the
biological effects of oil spills since the mid-1980s. After the *Exxon Valdez*
accident, he received contracts from the oil corporation, as well as from
other funders, such as the state of Maine. Over the years he had come to
regard the government scientists at N.O.A.A. as rent seekers who per-
petuated a narrative of persistent oil pollution in order to justify their
professional funding and projects. "It's like the Arabian Nights—if you run

out of stories, you get your head cut off," he said. "They kept doing research long after it would do any good."[13]

It was with Page's collaboration that ExxonMobil began to deliver the Freedom of Information Act (F.O.I.A.) requests to the Auke Bay Laboratories at a rate sometimes as high as four per week. Whenever Short or Rice made a public presentation of their findings, "an Exxon lawyer, biologist or chemist would be in the audience," and the corporate-affiliated scientists would sometimes stand up to make "an out-and-out attack on our work," Rice recalled. At one conference in San Diego, Rice had heard enough; he "got up and told them I thought it was a classless act. People attending were shocked—it's not something you normally see."

Pete Hagen, a biologist who arrived at Auke Bay as the program manager for Exxon Studies, a title that referred to N.O.A.A.'s research into the *Valdez's* impact, found that his views about the scientists working for ExxonMobil hardened as time went on and as F.O.I.A. requests accumulated on his desk. "They may want to wear down government scientists," he reflected. "Beyond harassment, we don't know what Exxon's motivation is." His thinking about the oil corporation and its allies in the scientific community "has become more extreme," he admitted. He felt that their willingness to bend data to serve the corporation's legal and business aims was "not too dissimilar to what the tobacco industry went through, or the lead industry. . . . Sometimes you win by persistence." For his part, Jeep Rice regarded ExxonMobil's tactics as "legal, but just immoral." Jeffrey Short resented ExxonMobil's drive to "access data before we'd even published it, which put us in the position of giving data before we interpreted it—giving them, in theory, the chance to write up papers before we did."

For David Page, too, the arguments about science in Prince William Sound grew personal. He found the N.O.A.A. scientists' responses to his criticisms of their research to be "shrill . . . I wasn't accusing them of fraud. I was just saying my observations were at variance with what they were claiming." In fact, his published essay did come close to implying that Short's team had faked its hole digging, but after the initial, accusatory exchanges, Page did not return to that charge. He supported Exxon's Freedom of Information Act demands because Jeffrey Short "rarely presents data" and "the only way" you can get detailed information underlying his studies is to make a formal legal request. "F.O.I.A. is not harassment; F.O.I.A. is to find out important information that government agencies aren't willing to let become public," Page said. As to Short's concern that ExxonMobil was seeking raw data in order to advance arguments in public before the government scientists could, he said, "The record shows that our requests were made well after published reports were made in various venues, often several years or more after the studies were done."[14]

Science is innately uncertain. Its progress has been marked again and again by the defiance of settled wisdom by independent-minded mavericks, from Galileo Galilei to Charles Darwin. It can be difficult even for excellent scientists to distinguish between a revolutionary new insight and plain foolishness. Vested interests—governments, clergy, or private corporations—have long sought to control and manage the policy implications of scientific findings. Only in an environment of free debate can the best scientific facts and interpretations eventually win out. Even where new facts affecting the public welfare become well established, as had occurred with the research into global warming by 2001, it does not follow from scientific logic which public policy response is the best one; in the case of climate change, the economic costs of full and rapid remediation would be high. Governance and economics are not hard sciences, despite the contrary aspirations of some of their theorists and practitioners.

Yet, the environment in which ExxonMobil and the Bush administration devised parallel approaches to managing science and public policy in the age of oil spills and global warming was influenced by several factors that Darwin would not have recognized. One was the prominence of lawyers and their win-for-the-client mind-sets. The tobacco industry's near bankruptcy had demonstrated that not even talented lawyers could overcome terrible facts in a product liability matter. Yet that example had also shown how industry funding and purposeful, subtle campaigning could profitably delay a legal reckoning for a dangerous product through the manipulation of public opinion, government policy, and scientific discourse.

The scientific facts about oil pollution and climate change that Exxon-Mobil and its political and intellectual allies in Washington had to manage as the Bush administration took office were nowhere near as daunting as those that confronted the tobacco industry when the dangers of smoking were publicly recognized in the early 1960s. By comparison, the public health effects from the burning of fossil fuels were often indirect. The American economy's dependence upon oil and gas was not the product of some clever marketing campaign, as cigarette smoking arguably was, but was embedded in technological and industrial evolution.

When regulators or lawsuits challenged ExxonMobil's liability on environmental matters, the corporation turned fiercely combative—Irving's internal protocols provided for rapid intervention by ExxonMobil's law department, which spent large sums on the most talented and aggressive outside litigation firms. "They took a very hard line on the legal issues," explained a member of the corporation's board of directors. "It's very much a take-no-prisoners culture." From Washington to Houston to capitals worldwide, ExxonMobil executives internalized the corporation's attitude toward lawsuits of all kinds: "We will not settle just to avoid a struggle; if we believe we are in the right, we will use our superior resources to fight and appeal for as long as possible, and when the case is over, your house may no longer be standing. Think twice before you take us on." ExxonMobil's spokespeople and lobbyists regularly expressed dismay that the scientific findings they presented about the *Valdez* cleanup, climate, chemical regulation, and other public policy issues were not accepted by journalists, judges, and politicians as fully credible.

David Page knew that the government scientists thought of him as a corporate shill, and he felt insulted by that accusation. "It's not about corporate America," he said. "To compare Exxon to a tobacco company is totally outrageous. They are two very different things. I will tell you, my livelihood was teaching students chemistry and biochemistry. I didn't need to work for Exxon or anybody else. If I thought for a minute that I was being asked to say something that wasn't true or to hide information or act in [an] indefensible way at all, I would have taken a hike and not had any further relationship. I don't hold my nose when I'm talking."

Jeffrey Short ultimately quit his government job in part because of the distractions caused by the corporation's unrelenting Freedom of

Information Act requests. "We're all scientists—we didn't sign up to do that," he said.

Mandy Lindeberg could think of only one positive aspect of her experience as an ExxonMobil adversary. Knowing that every field note she and her colleagues made would be scrutinized by corporate litigators and scientific consultants, she said, "forced us to be very good scientists."[15]

Six

"E.G. Month!"

Equatorial Guinea seemed like a place that Gabriel García Márquez had invented, a former American ambassador once remarked.[1] A thumbprint on the west coast of Africa, the entire country consisted of an offshore island, where the capital of Malabo was situated beside a cliff-walled harbor, and a sliver of land on the continental shoreline. It had been one of Spain's few African colonies. In 1968, Generalissimo Francisco Franco, Spain's then-septuagenarian dictator, granted independence to a government led by Francisco Macías Nguema, an anticolonial politician. Macías turned out to be a depraved mass murderer. He imprisoned, tortured, and killed his opponents by the score. He closed schools, campaigned against intellectuals, and burned boats to prevent his people from fleeing his realm. His security guards executed 150 people in a sports stadium on Christmas Day while loudspeakers blared "Those Were the Days." In Malabo, Macías maintained an active torture chamber in Black Beach prison. He increasingly walled himself away on the slice of mainland Africa where his Fang ethnic group predominately resided. He descended

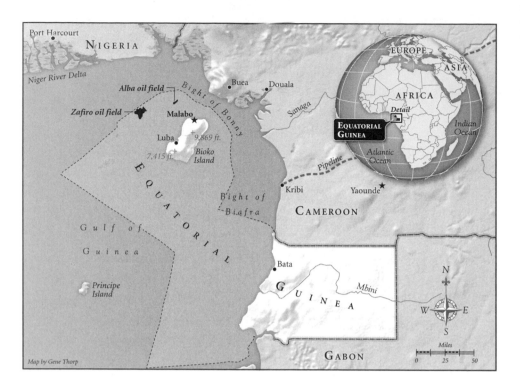

Map by Gene Thorp

into paranoia and lashed out at anyone who challenged him. Roughly a third of Equatorial Guinea's small population would die or manage to escape during his reign.[2]

On August 30, 1971, Lannon Walker, a diplomat at the United States embassy in neighboring Cameroon, telephoned his colleague Len Shurtleff to report signs of trouble emanating from the nearby American embassy in Malabo. Walker reported that Al Erdos, the American chargé d'affaires in Malabo, might also be going insane.

Erdos had been sending strange cables in recent weeks. Now he had come on the shortwave radio to report on some sort of Communist plot involving one of his American colleagues, whom he had tied up in a vault in the chancery. Walker asked Shurtleff to fly to Malabo to investigate. The diplomat arrived that evening by chartered aircraft. At the tiny embassy— little more than a rented house—Erdos, visibly distraught, pulled him aside. He announced, "I lost my cool. I killed Don Leahy."

He was referring to the embassy's administrative officer. Inside the chancery Shurtleff found scattered papers and spattered blood. A woman's scream called him to an interior room. There he found Mrs. Leahy kneeling over the body of her dead husband. He had been stabbed to death with a pair of embassy scissors.

An autopsy showed that Leahy had semen in his trachea, suggesting that love or sex had been an issue between murderer and victim. At his subsequent trial in a Virginia federal court, Erdos entered an insanity defense; his lawyers blamed Equatorial Guinea's menacing tropical dictatorship for having driven him mad. The jury convicted him of manslaughter.[3]

The case established a tone in U.S.–Equatorial Guinean relations that would persist for years to come. The host government accused one of Erdos's successors of sorcery and expelled him; the ambassador in question, John Bennett, had persistently raised concerns about the country's human rights record. The United States shut its Malabo embassy, pleading budget constraints. Thereafter it serviced Equatorial Guinea by airplane from its embassy and consulate in Cameroon.

Equatorial Guinea was perhaps the most politically toxic oil property Lee Raymond had acquired from Mobil. Given Exxon's reserve replacement challenges, however, it was hard to be picky. Resource nationalism in the Middle East had driven all of the Western majors to Africa in search of bookable reserves. Exxon had had success on its own in Angola, but Raymond would mainly be dependent on Mobil's legacy properties if he wanted to share in Africa's emergence after 2000 as an increasingly important oil play. Some of the contracts available to Western oil companies in West Africa, such as in Nigeria, could be restrictive. Equatorial Guinea was a case where the upside was more attractive, financially—as it had to be, given that, on the world's political risk charts, the country presented an extreme case of uncertainty. By 2001, ExxonMobil operated oil platforms about forty miles offshore, where workers on four-week rotations pumped steadily rising amounts of crude—more than 200,000 barrels per day and rising, or about 8 percent of ExxonMobil's worldwide production of oil and gas liquids that year. Mobil had negotiated a contract with Malabo's inexperienced government in which it secured the right to recoup its investment expenses from oil sales in the early years of production, paying Equatorial Guinea's government an initial royalty of only about 10 percent. The principle that Mobil should be able to recover its costs early was typical of deals designed to protect international oil companies from political risk, but these specific terms were favorable. Now oil prices were rising, and the project looked likely to pay off big to both parties.[4]

ExxonMobil's headquarters and residential compound in Malabo stood beneath a towering dormant volcano. A large population of monkeys inhabited the mountain's rain forests; they were shy because humans had long hunted them as food. In the evenings heavy tropical clouds often skirted the volcano, which had two peaks, like a double-humped camel. Lightning flashes and quiet rumbles of thunder added to an air of ominous majesty. The ExxonMobil refuge contained stucco buildings with Spanish red tile roofs—residences, offices, and recreation facilities, including a swimming pool. It was not particularly luxurious—certainly not as comfortable as the burgeoning Marathon Oil waterfront compound across the bay, which housed more workers than ExxonMobil's did and had been laid out for tennis courts, basketball courts, squash and racquetball courts, a clubhouse, and a restaurant. At night, Marathon's gas flares and the white safety lights at its liquefied natural gas and methanol plants illuminated the dark water that spread out beneath ExxonMobil's smaller facility.

The second-generation dictator who oversaw ExxonMobil's inherited contract was Teodoro Obiang Nguema, a nephew of Macías's and a brigadier general trained at a military academy in Spain. In 1979, Obiang had led an uprising against his uncle and seized power. He arrested Macías and assigned mercenary bodyguards to execute his uncle by firing squad.[5]

"When I came to power, the place was completely destroyed," Obiang remembered. "No electricity, no roads. The schools were closed."[6] Few reliable statistics were kept for the country by its government or international agencies, but per capita income was perhaps one hundred dollars per year. Hunger stalked the forest villages, less than a third of the population had access to safe drinking water, mothers and babies commonly died in childbirth, and life expectancy was less than fifity years. Because of the depredations of the Macías years, the country's cocoa economy, once modestly successful, no longer existed. Obiang settled into power in a less wanton but no less ruthless manner than his predecessor. He turned ministries, businesses, and land over to his family members and through them constructed layers of internal security, strengthened at the inner core by his palace guard of salaried Moroccan soldiers. Black Beach prison remained open, and the torture techniques of its jailers and political prosecutors did not much change, according to one human rights investigation after another.

Still, Obiang kept the United States satisfied about his foreign alliances. Unlike Macías, he sought to work as a professional authoritarian, in the manner of those who led neighboring nations in West Africa. At business meetings Obiang usually turned up in a dark, tailored suit with a pocket kerchief; he could be coherent, direct, and even sophisticated, if also persistently obtuse about the precepts of good government. To interact more successfully with his French-speaking neighbors, Obiang took on a French tutor at his palace. He played tennis regularly and jogged along the rutted, red clay road between the airport and Malabo's elegant but water-streaked colonial plazas. He danced through the night at parties and drank copiously.[7] He developed cancer and eventually traveled to the United States about four times a year for treatment. The scope and seriousness of the disease was a closely guarded secret, but Obiang, as rugged as a crocodile, seemed to overcome it. His regular medical travel to the United States and a deep antipathy toward France and Spain turned him gradually into an unrequited friend of America's. Oil, he hoped, might persuade Washington to embrace him.

Equatorial Guinea's territorial ocean waters encompassed some of the same geology that had much earlier enriched Nigeria and Gabon with oil dollars. Obiang provided exploration leases to Spanish oil companies during the late 1980s. It was relatively early in the development of offshore oil technology, and the Spaniards reported they could find nothing. "Thanks to the American embassy" in Cameroon, Obiang recalled, Joe Walter, an irrepressible Houston wildcatter, agreed to take a second look.

In 1991, tiny Walter International discovered Equatorial Guinea's first oil, in the Alba field. Obiang interpreted the news as evidence that Spain had deliberately suppressed his country's economic potential, while the Americans seemed prepared to back him. Walter sold out its holdings as Mobil, Marathon, and Amerada Hess arrived to explore farther. Equatorial Guinea hired no outside lawyers or investment bankers to negotiate; Obiang's first-ever oil minister, Juan Olo, worked out the terms. Mobil acquired rights to the offshore Zafiro field, which, as it turned out, contained at least a billion barrels, or at least three times Mobil's entire annual worldwide production of oil and gas liquids in 1995.[8]

Mobil embedded itself in financial partnership with the Obiang government; it paid for land, office leases, and security services. The local companies it worked with had many ties to the president and his family. Under the Foreign Corrupt Practices Act, it can be illegal for American companies to make sidebar payments to businesses controlled by foreign government officials who are at the same time handing out lucrative contracts for oil. In later years, the Department of Justice questioned Mobil's deals with local firms, but ExxonMobil warded off the investigations by arguing that it had no alternative but to invest with the ruling family because there was no market for land or services that was not controlled by the Obiang clan. The Foreign Corrupt Practices Act, as interpreted by Justice, did not hold that some countries should be avoided altogether, only that American corporations should not act corruptly if they had a choice in the matter. Some of Mobil's and later ExxonMobil's payments to Obiang's regime covered scholarships for students and relatives selected by the president to study in the United States. The corporation also held a joint investment in a fuel services company; Obiang controlled the venture's minority partner, a company called Abayak, according to the findings of a U.S. Senate staff investigation.[9]

"The private American (especially oil) companies would not wish to be pulled into U.S.G. [United States government] efforts to combat human rights violations in Cameroon and Equatorial Guinea," reported a U.S. embassy cable written just after Equatorial Guinea's oil began to flow in earnest. "U.S. companies are aware of human rights violations . . . [but they] present themselves as 'ahead of the curve.'"[10] That seemed mainly a euphemism for corporate strategies of hunkering down, avoiding publicity about human rights and other controversial aspects of Obiang's reign, and staying as far away as possible from recurring State Department campaigns to reform Equatorial Guinea.

There were foreign policy episodes, such as in the Aceh war, when ExxonMobil leaned on State to intervene on the corporation's behalf. Equatorial Guinea provided a different imperative: The Bush administration's human rights campaigning in Africa was more likely to taint Exxon-Mobil in Obiang's eyes than to help the corporation's position as an oil contractor. ExxonMobil therefore adopted a low-profile posture of strict noninterference in Equato-Guinean politics, coupled with quiet advice to Obiang aimed at helping him improve his international reputation, which would redound to their mutual benefit.

Lee Raymond regarded the State Department as not particularly helpful to ExxonMobil, notwithstanding the example of the administration's intervention to stop G.A.M.'s targeting of the corporation. America's career diplomats did not understand international business very well, and some of them were outright hostile to large oil firms, Raymond believed. Where they did try to intervene in a commercial or contract matter, they often did not know enough detail to be constructive, and they did not appreciate the need for strict confidentiality, he concluded. Raymond could talk from time to time with his friend Vice President Cheney, who understood his issues, but the ExxonMobil chief executive had come to the view that as for the American foreign policy and government bureaucracy in general the best approach was, as he told his colleagues, "Don't talk to them." That did not mean ExxonMobil never asked State for favors; it meant only that the corporation's demands for Bush administration intervention were erratic, inconsistent, and influenced by Raymond's access to back channels with Cheney and other officials he regarded as sophisticated and reliable, such as Samuel Bodman, who would become secretary of energy during President Bush's second term. ExxonMobil and its handful of international American peers in the international oil industry "blow hot and cold," the veteran American diplomat

John Campbell wrote in one cable to Washington from West Africa. "For the most part [they] prefer to try to address industry-specific issues themselves. They may turn to the [State Department], only to back away from requests on further consideration. If their efforts fail to achieve a resolution or problems become more acute, they can quickly return demanding action."[11]

During the late 1990s, after Equatorial Guinea's big contracts were signed but before oil and cash flowed, Obiang traveled to Washington, D.C., for annual World Bank meetings. His country's decades-old struggles against poverty were about to end—he needed to think about how to manage the great sums that would soon come his way. Obiang was in some respects naive about global affairs, but it did not require an advanced degree in political science to notice that small, weak countries with huge amounts of oil tended, as Kuwait had done, to ally themselves protectively with the United States, a superpower with a thirst for hydrocarbons and a military large enough to deter any power that might bully its oil-supplying friends. By opening Equatorial Guinea's fields exclusively to American companies, Obiang hoped in time to coax Washington into strategic partnership. The president and his companions walked one afternoon past the grandiose main branch of Riggs Bank, at 1503 Pennsylvania Avenue, across from the Treasury Building and diagonally opposite the White House. The bank's gray ionic columns stood several stories tall and created the impression that this might be the American president's own financial institution—or, at a minimum, that it was deeply connected to the corridors of American power. "We should put our money here," Obiang told his companions. At the time, they still did not have much of a check to write. They opened a Riggs account with a $5,000 deposit.[12]

Two years later, in 1997, with Zafiro in production, $1.2 million a month began to flow into the Washington, D.C., bank. Riggs's executives woke up to the gusher they had struck. It kept growing.

"Equatorial Guinea has gone from being a very small, insignificant relationship to the largest single deposit relationship at Riggs," a manager named Ray Lund wrote to senior colleagues in 2001. With ExxonMobil now operating and profiting from Zafiro, Equatorial Guinea had $200 million on deposit and expected additional cash flow at a rate of about $20 million a month for the foreseeable future. "Where is the money coming from? Oil—black gold—Texas tea."[13]

Obiang had been denied high-level meetings with Clinton administration officials. Equatorial Guinea's human rights performance, he was told, was the obstacle to such access. Africa Global, a small Washington lobbying firm, advised Obiang that the election of George W. Bush as president of the United States presented an opportunity to rehabilitate Equatorial Guinea's reputation and to establish a deeper partnership in Washington based on oil interests. Obiang agreed to pay Africa Global hundreds of thousands of dollars to help him navigate the American capital and secure meetings at the highest levels of the new administration. One of his lobbyists secured an appointment at the State Department's Africa bureau on February 22, 2001, a few weeks after Bush's inaugural.

"Obiang has been waiting eight years" for the Democrats to leave office, said the dictator's representative. "He hopes he can now meet with senior levels of the new administration." After a State cable about the lobbyist's meeting circulated in West Africa, the American ambassador in Cameroon wrote to Washington to say that "we would be delighted if he [Obiang] were received at a higher level," although it would be better if "we [the American government] can get the credit instead of a lobbying firm."[14]

Henry Hand, a desk officer in the Africa bureau at Foggy Bottom, took a call a few days later from a Halliburton executive. The executive said his company "was being hounded by Africa Global to intercede with the Vice President's office" to obtain an audience with Cheney for Equatorial Guinea's president. "They declined to do so," Hand reported. On March 2, the desk officer rode over to Equatorial Guinea's threadbare embassy on 16th Street. Obiang's ambassador to the United States explained to him that Africa Global has "a very ambitious agenda" for the leader's upcoming private trip to Washington—the lobbying firm would be seeking meetings with Bush, Cheney, Secretary of State Colin Powell, and National Security Adviser Condoleezza Rice.

"As I left the embassy, oil company representatives were arriving for a meeting, having been convoked by the ambassador" to advocate for Obiang's access, Hand reported. Nor was Africa Global confining itself to the federal government: The desk officer reported two weeks later that Obiang had apparently secured a meeting with Washington, D.C., mayor Anthony Williams "who may declare March E.G. month!"[15]

The main practical item on Obiang's agenda was "the establishment of an [American] embassy in Malabo." But no meetings with Bush, Cheney, or Powell actually materialized in the days ahead. "He is reportedly irked," Hand recorded. "The energy companies are increasingly unhappy with Africa Global, which they feel is doing a poor job of getting across Equatorial Guinea's message. The lobbying firm has been very heavy-handed in leaning on these firms [ExxonMobil, Marathon, and Hess] in an attempt to get high-level meetings, after raising Obiang's expectations for such meetings to unrealistic levels."[16]

The best Africa Global could do, it turned out, was an under secretary of state—hardly an insult, but not a cabinet officer, either. Alan Larson, Bush's under secretary of state for economic, business, and agricultural affairs, who oversaw energy issues at the State Department, rode to the Equatorial Guinea embassy on March 19 to hear what Obiang had to say.

The president opened by stating that he would like to return to Washington on an official visit, so he could "convey his concern over the lack of a U.S. embassy in Malabo to the highest level of the U.S. government." His country, he continued, "had received much assistance from private American companies" and it was "unreasonable" that there was no embassy. He understood that there were budget issues vexing the United States, but federal tax revenues received from the American oil companies making profits in Equatorial Guinea were "more than sufficient to pay for a new mission." The decision to close the embassy in 1995, after the witch doctor incident involving Ambassador Bennett, was "based on erroneous human rights reports." State Department public reporting on human rights violations in Equatorial Guinea—which continued to highlight torture and detention of Obiang's political opponents, as well as the abysmal conditions at Black Beach—was misguided; it was the product of a temporary American diplomat in Malabo who had "no conception of the real situation. . . . Only officials posted in [Equatorial Guinea] would be able to understand the true situation."

Larson replied that the Bush administration was "very interested" in working with Obiang "on encouraging the growing bilateral business relationship." President Bush and Secretary Powell "had made it clear that promoting respect for human rights and democracy would be a continuing theme of our foreign policy," but the administration was nonetheless "prepared to work" with Equatorial Guinea's government. On the question of a U.S. embassy in Malabo, the State Department would be "reviewing the issue," Larson said.[17]

On Capitol Hill, among human rights activists, Republican-leaning global Christian groups concerned with governance and development in Africa, and the democracy-promoting enthusiasts of the neoconservative school, Equatorial Guinea was "the kiss of death," recalled a senior Bush administration official involved. But the oil companies joined Africa Global in pressing Obiang's cause. ExxonMobil, Marathon, and Hess worked through the Corporate Council on Africa, an industry trade and lobby group, to campaign at the White House and State for approval for a new U.S. embassy in Malabo. The oil companies argued through their Washington lobbyists that the American embassy in Malabo had been shuttered before the discovery of oil, when virtually no U.S. citizens resided in the country, whereas now there were upward of six hundred Americans living in Equatorial Guinea, shuttling in and out on rotation. Passport and visa paperwork had to be handled in Cameroon, and there was no permanent diplomatic liaison to address even routine business issues. Still, the Africa hands on the National Security Council, where the decision would ultimately be made, hesitated. They knew George W. Bush would be accused of selling out human rights for oil profits if the administration reopened the embassy.

Obiang wanted military training, too. His government had received a State Department license to hire Military Professional Resources International (M.P.R.I.), a government-connected security contractor based in northern Virginia, to improve the virtually nonexistent capabilities of

Equatorial Guinea's tiny coast guard and navy. It was unusual for State to approve any license for military training for a regime with a human rights record as bad as Equatorial Guinea's, but maritime defense work had been rationalized as necessary to protect huge American oil investments off-shore. Now Obiang wanted to expand M.P.R.I.'s training to include his military and internal security forces and to contribute to regional campaigns against maritime piracy and illegal fishing. Obiang had told Larson that he needed additional military assistance to "protect [Equatorial Guinea's] sovereignty and the U.S. investment."

He sent his foreign minister and energy adviser to Washington to explain that his request reflected Equatorial Guinea's "concern over security issues, particularly the safety of offshore oil installations, but also stems from the president's desire to emulate the United States in areas such as democratization and respect for human rights." The Bush administration stalled some more.[18]

The administration did allow Obiang a steady stream of official meetings when he visited America for cancer treatment or other private reasons. On September 7, 2001, Obiang again met Under Secretary of State Alan Larson. United States investment in Equatorial Guinea, all in the oil and gas sector, "is having a great impact on the country," Obiang pleaded. But he needed help. His country had been called the "Kuwait of Africa," he said, but with a mere 10 percent royalty rate on oil production in the early phase, "this does not accurately reflect the revenues that the government receives." The country was still waiting for ExxonMobil to cross the break-even point in the recovery of its investments, after which Equatorial Guinea's take would rise; the country was not yet as wealthy as it would be.

Larson agreed that the "investment of U.S. companies" in Equatorial Guinea "strengthens the bilateral relationship" with America. He warned, however, that the more prominent Equatorial Guinea became as a global oil supplier, "there will also be increased scrutiny from human rights groups around the world." There was nothing the Bush administration could do about that—it was a fact of global life in an age when the power of nongovernmental campaigners and media was increasing. Larson "encouraged Obiang to continue to work constructively" with Ex-xonMobil, Marathon, and Hess, "who have shown so much confidence in [Equatorial Guinea] to invest so much there." On the question of the royalty rate and other financial matters, Larson "assured" Obiang that the American companies would "deal squarely" with Equatorial Guinea.[19]

Al Qaeda terrorists struck Washington and New York four days later. Obiang was still in town, ensconced at the luxury Willard Hotel, which is located on Pennsylvania Avenue between the White House and the Capitol. He was smuggled out of the hotel through the garage. Terrorism fears now joined oil dependence as glue in the emerging U.S.-Obiang relationship. The Bush administration soon briefed Equatorial Guinea on "increased Al Qaeda operations under way throughout the world and the possibility that Al Qaeda might target petroleum facilities."[20] Obiang readily invited American diplomats to talk with him about security issues. His regime lived in perpetual insecurity, plagued by internal coup plots and menaced by much larger neighbors, particularly Nigeria, that might covet the country's oil wealth. Rising fears within the Bush administration that seaborne Al Qaeda–inspired terrorists might attack American offshore oil platforms would draw Malabo and Washington toward a security partnership, some of Obiang's advisers believed. Equatorial Guinea's production, moreover, was expanding by the month. The country pumped out just fewer than 300,000 barrels per day in 2002 and expected more than 350,000 barrels per day in 2003. This would mean, a State Department cable noted, that Equatorial Guinea "will become the third largest oil producer in sub-Saharan Africa, after Nigeria and Angola."[21] Hardly anyone had noticed the country's emergence on world oil markets. The Bush administration at last overcame its hesitations; the White House approved the embassy and prepared for its reopening.

George Staples arrived in November 2001 as the new American ambassador. He was a career foreign service officer, an African American who wore wire-rimmed glasses and who spoke Spanish, French, and Turkish.

Staples had divided his tours among the Carribean, Latin America, the Middle East, and the Continent. He had served a previous tour in Malabo, as a political officer, during the 1980s.

On January 23, 2002, Staples flew to Equatorial Guinea for the first time after a fifteen-year absence. The country he saw dazzled him—road construction under way, new hotels, the modern corporate enclaves of ExxonMobil and Marathon, and small supermarkets. Obiang remembered Staples from his earlier tour, when they occasionally bumped into each other while jogging on the airport road. They met now at the president's palace above Malabo harbor. The atmosphere in the receiving room was formal; bodyguards and aides stood by in attendance. The president greeted Staples warmly; they spoke in Spanish.

"America is our friend, versus Spain and some other Europeans who have other agendas," Obiang said. "The American companies made this possible," he said, referring to his country's boom. "The Europeans lied to us. We've never stopped believing this."

It was gratifying to see a country that he remembered as one of the world's poorest developing so rapidly, Staples replied. "You have to recognize that great wealth brings with it great responsibilities," he added. It was vital that Obaing use his wealth "in a responsible manner, and not waste it on foolish programs or see it disappear through corrupt practices that could destroy the country's reputation and erode its moral fiber."

Obiang declared that he was "determined" that Equatorial Guinea would not "become another failed African state." He elaborated: "Not a single West African oil state can be called a success. What has happened with all the oil money that came to Nigeria, Cameroon, Gabon, and Angola? When African leaders gather at Organization of African Unity meetings, they look to South Africa as the continent's engine of growth, when Nigeria and other oil states could have performed this function long before apartheid ended."

As to human rights, Obiang said he was "trying to develop a democratic system," but his countrymen "were not sophisticated and sometimes had a low tolerance for opposing opinions." When opposition politicians were "harassed and attacked," it was "not on orders from the government." The same was true of other human rights abuses made against his regime.

"Help us," he said.[22]

Like many diplomats, Staples was optimistic about the potential of the United States to improve a country such as Equatorial Guinea. Here it seemed more apparent than usual that American corporations and international nongovernmental organizations could lift the quality of governance. When he visited Malabo, Staples sometimes stayed overnight at the compounds of the several international oil companies, including ExxonMobil's; construction of international business hotels was under way in Malabo, but none had yet been opened, and the State Department had yet to lease its own housing. The oil industry executives he met during these stays "spoke highly of Equatorial Guinea's potential to become an African success story."[23]

The ambassador suggested ideas to his hosts, including ExxonMobil, about how they might directly contribute to Equatorial Guinea's development, outside of pumping oil, by flying in teachers to run seminars on business formation, accounting, and the like. Staples soon learned, however, that while ExxonMobil's local representatives were sympathetic and interested, they had trouble winning approval for such initiatives from headquarters. ExxonMobil, along with Marathon and other firms, did invest in malaria eradication in Equatorial Guinea, but the corporation shied away from anything that involved interaction with the country's politics or business classes. ExxonMobil's lawyers in the upstream division in Houston feared, for example, that if they trained young Equato-Guinean people and their businesses subsequently failed, the oil corporation might be somehow judged liable. To the American ambassador, this seemed careful in the extreme, but it did in fact reflect ExxonMobil's strategy in poor and volatile countries. The issues of concern to ExxonMobil were largely limited to the production of oil and the sanctity of contracts. "We are an oil company; we are not the Red Cross," as Andre Madec, an ExxonMobil executive who oversaw global community relations, once put it. "We don't want to be seen as the de facto administration."[24]

Obiang remained hopeful about the strategic partnership he could eventually build in Washington through oil and security. The president bought big houses for family members throughout the suburban Washington region and he traveled regularly to the capital and to New York. At the Riggs branch across from the White House, Obiang's aides arrived with heavy, bulging suitcases filled with plastic-wrapped hundred-dollar bills—up to $3 million at a time—and handed them over as cash deposits, which Riggs's account managers gratefully accepted.

In tandem with the oil companies and Obiang's lobbyists, Riggs's executives tried to burnish Obiang's reputation as best they could. "The president has requested to come to the bank to pay a courtesy call and brief us on developments in Equatorial Guinea," Simon Kareri, the bank's African-born account manager handling the Obiang and related accounts, wrote to Joseph Allbritton, the bank's chairman. "I would like to suggest that we recommend that the President hire a P.R. firm."[25]

Kareri briefed the bank's leaders on the political risk equations involving their idiosyncratic client. There was potential for coups or internal fracturing in Equatorial Guinea, and yet major American oil companies had "established a significant presence in the country and U.S. officials appear anxious to maintain good relations," he noted. "The country could be valuable to the U.S. from a geostrategic view, given its location in Central Africa and the abundance of oil."

Obiang passed through the branch's soaring columns and into a paneled conference room one summer day. The bank's senior executives listened as the president described his country's progress. Obiang mentioned that there had been "pressure" from other banking and financial institutions to move some of his business elsewhere, on the grounds that Riggs was not equipped to handle all of the nation's finances. However, "the President's regard and loyalty to Riggs is unquestionable because he has dismissed all possible suitors as 'speculators,'" Kareri's notes of the meeting recorded. As to the allegations about his human rights record, publicized by Human Rights Watch and others, Obiang told the bank executives that he had not reacted to "the innuendoes" about such matters. "The President clearly regards his engagement of such discussion as demeaning to his stature," Kareri noted.

Obiang did have one piece of business he wished to mention while he was visiting his bank: He requested a $34.4 million loan to purchase a presidential jet from Boeing Corporation, a 737-700 that would be outfitted with a king-size bed and gold-plated bathroom fixtures.[26]

Seven

"The Camel and the Jackal"

O n New Year's Day, 2000, about two months after he arrived in the capital city of N'djamena to serve as United States ambassador to Chad, Chris Goldthwait sat down to write a letter home to friends entitled "Is It Hopeless?" He was referring to Chad. Certainly it was a troubled place. Goldthwait was a senior officer in the Foreign Agricultural Service and a part-time novelist. He had been posted overseas before, but never as a full-fledged diplomat; the State Department dispatched him to Chad as part of an effort to recruit ambassadors from outside of its own ranks. During his first weeks in the country, he took the embassy boat, which had been acquired to allow the ambassador and his colleagues to escape to neighboring Cameroon in the event of coups or other violent unrest, out on the wide Chari river to look for hippopotamuses. He found a few, but he was more entranced by the families living in adobe brick huts—premodern-looking dwellings supported by mats and poles, their tin roofs "held down against the wind by concrete blocks or big stones." Goldthwait was single and entering the twilight of his professional life; he was a self-contained man, ready to travel and inspect his vast assigned territory. It didn't take him long to discover that "all aspects of life are starker here than at home—greed, poverty, hatred, disease, death, honor, friendship and love."

If Chad's history was "our standard for judgment," then there was little evident cause for encouragement. To seize the territory for the French empire, two colonial captains invaded from Senegal late in the nineteenth century. They left a trail of burned villages and decapitated bodies before their African troops mutinied and murdered them. After independence, a succession of coups, rebellions, incursions from neighboring Libya, French interventions, and American cold war–inspired covert maneuverings left Chad's eight million people in the grip of Hissène Habré, who arrested, tortured, and murdered several tens of thousands of his countrymen. One of Habré's French-trained generals, Idriss Déby, eventually overthrew him; the dictator fled to exile in Senegal with as much gold as he could load onto his escape plane. Like his predecessor, Déby trusted only his northern tribal kinsmen and "lavished upon them the lion's share of governmental largesse and responsibility," Goldthwait noted. "It has meant that a government with meager resources, degenerating early after independence into corruption, has come to be viewed mainly as a patronage system."

Chad's borders on international maps mark a landlocked expanse almost twice the size of Texas and breathtaking in its internal diversity. Its people speak 128 distinct dialects. The country's southern forests and agricultural lands, bordering Cameroon, receive as much rainfall as the American East Coast. In the north, nomads roam with camels and cattle through parched dunes and rocky crags. There are fewer than eighty miles of paved roads in the entire nation. Chad's poverty ran even deeper than Equatorial Guinea's.

Life expectancy at the century's turn was just forty-six years, according to the United Nations Development Programme; less than half the population was literate and only a third of school-aged children were enrolled in classes. The economy suffered from "poor to non-existent infrastructure, chronic energy shortages, high energy costs, a scarcity of skilled labor, limited understanding of the English language, a high tax burden, and corruption," according to a formal U.S. embassy assessment cabled to Washington. Five of Chad's six neighbors—Libya, Niger, Nigeria, the Central African Republic, and Sudan—were politically unstable. Idriss

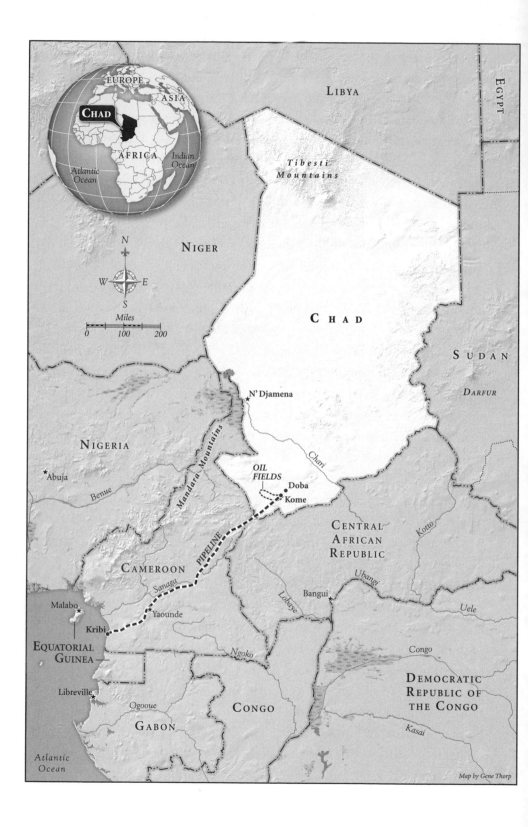

EUROPE
ASIA
CHAD
AFRICA
Indian
Ocean
Atlantic
Ocean

LIBYA

EGYPT

Tibesti
Mountains

NIGER

N
W   E
S

Miles
0   100   200

CHAD

SUDAN

DARFUR

N'Djamena

NIGERIA

Chari

Abuja

Benue

OIL
FIELDS
Doba
Kome

CENTRAL
AFRICAN
REPUBLIC

Kotto

Mandara Mountains

PIPELINE

CAMEROON

Sanaga

Ubangi

Bangui

Uele

Malabo

Lobaye

Congo

Yaounde

Kribi

EQUATORIAL
GUINEA

Ngoko

DEMOCRATIC
REPUBLIC OF
THE CONGO

Libreville

CONGO

Ogooue

GABON

Kasai

Atlantic
Ocean

Map by Gene Thorp

Déby, like coup makers worldwide, had promised reforms when he came to power, but he had settled into a self-protecting regime made up of relatives and cronies with apartments in Paris, ruthless palace guards, half-loyal regional militias, and a training contingent of about a thousand French troops and airmen on standby at N'djamena's airport. Déby had some skills: He was a thin, composed man with a general's sense of military maneuver and a tribal sheik's instincts for political balancing. Still, although the Chadian elites in the precarious capital spoke a "perfect French," Chris Goldthwait wrote, "there isn't anything behind it in the way of knowing what to do to solve problems."

Hopeless? The arriving American ambassador chose to believe otherwise. Democracy movements had swept even the poorest African countries during the 1990s. National economies had been growing across the continent, even in the midst of civil violence. By the time Goldthwait landed, Chadians had become "marvelous at saying what they know we want to hear from them about democracy, development, national reconciliation, etc.—sooner or later they will start to believe some of it themselves." He wrote to his friends:

> So no! It isn't hopeless, only incredibly difficult. More difficult than the miracles of development in East Asia. More difficult than the recovery of the District of Columbia in our own country. More difficult than the challenges facing South Africa or other African lands. But the people here will face the challenge because they have no choice. Is there anything we can do to help them?[1]

ExxonMobil employed considerably more geologists than political scientists. The complexity of Chad that attracted their attention involved the subterranean formations of the Chari river floodplain, particularly around the Doba basin. Exxon's local subsidiary—Esso Exploration & Production Chad, Inc.—had started exploration and development work in the basin in 1977. Its engineers confirmed significant deposits at Bolobo. The oil there was thick and sour, meaning that it was infused with sulfur and thus less valuable than the "light and sweet" crude blends that attracted the highest prices on global markets. (Those lighter blends were easiest to refine into gasoline and other fuels.) The viscosity of Chad's oil presented production challenges, as did the diverse and inconsistent strata of rocks and water beneath Chad's eroded land, which complicated efforts to pump oil to the surface.

The scientists who puzzled over Chad's subsurface worked mainly out of Houston, in the upstream division of ExxonMobil charged with buying, developing, and producing oil and gas. After the merger, to reduce head count and improve flexibility, Raymond oversaw a reorganization that moved many of these scientists into "skills" groups that could deploy worldwide, like geology special forces teams, to support local business units and projects. Their ability to accurately assess oil reservoirs improved year by year as faster computing and improved graphic and imagery software allowed for more accurate visualization of underground geology than in the past. There was a basic Geophysics skills group, with obscure-sounding subunits such as the Gravity Magnetics Group. There was a skills group called Geological Operations that brought scientific expertise to well and pipeline engineering. There were scientists grouped together to work on New Field Development and Mature Field Development, as well as a group specializing in Stratigraphy, a field of geology that studies how complex layers of rocks and sediments such as those beneath Chad's Doba floodplain change over time, influenced by water and weather. One of the most important scientific groups based in Houston was called Formation Evaluation. These were the specialists who bought, captured, and analyzed seismic and other data about where undiscovered oil and gas formations might lie beneath land or ocean floors. As Houston struggled to define an economically viable plan to lift Chad's thick, dispersed oil to the surface, its stratigraphers wrestled with geological problems tens of millions of years in the making.

Then there was the challenge of transport, even if Chad's oil could be coaxed to the wellhead. There was no profitable market for large volumes in Chad's neighborhood. The oil would have to be piped hundreds of miles overland to the Atlantic Ocean and then shipped to refineries in Europe or the Americas.[2]

During the cold war, despite his torture rooms, Hissène Habré had maintained cordial ties with the United States, which regarded him as a regional counter to Libya's leader, Moammar Gaddafi. The Reagan administration trained elements of Habré's notorious security service and provided the regime with tens of millions of dollars of military aid each year. When Habré visited Washington in June 1987, Reagan assured him, "Chad now knows it can count on its friends." The American president took note, too, of Chad's economic problems, particularly its recent "locust plagues."[3] Exxon's oil discoveries offered relief from such trials. Bolstered by the Reagan administration's alliance, Exxon outflanked French oil companies, negotiated with the dictator's aides, and, in 1988, produced the Convention for Exploration, Exploitation and Transportation of Hydrocarbons in Chad. It provided for a thirty-five-year compact among Chad's government, Exxon, and two partners; Exxon would be the lead operator, and eventually, the corporation secured Chevron and PETRONAS as partners. The terms were favorable to Exxon and its partners in comparison with typical contracts elsewhere: Chad would receive a 12.5 percent royalty on all oil produced, plus taxes equal to 50 percent of the consortium's net profits, which could rise to 60 percent if world oil prices soared. Chad's take of less than two thirds of revenue after expenses compared to rates closer to 90 percent in Nigeria.

The generous terms were required, the oil companies insisted, to compensate for the exceptional risks they would endure in Chad. No political order in the country was likely to last for thirty-five years. Exxon's negotiators addressed this conundrum not just by negotiating for favorable royalties; they also inserted into the 1988 contract what was known in the oil industry as a stability clause. Article 34, entitled "Applicable Law and Stability of Conditions," placed the terms of the convention beyond the reach of any Chadian law that might be enacted by any government of the future. The clause protected Exxon against political risk. That Exxon had the power to carve out rights trumping any future law passed by any future Chadian regime was perhaps not surprising in this instance; Exxon's 1988 net profits of $5.3 billion exceeded by several times the size of Chad's entire economy. Article 34.3 declared:

During the term of this Convention the State guarantees that no governmental act will be taken in the future, without prior agreement between the Parties, against the Consortium which has the effect either directly or indirectly of increasing the obligations or amounts payable by the Consortium or which adversely affects the rights and economic benefits of the Consortium provided by this Convention.

The language binding Chadians to Exxon's "rights and economic benefits" was strikingly broad—it could even be interpreted to mean that future governments in N'djamena might be prevented from broadening civic freedoms or permitting unions to organize if such changes raised the oil consortium's costs. More realistically, the stability clause provided a strong defense against any future Chadian coup maker's inclinations to raise taxes on Doba oil production. The contract was unambiguous about the parties' relative sovereignty: "In case of contradiction or inconsistency between this Convention and the laws and regulations of the Republic of Chad, the provisions of this Convention shall prevail, unless the Parties decide otherwise." When Déby presented the contract to his cabinet for approval, recalled Salibou Garba, then the country's minister for post and telecommunications, the president declared, "You don't have time to read this—and they need it in Houston." Even Déby "did not take time to go through it," Garba said. "Only later did he realize that the terms were not as favorable as he wanted."[4]

Rosemarie Forsythe, who rose to run ExxonMobil's global political department out of Irving, had been a precocious child. She graduated from Indiana University at the age of sixteen after studying the classics, Russian literature, and political science. She overcame the psychological burdens of prodigy and grew into a calm, professional woman with a knack for making herself useful in large organizations. Fascinated by Russia and its neighbors, she joined the State Department in 1987 and then moved to the National Security Council, where she became a traveling specialist in the political affairs of the new republics born from the Soviet

Union's dissolution. In the late 1990s, Forsythe left government to work for Mobil Oil as a political adviser, based in London. After the Exxon merger, she was summoned to build the combined corporation's political department at the Irving headquarters. She came to function as Exxon-Mobil's chief political risk analyst. She filtered and synthesized political assessments flowing to Irving from the corporation's far-flung field offices. She adjusted gradually to ExxonMobil corporate culture with some of her sense of irony intact; she told friends that the oil corporation's system for maintaining confidential information was far more severe than anything she had seen while holding a top secret clearance at the White House. (ExxonMobil so guarded its internal estimations of country-by-country oil reserves, for example, that when executives talked about that subject with outsiders, they used the published estimates of rival BP rather than reveal their own.)

Forsythe worked in a modest-size office on the ground floor of the Death Star; she enjoyed a window looking onto the campus's green lawns. Each April, when ExxonMobil conducted an annual multiyear strategic planning exercise, she integrated global and demographic forecasts into the plan. She also advised Lee Raymond and the Management Committee about the international political dimensions of investment and technology decisions.[5]

Chad presented an emblematic case of the challenges Exxon faced abroad. Much of the oil and gas ExxonMobil could hope to acquire and book as equity reserves by the year 2000 lay in weak states—countries that were too poor, underpopulated, or badly governed to produce and control their oil on their own. Many of these weak states lay in Africa.

In her presentations to ExxonMobil's Management Committee, For-sythe showed PowerPoint slides that divided the world's governments into three categories: democracies (blue), authoritarian regimes (yellow), and transitional governments (red). The latter were characterized by political instability. One slide showed that in recent decades the percentage of the world's known and estimated oil located in unstable countries had doubled to about half. (By comparison, much of the world's natural gas reserves lay in authoritarian nations, particularly Russia and Iran.) Forsythe's work also showed how demographics in the red and yellow countries—

particularly the "youth bulges," or the growing generations of young people swelling in the Middle East—suggested that even more instability in
oil-producing regions could be anticipated. (Large numbers of teenagers
tended to create havoc wherever they lived; her forecasts anticipated the
Arab Spring many years before it arrived.) Forsythe also produced maps
showing global piracy problems and threats to shipping lanes.[6]

An implication of her analysis was that if a company like ExxonMobil
wanted to continue to replace the hundreds of millions of barrels of oil
reserves it pumped and sold each year, and to show Wall Street that its
booked reserves were holding steady, the corporation would be drawn more
and more into poor and unstable countries, and it would have to find ways
to operate successfully in such places. As Vice President Cheney had once
remarked when he ran Halliburton, "The good Lord didn't see fit to put
oil and gas only where there are democratically elected regimes friendly
to the United States. . . . We go where the business is."

Exxon hired a former career foreign service officer with long experience in Africa, Simeon Moats, to serve as an Africa desk officer in the corporation's Washington office. Moats consulted with former colleagues at
State and at the National Intelligence Council to stay abreast of Chadian
affairs. Lee Raymond's Africa team also brought in Herman "Hank" Cohen,
an assistant secretary of state for African affairs during the George H. W.
Bush administration. He was hired as a consultant to train Exxon managers about the political and cultural issues they might face in Chad. Exxon
ran the Chad project out of an office in Paris and hired French nationals
and French-speaking Canadian executives to supervise its operations in
N'djamena. Many of the engineers and managers who did the day-to-day
work in the oil fields, however, were English-drawling Americans dispatched from Houston. Cohen set up classrooms in Houston and New Jersey; his students were "all white males," as he recalled it. He developed a
syllabus to instruct them "about the nature of the Chadian government,
how it worked, how to get along with them," and perhaps even more
challenging, "how to get along with French people." Cohen had known
Idriss Déby during the anti-Libyan campaigns of 1987, when Déby served
as Chad's chief of army staff; Cohen remembered him as "a tribal warlord,
basically," who drank too much and who struggled to share power and

wealth adequately even with members of his own tribe. Cohen warned Exxon's executives that there "would be a challenge to his rule at some point."[7]

The possibility of a coup d'état was not the only political risk the corporation had assumed in Chad. As with climate change and the management of corporate security in Aceh, Lee Raymond's decision to explore for oil in such a poor African nation had involved Exxon in a burgeoning global contest of ideas, this one concerning the social and political consequences of oil production in very poor countries. Rosemarie Forsythe's slides for the Management Committee implied that more and more of the world's oil happened to be located in unstable countries, more or less coincidentally. A growing body of academic research suggested that oil production was likely a *cause* of their instability.

Exxon's exploitation of Chad's oil involved both engineering audacity and big thinking about how to relieve poverty in Africa. To reach global markets, Exxon formed a consortium to build a 660-mile pipeline across Cameroon's forests to the town of Kribi, on the Atlantic Ocean. From there it would pipe the oil an additional 7 miles to an offshore marine terminal. To ensure that the project met global standards for the management of oil revenue and involved credible plans to relieve Chad's poverty, Exxon enlisted the World Bank, the Washington-headquartered institution funded by rich countries to support economic development in poor countries. During the mid-1990s, Exxon and the bank conceived an unprecedented plan: In exchange for loans from the World Bank's finance arm, Chad's government would be pressured to accept covenants requiring that it spend most of the royalties and taxes it received from oil production on health services, education, economic infrastructure, and other poverty alleviation programs. To ensure that Idriss Déby or others in his government did not cheat, the plan would require that Exxon route Chad's oil money through special bank accounts in London controlled by the World Bank.

For a cautious company run mainly by engineers, the Chad project's terms amounted to an extraordinary venture by Exxon into social engi-

neering and nation building. There was something about the starkness of
Chad's poverty that seemed to attract Exxon's engineers; they talked
about the country as a place that could be entirely remade. "We have the
opportunity of applying this model on a clean slate," explained Tom Walters, the corporation's vice president for oil development in Africa. "There
was no prior history of development to deal with." Foreign missionaries
of an earlier era imagined that they might improve Africa by imbuing its
people with Christian values of work and rectitude. Exxon's managers believed they might improve Chad's government by demonstrating through
their own example the benefits of corporate discipline and principled
consistency—the gospel of the Operations Integrity Management System.
"A big part of what we think we can bring is a lot of the ethical behavior
that we can portray to the government," Walters said. "We have been
working with this government now for a good ten years. . . . And we
do not back down, and I think that education is going to have dividends
over the long haul." Rex W. Tillerson, who was rising to prominence as a
leader in the corporation's international division, described the project as
"a clean sheet of paper" where Exxon had "the opportunity to put things
in place perhaps the way you'd like to see them carried out from the very
beginning."[8]

The corporation's ambition in Chad stood in contrast to its modest
local development projects and political quietude, bordering on complicity, in Equatorial Guinea. The difference had more to do with oil market
geography than with any deep-seated desire within ExxonMobil to reform
Chad. The corporation's Equatorial Guinea properties lay offshore and
could be shipped to markets without much need for ExxonMobil employees to involve themselves in the country, except to go and come from
the airport. As long as Obiang's rule was stable, ExxonMobil could keep
a low profile and pump oil. In Chad, the corporation's oil was stranded
inland. Constructing a pipeline to the Atlantic Ocean inevitably meant
the political visibility and risks of the project would be elevated. Land
acquisition, population resettlement, cutting down trees, and environmental protection plans were sure to attract local and international scrutiny
from the start.

By recruiting the World Bank as a partner, Exxon's leaders shrewdly insulated themselves from many of the project's most daunting reputational risks, particularly those arising from the objections of environmentalists and nongovernmental organizations. The World Bank's technocratic experts in poverty alleviation and development—not the oil corporation—would design and implement the plan to manage Chad's oil revenue as a public trust. "The notion that ExxonMobil should be telling the Government of Chad how to spend its money—like Shell telling the U.K. government how to spend its money—wouldn't go down well," Lee Raymond observed. The World Bank, however, had the mandate.

"The biggest thing this company can bring to some of these countries is the opportunity to see capitalism and the free market work," he said on another occasion. "Am I comfortable with everything the government of Chad does? No. Am I comfortable with the concept that we're now going to give the Chadian people an opportunity to improve their lot through economic development? Extremely comfortable."[9]

The project's New Jerusalem ambitions reflected the World Bank's evolving priorities. President Clinton had appointed James Wolfensohn, a multitalented Australian investment banker, cellist, and fencer, as the bank's president. Wolfensohn inherited an institution under increasing criticism from European board members and antipoverty campaigners for its reliance on the financing of large infrastructure projects—dams, highways, and the like—that did little to stimulate private investment or meet the human needs of poor people. Wolfensohn traveled the world seeking new ideas to address these concerns. The Chad revenue-management plan presented an opportunity to experiment with a new conditional-lending model. It became "the child of James Wolfensohn," recalled Jane Guyer, an anthropologist who advised the bank.[10] Exxon seemed an unlikely corporate partner for a nation-building project of such visibility and ambition. Lee Raymond, however, had confidence in Wolfensohn. In his earlier career as an Australian merchant banker, Wolfensohn had worked with Raymond on some of Exxon's oil and gas deals. Later, as Exxon's chief executive, Raymond looked at the Chad property and realized there were only three choices: Do nothing with the oil, which would be unfortunate;

ship it north, through Libya, which looked impossible, politically; or sign up to the World Bank's plan to improve governance and raise living standards in two very poor countries. Why not try?

Exxon's plan for Chad offered to break the pattern of oil-related corruption and violence; it offered a way to use public development funds to stimulate private economic activity in Africa; and it seemed to offer a credible reply to the bank's critics. Persuaded that it had a choice between the bank's terms and no oil revenue, Chad's parliament passed Law 001 in 1998, and Déby signed it the following year; it pledged 10 percent of the country's future oil revenue to a "future generations" fund, and 80 percent of the remainder to "priority sectors" such as education, health, agriculture, and the environment. In addition to parliament, a "college" of civil society groups would provide oversight of these investments.[11]

The plan offered an answer to the "resource curse," a syndrome described by economists and political scientists. The curse referred to evidence that when poor countries became suddenly rich in oil or minerals, they could often expect to go backward rather than forward. In 1993, the British economist Richard M. Auty published *Sustaining Development in Mineral Economies: The Resource Curse Thesis*. (His work drew on earlier economic analysis about the "Dutch disease," which referred to the distortions that took place within the Netherlands's economy after a major natural gas discovery.) Essentially, the resource curse described a condition within governments similar to what happens to many individuals after they win the lottery. When nations became enthralled by the short-term riches offered by a finite national resource, capital and talent often migrated away from more productive and self-sustaining economic sectors such as agriculture. In 1997, the American political economist Terry Lynn Karl applied these insights to oil development in poorer countries; her book, *The Paradox of Plenty: Oil Booms and Petro-States*, used the example of Venezuela to show that weak governments made rich by oil were prone to corruption and underinvestment in agriculture. These countries also sometimes attracted violence among internal factions competing for control of the engorged national bank vault. Philippe Le Billon of the University of British Columbia reviewed twenty separate studies of the resource curse and found that while "oil broadly correlates with higher risk of conflict," countries with high wealth per capita faced less danger, even where, as in Saudi Arabia, corruption was pervasive. The highest risk scenario, Le Billon found, was one involving "onshore production, institutionally weak central government, generating low rents per capita, with high level of dependence on the oil sector."

That described Chad. Even before the Doba oil flowed, a southern rebel group, the Armed Forces for a Federal Republic, known by its French acronym as F.A.R.F., sprung from nowhere to challenge Idriss Déby's regime for control of the region around the Exxon fields. Déby violently suppressed the F.A.R.F. and its suspected supporters. "The Chadians came in and were quite rough, but they terminated it," recalled an oil industry security adviser who monitored the campaign. Exxon stood by as the oppressive violence occurred; security was the responsibility of Chad. "There were regular exchanges between embassy and Exxon officers when human rights abuses occurred" during Déby's counterinsurgency campaign, noted a State Department cable, but these meetings did not consider "actions Exxon could possibly undertake in reaction to the abuses."[12]

By the late 1990s, the resource curse thesis had become conventional wisdom among liberal-leaning development economists and policymakers, including some at the World Bank. As a decision about Exxon's oil project neared, antipoverty groups with long experience in Africa, such as Oxfam and Catholic Relief Services, argued that the World Bank was moving too hastily. Their researchers agreed in principle that ordinary Chadians deserved the benefits of their oil wealth, but they doubted that the fragile, violent, embryonic half democracy over which Idriss Déby presided could offer the oversight of oil revenues imagined by the bank. These nongovernmental researchers chafed at optimistic assessments published by the bank, which seemed to them to ignore recent events such as Déby's suppression of the F.A.R.F. "You really have to examine: What is the nature of the state?" said Ian Gary, who monitored the deal at the time for Catholic Relief Services. "You can't just see what you want to see."[13]

The Chad compact also suggested a double standard for Africa. The World Bank and the Clinton administration did not question the right of Saudi Arabia to use its oil revenue to enrich a corrupt royal family. They did not seek to control social spending in the new oil-producing states of

Kazakhstan or Azerbaijan, despite the myriad ways those governments failed to serve their people. Why single out Chad? Why single out Africa? "The oil project is frankly the only hope on the horizon for the nearly eight million people of Chad," argued Donald Norland, a former American ambassador to the country who used some of his retirement years to campaign for the oil project. "It seems particularly unconscionable, therefore, for outsiders surrounded by modern comforts to oppose the project and thereby condemn the people of Chad to a kind of pristine poverty."[14]

It seemed Idriss Déby sent a cousin, Tom Erdimi, to Houston, where Erdimi took an office at Exxon's upstream division and worked as Déby's liaison to the corporation. Erdimi helped to manage and push Déby's negotiations with the bank. The president also appointed Ahmat Hassaballah Soubiane as his ambassador to Washington. Soubiane was a political activist who believed the oil project would be good for Chad in the long run, but who doubted his own president's sincerity. As the negotiations reached a climax, Déby continued to resist the World Bank's oversight terms. He "talked about a plot against the sovereignty of Chad" and urged his ambassador to fight. Privately, however, Soubiane "felt very strongly" that the bank and its "instruments of control and surveillance" would help protect his country against "Déby's desire to do whatever he wanted with the oil revenues." Soubiane worked to bring Déby around to a compromise.[15]

In the spring of 2000, Soubiane met on a Saturday with James Wolfensohn at the bank's headquarters, five blocks from the White House. They argued about Déby's stubbornness. Soubiane promised to push his president harder. He described how Chadians had come to see the international arguments about their country's oil. Northern nomads in Chad tell a creation story, "the camel and the jackal," he explained: The camel was created first, followed by the jackal. The jackal then noticed that the camel was a weaker animal and so it followed wherever the camel went, waiting for it to fall down dead. "Chadians remain skeptical and they believe that Chad's oil has become the camel," Soubiane said. "The jackal, since the beginning of the world, has been waiting to watch it fall down."[16]

Treasury Secretary Lawrence Summers made the Clinton administration's policy decisions about World Bank projects; he took advice from

Timothy Geithner, who then ran Treasury's international division. After two years of interagency reviews, Geithner and Summers were united in their support for Exxon's plans. "Clearly, the project has significant risks," Summers wrote to the U.S. Catholic Conference, a religious group that had questioned the project. "However, the Bank and borrowers have made a serious and credible effort to deal with these inherent risks, to incorporate lessons learned from failures of the past, and to set a higher standard."[17]

The World Bank at last approved the deal. Déby's political party organized a mass gathering in the streets of N'djamena. The president appeared before the crowd and accepted his people's congratulations. "The city burst into celebration," Chris Goldthwait wrote home. "It was to a degree organized, but there was genuine feeling in the crowd that gathered at the parade grounds opposite the Presidential Palace. And there was spontaneity in the horn-honking motorists who sped around town. The President walked over . . . relaxed and clad in a polo shirt for the evening warmth. The most important economic project in the country's history would go forward."

At an inauguration ceremony, Déby seated the ExxonMobil delegation in front of the diplomats before the stage; "the Saudi charge, always sensitive to such matters, complained." But Goldthwait felt that he "couldn't fault the Chadians this little breach of protocol—they know where their bread is buttered."

ExxonMobil's oil deal in Chad signaled the shifting sovereignties of a rising era in which formal governments were losing relative power. A warlord running a teetering state surrendered prerogatives of his office in exchange for the private capital and cutting-edge technology he required to strengthen his reign. A multilateral lending institution brokered the agreement and afterward contracted with the London office of a global bank, Citigroup, to manage and control most of the revenues due to the warlord's government. Oxfam, Catholic Relief Services, Global Witness, and other worldwide antipoverty campaigners organized conferences at which they taught Chadian civil society activists how to secure their rights. ExxonMobil, having conceived and financed the oil project in the first instance, and having achieved its business aims after more than two decades of effort, now moved to produce oil on a schedule of its choosing and under contract terms that enshrined its rights ahead of those of the Chadian government.

The United States embassy in N'djamena transmitted a "Sensitive" cable to Washington that candidly described its position in relation to ExxonMobil, when the embassy sought to advance American policies on human rights:

> Exxon has been an operator in Chad for almost 30 years now and is self-sufficient working in this environment. To date they have not come to us for advice on how to confront particular human rights situations or how to advance specific political agendas. . . . It is our impression that on a number of fronts, they would probably prefer to keep a certain distance from the embassy. . . . Our experience has been that Exxon has not felt the need to request the presence of senior embassy officials during meetings with government officials.[18]

And why should it be otherwise? ExxonMobil's investments in the Chad-Cameroon oil project would amount to $4.2 billion. Annual aid to Chad from the United States was only about $3 *million*. "Only this pittance, in one of the ten poorest countries of the world," Chris Goldthwait wrote.

As Exxon's work around Doba expanded, Goldthwait traveled through the oil area and marveled at "just how much control the government [of Chad] has ceded to Esso over what happens in the south, almost a loss of sovereignty!" The company "calls the shots" in four of the country's most populous administrative regions, he noted. Oil personnel managed security by suggesting travel routes and accompanying Chadian military and paramilitary patrols; they controlled local roads and ran their own satellite radio network reaching to the capital; and their spending on employment and construction in and around their oil fields dwarfed that by Chad's government and foreign donors. Goldthwait observed: "Esso coordinates every step of the way with central and local authorities, but insists on what it needs; Esso rules the south."[19]

———

William Foltz, a Yale University professor of African studies and political science, and the author of *Arms and the African*, visited N'djamena and delivered a lecture; Goldthwait went along to listen. The ambassador had been reading and reflecting on *Le Mendiant de l'Espoir*, or *The Beggar of Hope*, a novel by an imprisoned Chadian writer, Ali Abdel-Rhamane Haggar, who described the "root divisions" of Chadians, particularly "tribalism and corruption." Foltz set Chad's travails into the wider context of Africa's transition from an earlier postcolonial era of periodic coups to "today's much bloodier fighting in civil war," a transition aided by the availability of weapons, the spread of dislocated refugees, and the ability of African governments to sell off resources to "finance the purchase of more weapons." As he listened, Goldthwait jotted down other factors that the professor had not mentioned: "Ethnicity, a tradition of fighting, the fact that people have very little to lose, corruption, lack of results from government's actions, lack of faith in democratic institutions."

"The gap that bothers me the most," Goldthwait wrote home, "is in our development model for a country as poor and isolated as Chad. We know what works at the grassroots to help farmers or herders take the first small step beyond subsistence. . . . And we can identify a handful of critical big-ticket infrastructure needs like paved, all-weather roads or reliable supplies of water or power. But this leaves a huge void before the country achieves sustainable economic growth—and we don't know how to bridge it."[20]

Shell and the French oil giant Elf Acquitaine originally joined Exxon-Mobil in the Chad project. Then an Elf chief executive, Loik Le Floch-Prigent, was indicted on embezzlement and international bribery charges; he was accused of skimming corporate funds and using some of them to pay off African and other overseas officials to win oil deals. ("It was like that before me," Le Floch-Prigent claimed. "And my successors are doing it as well.") Lee Raymond and his Dutch senior aide, Rene Dahan, once dined with Le Floch-Prigent in Paris and walked away shaking their heads. "I'm not doing that again," Raymond barked at Dahan. ExxonMobil had its flaws, but outright crookedness was not one of them.

The scandal weakened Elf, led ultimately to a merger with Total, and forced ExxonMobil to find new partners in Chad. The Chad project had by now lasted three decades without producing a barrel of oil; it had turned out to be "a good example of long-term persistence, but a bad example of getting things done quickly," as Harry Longwell, Raymond's upstream lieutenant, put it dryly. The corporation finally roped in Chevron and PETRONAS of Malaysia as new nonoperating partners.

Chevron paid Déby's government a "signing bonus" of about $25 million. The payment was not subject technically to the World Bank's rules, but Déby had rashly pledged in public that he would apply the bank's formula to all of his revenue. He changed his mind. There were rumblings of armed rebellions in the west, along his chronically troubled frontier with Sudan, near Darfur. Déby felt he needed to shore up his regime's defense; he invested about $4.5 million in new military equipment before 2001 was out.[21]

After years of absence from Chad following the cold war's end, the Central Intelligence Agency reestablished a station in N'djamena around the time that ExxonMobil decided to go forward with the oil project. The threat of terrorism gradually drew its officers back. Even before the September 11 attacks, Al Qaeda menaced Africa. Osama Bin Laden, the group's emir, lived in Khartoum, Sudan, between 1992 and 1996. He funded violence in Somalia and Ethiopia and built up cells in Kenya and Tanzania. The vision of pan-Islamic risings Bin Laden articulated reached the lightly governed regions of the Sahel, including northern Chad. From Sudan, Al Qaeda financed weapons shipments across the desert to sympathizers in Libya and Algeria. The reopened C.I.A. station in N'djamena was partially intended to address this threat. The station was "declared" to Déby's government, meaning that the intelligence officers posted to Chad were known to their counterparts in the Chadian security services. The station evolved into a security liaison operation through which the United States could covertly provide training and equipment to counterterrorism units created by Déby for that purpose.[22]

After the September 11 attacks, the C.I.A. and the Pentagon's Joint Special Operations Command no longer faced budget constraints; they charged into African capitals long neglected. At first, the coordination between the agency and the Pentagon in Chad and elsewhere in the Sahel was lacking. Deeply compartmentalized secrecy was the operational norm at both the C.I.A. and within the Special Forces. In the early months of 2002, the C.I.A.'s Africa division sometimes learned what the Pentagon was up to only through the reports of its paid local agents. The Pentagon lacked the language skills—particularly French—to set up new Special Operations liaisons with host governments like Chad's without using translators. The translators would then be recruited as spies by the host government, so its leaders could keep track of what the Pentagon was planning. The C.I.A., running its own security liaisons, would hear about the Pentagon operators through the same translators. Once, in Ethiopia, local prostitutes were videotaped servicing American clients—the clients turned out to be Special Forces operators who were in the country without the knowledge of the local C.I.A. station, which then had to smuggle the Pentagon personnel out of the country before they were arrested. Gradually, the coordination improved. General Stanley McChrystal, a graduate of West Point, took charge of the Pentagon's Joint Special Operations Command, which ran the operations against Al Qaeda in Africa. During the same period, after the September 11 attacks, C.I.A. director George Tenet appointed Mel Gamble, a longtime operations officer who had served in South Africa, Nigeria, Kenya, and Liberia, to run the Africa division in the C.I.A.'s Directorate of Operations. McChrystal and Gamble worked together on a series of covert counterterrorism initiatives in Chad and elsewhere in the Sahel.[23]

"Terrorism was a rising star," recalled Karen Kwiatkowski, then a lieutenant colonel working on Africa policy at the Department of Defense in Washington. "Oil, as the price increased, was a rising star." It was natural for those, like her, who were charged with crafting Africa strategy at the Pentagon to take their cues from the White House. "And certainly," she recalled, "we knew what Cheney liked. . . . We didn't understand the oil economy. What we knew is that these are ways to make your department more relevant to the national mission and the national priorities."[24] The Pentagon, seeking funding and relevance, assumed that the Bush administration would support a focus on counterterrorism and oil supply security.

The National Intelligence Council had forecasted as the Bush administration arrived in office that West African oil would make up 25 percent of American imports by 2015. In the first years after September 11, there was a gathering sense within the intelligence and defense bureaucracies that instability in the Middle East required paying greater attention to oil-producing regions elsewhere. At a 2002 symposium entitled "African Oil: A Priority for U.S. National Security and African Development," Walter Kansteiner, the Bush administration's assistant secretary of state for African affairs, told the audience, "As we all start looking at the facts and figures of how many barrels a day are coming in from Africa, it's undeniable that this has become a national strategic interest for us."[25]

In Chad, the Africa division of the C.I.A. instructed the station chief to tell President Idriss Déby, in effect, "All the [oil] money that's coming in—you have to do a better job of helping your people." Yet Déby might be forgiven for believing that oil and terrorism mattered more to Washington than good governance.

Soon after McChrystal took charge of the Joint Special Operations Command, elements of an Algerian-based Al Qaeda affiliate, the Salafist Group for Preaching and Combat, crossed Chad's northern border and entered the country. The C.I.A. and the Pentagon turned to Déby for military assistance. The Pentagon's European Command used reconnaissance aircraft to pinpoint the Islamist cell's location. The C.I.A. station chief asked Déby to send in his own ground forces to attack. The overall message was, as one individual involved put it, "If you can help us on this, we'll help you in a lot of ways."[26]

In the political economies of African strongmen, the World Bank was potentially useful; the C.I.A., on the other hand, was where the world really turned. Salibou Garba, the Chadian opposition leader, believed that by 2002 or 2003, as ExxonMobil's oil began to flow in earnest and as counterterrorism rose as an American priority, Idriss Déby had concluded that his own security must be his overriding priority. His cooperation with the C.I.A. could reinforce his rule by making him indispensable to the Bush administration's global counterterrorism campaign. The World Bank's priorities of governance and social investment offered no comparable benefits and might cause him to lose his political grip if pursued too vigorously. "He was like a driver who ran over a pedestrian and just kept driving," Garba said.[27]

As he prepared to depart his ambassadorship after an extended tour, Chris Goldthwait tried not to dwell unduly on "the Great Frustration," as he called it, which arose from the fact that he could point to relatively few concrete achievements from conventional development aid or diplomacy that would help ordinary Chadians—particularly agricultural projects, in which he had long specialized. "The U.S. government isn't really a player in Chad's economic development," he wrote. "Beyond the general strictures of our Africa policy, the U.S. has one specific interest in Chad, the oil project, and one more amorphous one, regional stability. . . . Frankly, I could do a lot more and keep a lot busier running around town to see folk. But to what point? More cables won't help anyone back in Washington."

He held on to hope for the many striving Chadians he had come to know on his travels across the country. "From a purely economic view, the future is now," he wrote in his last letter. "The oil project has been producing. . . . Per capita income is growing 10 percent annually, solely on the basis of local spending by the consortium. Cell phones and Internet connections are booming. New construction is visible all over N'djamena.

"The great lingering question is whether the government of Chad will indeed adhere to the oil revenue management plan and whether the income will benefit the people. I'd say the odds are good, but it's still a crapshoot."[28]

ExxonMobil judged the risks it had taken to pump oil in Chad to be a success: "On target to reach full production status as scheduled." The corporation had so far drilled 115 oil wells in its southern zone; it had paid out millions of dollars in compensation to local farmers for land rights, and it continued to dole out about $25 million in wages to local employees annually, a large infusion of cash to the country's economy. ExxonMobil "has had and will continue to have significant positive economic impacts in Chad," the corporation declared.[29]

These were still early days; hope was in the air. A few years on, the World Bank's assumptions about Chad's potential as a socioeconomic experiment would be more fully tested, and the distribution of costs and benefits from oil production would be clearer.

Eight

"We Target Oil Companies"

F rank Sprow, the ExxonMobil vice president in charge of safety and environmental issues at the corporation after 2000, was a trim, athletic man who stood about six feet one inch tall. He had worked at Exxon since the 1960s; as he reached the corporation's executive ranks, he wrote and oversaw its rigorous post-*Valdez* safety rules. He was also, in his private life, a thrill seeker.

In off-hours and on weekends and vacation days, Sprow sometimes went hang gliding. He tried extreme skiing. He raced motorcycles. He raced midget sports cars at Soldier Field in Chicago. He flew stunt helicopters and went parachuting. One of his friends owned a stunt plane, a mini Messerschmitt, and on the weekends the two of them occasionally went for a ride—they would fly straight up, cut the engine, and dive. To manage his hobbies while carrying out his professional role as the chief enforcer of ExxonMobil safety codes so pervasive that they encouraged employees to confess paper cuts, Sprow mainly chose to keep his off-campus activities to himself.

At one stage he developed a particular passion for bicycle racing. He competed in Europe and the United States. Racing proved hazardous, however; he was hospitalized a number of times after accidents. One day Sprow was lying in a Princeton, New Jersey, hospital bed, nursing a broken collarbone, when the telephone rang. An ExxonMobil colleague who was visiting him, and who was aware of his friend's private passions, answered the call. It was Lee Raymond. Sprow shuddered a little and took the receiver.

"You're probably thinking that I'm calling to ask how you're doing and to wish you a speedy recovery," Raymond said. "The reason I'm calling you is to tell you that this is your last injury as an employee of Exxon. Do you understand that?"

"Yes, I understand," Sprow managed. Raymond hung up.

Sprow gave up bike racing and entered triathalons; he managed to avoid additional injuries, at least of the type that led to missed workdays and unfavorable safety statistics. He did not slow down much, however. He typically started his days by running ten or twelve miles. After the Mobil merger, despite a general awareness of Sprow's hobbies, Raymond appointed him to run Safety, Health, and Environment for the worldwide corporation.

Sprow had shown that he could deliver measurable improvements in ExxonMobil's safety and operational performance. Alongside Raymond, he proselytized the belief that zero accidents and defects was achievable, and that a fanatical devotion to safety in complex operational units such as refineries could lead to greater profits because the discipline required to achieve exceptional safety goals would also lead to greater discipline in cost controls and operations.

Sprow had studied chemical engineering and physics at the Massachusetts Institute of Technology, and he earned a doctoral degree in chemical engineering at the University of California at Berkeley. His advanced knowledge of chemistry gave him special credibility with Raymond, a fellow doctoral degree holder in the field. So, too, did Sprow's ability, in a succession of management assignments, to break through Exxon's habitual, bureaucratic resistance to change. Even so, along with many other managers, he occasionally suffered Raymond's withering scorn. "He's a pretty brusque guy," Sprow conceded. "That's probably an understatement. I have been cursed at by him a few times, and perhaps demeaned. . . .

There were a few times when I came home and told my lovely wife, 'This is probably going to end soon.'"

After taking charge of worldwide safety and environmental matters at ExxonMobil, Sprow became interested in dangerous game hunting—a hobby at least a little more in keeping with the masculine corporate culture. Sprow sought more challenging landscapes than ExxonMobil's sprawling Texas hunting ranch; he traveled to Africa periodically to shoot lions and rhinos. This bolstered his resolve when facing down Raymond.

Sprow served as one of Raymond's key lieutenants on climate change matters. Ken Cohen and his public affairs shop, in tandem with the K Street office in Washington, oversaw contributions to free-market advocates who published, spoke out, and filed lawsuits to challenge policies designed to reduce greenhouse gas emissions or assess the long-term impact of global warming. On the ExxonMobil organization chart, however, climate policy fell within Safety, Health, and Environment because it was, like oil spills or air pollution, an environmental issue that required worldwide corporate edicts and continuous management attention. Sprow was among those, like Raymond, who believed that "some of the environmental groups, with relatively little science behind it, at least in the early days," had "rallied way over to the side that we just need to flat move away from fossil fuels." He agreed with Raymond that poorer governments should be very cautious about the economic damage they might do to their struggling populations if they moved to tax or limit carbon-based energy sources too quickly or severely.

By 2002, however, it had become apparent to Sprow and other senior executives that ExxonMobil had, at a minimum, a communications problem surrounding its climate policy advocacy. The corporation was being vilified, and it seemed, at the same time, to be undertaking no constructive activities to address or even acknowledge the risks climate change might pose. Early in 2002, the Bush administration had developed a strategy for fending off critics of its own: The administration pledged to invest in scientific research, while resisting any legislation or treaty that would tax or limit carbon-based energy. Around the same time, Sprow pitched Raymond on a similar positioning for ExxonMobil.

"Okay, we cannot prove that climate change is being driven by man-made activities," Sprow said. Also, even if the burning of coal and gasoline was warming the planet, there were no "cost-effective steps" that would ease the problem, at least not "steps that are available now." Thus far, he was in lockstep with Raymond's views.

The science of climate change might contain significant uncertainties, he continued. Yet the uncertainties also cut in the other direction: It was at least *possible* that global warming was indeed under way and that it could create large disruptions in the future. Sprow put a question to Raymond: "Is it not the case that the risk of climate change is high enough that responsible efforts . . . to mitigate risk would be worthwhile?"

Raymond did not denounce him as an idiot, which was a start. The two of them, other senior ExxonMobil executives, and members of the corporation's board of directors who oversaw environmental and public policy "talked about that a long time." They decided during 2002 that ExxonMobil should invest in research that might bring breakthrough clean energy technologies forward, without altering the corporation's opposition to greenhouse gas regulation—a policy evolution that exactly mirrored the Bush administration's approach.[1]

ExxonMobil wanted "partners with a faultless reputation" whose work on alternative energy technology could burnish the corporation's reputation in the field, Sprow recalled. Although he remained skeptical of what he regarded as General Electric's favor mongering in Washington, Raymond reached out to Jeffrey Immelt, G.E.'s chief executive and the successor to Jack Welch, to explore a joint venture: Perhaps ExxonMobil could benefit from some of G.E.'s carefully honed image making. ExxonMobil, G.E., Toyota, and Schlumberger eventually agreed, late in 2002, to provide $225 million in funding to a new Global Climate and Energy Project, housed at Stanford University; ExxonMobil's contribution would be $100 million.[2] The project would conduct basic research, overseen by independent scientists, into alternative energy technologies promising enough to address the huge demand for energy worldwide, while at the same time offering pathways to reduce greenhouse gas emissions.

ExxonMobil had also funded computerized climate modeling work at the Massachusetts Institute of Technology. The Stanford project would provide the corporation with something new and credible to point to in response to accusations that it put its profits before the public welfare. The Stanford investment soon became a familiar element of ExxonMobil PowerPoint slides and media talking points, but it would prove to be a limited basis for enhancing the corporation's reputation on the issue. For one thing, Lee Raymond had not abandoned his fundamental doubts about whether the planet was warming at all. Nor was he prepared to yield the issue to environmentalists. Like his friend Dick Cheney, Raymond believed that the best way to defeat or at least contain an enemy was to remain clear-eyed and unsentimental about the enemy's intentions.

G reenpeace's penchant for direct action, guerrilla theater, and civil disobedience had attracted more than two million members since its founding in the counterculture coffeehouses of Vancouver during the Vietnam era. Greenpeace now ran its global campaigns from Amsterdam and maintained offices in more than three dozen countries; its portfolio of issues had expanded to protecting whales and warning of the dangers of climate change.

The 2000 election had forced a reevaluation of Greenpeace's strategy in the United States. With Al Gore out and George W. Bush in the White House, "Climate [was] suddenly on shaky ground," recalled Kert Davies, the organization's Washington-based research director, who had arrived that year from a group called Ozone Action. The question for Greenpeace was how to best launch a worldwide campaign targeting Bush. "The first instinct was, 'He's Big Oil,' so we target oil companies." Greenpeace began with a generalized campaign highlighting "oil addiction" and distributing posters of Bush holding a gas nozzle like a weapon. That campaign soon evolved into one that explicitly targeted "Exxon as a proxy for Bush . . . a proxy for corporate power over government in the United States."[3]

Davies played a prominent role in shaping the new campaign. He knew the art of media advocacy; he could remember the "killer quotes" he had placed with particular reporters from major newspapers to advance his messaging. He was a tall, thin man who biked to work at Greenpeace's office, a warren of desks made from recycled materials, in the Chinatown section of Washington, D.C. Davies had grown up in central Philadelphia, a child of "campers and hippies." His father was an architect, and his mother was a computer programmer and elections judge. In the midst of the Reagan administration, he studied environmental issues at Hampshire College, whose reputation for crunchiness was the subject of a recurring skit on *Saturday Night Live*. After graduation, he backpacked around the world and earned a master's degree in environmental studies at the University of Montana before taking up green campaigning as a formal profession.[4]

He had substantial experience by the time the ExxonMobil challenge fell to him. His thinking was forged as well by Greenpeace's rigorous internal culture. The organization engaged in ruthless internal reviews and self-criticism in regard to the effectiveness of its advocacy; Davies repeatedly had to defend his choices before colleagues in meetings and in conference calls, some of whom were no less direct than Lee Raymond in their cross-examinations of his decision making. He proved persuasive: ExxonMobil seemed the best target for Greenpeace's climate work during the first Bush term, because "it was the biggest and the ugliest and . . . had said the worst things about climate." Lee Raymond, in particular, seemed a gift.

Davies's strategy was to "pillory this company," document its "wrong behavior" on climate, and "force other companies to run away from that model." In July 2001, Greenpeace released "A Decade of Dirty Tricks: ExxonMobil's Attempts to Stop the World [from] Tackling Climate Change"; an unflattering photograph of Lee Raymond, jowly and thick-lipped, stared out from the report's cover. The next year, in "Denial and Deception: A Chronicle of ExxonMobil's Efforts to Corrupt the Debate on Global Warming," Greenpeace documented the corporation's funding for proxy groups that raised doubts about climate science. In an introductory note, the executive director of Greenpeace described ExxonMobil as "now standing in the path of history." In its environmentalist opponents, he continued, the corporation faced a force akin to "the movement led by Gandhi to free the Indian people, the U.S. civil rights movement, and

Solidarity's defeat of Polish communism. The goliath ExxonMobil may provide the perfect catalyst for much needed reform of corporate power. . . . Rave on."

Greenpeace did not fantasize that it could change ExxonMobil or curtail its profit making; the campaign Davies oversaw did not rely on calls for formal boycotts against ExxonMobil products. "Those objectives would be too hard to reach," Davies said. Instead, the idea was to show the world—and other multinational corporations—"what it means to be wrong on climate change" and to drive other chief executives away from Lee Raymond's position. The real targets of the campaign, Davies said, were the "BPs and the Fords and the Coca-Colas" that were iconic in their markets and led by executives less determined than Raymond to challenge climate science and the Kyoto Protocol.

ExxonMobil's corporate culture prescribed "very careful action," as Davies put it. This meant that Irving executives would be slow to react to Greenpeace's provocations and accusations, and when they did react, their responses would often be so heavy-handed that they would only reinforce the Greenpeace message. "They basically held their ground," Davies said. "They didn't give an inch on the policy. They didn't try to moderate their voice at all."[5]

Greenpeace and allied researchers filed Freedom of Information Act requests seeking documents about early corporate contacts between major oil companies and the Bush administration. Davies already knew that ExxonMobil funded small think tanks and science policy groups in Washington that issued reports and statements arguing that climate change was unproven and of no public concern. The documents produced to Greenpeace under the F.O.I.A. request turned up additional briefing memos and PowerPoints prepared for the White House by ExxonMobil lobbyist Randy Randol and the corporation's astrophysicist Brian P. Flannery. "Gaps and uncertainty in observations and scientific understanding of critical climate processes limit current ability to predict the rate and consequences of future climate change," Flannery wrote in a March 2002 memo. An accompanying PowerPoint urged the administration to "avoid near-term caps on CO_2 emissions" and "expand nuclear energy."[6]

The Bush White House did not require ExxonMobil's bullet points to adopt these recommendations; a majority of the president's cabinet and the Republican leadership in Congress already shared many, if not all, of Lee Raymond's views. Davies and his Greenpeace colleagues believed, however, that the documents they discovered established some degree of cause and effect: that ExxonMobil and other politically influential fossil fuel companies had, in effect, through their campaign contributions, purchased the Bush administration's climate policies, and then reinforced this achievement by sowing public doubts about climate science through the systematic funding of proxy groups.

ExxonMobil had, in fact, self-consciously invested in the dissemination of doubt about climate change. Under Lee Raymond, ExxonMobil had persistently funded a public policy campaign in Washington and elsewhere that was transparently designed to raise public skepticism about the science that identified fossil fuels as a cause of global warming. ExxonMobil ran some aspects of its campaign clandestinely; that is, it did not initially disclose the full scope and purpose of contributions it made. ExxonMobil's opponents were guilty of lumping together the corporation's support for small, havoc-making groups focused heavily on climate issues with ExxonMobil's support for legitimate, well-established conservative and free-market research institutes such as the Cato Institute for Public Policy Research, the American Enterprise Institute, and the Heritage Foundation.[7]

What distinguished the corporation's activity during the late 1990s and the first Bush term was the way it crossed into disinformation. Even within ExxonMobil's K Street office, a haven of lifelong employees devoted to the corporation's viewpoints and principles, an uneasy recognition gathered among some of the corporation's lobbyists that some of the climate policy hackers in the ExxonMobil network were out of control and might do shareholders real damage, in ways comparable to the fate of tobacco companies. The more it went on, and the more Greenpeace and other activist groups exposed ExxonMobil's more clandestine investments, the more it became clear that the corporation was taking on risk. If ExxonMobil were ever judged in a courtroom to be cooking science to appease Raymond's personal beliefs about warming issues, it could be devastating. The corporation was not alone in its support of fringe activists on climate, but it persisted longer than many other business groups and individual Fortune 500 corporations. The available record suggests that ExxonMobil was more aggressive than all but a handful of peer companies during this period, despite the fact that the corporation, because it produced no coal, did not belong to the energy industry's most vulnerable sector if restrictions on carbon fuels were enacted.

Raymond regarded the groups he supported as entirely legitimate participants in mainstream scientific debate. The credibility of this claim seemed increasingly doubtful, however. *Science* published a review of 928 peer-reviewed papers on climate science written during the late 1990s and the first Bush term; none of the papers, the survey's author found, "disagreed with the consensus position" about the probable man-made causes and dangerous trajectory of climate change—an assessment Raymond and his allies rejected.[8] ExxonMobil funded a few institutions that supported the consensus position on climate science, but the corporation's allies in climate science advocacy were more aligned with James Inhofe, the United States senator from Oklahoma, who asked on the floor of the Senate, in 2003, "With all of the hysteria, all of the fear, all of the phony science, could it be that man-made global warming is the greatest hoax ever perpetrated on the American people? It sure sounds like it."[9]

These investments in skeptics of the scientific consensus coincided with what at least a few of ExxonMobil's own managers regarded as a hypocritical drive inside the corporation to explore whether climate change might offer new opportunities for oil exploration and profit. One of ExxonMobil's most accomplished earth scientists, Peter Vail, had won acclaim for his insights into how changes on the earth's surface affected ocean levels and other geological shifts. Vail had developed a calculation known as the Vail curve to describe some of these ocean events. In the ExxonMobil upstream division in Houston, scientists in charge of finding new deposits of oil and gas began to explore whether Vail's scientific insights might give them a leg up in exploration by allowing them to predict how climate change—if it did materialize—might alter surface and ocean trends and lead the corporation to new oil finds. "So don't believe for a minute that ExxonMobil doesn't think climate change is real," said a former manager involved with the internal scientific review. "They were using climate change as a source of insight into exploration." This work remained unpublicized.

A raiding party of about four dozen arrived at ExxonMobil's headquarters in Irving, Texas, shortly before eight in the morning on May 27, 2003. The raiders divided themselves into three units. The first group pulled up at the main entrance in two panel trucks marked "ExxonMobil Global Warming Crimes Unit." They scrambled out, blocked the driveway, and chained themselves to the trucks in front of the gate. A second team wearing business suits and toting briefcases arrived at the maintenance gate at the rear of the one-hundred-acre campus. They cut a lock and drove inside in two rented Jaguars. Two vans pulled up at the delivery gate carrying the third unit of attackers. Most of that group had dressed in tiger costumes, mocking Exxon's old "Put a tiger in your tank" advertising slogan. They unpacked a ladder and a raft and climbed over the gate, chaining it shut behind them. Some of them dragged their raft to the pond within view of the executive suites and set themselves afloat. Other tigers climbed onto the roof, where they unfolded a banner that declared ExxonMobil's headquarters to be a global warming crime scene and tossed around a balloon designed as a globe.

The protesters wearing business suits drove their Jaguars across an unpaved road, entered the employee garage, and found their way inside the headquarters building. They fanned out and offered spontaneous lectures and leaflets on climate change to bemused ExxonMobil executives and staff. Two activists in tiger suits also made it inside the headquarters. Some of the oil corporation's employees thought a terrorist attack might be under way. The Irving Fire Department eventually brought in one of its ladder trucks to remove the tigers on the roof.

It had taken Greenpeace three months and several tens of thousands of dollars to plan the raid. It did so in strict secrecy. The group's activist coordinator, Maria Ramos, had dispatched a recruiting notice seeking those who might be interested in "challenging the world's largest company . . . and engaging in guerrilla tactics." One of the volunteers, Anne Nunn, traveled from Australia, where she had recently completed a raid on a Mobil tanker off the Australian coast. The Irving raid was timed to influence news headlines before ExxonMobil's 2003 annual share-holder meeting; it succeeded. "If you're a fringe, radical organization like Greenpeace," said Tom Cirigliano, of the corporation's public affairs department, "you need a target, you need an enemy, and you need a villain."[10]

Increasingly, inside ExxonMobil, the corporation's image strategists reflected upon whether they could find some way not to play the role Greenpeace had assigned them. The raid provided additional evidence, if more was required by this time, that Lee Raymond's visibility on the climate issue had drawn extraordinary attention that was unlikely to dissipate. The attention had reached a point where it was undermining Raymond's own cause. ExxonMobil's many allies in Washington who opposed the Kyoto Protocol on economic and fairness grounds—utilities, carmakers, free-market conservatives in academia and journalism—found themselves tarred by the accusation that all of their arguments might be merely a front for the oil industry's largest corporation.

ExxonMobil executives consoled themselves by saying that the Greenpeace campaign was just part of the price of doing business in the modern oil industry. In the marketplace of nonprofit fund-raising, Exxon-Mobil's notoriety offered an attractive opportunity for environmentalists to raise money by promising to hold Big Oil to account, and there was nothing much they could do about that.

The corporation's executives did not have to passively accept Greenpeace's assault, however. After the Irving raid, ExxonMobil approached its Greenpeace problem as an aggressive litigator would. The corporation encouraged the Dallas County district attorney to prosecute fully the Greenpeace protesters who had participated in the Irving action. Exxon-Mobil also sued Greenpeace and thirty-six individuals who had been arrested on its campus. By threatening fines and jail terms, the corporation eventually won a seven-year standstill accord in which Greenpeace agreed not to commit any crimes while campaigning against the corporation. As a result of this settlement, the group's anti-Exxon campaign migrated from newsmaking direct action and civil disobedience into online publishing.

Investigators from the Internal Revenue Service turned up at the Chinatown office in Washington to conduct an audit. A small nonprofit group called Public Interest Watch had raised questions with the I.R.S. about whether Greenpeace was compliant with federal laws governing groups that received tax-deductible contributions. Greenpeace passed the audit and opened its own investigation of Public Interest Watch.

The group's tax form, filed about two months after the activists in tiger costumes had scaled the Irving headquarters' roof, showed that a single donor was responsible for $120,000 of Public Interest Watch's $124,000 in annual revenue: ExxonMobil Corporation.[11]

A s Raymond battled Greenpeace, the international oil company he most admired after his own, Royal Dutch Shell, stunned stock market investors by revealing that it had overstated its true proven reserves of oil and gas; the company eventually calculated that it had puffed up its holdings by 4.35 billion barrels of oil, an amount equivalent to more than a fifth of ExxonMobil's total proved reserves worldwide. Three top Shell executives resigned. The scandal made plain that the pressure on the very largest international oil companies to replace reserves in the era of resource nationalism had become so severe that it could induce grotesque distortions.[12]

Shell's revelation galvanized regulators at the Securities and Exchange Commission in Washington to look at reserve counting and reporting practices by major American oil corporations. That review in turn brought fresh attention to a practice ExxonMobil had gotten away with for many years: The corporation still claimed each winter in press releases and in Wall Street presentations that it had an unbroken record, dating back to 1993, of replacing, through the discovery and purchase of new reserve additions, at least 100 percent of the oil and natural gas it pumped or otherwise disposed of each year. But the assumptions ExxonMobil used in making these public claims did not conform to S.E.C. regulations—

and the commission and its staff had done nothing, under either the Clinton or Bush administration, to force ExxonMobil to modify its public statements.

To protect stock market investors from oil operators that inflated their reserves to boost their share prices, Congress had mandated in the Securities Act of 1933 and the Securities Exchange Act of 1934 that Washington regulators oversee how publicly traded companies reported their numbers. The S.E.C. had later issued detailed regulations, under Rule 4-10, about how a corporation such as Exxon should count its oil and gas holdings and report them in mandatory S.E.C. filings. Among the regulations: To report proved reserves, a company had to show there was a "reasonable certainty" that the reserves would be "recoverable in future years from known reservoirs under existing economic and operating conditions." That meant, too, a company had to be able to transport the oil to market by sea or pipeline at a profit-making cost. This was obviously an imprecise standard—the reserves being counted were by their nature difficult to measure scientifically, so oil companies retained, by regulatory design, some discretion to decide what was proved and what was not.[13]

To calculate the economic viability of reserves, companies were required to mark oil prices on the last date of every year. Also, certain forms of oil, such as bitumen or oil sands extracted by techniques that resembled mining, could not be counted. The latter rule remained the main reason ExxonMobil's public claims about reserve replacement differed from the disclosures it made officially in S.E.C. filings. If ExxonMobil had not disregarded the S.E.C. oil sands rule, it would not have been able to boast of an unblemished record. "This marks the tenth year in a row that we've exceeded 100 percent reserves replacement," Raymond declared in a press release disclosing the corporation's 2003 results. Yet that was true only by using ExxonMobil math. According to S.E.C. rules, the corporation had replaced reserves fully in only two of the previous four years. And the fudging involved issues that were material to investors: As Raymond put it himself, "Continued high-quality additions to ExxonMobil's resource base are the foundation of our long-term profitable growth."[14]

As Wall Street focused in on reserves and as ExxonMobil implemented O.I.M.S., Raymond tightened and made uniform the corporation's reserve counting rules. The system's stated goals included objectivity and rigor: "A well-established, disciplined process driven by senior-level geoscience and engineering professionals . . . culminating in reviews with and approval by senior management. Notably, no employee is compensated based on the level of proved reserves." Such bonus incentives for managers involved in reserve counting had apparently contributed to Shell's overstatements; ExxonMobil eschewed them even before the Shell scandal broke.[15]

By all accounts, Raymond genuinely wanted order and accuracy. In financial management, for example, to complement O.I.M.S., he created the Controls Integrity Management System, or C.I.M.S., a financial audit and risk management system designed to identify and root out managers who cut corners, massaged revenue reporting, or fiddled with expense accounts. Many corporations tolerated split contracts or other gray-area accounting practices designed to smooth out quarterly earnings reported to the public, so that shareholders might see a picture of stability. At ExxonMobil under Raymond, such accounting manipulation could be a firing offense. Raymond would also order employees terminated over tiny expense irregularities.

On oil and gas reserve counting, however, the ExxonMobil system tolerated more flexibility. S.E.C. rules effectively allowed oil companies to manage the timing of announcements of new proved reserves. This helped ExxonMobil control when new proved reserves were revealed publicly, and by doing so aided its effort to portray a steady story of year-by-year reserve replacement for Wall Street. ExxonMobil's internal rules held, for example, that for reserves to be counted as proved, the corporation's management must have authorized investment for their development. This meant, as a practical matter, that the Management Committee could adjust the annual timing of proved reserve additions by synchronizing investment decisions. "The key to reserves, among other things, is when you can actually book them," an executive involved with the process said. "You can't and you shouldn't book reserves until you've really made an investment to develop the reserves. So consequently, if you've figured out a way to manage the system properly, consistent with the continuity of capital budgeting, you can have a pretty smooth reserve identification over time." As the Wall Street analyst Mark Gilman put it: "If you are conservative about when you book reserves during the development of projects, in effect you create an inventory going forward" that can be declared as new proved reserves as the timing of reserve replacement announcements requires in order to present a smooth picture.[16]

There would be nothing illegal or even improper about this managed timing if it were linked to a rigorous internal counting system, such as ExxonMobil possessed; if the counting conformed to S.E.C. rules; and if the results were communicated honestly to investors and the public. Characteristically, ExxonMobil's internal system might have been the most rigorous in the industry, although neither the S.E.C. nor any other regulator had the capacity to confirm that through auditing. Also characteristically, the corporation rejected the precepts of government regulators and communicated in public on its own terms.

Besides the S.E.C. rule that prohibited the counting of oil sands, the other regulation that galled Raymond was the one dictating how a corporation should determine whether proved reserves were economically viable. The S.E.C. held that the viability of reserves should be judged against prices on the last day of the year, December 31. Raymond thought that was dumb: ExxonMobil ran its business by thinking about price ranges and averages, not one arbitrary day's price. So in public and to Wall Street, he used his own system of average prices to calculate whether reserves should be counted.

After the Shell scandal, the S.E.C. issued new "guidance" to Exxon-Mobil, saying that it should use the year-end pricing rule, at least in commission filings. Raymond now grudgingly reported the S.E.C. number alongside his own. For the first two years, the consequences of following S.E.C. regulations were highly unfavorable, amounting to a total reduction in ExxonMobil's proved reserves of oil and gas liquids of 1.27 billion barrels. Even as he noted the S.E.C.-mandated numbers, Raymond went right on issuing press release claims that ignored the rule. With other industry executives, he also initiated a Washington lobbying campaign to have the rules changed.[17]

In some later years, if ExxonMobil had followed the S.E.C. rules, the corporation's reserve replacement figures would have been higher than under its own system of counting—it would have looked better. The corporation ignored the advantageous swings in the same way that it ignored the disadvantageous ones. Overall, the two rules ExxonMobil seemed to find most onerous—the oil sand prohibition and the pricing formula—deprived the corporation of the claim that it had smoothly replaced reserves, year after year, since Raymond became chief executive. The picture under S.E.C. rules, instead, although it was not one of disastrous decline, was one of volatility, which implied a degree of change and insecurity in the reserve arena. This in turn happened to reflect a broader truth about the challenges confronting major international oil corporations.

Raymond certainly had a case on the merits: The S.E.C. prohibition of oil sands could be regarded as outdated, and the pricing rule could be dismissed as too arbitrary. Investors might benefit from revisions that better aligned S.E.C. regulation with the rising importance of oil sands. ExxonMobil's position would have been more defensible, however, if it had communicated to investors and the public more forthrightly, in alignment with S.E.C. regulations, while it petitioned for regulatory change. "Exxon seems incapable of simply stating, 'We did not replace all of our reserves this year, but we have a heck of a lot of reserves anyway, and we convert them into cash at a more efficient and consistent rate than any of our competitors,'" wrote Steve LeVine, the journalist and analyst who first called attention to the gaps in ExxonMobil's public reporting. "Nope, it has to say that it's replaced its reserves for [ten years straight] when, legally speaking, it hasn't."[18]

The dodge reflected on how the need for reserves pushed the corporation toward higher-risk frontiers: Chad, Equatorial Guinea, and deep ocean waters where drilling technology and safety procedures had to be reengineered on the fly. The reserve replacement conundrum pushed the corporation, too, toward unconventional resources like the Canadian oil sands, where innovation and uncertain new drilling techniques would be required to make money at the rate ExxonMobil executives and shareholders expected and where the environmental risks were higher than normal. It also led Lee Raymond to reflect regularly on whether it might be possible to find a new way back into the huge oil zones where abundant crude could make reserve-counting rules irrelevant, as they had been for Exxon and Standard Oil for so much of the twentieth century, when the corporation was awash in equity oil in the Middle East.

Above all, any global oilman thinking big coveted access to Saudi Arabia.

Nine

"Real Men—
They Discover Oil"

Lee Raymond had always believed that he could rationalize the $81 billion Exxon paid for Mobil by driving operating costs down enough to justify the combination for shareholders. Assessing the true long-term value of Mobil's sprawling oil and gas assets was difficult, however—and that would determine the strategic payoff from the merger. The long-term value of Mobil's holdings would be a function of many factors—not just how much oil and gas actually lay in the ground when all the wells were drilled, but also the evolution of global markets, geopolitics, and the advent of new technologies that might unlock value from reserves previously thought to be worthless. After the merger, Exxon's geologists, engineers, and marketing specialists tore through Mobil's business divisions on a quest to understand the assets they had taken on. Gradually, they came to appreciate the astonishing value of one asset they had not comprehended adequately at the time of the merger deal: Qatar's North field.

Raymond would eventually quip in private that the North field alone was probably valuable enough to justify the full Mobil merger price, and that everything else that came with the company—all its oil and gas fields in Africa, Asia, and the former Soviet Union—were a bonus. That was an exaggeration, intended in jest, and yet "it would be fair to say that we did not totally appreciate what the scale of it might be," Raymond recalled.[1]

Qatar protruded into the Persian Gulf from the desert landmass of Saudi Arabia; on maps, it looked like a small spruce tree. It was a featureless, flat, barren, sandy, humid kingdom without oases or other natural greenery. At the turn of the twentieth century, Qatar's native population of impoverished fishermen, pearl divers, and Bedouin Arab herdsmen numbered perhaps five or ten thousand. Even by comparison with the other sparse, isolated emirates of the Arabian peninsula—Saudi Arabia, Kuwait, Bahrain, Oman, and the United Arab Emirates—Qatar had been a backwater. A single family, the Al-Thanis, had ruled the peninsula since 1825. Japan discovered a method for synthesizing pearls during the 1930s, which caused a crash in the global pearl market, leaving Qatar even more isolated and poor. Around the same time, the emirate's Persian Gulf neighbors discovered and pumped oil, but Qatar lagged. It had been endowed with more natural gas than oil and it lacked the leadership and skills to exploit either profitably. The Al-Thanis feuded among themselves; in 1995, one of the king's sons, Hamid Bin Khalifa Al-Thani, overthrew his father bloodlessly. As late as 1990, the emirate remained a ramshackle, underdeveloped place, whereas in oil-engorged Saudi Arabia booming revenue after the 1970s paid for California-style freeways, industrial ports, airports, skyscrapers, ornate princely palaces, and shopping malls.

Geologists knew that Qatar's North field held natural gas—lots and lots of gas. It held so much gas that it was not easy to estimate the full amount accurately—800 trillion square feet eventually became a common estimate, the equivalent of more than 130 billion barrels of oil. By comparison, Mobil's highly lucrative gas field in Aceh, Indonesia, held only about 17 trillion square feet, equivalent to just under 3 billion barrels of oil. For all practical purposes, the size of the North field was infinite; it would last for generations, probably beyond the point when fossil fuels would be a dominant source of energy supply for the world economy. After the Mobil merger, Lee Raymond organized a natural gas task force. The paradox its members confronted was that while the North field's abundance was assured, little had been done to develop it profitably. Why had other corporations failed, and what might ExxonMobil do differently?

The natural gas industry differed from the oil business in that, during the postwar period, the main challenge had not typically been the search for new fields. The problem instead was to profitably exploit the largest natural gas reserves that were known to exist but were geographically "stranded," that is, physically disconnected from commercial markets. Pressure and heat formed and trapped natural gas beneath the ground by processes similar to those that formed oil. Much of the world's gas was mixed up with, or "associated" with, oil deposits. Qatar's North field was a mother lode of "nonassociated" or freestanding natural gas. There were a handful of proven, concentrated areas of large nonassociated gas reserves in the world: in Qatar, Iran, and Russia. The latter two could use some of their gas domestically, and Russia exported gas to Eastern Europe, where it was a critical source of heat and electricity. Qatar's gas, on the other hand, was sitting thousands of miles from any customers that might burn it. It would be prohibitively expensive and politically impractical to connect Qatari gas by pipeline to large population centers in Europe or Asia.

As an energy source, gas had many attractions. It could heat homes, cook food, power turbines to produce electricity, fuel automobiles if the cars were configured properly, and be used to make chemicals and other industrial products. Gas also emitted considerably fewer greenhouse gases than oil or coal when burned. Qatar's case illustrated one of gas's major liabilities, however: Its form made it difficult to transport. Oil was a remarkably easy fuel to move around. It sloshed easily into storage tanks; it streamed cooperatively down pipelines; it poured smoothly into supertanker holding bins; it poured out again into refinery pipelines; it flowed out the other side of a refinery as gasoline; and it spilled into tanker trucks for delivery to retail stations. Gravity was oil's friend. The natural tendency of gas, on the other hand, was to dissipate into the air; gravity was its enemy. Engineers could design systems to transport gas by pipeline easily enough, but for many decades that had been the only practical way to move it from a field where the gas was pumped out of the ground to facilities where it was burned. This meant gas-fired electricity plants, for example, had to be located within economical piping distance of a gas source, whereas an oil-fired plant could use oil from halfway around the world.

For Qatar, rich in gas but bereft of oil, in the first decades after the Second World War, all this had amounted to an equation that kept the emirate locked in poverty. The only semimodernizing economies within easy pipe distance—Saudi Arabia, Iran, and Iraq—had plenty of their own gas. Qatar also lacked even the basics of a manufacturing economy of its own, such as freshwater and a skilled workforce.

It had been known since the early twentieth century that, as a matter of chemistry, natural gas could be converted into a liquid and then, after transport, be reconverted into a gas for burning. This process might solve the problem of a stranded-gas holder like Qatar: Its gas could be turned into liquid, loaded into oceangoing tankers, shipped to populated markets, and then reconverted into gas for commercial use. The technology to accomplish this conversion and reconversion at a large scale was unwieldy, however. Britain and Algeria signed the first major commercial liquefied natural gas contract in 1961. A huge refrigeration plant in Algeria cooled that country's stranded gas into a liquid; ships carried the liquid gas to Britain; and a reconversion plant turned it back into a fuel for electricity. Indonesia soon moved into the L.N.G. industry with energy-starved Japan as a customer; Mobil became the operating partner in Aceh. For years the profitability of Mobil's L.N.G. business in Aceh was an exceptional success, however. It relied, effectively, on Japan and South Korea, which were industrializing very rapidly but had few hydrocarbons of their own; they were willing to pay high prices for secure L.N.G. supplies. The technology Mobil employed to fill these contracts was very costly, and it seemed that it would be some time before those methods would be economical enough to deploy worldwide.

Exxon had a troubled history in the L.N.G. business before the Mobil merger. The corporation had built, relatively early on, an L.N.G. plant in Libya and a reconversion terminal at La Spezia, Italy. The Libyan plant proved to be balky and trouble-prone. In the early 1970s, Exxon's Italian subsidiary became embroiled in scandal when the unit's president, Vincenzo Cazzaniga, was accused of setting up a web of hidden bank accounts to funnel almost $50 million of Exxon's revenue to Italian political parties—including a small amount to the country's Communist Party—to win tax and other favors. Exxon eventually entered into a consent decree with the Securities and Exchange Commission over the matter; the affair soured the corporation's executives on their Italian subsidiary, and their L.N.G. investments languished.[2]

Mobil had stumbled into its gas partnership with Qatar during the 1990s. A substantial number of the oil industry's big success stories were the product of luck, not brains. Oil executives had flown in and out of Qatar for years, but none of them could think of how to commercialize the North field. Royal Dutch Shell led the global L.N.G. business by the 1990s. Mobil was a second-tier but significant player, because of Aceh. Shell negotiated access to the North field but pulled out in a dispute over financial terms. British Petroleum and the French giant Total moved in afterward and negotiated to build an initial pair of L.N.G. "trains," the industry term used to describe the giant refrigeration complexes that converted gas into liquid form for sea transport. The consortium struggled with some of the technical challenges; British Petroleum pulled out. "They had it on a golden plate, but they rejected it," Abdullah Bin Hamad Al-Attiyah, Qatar's energy minister, remembered. The Qataris realized there was hardly anyone else in the global oil industry but Mobil who could do the work they wanted. "They came immediately," recalled Al-Attiyah.[3] Lou Noto slipped into the Total deal as a partner, but he also won the exclusive right to build future Qatari gas trains. He structured a long-term sales contract for Qatari gas with South Korea as the customer and handed off the whole project to Lee Raymond at the time of the merger.

The North field challenge played to Exxon's strengths: budget- and performance-conscious management of gargantuan engineering projects, combined with profit-maximizing financial planning. In Mobil's L.N.G. group Exxon also acquired technical expertise it otherwise lacked. After 2000, ExxonMobil committed to multibillion-dollar investments to develop huge new L.N.G. gas trains from the North field as an exclusive 25 percent partner with Qatar Petroleum. Its engineers found that the emirate's natural gas was of unusually malleable quality—relatively easy to liquefy or to process to separate out other industrial products. This made it cheap to produce. The projects Raymond authorized in Qatar were designed to be profitable if the natural gas they produced sold at just three dollars per thousand cubic feet. Within a few years, prices soared as high as fifteen dollars. ExxonMobil's direct gas sales from Qatar took place under long-term contracts, so the corporation did not reap all of the benefit of this windfall on spot markets, but its gas-derived profits soared nonetheless. Also, the corporation's share of profits from auxiliary products manufactured in Qatar, referred to as gas liquids, would soon exceed $1 billion annually. Only ExxonMobil, Raymond boasted to Wall Street analysts, had figured out how to unlock the value of Qatar's bounty.

Gas figured increasingly in the search by ExxonMobil to replace the oil and gas reserves it pumped and sold each year. The sheer scale of ExxonMobil's reserve replacement challenge—its need to find and book oil and gas in equivalent or greater amounts to that which it pumped out and sold—now meant that the corporation "had to find a Conoco every year," as Raymond put it.[4] (In 2001, ConocoPhillips had worldwide revenues of almost $40 billion.) The scale problem was genuine, but it also sounded more and more like an excuse—nobody had forced Exxon and Mobil to merge, and Raymond had advertised the combination as full of strategic advantage. In any event, ExxonMobil's total portfolio was shifting away from oil toward gas. In 2002, the corporation pumped slightly less oil than it did in 2001; in the first half of 2003, oil production fell slightly again. For Wall Street, ExxonMobil counted oil and gas reserves as a single number, as "oil equivalent barrels." Analysts converted gas reserves to equivalent barrels of oil with formulas accounting for energy content and price. Yet the truth was that gas was less profitable than oil, equivalent barrel by equivalent barrel. Oil prices averaged about 30 percent more than natural gas on an energy equivalent basis after 1995, and the United States Energy Information Agency projected that this gap would widen into the future. Gas production could be more costly. Customer markets were less flexible, less interconnected. Yet because of resource nationalism and the depletion of accessible supplies in the United States, oil was harder and harder for ExxonMobil to own. The slow migration of ExxonMobil's reserves from oil to gas did not show up clearly in the numbers the corporation reported to Wall Street—and certainly not in the numbers it emphasized in public and investor presentations—but over time, the higher proportion of gas investments could threaten the corporation's impressive record of profitability.

Raymond lobbied in Washington to ensure that the United States had enough big import terminals to handle liquefied natural gas ships. Forecasts by the Bush administration's analysts at the nonpartisan Energy Information Agency suggested that the United States had only about twenty years' worth of natural gas supply left under its soil, government analysts then believed. America would soon need to import gas just as it already imported oil. In the United States, in 2003, gas supplied about a quarter of the country's energy supply, to generate electricity, heat water and homes, and fuel industrial processes.[5] ExxonMobil supported a National Petroleum Council study in 2003 that made recommendations to the Bush administration to expand the industry.[6]

Raymond had developed a friendship with Federal Reserve chairman Alan Greenspan. The men had gotten to know each other while serving together briefly on the J.P. Morgan board of directors, and then stayed in touch. Raymond impressed his analysis about natural gas on the Federal Reserve chairman: The American economy needed planning to build the facilities to import and reconvert liquefied natural gas in the future. ExxonMobil's economic forecasters in corporate planning reported to the Management Committee that they expected the global L.N.G. market to double by 2010. ExxonMobil was busy investing in that market worldwide. The global sales force in the corporation's gas marketing division finalized a contract to ship two billion cubic feet of liquefied gas from Qatar into the United Kingdom, for example. Raymond educated Greenspan about the coming shape of the emerging global L.N.G. market. Without telling Raymond in advance that he intended to go public, Greenspan testified before Congress, highlighting America's coming gas deficit as a strategic issue for the American economy. Preparing for an L.N.G. world would require construction of large import terminals that carried environmental and safety risks, but the thrust of Greenspan's testimony was that America's gas deficits would demand such risk taking. Greenspan's friendship with Raymond was not well known, but one analyst aware of the relationship remembered reading Greenspan's unusual testimony about natural gas markets and thinking, "He's giving Raymond's testimony!"

As it turned out, the natural gas market in the United States was one of the few industry subjects that Lee Raymond had misjudged. "Gas production has peaked in North America," he declared at an industry conference in 2003. America's only large, unexploited deposits of gas were in Alaska, stranded from commercial markets in the Lower 48 for lack of a pipeline. Even if a pipeline were built, Raymond continued, he expected total American gas production to decline, "unless there's some huge find that nobody has any idea where it would be."[7] In fact, such a find, of sorts, was coming by the decade's end, and it would transform ExxonMobil's strategy within the United States. Lee Raymond just did not see it coming. Hardly anyone else did, either.

A bdullah Bin Abdul-Aziz, the crown prince of Saudi Arabia, was in his mid-seventies at the time of the ExxonMobil merger. He moved among manicured, well-watered palace complexes the size of some college campuses. There was one palace in Riyadh, the Saudi capital, and another in Jeddah, and another in the desert where Abdullah bred Arabian horses. The prince kept an unusual schedule. He slept in two four-hour shifts, one between 9 p.m. and 1 a.m. and a second between 8 a.m. and noon. In the hours between he swam for exercise and did office work. He was a goateed, barrel-chested man with a serious and penetrating gaze.[8]

He had much to contemplate. His older half brother, King Fahd, had been incapacitated by a stroke in 1995. The Saudi royal family was too decorous and divided to remove Fahd from power formally, despite his incapacitation, so Abdullah ran the country as de facto king, but he was constrained by shifting family and ministerial factions. Abdullah felt that his kingdom needed to modernize its economy and its education system. Saudi Arabia imported too much of its skilled labor from Asia and Europe while employing its native sons in do-nothing government bureaucracies and religious institutions. The state oil company, Saudi Aramco, which had been owned in part by Exxon and Mobil before nationalization during the 1970s, was so bloated that it employed about three quarters as many people to operate within the kingdom as ExxonMobil did to operate worldwide. The Saudi regime needed to create jobs for its restless population of young men, but even with the inefficiencies that resulted, Saudi Aramco was a rare bright spot in the Saudi economy in that many of its homegrown employees and engineers were professionals who could work to international standards. In many other bureaucracies in the kingdom, too many Saudis lacked the skills and leadership to compete in the global economy. If the royal family did not do something to change this before its oil was depleted, then a common fatalistic aphorism among the Saudi elite—We started on camels; we acquired jets; we will return to camels—might well be borne out.

By 1998, seeing what Qatar had undertaken with its massive gas-fed industrial complexes, Abdullah decided to leapfrog beyond Saudi Arabia's dependence on oil sales into a more sustainable, job-creating future. The key to his thinking was natural gas.

That year, in the autumn, while on his first state visit to America as regent, Abdullah invited the chief executives of the seven largest American and European oil companies to the McLean, Virginia, mansion of the cigar-chomping Saudi ambassador to Washington, Prince Bandar Bin Sultan. Lee Raymond and Lou Noto attended. It was awkward for them because at the time they were in the advanced stages of merger discussions known only to them and a few dozen others involved in the talks. They agreed to act as if nothing unusual was going on.

It was extraordinary for all of the executives of the largest oil corporations to gather in one place with the head of state of an oil-rich country. In an Arabian-style *diwan* setting of cushioned chairs and couches, overlooking the Potomac River, the meeting began stiffly; it suggested the formal, tensely competitive atmosphere of a meeting of the heads of competitive crime families trying to divide up casino building rights. Abdullah invited the oil chiefs to speak about how they might work with Saudi Arabia's natural gas resources if they were invited back to the kingdom as investors for the first time in more than two decades. This was an enormous opportunity for all of the executives present—Abdullah's gas initiative could not make up for the economic pain of oil nationalization, but it offered a rare chance to reenter the kingdom with a big play, and who knew where that might lead.

Raymond began. He talked about the size of Saudi Arabia's presumed gas reserves and outlined how Exxon might be able to exploit them. Each of the other executives spoke similarly until the circle came around to Noto: Little had changed since Mobil was the smallest partner in Aramco, he joked. He was still the last in line.

Saudi foreign minister Saud Al-Faisal, an enthusiast of Abdullah's plan, sat quietly in the room; he was a favorite of Raymond's and other American oil executives because he was pragmatic, competent, comfortable in the West, and interested in forging new pathways to industrial modernization at home. Also present was the kingdom's oil minster, Ali Al-Naimi, a nonroyal who had ascended through Saudi Aramco's ranks. Al-Naimi looked on Abdullah's outreach to international corporations with suspicion; the initiative could encroach on the prerogatives of Aramco, which Al-Naimi oversaw. As Raymond, Noto, and other chief executives spoke, Al-Naimi "looked like he had eaten a sour lemon," one person who attended recalled.[9]

To Raymond, there appeared to be very few places on the planet with enough oil and gas resources to make a material difference to the revenue and profit picture of Exxon. Chad was a welcome play, but it was hardly an "elephant," as exploration and production geologists called huge oil and gas fields. Raymond could count on one hand the countries with enough proven oil and gas reserves to lift Exxon's equity holdings and address its reserve replacement challenges in a serious way: Russia, Iran, Iraq, and Saudi Arabia. Two of them—Iran and Iraq—were entirely closed off to Western investors. If Saudi Arabia was even hinting at the possibility of reopening its reserves—even if it involved only natural gas, for now— Exxon had to try to make it work, Raymond believed. The loss of Saudi oil when the royal family nationalized Aramco in 1975 had been a blow to Exxon's oil and gas production volumes from which it had never recovered. The expropriation had followed repeated and phlegmatic negotiations in which Exxon's Clifton C. Garvin Jr. had played a leading role. Not for the first or last time, the Saudis had exasperated an American negotiator with their opaqueness, delays, and changing terms: "I have to say I can't figure out what they want," Garvin declared at one stage. "We keep leaving pieces of paper detailing how we can work with them, and they keep asking for more talks." Raymond felt there was little choice, however, but to try again.[10]

Abdullah's vision was to allow foreign corporations such as Exxon to develop freestanding gas fields in exchange for their commitment to use the gas to fuel industrial projects such as water desalination plants, electricity generation, and petrochemical manufacturing. These multibillion-dollar projects would create skilled jobs for Saudis while addressing chronic infrastructure and electricity problems in the kingdom. The projects would also allow Saudi Arabia to stop wasting its oil on electricity generation. Most of the world's economies had stopped burning fuel oil to make electricity decades earlier; it was a dirty method and economically irrational, because the oil fetched greater sums at refineries where it could be made into gasoline or jet fuel. Saudi Arabia still burned off an astounding 200,000 to 300,000 barrels of oil a day to power its heavily air-conditioned cities, a figure that would soon rise toward 800,000 barrels a day—and that production counted against the kingdom's quota as a member of the Organization of the Petroleum Exporting Countries cartel.[11] By using natural gas instead, the kingdom would earn more revenue overall. ExxonMobil already operated large, profitable refining and chemical plants in the kingdom that it had agreed years earlier to construct and operate in exchange for preferential access to Saudi crude. With this new natural gas opportunity Lee Raymond could expand and diversify Exxon's position in Saudi industry.

Saud Al-Faisal led the gas negotiations for Abdullah. They proved, unsurprisingly, to be long and complicated. As they dragged on, September 11 became a factor. The attack and its aftermath sowed U.S.-Saudi relations with mutual resentments and mistrust; at night, in their palaces, Faisal and other senior Saudis tuned in to American satellite news programming, whose presenters and commentators increasingly seemed to them to be engaged in anti-Saudi race baiting. When Al-Faisal visited the White House, Bush administration officials, including Raymond's friend Cheney, urged the foreign minister to take stronger action in response to evidence that Saudi clerics and businessmen were financing Al Qaeda. Al-Faisal had attended Princeton University; it pained and angered him to be spoken to as if he were some sort of double-dealing international criminal.

Raymond sympathized with Al-Faisal. The ExxonMobil chairman had been visiting Saudi Arabia since the early 1970s and had come to know Al-Faisal well. He shared the Bush administration's outrage over the September 11 attacks, but increasingly, he felt uneasy about the hard line taken by Cheney. Raymond told colleagues he feared that an American overreaction could destabilize the Persian Gulf region. The Bush administration seemed not to understand, in particular, the importance of the Sunni-Shia sectarian divide, Raymond said. Saudi Arabia's Sunni royal family lived in deep anxiety about the expansionary ambitions of Iran's Shia-led revolutionary government. There was a restive Shia population within Saudi Arabia, and Iraq's people were mostly Shia; if the region were destabilized, Iran might emerge stronger. In any event, after September 11, there seemed to be a widening gap between how the Saudis analyzed the region's challenges—they placed a strong emphasis on the sectarian issue and Iran—and the way the Bush administration saw them, intently focused as it was on Al Qaeda and global terrorism. As Raymond and his colleagues negotiated for access to Saudi Arabian gas reserves, Exxon-Mobil found itself straddling the chasm that opened between Washington and the Saudi regime. Its executives believed Al-Faisal to be a reliable friend and partner of the West, but also a realist about the Middle East. As it became clear that the Bush administration intended to invade Iraq, against Saudi advice, Al-Faisal told his anxious ExxonMobil colleagues, "It's inevitable. There's nothing I can do."

Abdullah appointed ExxonMobil as the lead partner in two of the three gas projects he initially approved. Abdullah staged a ceremony in Jeddah for about three hundred people at which the crown prince, resplendent in robes, held court to congratulate the ExxonMobil team: "*Mabruk!*"

Raymond selected Ralph Daniel Nelson, a longtime Mobil executive with extensive experience in the Middle East, as his point man—lead country manager, in the ExxonMobil vernacular—in Riyadh. Nelson was a Naval Academy graduate and former U.S. Marine infantry officer who had served in Vietnam during the late phases of the war. He was a tall, silver-haired, broad-shouldered man. He could handle Raymond's intimidations and he conformed to Saudi expectations—born of *Dallas* and other prime-time soap operas relayed by satellite—of what American oil executives should look and sound like. Nelson had years of experience in Qatar and the Gulf region and he knew the natural gas industry from previous work for Mobil. With Raymond behind him, Nelson pressed for deal terms that would produce returns for ExxonMobil of more than 16 percent on capital invested. A successful deal would deliver as much as $15 billion in investment to the kingdom.[12]

Nelson dined monthly with Saud Al-Faisal at the foreign minister's relatively modest (by the standards of Saudi princes) Riyadh home. Five or six days a week, Nelson and Raymond conferred by telephone about the Saudi project, punctuated by face-to-face meetings in Irving. Nelson grew into a mysterious and somewhat feared figure in ExxonMobil's executive ranks, by virtue of his unusual access to the chairman; he was the only lead country manager who worked directly for Raymond.

Terrorists struck Saudi Arabia sporadically after the September 11 attacks. Slightly before midnight on a dark Riyadh night in 2003, a car pulled up to the security station of the Al-Hamra Oasis Village, a 404-unit residential compound favored by Western professionals. As the guards began to open the compound's formidable gate for the car, whose driver they recognized as a resident, an unfamiliar Toyota sedan and GMC Suburban truck turned into the entrance. The vehicles were moving suspiciously fast. The guards scrambled to shut the gate, but were foiled by a spray of bullets shot from the Toyota's windows. As the guards fell, the cars forced their way into the grounds and proceeded to the swimming pool, where a residents' party was in progress. Four men armed with AK-47s sprang from the Toyota and mowed down as many guests as they could before continuing to the compound's villas. The gunmen banged on doors and mercilessly shot those who emerged. "I will kill them all!" one gunman cried.

When they had restored order, officials reported at least thirteen dead and dozens injured. Among those harmed were two ExxonMobil employees and one of their wives, who was pregnant at the time.[13]

Raymond and the Management Committee at headquarters set up a corporate security team to assess the vulnerability of employees and assets worldwide, in light of Al Qaeda's terrorism. Saudi Arabia was a place of obvious risk. Michael Shanklin, a former marine and Central Intelligence Agency case officer from the Watts neighborhood of Los Angeles, who now worked for ExxonMobil Global Security, traveled to the kingdom. He developed a security plan in consultation with Mohammed Bin Nayef, a powerful royal family member at the Saudi Ministry of the Interior. The local C.I.A. station relayed intelligence that Nelson himself was an Al Qaeda target. Senior executives at Irving proposed evacuating Nelson and the rest of the corporation's staff in the kingdom; Nelson resisted. "An evacuation will kill our venture potential," he argued. Although there were fierce internal debates over the question, most of the corporation's employees remained.

For ExxonMobil, the big question remained whether Saudi Arabia had enough freestanding natural gas to justify the risks to employees. ("Associated" gas, intermingled with oil, was too complicated to produce for the purposes the crown prince had in mind.) ExxonMobil still had libraries full of field data from its time as an Aramco partner, before nationalization. The corporation even employed geologists and engineers who had worked for Aramco in that era. Raymond and other executives polled them and discovered that they were skeptical about Abdullah's hopes. "Our explorers and these guys who worked in Aramco were very doubtful that there would ever be significant reserves sufficient to really support the kinds of projects" that Crown Prince Abdullah and Foreign Minister Al-Faisal envisioned, Raymond recalled. The massive industrialization they outlined would require "an enormous amount of gas. . . . Our people kept saying, 'No, it's not going to be there.'" ExxonMobil's biggest prospect was a structure called Tukhman in the kingdom's South Ghawar field. The more the corporation's geologists scrutinized it, the more doubtful they grew.[14]

Raymond and Nelson eventually advised Al-Faisal that if the kingdom wanted to find enough freestanding gas to fuel the projects it had outlined, the partners would have to move into territories previously set aside for Saudi Aramco. But Abdullah proved unable or unwilling to do this. Instead, "what they wanted to do was build the kinds of projects" Abdullah had proposed "and then find the gas," Raymond recalled.

"No way," Raymond told his colleagues. "We are going to end up with some projects where the only financial motivation behind them is to produce the gas—and if the gas isn't there, then we are just going to end up with a bunch of albatrosses." For their part, Saudi negotiators felt that the midteens profit margins demanded by ExxonMobil and other corporations were too high and that the Big Oil executives were not willing to take enough risks.[15]

Saud Al-Faisal owned a home in Beverly Hills; one of his neighbors was the actress Drew Barrymore. As the negotiations foundered in 2003, he summoned Raymond and Al-Naimi, the oil minister, to his home.

Raymond announced: "I think I ought to pull out of this deal. There's not enough gas to drive the process forward—it can't work this way. You're asking us to drive an Abrams tank with a Toyota engine."

Al-Naimi challenged him; Saudi Arabia had plenty of natural gas, he believed, more than enough to fuel profitably the projects Abdullah had in mind. "Lee, I think your people aren't being very honest with you." He implied that ExxonMobil's geologists and executives were underplaying the potential of the deal to gain an advantage while negotiating financial terms.

Raymond exploded. "Ali, you can insult the hell out of me—I don't care what you say about me. But when you start screwing with my people, that's another matter."

He was so hot that they had to call a break. The ExxonMobil team stepped outside on a deck, overlooking Drew Barrymore's yard. "I wish that hadn't happened," Raymond said. "Do you think I overreacted?" he asked. Still, "I couldn't let him insult my workforce."

They went back inside; the mood was calmer. But the Saudi gas initiative was officially dead. At a later meeting, Al-Naimi handed Nelson a letter, one he would also give to other consortia members, canceling their rights to negotiate.

Raymond eventually learned that Aramco had actually started drilling in the areas ExxonMobil had evaluated; the Saudis apparently wanted their own evidence about how much gas was really in the ground. Raymond was irate; this is not how partners operated. He blamed Naimi. "If that's the game, you can count us out," he told Al-Faisal.

Raymond also wrote to Abdullah to ask if Naimi's actions truly represented the crown prince's position some four years after the hopeful initial convening in Virginia. Naimi soon eliminated all doubt by redesigning the project and bidding it out to new corporate partners. The areas ExxonMobil said were dry turned out to be dry. Five years of effort had come to nothing.[16]

Lee Raymond ruled over ExxonMobil in the manner of an emir. During the difficult years of restructuring, he had worked very closely on the Management Committee with two key aides. Harry Longwell, a garrulous southerner, ran the upstream. Rene Dahan, the Moroccan-born Dutchman, supervised the downstream operations. Dahan might have been a candidate to succeed Raymond, although he was a little on the older side of the ideal age range. In any event, he decided to retire early and return to Europe, in 2002. Longwell was essentially Raymond's age, too old to be considered as his successor. By the time the Saudi deal fell apart, Raymond was approaching sixty-five, but showed no interest in retirement. Increasingly the outside members of the corporation's board of directors regarded the lack of a clear succession plan with concern. Hardly anyone at ExxonMobil stayed on beyond retirement age. Lawrence Rawl, Raymond's predecessor, had retired at sixty-four. "He wanted to stay longer," a director remembered. "The board was a little uncomfortable with it."

After the dust settled from the absorption of Mobil, the board had come "to a fairly clear view that, because of the merger, the people who were likely to succeed Lee" were not in position, recalled an executive involved, because these younger candidates were still out leading operating divisions and had not spent enough time at headquarters or interacting with the outside corporate directors who would be responsible for the final choice. A successor needed to be in his early fifties to have a chance

to lead the company for an extended time. (There were no women anywhere near in contention for the top job at ExxonMobil.) Younger candidates "should have been brought in much, much earlier, to sit around the
table," the executive who watched the succession process recalled. But
Raymond argued that he needed to keep the most talented younger leaders out in the field, to make sure that the reorganization following the
Mobil merger took place properly. "The rationale was that we'd lost two
years" in developing successors because of the merger.

The board had therefore agreed to extend Raymond's tenure beyond his scheduled retirement in 2003. At the same time, the directors
told him, in essence, "We need to bring these people in."[17] Raymond
named two promising younger candidates, Rex Tillerson and Edward G.
Galante, to coequal jobs at headquarters. Tillerson was a Texan who had
spent much of his Exxon career in the upstream exploration division.
Galante was a New Yorker who had risen on the downstream side. His
upstream experience seemed to give Tillerson a built-in advantage, because at ExxonMobil, as a director put it, "real men—they discover oil."
A few members of the board felt, as the director recalled, "there was just
no question that Rex was going to be the successor. He came from the
discovery side." Yet Raymond had spent time during his rise running
downstream facilities and was not an upstream oil hunter by specialty.
Other Exxon chiefs before him had also emerged mainly from downstream careers, including Cliff Garvin, Exxon's fourteenth chief executive.
(Raymond was the sixteenth.) In years past, the profitability of the upstream had subsidized downstream operations, which often struggled to
break even or eek out modest returns. Raymond had insisted that the
downstream businesses had to stand on their own; Galante had been part
of this successful transformation. Still, because some members of the
board of directors assumed Tillerson would prevail, largely because of his
command of the big oil and gas portfolios abroad, they questioned Raymond's motivations. "My concern was that it was kind of a charade to buy
him [Raymond] more time," a director said.

Once a year, on a Tuesday afternoon in October, Raymond organized
a special meeting of the board. At this session, Raymond was the only
ExxonMobil executive in attendance. Raymond provided reviews to the

outside directors of the performance and potential of his most senior executives. "It was always the case that the possible successors were not ready yet," an executive who heard Raymond's briefings recalled. Raymond would tell the board, "Maybe in eighteen months or two years." The directors would "talk amongst themselves: 'This could go on forever.'" They made "several efforts" to raise the matter with Raymond, "and [they were] rebuffed."[18]

On paper, Raymond worked for the board; in practice, he controlled his directors carefully. "The board wasn't able to impact management very effectively," a director recalled. "They were a group unwilling to challenge the status quo. . . . That is one of the few boards I know where the whole is less than the sum of the parts."

There was very little free-flowing discussion at board meetings. Raymond's remarks, presentations by other senior executives, and votes on board resolutions were written down well in advance and read out from sheets of paper. Board committee meetings could be a little looser. Even there, ExxonMobil executives listened carefully if outside directors asked hard or challenging questions and then reported back to Raymond—the offending director would soon be smothered with attention, to deflect the concerns he had raised. Once, after a director spoke up to defend one of Raymond's policies during a committee meeting where Raymond was not present, the chairman approached him to say, "I'm really glad you spoke up in that committee." The director was taken aback: "I was just amazed that he had that kind of intelligence; it was very revealing to me."

The approach was a throwback to the way board meetings often had been run in corporate America during the 1960s and 1970s. "The world had changed, but they had not," the director recalled. As oil prices rose, the corporation's financial and operating performance was so strong that there were few big issues that the directors felt they needed to intervene about. "That was a little bit of the board's problem," an executive recalled. "The company was so successful, it was kind of hard to argue. There was a little bit of a prisoner's dilemma."

On the question of Lee Raymond's successor, about all the board could accomplish was to push Raymond to bring Tillerson and Galante to more board meetings, to show off their skills to the directors. They each made presentations at the board's retreat in Scotland in June 2003, in the midst of Raymond's intensifying, failing negotiations with the Saudis. Gradually board members got to know the pair better. The formal presentations they made during meetings were heavily scripted and revealed little, but afterward, at lunches and dinners, the two would sit with outside directors and engage in more informal banter. Also, the corporation periodically organized board field trips to ExxonMobil divisions or operating sites, travel that also allowed Tillerson and Galante to interact with directors spontaneously. At the personnel review sessions with the board each October, however, Raymond's message did not change: "We need more time to see them perform."[19]

Ten

"It's Not Quite as Bad as It Sounds"

Raymond had promoted Ken Cohen to vice president of public affairs at the time of the Mobil merger. His predecessor, Tony Atkiss, had been "a workmanlike hand, but I think the perception was that we needed to have a little different view," an executive involved recalled. At the time, Cohen was a forty-eight-year-old in-house lawyer with no direct professional background in lobbying, media, or public relations. "There wasn't a long list of candidates," the executive remembered. "Being the head of public affairs for Exxon is probably not viewed as one of the more desirable jobs in the world. . . . The Exxon culture just doesn't generate a lot of people who would a) be very good at it and b) like it. . . . The perception was that Ken was reasonably articulate and would bring a different view."

Exxon in-house lawyers had founded the public affairs department; the corporation was heavily regulated in the United States and abroad, so its lawyers often had the most direct knowledge of the public policy issues that affected the bottom line. Still, Cohen did not possess an obvious background for corporate image management. In his previous position, he had served as a senior lawyer at Exxon Chemical Company, a division that litigated continually with regulators and environmental campaigners. Cohen had been conditioned to think that anything he or other Exxon executives said in public could be used against them in a court of law. He inherited a public affairs department whose core media strategy was to say "no comment" in fifty different languages, as an in-house joke had it.

Cohen kept a black notebook binder on a shelf in his second-floor office on the Irving campus. He distributed copies to members of Exxon-Mobil's Management Committee and other senior executives. The first page carried the title "Public Policy Issues." A list of about two dozen subjects followed, from climate change to energy pricing to alternative fuels. The Management Committee updated and approved the list annually as part of ExxonMobil's Srategic Planning process. Cohen's aides wrote a summary of the current policy debate under each subject heading and then a description of ExxonMobil's position. Cohen also reviewed the notebook's contents with a special public policy committee of the corporation's board of directors, which met nine times each year.

If the corporation's management ever changed its thinking about a public policy matter, the change would be vetted through the Strategic Planning process and then inscribed into Ken Cohen's notebook and reviewed with the board committee. "Issues managers" and "issues teams"—some in Irving, some in Washington, some in Fairfax, and some in Houston—monitored and updated the listed public policy debates throughout the year, reporting to Cohen and, through him, to the Management Committee. The issues managers ensured that the language used worldwide to describe ExxonMobil's position on any policy question was consistent. The language on PowerPoint slides presented to Chinese Communist Party functionaries in Beijing about, say, the regulation of ethanol was essentially the same as the language on PowerPoint slides left behind in the offices of first-term congressmen in Washington. ExxonMobil lobbied the same way it ran refineries—it employed a top-down, Global System, vetted at the highest levels of the corporation, and it expected all of its managers to follow that system exactly.

The issues management system, as it was known, displayed ExxonMobil's signature internal discipline, but it could also be as rigid, slow, and inflexible as a Soviet five-year agricultural plan. The specific policy positions that emerged in the final draft of the annually updated black binder were sometimes vague and abstract, and might be of limited use to a former senator under contract to the ExxonMobil K Street office to sway votes in a Capitol Hill legislative conference. Legislative deal making was a fluid, adaptive process, involving more improvisation than Exxon's engineering-led culture often knew how to manage. Public affairs dispatched an internal newsletter on the fifth of every month describing what political and policy developments it had been monitoring since the tenth of the previous month; its analysis was often out of date. In the lobbyist's art, personal relationships and spontaneous compromises shaped success, not prepackaged analysis.[1]

Ken Cohen had grown up in a placid Midwest community in the postwar era as one of four sons in an achievement-oriented, education-driven family. His grandfather had fled anti-Semitic pogroms in Ukraine and immigrated to the United States; the relatives on which he relied drew him to Illinois. He settled in iconic Peoria and opened a dry goods store. Cohen's father attended Northwestern University and became a medical doctor. Ken also enrolled at Northwestern, but left after his third year and enrolled at Baylor University's law school, in Waco, Texas; Baylor was one of a relatively small number of law schools willing to accept students who had not finished their college degree. Cohen loved the law, particularly its pedagogy, but he concluded that he should practice before taking up a teaching career. He joined Exxon in 1977, at twenty-six, and never left.

He was a mild-looking man of modest height with a full head of salt-and-pepper hair, which he parted near the middle, a choice that gave him a slightly old-fashioned air. He owned a condominium in the upscale Turtle Creek area of Dallas. He was married to a former colleague from the Exxon public affairs department, Darcie A. Bundy, who in midlife entered into the interior home design business in Dallas and in the town where the couple summered, Kennebunk, Maine. Cohen did not think of himself as possessing a political ideology, but as he stayed and rose at Exxon, he adopted the corporation's official skepticism toward regulators and its preferences for free-market policies. He was quiet and calm, professional, but intensely competitive. He could master a new subject quickly and he organized concise briefings about public policy issues for his colleagues in management. Like many in the corporation's upper ranks who had lived through the post-*Valdez* years, Cohen had developed a defensive posture. The attitude he and his colleagues projected came across as arrogance, but if it was that, it was conditioned by mistrust of outsiders, especially of environmental campaigners and journalists. "We've had our hearts broken so many times," Cohen told his colleagues.[2]

Each year, Lee Raymond held a private retreat for ExxonMobil's most senior executives. At the first such conference after the merger, Raymond told the assembled, "ExxonMobil is different than either Exxon or Mobil—it will occupy a different position in the economy and the industry." Raymond was referring obliquely to his intention to use the merger to reinvent Exxon. It would have been easiest for Raymond to have managed the absorption of Mobil by just taking the acquired company apart and then bolt the pieces onto Exxon's existing divisions, while seeking cost efficiencies. Raymond had rejected that approach. He intended to use the merger to restructure Exxon itself. "Everybody is going to have a new job," he announced. As the merger proceeded, more people from Exxon than from Mobil departed or were let go, accounting for their relative size before the deal. With his withering inquisitions and his relentless push for financial excellence, Raymond managed to some extent by using insecurity as a motivation tool. It was up to the senior department heads to interpret his dogma about ExxonMobil's being different from either of its predecessor companies; Raymond would judge the result. In public affairs, Cohen understood the chief executive's edict as an invitation to rethink the corporation's brand position.

ExxonMobil spent very little money on advertising or image building for a company of its size. Typically, in previous decades, oil companies had mainly invested their advertising and marketing budgets in their highly competitive retail gasoline businesses. Exxon's version of this marketing campaign had centered on the "Put a tiger in your tank" slogan, until Kellogg's, the corporate home of Tony the Tiger, of Frosted Flakes renown, persuaded Exxon to drop the tiger as part of a trademark infringement settlement. (The Greenpeace protesters in tiger suits were out of date.) The truth about gasoline, however, was that it was basically all the same. Each of the major oil companies blended in a few unique chemical additives to improve the fuel's performance and then spent large sums of money on ads promoting the supposed superiority of its magic formula. Convenience stores, credit cards, and membership point schemes later enhanced these campaigns for retail market share. "We're drivers too" was the bland slogan that ExxonMobil adopted after it lost its tiger.

Given the corporation's business performance—it returned to the top of the Fortune 500 in April 2001 and earned $15 billion in profits that year, more than any corporation in America—it was not obvious what strategy of image building ExxonMobil required. After the merger with Mobil, Raymond cut about twenty thousand jobs and reduced operating costs by a further $8 billion after initially promising investors that he would save only about $3 billion. The extra $5 billion in savings partly reflected what Raymond had believed all along: that it would be possible to achieve more cuts than he had advertised publicly when the merger was announced. After the deal closed, Raymond told his top two executives, Harry Longwell and Rene Dahan, that he wanted the combined company, within three or four years, to be no larger than Exxon had been before the merger, without counting the additional people who would be necessary to run new refineries or chemical plants. Raymond's goal reflected his long-standing conviction that Exxon's management was underutilized and had the capacity to take on harder problems.

Raymond's financial strategy was clear, but it was less obvious how spending on public policy, image advertising, and lobbying could help. Raymond had made up his mind about global warming policy and environmental policy, and he was not about to revisit his thinking on those issues to appease activists or uncompromising Democrats. In any event, big, highly profitable oil companies were not likely ever to enjoy wide enthusiasm in the populist-influenced United States—or anywhere else, for that matter. Worldwide, Cohen's private opinion surveys showed, ExxonMobil enjoyed a top reputation among oil companies in only one country: Singapore, a tiny, authoritarian, free-trading state where the corporation was a big investor, and a country whose top-down conformist culture resembled that found in Irving.

Ken Cohen and his public affairs team researched the history of Standard Oil and successor companies' ad campaigns during the twentieth century and they reviewed the public opinion results those campaigns achieved. The numbers fluctuated between terrible and tolerable, but Cohen's group could find no evidence of a golden age of oil company popularity.[3]

ExxonMobil could not afford to accept low public esteem as a given, however. Engineering and scientific talent in the United States was in high demand. The corporation competed with other super-majors to recruit the most talented geophysicists and geologists at the world's top schools. Scientists who agreed to join entered a two-year training program—a kind of free graduate school in which ExxonMobil invested tens of thousands of dollars to advance their skills in applied settings. The children of the baby boomer generation had been reared during an age of environmentalism, and it would be difficult to recruit and retain the best and brightest of them if working at ExxonMobil seemed a morally compromised choice. "There's a big population of liberal young folks in the company," a former manager recalled. These left-leaning employees wrestled among themselves with how they could be liberals and still work at ExxonMobil. Some of them pushed for paternity benefits and nursing rooms at the office—and succeeded. But these were incremental achievements and the ambivalence remained.

The difference between excellent and mediocre geologists could be the difference between finding oil and failing to do so. Scientists could be an independent-minded lot. Young ExxonMobil geologists often received "poaching" recruitment offers after four or five years of employment. Salaries at ExxonMobil were modest by industry standards, except for the very best performers, who were well rewarded. British Petroleum's pay scale, for example, meant that it could often offer significant raises to geological scientists, $20,000 or more annually. After the merger, ExxonMobil became concerned about unusually high attrition rates in these talent wars; the corporation seemed to have particular difficulty holding on to women. Senior management organized "listening post" meetings with invited groups of department leaders and asked them, "Are you proud of your experience" at the corporation? "What do we need to do to keep you here?" Some of the rising managers were frank: "Look, we have a really bad reputation as a company." Informally, "the employees talked about it all the time," a manager who participated recalled.[4]

There were other measurable costs to being despised. Late in 2000, an Alabama state jury deliberating a civil fraud case handed down a $3.42 billion punitive damages verdict against ExxonMobil for allegedly cheating the state out of natural gas royalties. Ken Cohen and other lawyers at the company were adamant that the verdict would be thrown out on appeal (and it was), but the case offered a reminder, if one was needed after the *Exxon Valdez* jury trial, that public skepticism toward the company could easily present itself in a courtroom in the form of hostile jurors.

Cohen's group commissioned public opinion surveys, opinion leader focus groups, and other elaborate research endeavors designed to map the ways in which ExxonMobil was hated. Engineers ran the company, and only numbers could persuade them. Even so, some of the corporation's executives, including Raymond, received the insights from these "scientific" surveys and focus groups with undisguised skepticism about their usefulness.

If a survey reported progress in ExxonMobil's reputation in comparison to, say, Chevron, Raymond would comment, "Just remember: Chevron is having the same meeting today and their guys are saying exactly the opposite of what you just said."

The more he learned about the details of public opinion surveying— how, for example, survey questions had to be asked exactly in the same form for the results to be truly comparable—the more Raymond doubted its validity as science. He did not want to ignore it entirely, and he budgeted for continual opinion surveys, but he declared that it would not drive his decision making. "If you start to play to that kind of thing, where does that end?" All oil companies saw their reputations rise and fall mainly as a function of whether retail gasoline prices were high or low, Raymond believed; the rest was just noise.[5]

Cohen accepted that the public's sense of being vulnerable to volatile gasoline prices and, more generally, to the political power of big oil corporations, was nearly universal. The implication, he concluded, was that ExxonMobil should seek to be credible rather than popular. The corporation's communications should be clear, consistent, and fact based. Exxon-

Mobil's leaders and employees should accept that there would be many people who did not like what they had to say about public issues, but some of the company's skeptics might nonetheless be persuaded to accept that the corporation's positions were the product of empirical analysis. Global oil production was indispensable to the American economy—in that respect it was very different from, say, the manufacture of addictive tobacco products or the promotion of a particular image of a youthful lifestyle through the sale of soft drinks. The resurrection of ExxonMobil's reputation could not be regarded as a marketing goal, equivalent to the establishment of a new soft drink brand, Cohen told his colleagues; it should be seen as an outcome of consistent, credible communication, in the face of predictable and persistent public skepticism.[6]

D uring the Aceh crisis, Robert Haines, the manager of international government relations who worked for Cohen out of the Washington office, held regular off-the-record meetings with representatives of Human Rights Watch to hear the organization's concerns and to try to persuade its investigators that ExxonMobil was doing everything it could to control the abuses of the Indonesian military. Haines used the sessions to present ExxonMobil's brief about the conflict: The corporation was a guest in the country; its security arrangements were a requirement of its contract; ExxonMobil exercised no control over the T.N.I.; and the corporation did not condone human rights violations. Yet Cohen and Lee Raymond had still not signed the Voluntary Principles on Security and Human Rights developed in 2000 by the Clinton administration and the Blair government in Great Britain. They may have wondered whether the Bush administration would abandon the initiative, but as it turned out, Bush developed a strong interest in human rights issues, and in 2002, his administration formally embraced the Voluntary Principles as official policy. Around that time, Ken Cohen decided that he should revisit ExxonMobil's own decision about the Voluntary Principles and begin to think more deeply about how the corporation could improve its credibility on human rights questions.

On one of his regular trips to Washington, Cohen invited Mike

Jendrzejczyk, the director of Asia advocacy at Human Rights Watch, and his colleague Arvind Ganesan to dinner at the Four Seasons hotel in Georgetown. Ganesan was a lawyer; Jendrzejczyk, as one of his colleagues put it, had "the charm of a con man, the energy of a five-year-old, and the persistence of a used-car salesman." That was not exactly the personality profile cultivated by ExxonMobil in its lifelong employees, but Cohen had learned to accept human rights and environmental activists as he found them. The violence in Aceh was continuing. Cohen refused to discuss the specifics of ExxonMobil's natural gas operations there; the lawsuit filed on behalf of some of the province's torture victims was now pending in federal court, and Ganesan and other activists had learned that if they raised Aceh specifically in private meetings with Exxon officials, they received silence and a "meeting over" look. On global human rights more generally, however, Cohen saw the Four Seasons dinner as an opportunity to open a dialogue. His message was, as Ganesan recalled it, "We're doing a great job, but we think we need to be more engaged."[7]

Cohen invited Ganesan to speak at an off-site retreat for about seventy-five rising ExxonMobil managers. The executives were senior enough to have earned an internal designation as "gold-" or "platinum-" level leaders; their reward was a multiday public relations boot camp at a corporate retreat center in Norwalk, Connecticut. Other corporations retreated to semitropical golf resorts; at ExxonMobil, there was Norwalk. Ganesan traveled there on October 27, 2002. He had attended corporate retreats before, but the ExxonMobil event struck him as "one of the strangest" in his experience. He felt that he was in a classified facility. He was not allowed to enter the meeting room until his speaking time arrived and he was ushered out immediately afterward; the atmosphere seemed "clinical." Later, he was told that he had been invited not only so that the managers could hear his views about oil corporations and human rights, but also so that he could provide a kind of live exercise in how to deal with a social activist.[8]

Ganesan reviewed for the assembled managers the history of Human Rights Watch. He described why he and other activists believed that oil production hurt poor countries more than it helped them: "Energy wealth does not necessarily lead to better standards of living, increased democratic participation in government, or a better climate for human rights. Instead, economic, social, and political conditions may stagnate or even deteriorate." He described the appalling human rights records of the governments that were major oil and gas producers, starting with Saudi Arabia, a business partner of ExxonMobil's in the refinery and chemical industries. The Saudi monarchy, Ganesan pointed out, "executes prisoners, engages in torture, curtails due process rights, and uses barbaric forms of punishment such as amputations and beheadings." Nor was its record unique. Seven members of O.P.E.C.—Algeria, Iran, Iraq, Kuwait, Libya, Qatar, and the U.A.E.—were "undemocratic with poor human rights records and limited economic diversification." Three others—Indonesia, Nigeria, and Venezuela—were "nominally democratic but plagued with widespread corruption and poor human rights records." New oil exporters such as Angola, Azerbaijan, and Kazakhstan were becoming "models of corruption, mismanagement, and human rights violations." ExxonMobil operated in many of these countries and collaborated with the governments Ganesan found so wanting. He then turned to the reputation of ExxonMobil itself. He had decided not to spare the managers' feelings.

"I have interacted with ExxonMobil for at least the last five years, and found them to be hostile and unproductive prior to this current effort," he said. "ExxonMobil seemed like an arrogant, opaque company that was hostile to social responsibility and preferred to go its own way." He continued: "This is not just my perception of the company, but shared by every NGO and many others. Several company representatives have come to me over the years and have justified their companies' actions or inactions by saying, 'At least we're not ExxonMobil.'"

He did finish on a note of aspiration. There was "another widespread perception" of ExxonMobil, namely, that "once it decides to do something" it will "do it better than anyone else in its industry." If the corporation would seek to improve its human rights record in a serious way, its leadership in the international oil and gas industry "could have very beneficial effects." ExxonMobil should expect, however, "a considerable amount of skepticism" if it tried to change its ways.[9]

Ken Cohen did not invite Ganesan back to his retreats. But he did not give up on his outreach campaign to Human Rights Watch. ExxonMobil formally signed up to the Voluntary Principles and gradually began to implement them.

Cohen also turned to Bennett Freeman, the former deputy assistant secretary of state for democracy, human rights, and labor who had helped conceive the Voluntary Principles before returning to a corporate consultancy practice. Freeman, too, attended Cohen's off-sites at Norwalk. He respected Cohen's professionalism. He regarded himself as a constructive critic of ExxonMobil, but also a sophisticated thinker about corporate responsibility who was not innately hostile to multinationals. When he appeared at Norwalk conferences or at private "opinion leader dialogues" with ExxonMobil executives elsewhere, Freeman usually broke the ice by remarking that the corporation's human rights performance reminded him of what a critic once said about the music of Richard Wagner: "It's not quite as bad as it sounds."[10]

Over the years, Lee Raymond had told colleagues that he considered Royal Dutch Shell to be Exxon's most formidable competitor. Royal Dutch had weaknesses, in Raymond's estimation: a mind-boggling system of split Anglo-Dutch governance, a retirement age of sixty years that created disruptive turnover in corporate leadership, and a thick bureaucracy. Yet Royal Dutch maintained a greater focus on operations and project discipline than many other oil companies, Raymond told his colleagues. Exxon partnered in oil and gas operations with Royal Dutch more than any other company.

In comparison, Raymond and other ExxonMobil executives did not hide their disdain for BP. Increasingly there was a competitive edge to the rivalry. After the dust settled on the Big Oil merger scramble of the late 1990s, ExxonMobil and BP emerged as the nearest equals in size and global ambition, together at the head of the global rankings for shareholder-owned oil corporations. Raymond had admired one of BP's previous chief executives, David Simon, but he told colleagues that in general, he found the corporation to be bureaucratic, undisciplined, and unreliable.

Raymond was also no Anglophile. ExxonMobil's operations in Britain had frustrated him. The corporation ran refineries and retail gas sta-

tions in the United Kingdom under the Esso brand. In the O.I.M.S. era
these divisions had not measured up very well. Raymond traveled to
London and complained to his British subordinates: "You guys are really
great in poetry. But getting up every morning at 6:30 a.m. and saying,
'Okay, we are going to have the morning meeting—what's going on in the
refineries?'—that's just not in your skill set." He extrapolated the flaws he
perceived at Esso to explain the enduring worldwide management weaknesses he saw at BP.

BP began to annoy Exxon in the environmental lobbying arena, too.
By the end of the 1990s, more of British Petroleum's assets were located
in the United States than anywhere else; American public policy was critical to the company. John Browne, however, did not think about industry
issues as Lee Raymond did. To ExxonMobil's executives, he seemed to be
more of a financial engineer than an operations man. Browne was also in
tune with the transatlantic center-left politics of the late 1990s. He enjoyed a strong relationship with the newly elected British prime minister,
Tony Blair. He had easy access to Bill Clinton's White House; he was
exactly the sort of big business leader Clinton-era Democratic politicians
often seemed to value—a thoughtful globalist willing to endorse the principles, at least, of the mainstream environmental, human rights, and public health movements. Browne spoke early about the importance of global
warming. He rebranded his company as the letters BP and eliminated all
abbreviated and other reference to British Petroleum. He approved the
marketing slogan "Beyond Petroleum." The corporation's marketing team
chose a green-and-yellow logo that looked like the sun, as if BP were
moving decisively out of the oil and gas business and into solar power. An
ExxonMobil executive at the corporation's British affiliate took a photograph of a BP retail gas station with a windmill on top and sent it to Lee
Raymond with a note: "This is our competitor."

"Oh," Raymond said, dismissively. This is just a public relations strategy, he said. There is no substance to it; don't overreact.[11]

BP did invest in some solar manufacturing in India, China, Australia,
and the United States, where its plant was located in Frederick, Maryland,
a convenient drive from Washington, D.C., and thus an optimum site for
tours by members of Congress or their staffs who might be interested in

alternative energy. Yet the scale of BP's solar investments was minuscule in comparison with its oil and gas operations. The investments were understood within the corporation, according to one former senior executive, as justifiable not so much on business as on marketing grounds—BP Solar returned more to BP in favorable reputation than comparable sums spent on conventional corporate image advertising ever could.[12]

"The oil industry is already detested by people who think we're indifferent to the environment," Browne explained. "We must persuade our ultimate customers that this isn't true." Smog and other pollution from oil-derived fuels meant that customers "can see it and they can feel it and they can smell it. And they look at oil companies and say, 'You brought us this.' And we don't want to be in that position."[13] Ken Cohen occasionally seethed in private conversation about BP's image makeover. First, he pointed out, BP remained fundamentally an oil and gas company—one of the largest in the world. Of course it was "in that position"; how could it pretend otherwise? By 2002, Cohen had also assembled an issue-by-issue chart showing that on public policy controversies from climate change to human rights, the recommendations of BP and ExxonMobil were little different. Yet the public's impression was that the two companies had diametrically opposed approaches to climate change and corporate responsibility. As a recently minted public affairs strategist, Cohen could appreciate, in professional terms, Lord Browne's achievements. He knew, too, that BP had the advantages that ExxonMobil lacked—it was not burdened by the high negative ratings caused by the *Valdez* spill and therefore had much greater scope to reinvent itself in the public mind. BP executives and public affairs strategists looked on ExxonMobil the way many of its competitors did: as self-isolating, stubborn, inscrutable, and behind the corporate times. ExxonMobil executives rationalized their poor reputation, when compared with some of their industry peers, by assuring themselves that they conducted business ethically and operated safely and with financial discipline. They even took pride in their self-image as a corporation that did not try to pretend to be something it was not. Yet Cohen recognized that BP had accomplished something improbable—the cost-effective greening of an oil company. When he was in a more generous mood, he told his colleagues, "Hats off to them."[14]

The decisions Browne took at BP were not merely cosmetic. In 2002, the corporation announced that it would no longer fund "any political activity or any political party," a form of neutrality that ExxonMobil could not claim. Browne eventually extended BP's corporate benefits to the gay and lesbian partners of its employees; Lee Raymond declined to do so. On climate, however, BP dodged and wove during the first Bush term. "The science of global warming is unproven," Browne said in 2001, a formulation not much different from Lee Raymond's. "I question whether it will ever be proven. But there is a risk there," Browne said. This risk was enough to "begin to take steps to begin to make a difference." Still, the danger was not large enough to justify the costs and the global bargain contemplated by the Kyoto Protocol: The treaty was a "bridge too far," Browne said. Only very gradually would BP shift toward acceptance of the cap-and-trade system, a regulated, government-imposed marketplace that emerged in Europe to control carbon dioxide emissions and help governments there attempt to keep their commitments under Kyoto. He searched for "the right level of transparency or openness in order to build rather than to undermine trust in a world of suspicious media and single issue N.G.O.s."[15]

There were some public policy matters where not even the most creative corporate policies or public relations campaigns could make much difference, however. The invasion of oil-laden Iraq was one.

Eleven

"The Haifa Pipeline"

On February 11, 2003, Douglas Feith, the Bush administration's under secretary of defense for policy, appeared before the Senate Foreign Relations Committee, where he argued that the Iraq War, if it arrived, would not be a war for oil. "All of Iraq's oil belongs to all the people of Iraq," Feith said. The Bush administration had "not yet decided on the organizational mechanisms" through which the Iraqi oil industry might be restructured after the overthrow of Saddam Hussein, but he felt that he should "address head-on the accusation that, in this confrontation with the Iraqi regime, the Administration's motive is to steal or control Iraq's oil." That charge was commonly made, but it was "false and malign."

By the time of his Senate testimony, Feith had already become a punching bag for opponents of the Bush administration and its foreign policies. He was a tall, extroverted man with a mop of graying hair and round wire-rimmed glasses. His articulate self-confidence was of the type associated with student council vice presidents, and it grated on some people similarly; General Tommy Franks, then in command of all U.S. military forces in the Middle East, told colleagues at the time that he considered Feith "the fucking stupidest guy on the face of the Earth." (Franks's Pentagon colleagues debated his own acumen.) A lawyer in private practice before joining the Pentagon at the request of Secretary of Defense Donald Rumsfeld, Feith proved willing, at the least, to argue like a litigator about the rationales for a U.S. invasion of Iraq.

He told the senators that the United States had no historical record of stealing other countries' resources through war. "We did not pillage Germany or Japan; on the contrary, we helped rebuild them after World War II," he said. After Desert Storm, the U.S.-led campaign to liberate Kuwait from Iraq, which prevailed in 1991, "we did not use our military power to take or establish control over the oil resource of Iraq or any other country in the Gulf region."

The idea that the Bush administration would take on the human and financial costs of overthrowing Saddam Hussein's regime in Iraq for the sake of grabbing that country's oil did not make logical sense, Feith continued. "If our motive were cold cash, we would instead downplay the Iraqi regime's weapons of mass destruction and pander to Saddam in hopes of winning contracts for U.S. companies," he said. "The major costs of any confrontation with the Iraqi regime would of course be the human ones. But the financial costs would not be small, either. This confrontation is not, and cannot possibly be, a moneymaker for the United States. Only someone ignorant of the easy-to-ascertain realities could think that the United States could profit from such a war, even if we were willing to steal Iraq's oil, which we emphatically are not going to do."[1]

In the weeks to come, Bush administration cabinet officers and independent analysts would endorse Feith's position that the war had, as Defense Secretary Donald Rumsfeld put it, "literally nothing" to do with oil. The administration published its war aims; these made no mention of energy or economic issues. The invasion's stated goals were to eliminate Iraq's weapons of mass destruction, end the threat Saddam posed to neighboring governments, stop his regime's internal tyranny, cut off his links to terrorism, maintain Iraq's territorial integrity, liberate Iraq's people, and create a democracy. It was true that Saddam's capacity to threaten the world was in part a result of the cash he received from oil sales; in that limited but important sense, the administration's war aims could be said to be about oil. It could also be argued that the United States would not have incurred all the risks and costs of invading Iraq if the country did not have large oil reserves and therefore an innately important place in the global economy and regional power balances. But that was different from arguing that the United States intended to launch a war *for* the purpose of acquiring Iraq's reserves.[2]

What would it mean, in any event, for the United States to "steal" Iraq's oil? The question itself illuminated America's dysfunctional search for a national understanding of "energy security." The United States formally owned oil only to operate government vehicles and aircraft, and to fill a 700-million-barrel strategic petroleum reserve. The government amply met these needs by purchasing oil on the open market. The American economy required about 12 million barrels of imported oil every day in 2003, but these supplies were purchased from private and government-owned oil producers around the world; invading Iraq wouldn't change that market much, except perhaps unfavorably, from an American perspective, by raising prices through the disruptions caused by war. It was possible to imagine that President Bush might wage war as a conscious or unconscious proxy for the interests of American-headquartered oil companies, notwithstanding the fact that most of these companies were global in scale, employed more foreigners than Americans, and paid more taxes to overseas governments than to the United States Treasury. Yet even if the Bush administration were thoroughly infused by such corporate-inspired perfidy, invading Iraq did not seem like an especially cost-effective way to help ExxonMobil, Chevron, or Conoco expand their booked oil reserves. In an essay published on the Iraq War's eve, the oil analyst Daniel Yergin argued that even a "liberated" Iraq might be reluctant to allow much direct participation in its oil sector by American firms, because of the prevalence of resource nationalism among Arab populations. He cited the example of Kuwait: "After the 1991 Gulf War, a liberated and grateful Kuwait announced that it would open its oil industry to foreign investment in order to boost production. Eleven years later, that still hasn't happened, owing to nationalistic opposition."[3]

Perhaps, then, the invasion of Iraq would be a war for oil in a geopolitical sense, for the purpose of increasing Iraqi oil production from the moribund levels of the Saddam Hussein era, and by doing so reducing world oil prices and America's dependence on Saudi Arabia. In the run-up to the invasion, a few conservative thinkers floated versions of this rationale, inflamed in part by evidence of Saudi Arabia's support for Islamic radicals such as those responsible for the September 11 attacks. But although Iraq had large untapped reserves of 115 billion barrels or more, its daily production amounted to only 2 or 3 percent of the world's total. After a U.S.-led invasion, even if all went well, it would take a decade or more to double Iraqi production to 6 million barrels per day, and even then Iraq could not hope to challenge Saudi Arabia's dominance as the world's most influential "swing producer" and price setter in oil markets, a producer able to raise or lower output as market conditions demanded. Saudi production capacity would still likely be twice that of Iraq's.[4]

Nonetheless, as the war befell them, even Iraqis with a sophisticated understanding of the global oil economy remained suspicious about American motives. They did not believe, necessarily, that the United States intended to steal their country's reserves directly, but they regarded Iraq's oil as an essential context for the American invasion. History influenced them; without question, oil grabs had shaped Western intervention in the Middle East in the past. "The First World War was not about oil," said Tariq Shafiq, who would help to draft Iraq's postinvasion oil law. "But the loot to the victorious winner was the oil concessions in the East. The intention was not there, but that was the obvious outcome. Today with the oil being really the core of our civilization . . . you would expect that oil was a factor."[5]

From its thunderous opening salvos in the early hours of March 20, 2003, the American-led invasion of Iraq did unfold in ways that exacerbated such doubts, particularly among Iraqis. On their initial drive to Baghdad, American tanks and Jeeps refueled at depots called Exxon and Shell. The decision to choose those code names might be dismissed as the tone-deaf error of midlevel staff in the Pentagon's bureaucracy. It proved to be a signal of a deeper and persistent ambiguity.

Talking points written at the National Security Council and handed out to American officials charged with making contact with Iraq's oil bureaucrats during the early days of the invasion instructed them to emphasize, "We're not here for the oil; the oil belongs to the Iraqi people." Paul Bremer, the head of the Coalition Provisional Authority, or C.P.A., and the de facto regent of the country until 2004, declared that Iraq's "natural resources should be shared by all Iraqis" and that revenues from the sale of oil should be placed in transparent bank accounts to create a "humane social safety net" for the Iraqi people. In private, however, officials within Bremer's occupation authority wrestled over the "organizational mechanisms," as Douglas Feith had put it, that would govern Iraq's postinvasion oil industry.[6]

Standard Oil first invested in what became the Iraq Petroleum Company in 1928. By the 1960s, international oil companies, including Esso, the ExxonMobil precursor, still owned a share of Iraq Petroleum. Iraq later nationalized its oil industry and organized state-owned firms, akin to Saudi Arabia's Aramco. In its heyday, the flagship Iraq National Oil Company and its affiliates were highly professional, led by Iraqi engineers trained in the United Kingdom and the United States. Under Saddam Hussein, however, the state-run oil complex atrophied. By the time of the U.S.-led invasion, a few aging technocrats held Iraq's oil infrastructure together with proverbial gum and paper clips. The complex's maintenance problems ran so deep, "you could have brought the whole of ExxonMobil out there and they wouldn't have been able to operate that thing worth a damn," said Philip J. Carroll Jr., a former president of Shell U.S.A., who was appointed by Secretary of Defense Donald Rumsfeld to serve as Paul Bremer's first senior oil adviser.[7]

Even before the American invasion, it was clear, at least to some Iraqi exiles and American war planners, that a post-Saddam Iraqi government would have to consider whether to invite international oil companies to invest and help solve these deep-seated infrastructure problems. If a "liberated" Iraqi government wanted to draw on large sums of international capital to revitalize oil production, it would probably have to give up at least some equity oil reserves in return, by signing production-sharing contracts with international oil majors or through outright privatization. And yet allowing foreign companies to own Iraqi oil would undermine the Bush administration's public narrative that the war would not reduce Iraq's sovereign control of its natural resources. A desperate Saddam Hussein, toward the end of his time in power, had signed production-sharing contracts with Russian and Chinese companies, but those agreements had never been implemented. Otherwise, no Iraqi government had allowed outside oil ownership in four decades. Some financially and politically weak nations elsewhere still accepted production-sharing contracts—Azerbaijan, Indonesia, and Chad were among them—but such deals typically generated controversy, and they had essentially been banished as a contract genre in the Middle East.

Bush administration war planners anticipated this dilemma as they worked in secret before the conflict. The Oil and Energy Working Group of the Future of Iraq Project, a State Department planning body, noted in a paper written early in 2003 that postwar Iraq would require foreign investment "on the terms that best, rapidly and significantly increase [oil and gas] production," but that Iraqi privatization schemes or production-sharing contracts could "engender opposition from those who see this as selling out to foreign oil companies."[8]

Some free-market conservatives within and around the Bush administration saw no reason why a post-Saddam Iraqi government should feel embarrassed about trading some oil reserves for access to foreign capital and technology. In their view, all countries were better off if they privatized their economies to the greatest possible extent. "Privatization works everywhere," a paper published by the Heritage Foundation in 2002 declared. Its authors urged the Bush administration to work with Iraqi opposition leaders at once to "prepare to privatize government assets" after Saddam's overthrow.[9]

In Baghdad, immediately after the invasion, Thomas Foley, a business school classmate of George W. Bush's, organized a cell of privatization enthusiasts inside Paul Bremer's C.P.A.; Foley and his colleagues pushed plans for a "broad-based, mass privatization program" even before a transitional Iraqi government could be established. Iraqi technocrats who served as caretakers at the oil ministry in that chaotic spring and early summer of 2003 following Saddam's fall were stunned by the radicalism of some of the ideas the arriving Americans proposed. Pentagon planners suggested that Iraq should consider withdrawing from O.P.E.C. "This was part of the neoconservative view: Why have Iraq in O.P.E.C.?" recalled one American official involved. "'Let's break the cartel!'" The idea seemed preposterous to experienced Middle East hands, as such a proposal would only confirm ordinary Iraqis' worst fears about American intentions. State Department opponents of the proposal sought to dismiss it "out of hand," an official involved recalled. And yet, the idea "kept resurfacing."[10]

Pentagon officials also suggested that Iraq's oil ministry look into shipping crude down "the Haifa pipeline," as they referred to it. The pipeline had been constructed in 1934 to serve territory that eventually became part of the state of Israel. It ran from Iraq's oil-producing region around Kirkuk through Jordan to modern Israel's coastal city of Haifa. It ceased operations after Israel's birth in 1948, but it was marked on old maps.

After the invasion, Michael Makovsky, a member of the Pentagon's Iraq oil planning team under Feith, chaired weekly telephone conferences with American oil advisers in Baghdad. Late in the spring of 2003, Makovsky asked that inquiries be made at Iraq's oil ministry about the old pipeline's status. Was it operational? Could it be repaired or placed into service?

The assignment fell to Gary Vogler, a West Point graduate and former Mobil Oil executive who had entered Baghdad with the first wave of American civilians as part of the oil advisory team led by Phil Carroll. Vogler considered himself to be "politically naive." In the first weeks after the invasion, he established a strong working relationship with Iraq's interim oil minister, Thamir Ghadhban, a career ministry engineer who had been jailed briefly by Saddam but who had stayed and survived his reign. One day, Vogler traveled to the oil ministry and found Ghadhban at his desk, juggling telephones.

Vogler asked about the pipeline to Haifa. Ghadhban looked at him icily. "There are a lot of people in my organization, in the ministry, and throughout the country, who feel like the only reason why you guys came into this country is to get oil out to Israel," he said. "If I go out with a question like that, I'm only going to solidify their viewpoint."

"Forget I asked you that," Vogler said. "Don't follow up on it unless I ask you again."

The queries from Makovsky, in Washington, continued. Vogler resisted the questions, asking why Makovsky kept making such an issue of a pipeline that had never been discussed in prewar planning. The purpose of the questions seemed vague. "Put in writing what you need and why you need it," Vogler requested.

In an interview years later, Makovsky said he could not recall discussing the pipeline with Vogler, but he did remember being asked to review the pipeline's status by superiors at the Pentagon. "The Israelis were at one point interested in this at the beginning of the war," he recalled. "I was asked by an official to look at this." He investigated and wrote a brief report. "There were a lot of things I looked into that didn't go anywhere. I'm not aware of anyone in the U.S. government who was advocating building a line from Iraq to Israel. . . . I never advocated anything like that." Douglas Feith, too, said the idea surfaced with "lower-level Pentagon officials" and he "never supported the proposal."[11]

Makovsky clashed regularly with Vogel; he felt that the former Mobil executive was unreliable. A third American official who participated in the pipeline discussions in 2003, and who respected both Vogler and Makovsky, recalled that Makovsky's true purpose was to find an export route for Iraqi oil that would bypass Syria and benefit Jordan, not Israel. "Mike's view is that you can't have it go to Israel—he would like that, but he realizes you can't have that," this official recalled. Still, Makovsky was, in this participant's estimation, tone deaf. "What does he always refer to it as? 'The Haifa pipeline.'" Makovsky said later that aiding Jordan and undermining Syria was indeed the reason he was at times animated about the possibility of resurrecting the pipeline. He continued for years afterward to write articles supporting a pipeline route from Iraq to Jordan, arguing that it would create "an opportunity to export oil both to Asia, where demand is growing, and to Europe and the United States."[12]

Eventually, unhappily, Ghadhban made the inquiries demanded by his

American liaisons. He reported back with evident satisfaction that the pipeline in question barely existed anymore; it had not been maintained for decades and had been pulled apart in places by scavengers. There was nothing, realistically, that could be done for now.

These awkward early exchanges coincided with high-level reviews of how Iraq's state-owned oil industry should be restructured to attract foreign investment and improve production rates. Philip Carroll, the senior adviser, was a patrician and a deeply experienced peer of Vice President Cheney's in the oil business. After running Shell's American division, he had been recruited to turn around troubled Fluor Corporation, a government contractor and Halliburton competitor. At the time of the Iraq invasion, Carroll held a top secret security clearance from his Fluor days. He was a close social friend in Houston of George H. W. and Barbara Bush, the American president's parents. Carroll had reluctantly accepted a six-month Baghdad assignment at Rumsfeld's request; he considered it a call to national service that he could not refuse, although it would cost him about a half million dollars in foregone private sector compensation and require him to live in spartan conditions in Iraq's Green Zone.

Carroll waged a rearguard battle in Baghdad against the Bush administration's more radical privatization advocates; he made clear that he would resign rather than participate in a precipitous sell-off of Iraq's oil assets, according to one career intelligence analyst who worked with him at the time. Carroll, too, was a believer in free markets, but he knew the Middle East and he felt that the United States had no choice but to go slowly and defer to Iraqi decision making. Iraq's nationalism, coupled with the visible trends toward state ownership in the global oil industry, suggested that postwar Iraq should probably reestablish its state-owned oil company, at least as a first step. Carroll's personal view was that if he were running Iraq—"And believe me, I never want to do that"—he would build up a strong nationalized oil company and then later invite private international oil companies to invest as partners in Iraq's fields. That "mixed model" would free up Iraq's national revenue for "crying needs" in education, health care, and other social sectors. "If you bring in Exxon, with a very fat checkbook, they could basically throw money at something and get things done very quickly," Carroll said. Iraq would probably have to give up some oil ownership in exchange for such investment capital, but it would also gain access to the latest industry technologies and training. In any event, Carroll felt that he should not impose his private opinion on the interim Iraqi administration. He wanted to help Thamir Ghadhban and other key Iraqis "at least begin to be thinking" about their options for reorganizing their country's oil industry.[13]

He advocated approaches that might favor private oil companies in the longer run, however. On June 26, Carroll wrote a memo entitled "Future Policy Issues Concerning the Ministry of Oil," addressed to Paul Bremer. He noted that raising Iraqi production to its full capacity of about 6 million barrels per day might require as much as $30 billion or more in long-term capital investments. This raised the question of "when to invite new upstream discussions with prospective partners" from the oil industry. Carroll wrote that the next twelve months would not be "too early to start talks." He foresaw "an extended period to exchange concepts and to establish relationships."

Under the heading "Privatization in the Oil Sector," he continued, "Needless to say, this will be a very contentious issue within the Ministry, the government, and the population at large." He argued against a precipitous sale of stock in Iraqi oil enterprises. A trust fund for Iraqis, "along the lines of the Alaskan model," whereby regular cash royalties were paid to citizens, would be preferable. He also suggested that it "would be in the U.S. interest" to develop an educational program for employees of Iraq's oil ministry—specifically, "comers" who could be trained in the United States. Such a program "would not only meet the ministry's needs but [would] begin to build a group of future leaders who would have a taste of U.S. life."[14]

Carroll's memo accurately forecasted the Bush administration's oil policy in postwar Iraq. The policy was constructed to protect Iraqi decision-making prerogatives but also to account for enduring American interests, including those of U.S.-headquartered oil corporations. As Iraq's anti-American insurgency intensified after 2004, the Iraqi state's capacity to manage its own affairs gradually weakened. Iraq's potential to oversee oil production on its own, which would have been challenging in any circumstance, gradually became all but impossible. International capital and private oil companies would be required to rescue Iraq's position whenever the internal violence generated by the American invasion calmed. As Tariq Shafiq had observed about the First World War, access by Western companies to Iraq's oil did not need to be an explicit cause of the Bush administration's invasion to become an outcome.

D ouglas Feith had been thinking about global oil security issues since the first term of the Reagan administration, when he worked on energy policy as a young staffer at the National Security Council. He considered himself something of a contrarian on the subject. After the oil shocks and embargoes of the 1970s, it was common in Washington to think of "energy security" as a problem in which newly powerful Arab exporters could wield the "oil weapon" over vulnerable Western importers.

This was the political science model of oil security, as Feith put it. Oil supplies lay scattered around the globe, as on the board of a Risk game, and governments competed for advantage and control. The United States military developed contingent plans to seize oil fields in Saudi Arabia in an emergency, particularly if the Soviet Union moved against them. Hostile governments might squeeze the United States by withholding its oil, as the Saudis and other Arab producers had done over American policy toward Israel. A logical response to this embargo threat was the one taken by President Gerald Ford in late 1975, when he signed into law a bill that authorized, among other things, the construction of the strategic petroleum reserve, where the United States could store volumes of oil equivalent to several months of imports, to be released in the event of an embargo or supply disruption. The S.P.R., as it was known, provided the United States with a countermove against hostile oil producers in any contest of physical access. The reserve's existence presumably deterred oil-producing enemies from imposing embargoes. If one was imposed anyway, the S.P.R. provided Washington with time to pursue military or other interventions against the aggressor.

Feith concluded that this way of conceptualizing oil security was misguided. He was a young free-market thinker and he noted that neither economists nor oil industry executives saw the global oil market the way political scientists or naval blockade strategists did. In the economists' view, oil was a commodity just like any other commodity. As was true for cocoa or coffee, there was a single global market for oil. There were gradations of price for different types of oil quality, but fundamentally, oil's global price went up or down on the basis of worldwide supply and demand. The best way to visualize the market was to think of a global bathtub or pool of oil with spigots pouring into it from many different exporting countries and customers piping out supplies where they required them.[15]

Traders and speculators set the price of a new barrel of Iraqi or Russian or Chadian crude in Rotterdam's spot market or on futures exchanges in London, New York, and Chicago. In such a system no single oil producer could disrupt supplies or control prices very effectively for long, Feith believed. A major producer like Saudi Arabia or the O.P.E.C. cartel could attempt to withhold supplies and by doing so prop up prices temporarily, or try to punish a particular importer through a targeted embargo, but the forces of economic gravity in the pooled global market were likely to prevail over time. When global prices rose, as they did during the 1970s, the incentives to invest in new oil production also increased, and so new supply came on line, which in turn reduced global prices again—exactly as occurred during the 1980s, when Feith worked at Reagan's N.S.C. He noted that the collapse in the price of oil during the 1980s was precisely the opposite of what many political analysts inside the United States government had predicted. President Jimmy Carter's 1977 National Energy Program presumed that oil would become "very scarce and very expensive in the 1980s." In 1979, the Central Intelligence Agency forecasted, "The world can no longer count on increases in oil production to meet its energy needs."[16] Why were they wrong? Feith thought he knew the answer: The history of commodities was one of prices going up, supplies increasing, and prices falling back down again. Oil, fundamentally, was no different. The timelines required to bring new supplies on line were longer than, say, the planting of corn crops, but the underlying economic pattern was the same.

One implication of this analysis was that the geographical origins of a particular barrel of oil did not matter very much. Except for the issues of a particular barrel's quality—that is, the ease with which it could be refined—and transportation costs, global oil traders did not care whether a new barrel poured from a spigot in the Middle East, Latin America, Australia, or Africa. As it happened, half or more of global oil reserves lay in the Middle East, so that region's political stability and transport lanes would always command attention. However, in an economic sense, the Middle East's barrels were the same as all others.

Feith's views were reinforced by what happened in international oil markets after Islamic radicals seized power in Iran in 1979. Carter imposed a boycott on Iranian imports, but Iranian oil just found its way to European and other international traders. Those traders often resold Iranian oil on the spot markets. Other producers who had previously sold to Europeans now sold to American companies. Global supplies and prices proved to be resilient. Companies, not governments, decided where particular batches of oil were shipped for refining, on the basis of price, transport, and technical factors.

Within the Bush administration, by the time of the Iraq invasion, this free-market vision of a single, liquid global oil market had taken hold as a kind of quiet conventional wisdom—it was seen by some within the administration as a sophisticated basis for thinking about American oil security, as opposed to the misguided Risk board model of the 1970s. Rumsfeld and Stephen Hadley, the deputy national security adviser, as well as many of the economists who advised President Bush at the National Security Council and the National Economic Council, all had independently come to similar conclusions as Feith. Their consensus had clear implications for the Bush administration's energy policy: If oil constituted a unified worldwide market, and if the United States would be an importer in that market for an indefinite time, then it was in the interest of the United States to promote policies—free trade, open markets, low taxes, maximized oil production everywhere—that would fill the global pool with as much new oil as possible, and thus keep global oil prices low, to the benefit of the American economy. The alternative policy—the pursuit of "energy independence" by one means or another—was unnecessary, too expensive, and unrealistic. This was precisely what Lee Raymond believed, and what he had reiterated to Vice President Dick Cheney when they met in Washington soon after Bush took office. Cheney understood and agreed. Besides Cheney, Raymond told his colleagues at ExxonMobil, he had met only one other world leader who truly understood how global liquid oil markets worked, and what this implied for foreign policy: British prime minister Tony Blair. Blair joked that it was good that most politicians did not understand the oil markets because if they did, "they'll think they can do something about it."

Although Bush's national security team generally accepted Feith's vision of global oil, some of them offered partial dissents. Condoleezza Rice and Colin Powell were each struck by the destructive role of oil wealth in the development of African political economies; they credited aspects of the resource curse thesis. Paul Wolfowitz, the deputy defense secretary, also accepted the resource curse thesis and he expressed interest in the thinking of those such as former director of Central Intelligence James Woolsey and the journalist Thomas Friedman, who argued after the September 11 attacks that, even setting aside the challenge of climate change, America was arming its enemies by failing to wean itself from oil because oil exporters used their easy cash to challenge American interests.[17]

Oil security, it turned out, like other forms of economic security, lay in the eye of the beholder. One of the problems with Feith's arguments about seamless global oil pools forever replenishing themselves was that it required other world powers to act as if they shared his understanding.

China, crucially, did not; its leaders remained steeped in the political science model of oil power. Douglas Feith was a man of exceptional belief in himself, however. He was quite certain that his views of markets, history, and global oil security were correct. Feith's responsibilities included cochairing ongoing bilateral defense talks with Chinese counterparts. Soon after the invasion of Iraq, he decided to try to talk the Chinese government into changing its understanding of the character of the global oil market to conform to his own.

China became a net oil importer in 1993. That followed a much-hyped but failed search, in which Exxon had participated, for oil reserves in China's northwest Tarim Basin. "It was going to be the new Saudi Arabia and all that kind of thing," an executive involved recalled. What they found, however, were "basically dry holes."

By 2003, China had grown into the world's second-largest oil consumer, after the United States, and its oil imports were skyrocketing. Around 1999, China's Communist leadership coined a "Go Out" policy to encourage state-owned companies and diplomats to prowl the world for oil supplies that China could secure by long-term contract. Go out they did. Trade between China and Africa doubled between 2002 and 2003 to $18.5 billion; most of that increase described Chinese oil imports. Within a few years, China would invest $44 billion worldwide in oil projects, half in Africa.[18] Its methods struck American intelligence analysts as almost neocolonial—the Chinese government seemed to place a premium on physically owning oil supplies, in the belief that ownership would promote the country's long-term national security.

Stephen Hadley asked the Africa division of the Central Intelligence Agency for an assessment of China's oil deals in Africa. Were they a threat to U.S. national interests? The division's view was that "we didn't actually see them as that much of a threat," recalled an official involved in the review, "just an economic challenge." The C.I.A.'s analysts worried that China could displace U.S.-based oil companies from lucrative production deals in some African countries, but that was about the extent of their concern.[19]

David Gordon served as the Bush administration's chief national intelligence officer for economics at the time and participated in White House–led policy reviews. In the summer of 2001, Gordon had spent a month in China, steeping himself in the issues emerging from the country's rapid economic growth, "under the assumption that China was going to be the big economic intelligence story."

He was struck by China's "mercantilist approach to energy." Gordon subscribed, essentially, to Feith's view that the liquid, integrated nature of the global oil market meant there was no particular advantage for a country to "own" overseas supplies if it was a net importer; it was more efficient, economically, to purchase supplies as needed, unless a remarkably attractive long-term price contract was on offer. Gordon once gave a talk at a Chinese think tank in which he argued that American and Chinese energy interests "basically coincided." The two countries shared a need to have diverse global supplies, political security in the Middle East, and security on the open seas. The Chinese "sort of took it all down," but he could see that they did not really think that way.[20]

At the Pentagon, Feith found himself drawn into the Chinese conundrum. The Bush administration sought to mount pressure on Sudan's president, Omar Bashir, whose militias were responsible for a humanitarian crisis in Darfur, a separatist-minded province. Sudan financed itself with oil production and sent more than half of its oil output to China. The Bush administration wanted to persuade China to pull back from its Sudan contracts. Feith concluded that talking about a mercantile versus a free-market model of global oil in his bilateral military channel might help.

He commissioned a free-market economist, Benjamin Zycher, to create a presentation entitled "Historical Lessons from the World Oil Market" for a visiting Chinese delegation. Essentially, it was a nineteen-slide PowerPoint presentation summarizing what Feith believed he had learned from his days in oil policy research and in the Reagan administration.

The slides showed that U.S. government forecasts about oil's future availability had always been much too pessimistic. In 1980, the U.S. Energy Information Administration projected that there were only twenty-eight years' worth of proved oil reserves in the world remaining. Two decades later, the E.I.A. projected that there were still thirty-seven more years remaining. China did not need to lock up supplies with rogue countries like Sudan, damaging China's global reputation, Feith told the defense delegation, because there would almost always be oil available. Moreover, China had no reason to fear a supply disruption carried out for political reasons. The "embargo threat is empty," one of Feith's slides declared, because any attempt to cut off oil to a particular country would be overtaken by "ordinary reselling" elsewhere in the market, just as the United States had experienced after the Iranian Revolution. As an additional source of reassurance, the United States had urged China to build its own strategic petroleum reserve, so the Chinese Communist government would have at least short-term supply security, and therefore even less reason to feel anxious about the theoretical possibility of a future embargo.

On a slide headlined "Current Chinese Activities in the World Oil Market," the presentation noted, "It is clear that China views dependence on foreign oil unfavorably." Past mistakes by the United States, after the 1970s, however, "offer lessons for China today." The principal lesson, according to Feith, was this: "Dependence on foreign oil does not create vulnerability."[21]

His visitors from Beijing were not persuaded; China did not break its oil ties with Sudan. To the contrary, it was clear that China's leadership *did* believe that its dependence on foreign oil created strategic vulnerability, because oil might be a weapon in prospective competition with the United States during the twenty-first century.

The American-led invasion of Iraq, which Feith had also helped to author, hardened the fears of Beijing's Risk players. "They were very concerned when we went into Iraq," said Aaron Friedberg, a China scholar at Princeton University who served from 2003 to 2005 as a foreign policy adviser to Vice President Cheney. Right-wing Chinese, outside the government, said of the United States as the invasion unfolded, "They're putting their hands on the windpipe. They're occupying a major oil-producing country, solidifying their grip on our supply lines." Friedberg did not think the predominant view inside the Chinese politburo was quite so alarmist, but he and other Bush administration analysts could see that some patterns of China's overseas oil purchasing reflected its leaders' anxiety about the vulnerability of oil transport routes in the future—particularly on the seas.[22]

The United States Navy ruled the world's oceans and kept the seas open for all commerce, to support free trade. As a rapidly rising global power, however, did China really want to build an industrial economy dependent on oil supplies shipped from abroad that were vulnerable to interdiction by the U.S. Navy? In the event of a confrontation with the United States over Taiwan, for example, might not the United States use its naval superiority as a lever, threatening to cripple China's economy by blockading oil supplies? China could construct its own blue-water navy to challenge the United States, but that would take many years, entail great expense, and risk a draining competition with Washington. Some American analysts during the first Bush term asked why China could not just content itself to "free ride" on "the fact that the U.S. Navy is the only game in town," as Friedberg put it; that is, China could enjoy the economic benefits of secure ocean transport and allow American taxpayers to bear the price. But it was also obvious why this would not necessarily be appealing to the Chinese, looking to their future rise: "Think how we would feel if the situation were reversed."[23]

One alternative for the Chinese leadership was to maximize its access to oil and gas supplies that could be transported by land. This insight seemed to explain a thrust of Chinese foreign policy after 2000. There was some thinking among Chinese scholars and strategists that land-based empires seemed to last longer than those that were dependent upon the seas. The potential for land-only routes attracted Chinese strategists to Russian oil supplies—"They would like to stick a straw in Russia," was the way Friedberg put it—as well as to neighboring Southeast Asia, where China sponsored overland pipeline construction into Burma and Thailand.

At least some strategists in the Bush administration did think of China's dependency on seaborne oil imports as a source of potential vulnerability in the twenty-first century—just as right-wing Chinese analysts feared. In war game scenario planning involving flashpoints such as Taiwan, U.S. military planners did not always think a coercive oil blockade against China would be wise or necessary, "but they were content to see the Chinese anxious about it, because it might act as a deterrent," said one Bush administration official.[24]

Vice President Cheney seemed particularly interested in China's vulnerability to U.S. naval power. His experience of the global oil market while running Halliburton had left him with a deep understanding of oil's fungible nature. But his thinking about national security was influenced, too, by historical narratives about the rise and fall of great powers, and particularly the history of control of the seas, which had been critical to Great Britain and the United States, in succession, as a means to ensure the physical supply of commodities necessary for industrialization, including oil. Cheney read and admired *The Tragedy of Great Power Politics*, the 2001 book by the University of Chicago's John Mearsheimer. Mearsheimer predicted that the competition over security among world powers—

which had produced, during the twentieth century, the nine million dead in World War I, the fifty million dead in World War II, and the chronic violence of the cold war's proxy wars—would extend into the twenty-first century "because the great powers that shape the international system fear each other and compete for power as a result." The book's final chapter, entitled "Great Power Politics in the Twenty-first Century," reviewed the prospects for military and economic competition between the United States and China, including the potential of the U.S. Navy to squeeze China's oil supplies by controlling the straits around the Persian Gulf. In all, Mearsheimer's book was deeply pessimistic—a far cry from the free-market optimism of Douglas Feith. Cheney told colleagues that he liked the book until he reached the last chapter, where he thought Mearsheimer was a little too softheaded and hopeful that the Great Power struggles that inevitably lay ahead might be contained and managed so that they produced limited disruptions and damage.[25]

ExxonMobil ran its own war game scenarios about oil supply disruptions. The company's political risk analysts asked themselves what would happen in the most extreme case imaginable—for example, if Iran's 2.3 million barrels per day of oil exports were removed from world markets because of a war, while turmoil in Venezuela simultaneously removed another million barrels per day. The shock of losing 3.3 million barrels a day of exports would certainly create price spikes, and soaring oil prices could produce economic turmoil in the United States and Europe. Physical delivery of oil, however, did not look like a catastrophic problem, the planners concluded, at least not in a crisis that lasted less than six months. The strategic petroleum reserve would help cushion any disruptions. After a certain number of months, if the reserve ran low, governments might be forced to ration. But the extremity and unlikelihood of the imagined events needed to produce such a physical supply problem suggested to ExxonMobil's risk analysts that there was "an element of surge capacity in the global system," as the corporation's Rex Tillerson put it. The company's war gaming implied that the global oil markets had become more resilient and flexible than some analysts assumed. At the same time, the threat of disruption to physical supply from rogue states "is not very well understood, in terms of what would really happen," Tillerson believed.[26]

"The central reality is this: The global free market for energy provides the most effective means of achieving U.S. energy security," Tillerson said. "In the global market, the nationality of the resource is of little relevance. . . . Energy made in America is not as important as energy simply made wherever it is most economic." Punishing sanctions and uneconomic supply lockups such as those sometimes pursued by China did undermine American security, but only because the United States should be "enlarging this global energy pool, not dividing it."[27]

The economic interests of ExxonMobil and other international oil companies lay squarely in the realm of Douglas Feith's idealized vision of oil globalization, not in the realization of Mearsheimer's pessimism. ExxonMobil's oil and gas holdings were so widely scattered around the world that the corporation and its shareholders were at least as vulnerable to transport and production disruptions as China's government.

This understanding of risk shaped some of ExxonMobil's lobbying on foreign policy issues in Washington. The corporation promoted free-trade philosophies at every turn and opposed economic sanctions against oil producers, except in the most egregious cases; until September 11, ExxonMobil lobbyists in Congress had opposed oil sanctions imposed on Libya, Iran, and Syria. The corporation's economic self-interest on the sanctions issue was obvious, but the policy arguments mounted on PowerPoint slides by its in-house advocates were broad: The creation of a free-flowing global oil pool, one shaped to the greatest possible degree by market incentives, would promote American national security, ExxonMobil's representatives insisted.

A re people in your industry salivating over the possibility of gaining access to Iraqi oil?" an interviewer asked Lee Raymond as the war deteriorated.

"Oh, I don't think salivating, no."

"How would you describe it?"

"Well, I think there's a lot of caution in our industry about it. . . . I think everybody is interested, but you need to have security. You need to have a tax structure. You need to have a legal structure. . . . And there's no confidence that that's going to be there for some time. So I would describe it that we're all interested, because we know the resources are there. Whether or not there's the framework to develop those resources in an economic fashion—the jury is out on that one."[28]

There was more concern within ExxonMobil about the corporation's ability to keep replacing its reserves than Raymond let on. The numbers the corporation was reporting to Wall Street at this time were not impressive: At best, ExxonMobil seemed to be churning in place, finding only enough new oil each year to replace that which it had pumped and sold. A former manager recalled a 2004 meeting at which the message was: "We can't get enough new assets to replace our reserves." As a result, geologists and scientists throughout the upstream division took a fresh 360-degree review of reserve prospects worldwide. They revisited old assumptions. "They looked everywhere," the manager recalled.

Iraq offered a potential breakthrough in ExxonMobil's access to equity reserves, but Raymond counseled patience. ExxonMobil had owned oil in Iran and Iraq for decades during the twentieth century. It did not own oil in those countries in 2003, but twenty years into the future, as Tillerson put it, "we'll have another set of circumstances in some of those countries."[29] One attribute of a nation-state ExxonMobil lacked was an army or a navy of its own. China could "free ride" on the United States Navy's control of the open seas; Lee Raymond had no choice but to free ride on the Bush administration's efforts to subdue Iraq.

Raymond had been a friend of Vice President Cheney's for more than a decade. He was not immune, however, to the disillusionment that set in among many conservative Republicans as the Iraq War deteriorated and the Bush administration's overreach and incompetence in the conflict became increasingly exposed. As a global business leader whose corporate profits depended on international stability, Raymond identified more with Republican realists such as George H. W. Bush or his former national security adviser Brent Scowcroft than with the more idealistic activists and democracy promoters around Bush's son. ExxonMobil and its generations of home-bred executives felt they had learned long ago to deal with the world as it was, to bargain as needed with dictators and authoritarian emirs and revolutionary leaders. The transformational, Wilsonian streak in Bush's democracy promotion in the Middle East after September 11 increasingly discomfited Raymond. The ExxonMobil chairman still trusted Cheney and saw him frequently, and he admired greatly Bush's second Energy secretary, Samuel Bodman, a former oil industry executive. Throughout his cultivation of the Bush administration, however, Raymond purposefully kept ExxonMobil at arm's length from the administration's attempts to remake post–Saddam Hussein Iraq. It was not in ExxonMobil's interests to become tainted by failed nation-building projects in a country that held one of the world's largest unproduced oil and gas resource bases. American neoimperial ambition in Iraq might fail, but ExxonMobil's private empire had its own enduring interests, and these should not be rushed.

Midlevel State Department officials continually summoned ExxonMobil executives to meetings about the planned revitalization of Iraq's oil sector after mid-2003 and urged the corporation to open a Baghdad office. The Bush administration officials who ran these meetings seemed to believe that only security concerns stood in the way of ExxonMobil's making an immediate big push for Iraq oil contracts; the State officials tried to assure the corporation that security in Iraq would soon improve, even when the daily newspaper headlines suggested otherwise.

"Nobody at ExxonMobil wants to get killed for an oil well," Raymond's representatives explained at these meetings, "but our greater concern is political risk." Would there be an agreement guaranteeing the long-term presence of American troops in Iraq? What oil laws would a legitimate Iraqi government approve? Any investment ExxonMobil made in Iraqi fields would require multidecade commitments. Was Lee Raymond interested in Iraqi oil? Of course: Every corporation in the global oil industry was interested. But it was obvious that the Bush administration lacked the capacity to create conditions for ExxonMobil or any other serious international player to make a politically secure, economically rewarding deal in Iraq anytime soon.

Raymond had, in the meantime, joined the Bush administration as a partner in an oil play that might, in a single stroke, resolve Exxon-Mobil's reserve replacement conundrum for years to come. Of the four nation-states with oil reserves sizable enough to transform ExxonMobil's position—Saudi Arabia, Iraq, Iran, and Russia—one looked more plausible than the others. Quietly, without anything like the fanfare that surrounded Iraq, Raymond and his allies in Bush's first-term cabinet had developed an opportunity on a grand scale: in hopeful partnership with Vladimir Putin.

Twelve

"How High Can We Fly?"

On Wednesday evening, November 14, 2001, Vladimir Putin and his wife joined George W. Bush; the president's wife, Laura; and about two dozen friends and cabinet officials for dinner at Bush's ranch in Crawford, Texas. The guests wore jeans and boots; the menu included southern fried catfish. Putin stood to deliver a toast. Referring to his host, he said that the United States "was fortunate at this critical time in its history to have a man of such character at its helm." Bush returned the compliments. In June, when he had met Putin for the first time, in Slovenia, Bush said that he had "looked the man in the eye" and had discerned "a sense of his soul." Bush had been mocked for this claim, but in Crawford he again mentioned his respect for Putin's soul, and he told the Russian president that he sought to "transform the relationship between our two countries."[1]

After a dessert of pecan pie and ice cream, members of the group drifted outside to warm themselves around a fire pit. Putin sought out Don Evans, the president's close friend, who had served as the chairman of Bush's 2000 presidential campaign and was now his secretary of commerce. Evans had run an oil and gas company prior to joining the Bush administration. He told Putin that he appreciated the toast he had delivered about his friend the president. Putin raised the subject of oil and gas production. Evans figured that the Russian president, a former K.G.B. career officer, had probably read an intelligence file describing Evans's background. He welcomed the conversation, nonetheless; the new Commerce secretary was drawn to the prospect of a renewed energy partnership between Moscow and Washington. He and Putin talked about the possibilities.

"How has America accomplished so much in only two hundred years?" Putin asked him at one point.

Evans spoke about American freedoms, economic competition, innovation, and the rule of law. "People in America are good people," he said. "They wake up every morning trying to do the right thing. I don't care what kind of system you have, if the people aren't good, decent people, it won't work." But even moral citizens need a system of law and governance so that they will be treated fairly and, eventually, perhaps find financial reward, Evans continued. Putin had the opportunity to build such a system for Russia, as a legacy of his leadership, Evans and Bush believed. The United States could help by delivering technology, capital, and experience of the rule of law in international business.

In the oil industry, in particular, Russian requirements and American capabilities overlapped. Oil and gas production made up a large share of the Russian economy, so free-market reforms in the oil sector might provide a transformational catalyst for the entire country. Russia possessed at least 60 billion barrels of proved oil reserves, according to *Oil & Gas Journal*, and 1,680 trillion cubic feet of natural gas reserves, a quarter of the world's total and more than any other nation. For American oil companies, ownership of even a modest percentage of that bounty could be transformational.[2]

The fireside chat between Putin and Evans that night marked the start of an oil romance between the Russian and American governments, one that would soon draw in Lee Raymond and ExxonMobil. From the beginning it was an engagement marked by exceptional optimism on the American side, shadowed by a long history of mistrust rooted in the cold war—"old think," as Bush's ambassador to Russia, Alexander "Sandy"

Vershbow, put it in a cable to Washington.[3] The hypothesis Evans and others in the Bush administration pursued after the Crawford dinner was that a strategic campaign to deepen commercial ties between oil companies in the United States and Russia might transform Russia's internal politics, remake U.S.-Russian relations, and even alter the global geopolitics of oil. President George H. W. Bush had seen oil-for-friendship as a critical element of his campaign to build a partnership with Mikhail Gorbachev as the Soviet Union cracked up; Bush persuaded Gorbachev to endorse Chevron's pioneering, lucrative entrance into the Tengiz oil project in the then Soviet Republic of Kazakhstan. George W. Bush intended something similar with Putin, whose intentions as a political leader and as a sponsor of market-led modernization were at best enigmatic.

Bush did not see Russia only through the prism of oil, but he did regard energy policy as critical to his ambitions for his relationship with Putin. Bush's national security advisers, led by Condoleezza Rice, an academic specialist on Russia, and her deputy, Stephen Hadley, sought to construct a more cooperative relationship with Moscow, one that might coax Putin's government toward an embrace of democratic capitalism and diplomatic normalcy within Europe. They knew there inevitably would be tensions and disagreements that would drag them back toward rivalry. To prevent or at least limit that recurrence of old think, Bush and his advisers concluded that they needed to find three or four shared projects that would invest Putin and the Kremlin in a broad, pragmatic, ongoing pattern of U.S.-Russian cooperation. Space programs, and particularly the management of the International Space Station, offered one such pathway. Counterterrorism and preventing nuclear proliferation after the September 11 attacks also appeared to be promising. But the project with the greatest potential—the one that might literally invest the United States in Russia's success, and vice versa—involved Russia's oil and gas production. "I think all of us at the senior level thought this was a real moment to seize, from the president to Don Evans at Commerce to me to the vice president as well," recalled Spencer Abraham, then Bush's Energy secretary.[4]

Russia not only held the world's second-largest combined oil and gas reserves, after Saudi Arabia, it was rising again as an oil producer and exporter. The political chaos and economic freefall that followed the collapse of the Soviet Union had caused Russia's oil production to fall from a high of more than 10 million barrels per day to about 6 million in 1996, about half of which was exported.[5] As Putin took office on the last day of 1999, the industry had started to recover. Domestic consumption remained steady at about 3 million barrels per day, while exports were rising toward 7 million barrels per day and headed higher year by year. Most of the exports went to Europe, particularly to major economies such as Germany and the Netherlands. Yet Russian oil companies required new investment to reach their full production potential.

The question was whether the Bush administration's push for a strategic oil partnership with Moscow after the Crawford dinner was at all realistic. Energy policy specialists in the Clinton administration hadn't thought so; they had been working on the same issues of energy cooperation with Russia throughout the late 1990s and had concluded, as a former White House official put it, that "there was no discernible forward progress" toward free-market reform in the Russian oil sector under Putin, and "the pendulum was swinging markedly" against Western oil corporations that sought to own Russian reserves. If anything, from 1998 onward, there had been a "very, very clear trend in Russian policymaking toward the U.S. . . . of growing suspicion." The prize was so alluring that all of the major Western oil corporations persisted in Moscow, nonetheless. For all of Russia's traps—unpredictable politics, Mafia-like cliques in business and government who used murder and vendettas as negotiating tactics, prosecutors and judges subservient to shadowy powers, and a general absence of the rule of law—there was no other country on Earth with so much oil that offered even the pretense of a capitalist-friendly opening. The numbers—the challenge of annual reserve replacement, the scale of Russia's holdings—argued for risk taking.

"How high can we fly?" Ambassador Vershbow, a career foreign service officer who had long studied Russia, asked in a cable to Washington a few months after the Crawford dinner. Very high, he wrote. In developing a new strategic oil supply partnership in Moscow, the United States "will need to be sensitive to the reaction of other energy-producing states, both in the region (i.e., the Caspian) and outside it (notably Saudi Arabia)." Nonetheless, there was potential to change global economics and politics. Strategic energy cooperation between Washington and Moscow, he wrote, "is a deal. . . . Russia improves along market lines and continues to divorce politics from commerce, and we help it get what it wants— markets, investment, and stature. But the deal can be broader. If successfully developed, it can serve as a paradigm as well as a driver for cooperation and market-driven behavior in a variety of sectors, a trend that will only help Russia integrate into the world economy."[6]

The United States owned no oil companies, however. To forge the deal Vershbow had in mind, the Bush administration would need to encourage America's largest private oil firms, particularly ExxonMobil and Chevron, to cooperate.

Like a section of the American public, some in Putin's cabinet seemed to believe that the Bush White House made all the decisions for America's largest multinational oil companies, or vice versa. Putin himself seemed to understand how the American system worked, but some of his advisers projected what they knew about the Russian government's influence over its own oil and gas companies onto Washington, assuming it was the same there. German Gref, the Russian minister of economics, with whom Don Evans worked closely, would occasionally make points during their energy talks suggesting that he thought that Evans or Bush had heavy influence at ExxonMobil.

"Exxon writes their own checks and they are accountable to their shareholders—they're not accountable to me," Evans told Gref. "I don't write their checks for them."

Evans tried to explain that no matter what sort of agreement Washington and Moscow might forge about energy policy, no American oil company was likely to risk large amounts of investment capital if it feared that Russia's political and tax environment was unreliable. Evans quoted an aphorism often repeated by Bush's Treasury secretary, Paul O'Neill:

"Capital is a coward, and capital is not going to go any place that is un-friendly." His argument was undermined by the fact that all of the major oil companies—ExxonMobil, Chevron, Shell, BP, Total, and Statoil—had already risked capital in Russia, even though its political economy was dangerous and highly unsettled. But it was Evans's job to make the case for Russian legal, tax, and policy reforms. On specific transactions, he encouraged the Russian government to deal directly with the American corporations, particularly ExxonMobil and Chevron, and he always spoke highly of Lee Raymond.[7]

ExxonMobil ran an office in Moscow, headed by a former Australian diplomat, Glenn Waller. Rosemarie Forsythe, the former Clinton admin-istration National Security Council analyst, also advised Raymond during ExxonMobil's talks in Russia after 2001. The corporation, as ever, believed it was better off handling things in Moscow on its own. "Exxon operates independently," recalled Leonard Coburn, who worked at the Department of Energy on the Bush administration's Russian oil initiative. "They think they can do better than the [U.S.] government and often they do." Oc-casionally, Lee Raymond might turn to Vice President Cheney or another very senior official to give the Russians "a good kick in the pants," as Co-burn put it, about a specific tax or policy stalemate in Moscow. Otherwise, the corporation negotiated in private.[8]

ExxonMobil owned one significant oil interest in Russia at the time the Bush administration made its push to deepen cooperation. In the chaotic last months of the Soviet Union, investment bankers retained by elements of the dying regime presided over by Mikhail Gorbachev shopped around all sorts of natural resource deals; Lee Raymond had authorized bids on an exotic project in Russia's far eastern territory, near Sakhalin Island, just above Japan. The project dated to the 1970s, when Japanese corporations had lent money to the Soviet Union in exchange for exploration rights. But the deal went nowhere until the Soviet Union began to crack up and finally split apart. In the early 1990s, Raymond had called an old hand named Terry Koonce, who was running Exxon's oil operations in Calgary, Canada, and asked him to go to Moscow to "start to open the way in Russia."

"Man, I thought you sent me to Siberia when I went to Calgary," Koonce said.

"No, I was just getting you warm."

It was not until 1996 that ExxonMobil closed on terms for what became known as the Sakhalin-1 project. It was an undertaking that would test the corporation's engineering prowess like no other. Hurricane winds swept the Sakhalin region each autumn, and ice packs up to six feet thick built up during the winter. After the spring, melting ice floes threatened to knock down any offshore oil rig in their way. Exxon decided on a plan to drill a seven-mile horizontal well from mainland Russia underneath the ocean waters. It was the only well of its kind in the world. Sakhalin-1 suggested some of the appeal of Western oil technology and engineering skill to Russia. But to make the deal on terms acceptable to Exxon, Russia's government had had to set aside its nationalism and share oil ownership.

The Sakhalin-1 deal had been made at the nadir of the Boris Yeltsin era. Post-Communist enthusiasm about democracy and private enterprise had yielded to disillusionment and backlash as bankers and other new Russian oligarchs seized control of former Soviet state property in corrupt transactions. During this period ExxonMobil was one of three Western oil corporations to enter into production-sharing agreements, or P.S.A.s—contracts that allowed for direct oil ownership by the foreign firms and also sought to segregate arbitration rights in the deal from the morass of Russian law.

Rex Tillerson had followed Koonce as the lead ExxonMobil executive on Russia. He managed the Russia account from a base in Texas, flying over as necessary. Tillerson formed a friendship with the Sakhalin governor and decided to rope in a state-owned Russian oil firm as a project partner, so that ExxonMobil and the Russian government would be "on the same side of the table," as Tillerson put it later, if disputes over the project arose. ExxonMobil had connections at Rosneft, one of the smaller state-owned oil and gas companies, with about 10 percent of the country's reserves, a company widely regarded as a bureaucratic mess even by Russia's standards. Around 1995, Lee Raymond had met with Rosneft executives to talk about a possible acquisition of the firm. The Russian company's leaders had said they were willing to merge into Exxon, and "begged and pleaded" to be acquired, as an executive involved recalled it, but Raymond declined because even Rosneft's leaders seemed unsure about what their company legally owned. Rosneft's participation in the Sakhalin project provided some protection, and the legal and arbitration protections in the P.S.A. helped too, but the deal remained politically vulnerable in Moscow. Delays, arguments, and disputes over environmental issues, pipeline routes, and other subjects stalked Sakhalin-1 from its inception.

Under pressure, Tillerson applied the Exxon formula: no surrender. "We jacked this all the way to the top," recalled one of his colleagues. "We brought the issue up with the president [Putin] and we said, 'Look, we have got the contract signed, we are doing everything we are supposed to do—here are the rules. And these guys don't want to follow the rules. What are you going to do about it?'"

Putin offered to write out an executive order saying that Sakhalin-1 could proceed, but Tillerson refused. Putin did not have enough legal authority to satisfy ExxonMobil; Tillerson said he did not want to operate by decree, but by durable laws. Tillerson wanted to have "all the t's crossed and i's dotted exactly according to Russian law and regulation, and if we couldn't get it done, then we were not going to do it," the former executive remembered. Ultimately, after Putin "blew his stack" at ExxonMobil's affront, the Russian president agreed.[9]

In 2001, India's largest state-owned oil company bought into the Sakhalin deal in a transaction approved by Putin but which required ExxonMobil's acquiescence. ExxonMobil took its time reviewing the issue. At a White House meeting with President Bush, India's prime minister, Manmohan Singh, asked Bush to intervene. "Why don't you just tell them what to do?" Singh asked.

"Nobody tells those guys what to do," Bush answered.

This was the state of ExxonMobil's relations with Putin's regime when the Bush administration launched its ambitious energy diplomacy: Sakhalin had been successfully launched, but there had been a tough struggle over terms. Russia's negotiating culture of bluff and coercion seemed to suit ExxonMobil. The corporation and the country where it sought to deepen its investments had similar personality traits. ExxonMobil's executives convinced themselves that their earlier firmness had brought Putin around on Sakhalin. This self-affirming perception would shape Lee Raymond's attitude in the next round of negotiations, after 2001, when the stakes would be higher.

In early April 2002, Rex Tillerson arrived at the ExxonMobil terminal at Love Field in Dallas to board a jet to Moscow. Tillerson had turned fifty less than three weeks before. He had risen to the cusp of the top job at ExxonMobil partly on the strength of his work in Russia. The Bush administration's oil initiative placed ExxonMobil's business dealings in Russia in a new light. There was now the potential, if all went well, for ExxonMobil to acquire substantial equity oil in Russian fields, enough to make a major contribution to the corporation's reserve replacement requirements. There was no other place on Earth where the corporation enjoyed such full and explicit partnership with the White House in the pursuit of such large oil holdings—the sort of partnership that ExxonMobil executives often claimed they neither wanted nor needed.

President Bush had scheduled a visit to Moscow for May 2002 to meet with Putin and follow up on energy diplomacy, among other subjects. The administration hoped at this Moscow summit to announce new agreements between ExxonMobil and Russian oil companies, as well as between Chevron and Russian firms. They would not be large deals, but the announcements would display momentum. Tillerson rehearsed for negotiations with Putin, role-playing with a colleague.[10]

On the eve of Bush's arrival in Moscow, Tillerson announced a $140 million contract with a Russian shipyard to upgrade one of ExxonMobil's Sakhalin production platforms. Don Evans attended the contract-signing ceremony in May and declared that direct U.S. investment in the Russian oil industry would soon be increasing. A joint statement issued by Putin and Bush called out the "successful advancement of the Sakhalin-1 project" and drew similar attention to a Chevron pipeline agreement. Putin and Bush welcomed the "implementation of more projects in the fuel and energy sector . . . on the basis of Production Sharing Agreements and other frameworks."[11] It was unusual for an American president to put his name behind a particular contract genre, but unlike in Iraq, the Bush administration had decided that American foreign policy would embrace and promote the direct ownership of Russian oil by U.S. corporations.

Ambassador Vershbow cabled to Washington that the ExxonMobil signing ceremony had highlighted "the importance and sanctity" of ExxonMobil's investments in Russia, as well as the "tangible benefits for the Russian economy of cooperation in energy. U.S. and Russian companies are negotiating a slew of new potential projects that could add more meat to the bones of this framework."[12]

How, realistically, could ExxonMobil acquire direct ownership of Russia's oil and gas reserves? The sell-off of Soviet assets during the 1990s had given birth to a new generation of private sector Russian businessmen—entrepreneurs, bankers, former security and intelligence officers, former apparatchiks, and opportunists of all stripes. Oil and gas plays were big draws: The dollars involved were enormous, and just a 1 percent "point" as a commission on a deal or as a discount on a contract to sell oil abroad could make a man wealthy very fast. These Russian oilmen were not generally ideologues who shared George W. Bush's vision of a new order in global oil, in which Russia would participate on free-market contract terms of the sort promoted by the International Monetary Fund. But they did see opportunity in the eagerness of Lee Raymond and his counterpart at BP, John Browne, to own a piece of Russian reserves. By 2002, one of these tycoons, Mikhail Khodorkovsky, was in a mood to deal.

Khodorkovsky was just thirty-nine years old when he opened discussions with ExxonMobil in 2002. The fortune he controlled was estimated to be about $8 billion. In the fashion of the day, he kept his short hair combed forward in a style suggestive of ancient Rome. He was a handsome man with brown eyes and youthful energy. It required boldness to seize and build a great fortune from the ruins of the Soviet economy; in that respect, Khodorkovsky was like a dozen other Russian billionaires who had emerged from the 1990s. He was more politically ambitious than some of his peers, and as ruthless in business as any of them, but more recently, he had begun to profess a newfound appreciation for shareholder rights and even international market ideals.

He had grown up in Moscow, served as a Communist youth leader, and had graduated from a technical institute in the mid-1980s, just as Mikhail Gorbachev consolidated power and introduced reforms. He and his college friends worked with computers, symbols of modernization, and they joined one of the experimental cooperatives permitted under Gorbachev's new economic policy. When Gorbachev allowed private banking for the first time, in 1987, Khodorkovsky and his partners—then in their midtwenties—formed a bank called Menatep. They accumulated rubles as communism collapsed. Radical reformers around Boris Yeltsin issued 150 million vouchers with which Russians could supposedly buy shares in former Soviet assets. Many people dumped the vouchers; Menatep bought them up. By 1995, Yeltsin's administration was running out of money, in part because bankers like Khodorkovsky were not paying their taxes. Menatep helped to mastermind a "loans for shares" scheme with Yeltsin. In exchange for keeping the Russian government solvent and assuring Yeltsin's reelection, Khodorkovsky and his partners paid rock-bottom prices for state oil and gas properties. They spent about $300 million for a recently assembled energy corporation called Yukos, whose underlying oil and gas reserves were worth about $5 billion. Khodorkovsky was not only opportunistic, he was ruthless; a few years later, during the Russian currency and banking crisis of 1998, he shifted assets around to protect his fortune, while leaving depositors and investors, including some of the world's most respected investment banks, with large losses.[13]

As Russia's economy recovered in 1999, Khodorkovsky adopted a new strategy. All of the billionaires who had made their fortunes as he had claimed legitimacy, but they remained vulnerable politically, in part because many Russians justifiably believed they had stolen their wealth. Vladimir Putin spoke about advancing free-market reforms and fashioning a deeper democracy as he assumed the presidency in late 1999, and he accommodated the business oligarchs, as they were known, in exchange for their agreement to keep out of politics. Yet Putin also represented a reviving elite of former K.G.B. officers, military officers, and Soviet-era bureaucrats who might seek to challenge the private sector billionaires for power and the control of state wealth.

According to Vladimir Milov, who was deputy energy minister early in Putin's administration, Khodorkovsky's conflicts with Putin dated to the 1990s' wild scrambles for state assets. "By the time Putin became president, the relations between his group . . . and the Menatep group really had been tense." They had struggled over oil privatization deals during the 1990s and had evolved into rival camps influenced by former K.G.B. officers maneuvering for profits and power: "Yukos was the company which was, I believe, the most infiltrated by former K.G.B. officers out of all Russian business structures," Milov said. That was a difficult matter to quantify, but from the time Putin became president, his own group of former security men "saw an absolutely real challenge from Menatep" rivals.

Khodorkovsky sought to protect himself after 1999 in part by transforming Yukos from an opaque amalgamation of dubiously acquired property into a transparent multinational corporation that adhered to international accounting standards. His full motivations and intentions were difficult to discern: He was a man in a hurry with a record of questionable business dealings; he seemed ambitious politically; and he increasingly flew to London and Washington and Paris, where he delivered speeches, built networks, and maneuvered for gain. He now positioned himself as the leading practitioner of "normalized" democratic capitalism in Russia. He moved to develop two highly ambitious projects simultaneously: a merger between Yukos and Sibneft, another post-Soviet oil giant, which would create one of Russia's largest oil companies, and a sale of part of Yukos to a Western oil corporation. That would diversify Khodorkovsky's political risk by aligning Yukos with the interests of America or Britain or France, depending on whether it was ExxonMobil or BP or Total that invested in Yukos. "He probably rushed too early for this," recalled Milov. "I think he also realized, because of his personal conflict with Putin, he was maybe running short of time." A deal involving a Western buyer would also allow Khodorkovsky and his founding Menatep partners to take cash out of the business and legitimately move it abroad.[14]

Khodorkovsky needed American and European specialists to help him with this corporate metamorphosis, to spruce up Yukos's books so that someone like Lee Raymond could audit them and not walk away shaking his head, as he had earlier done over Rosneft. Khodorkovsky recruited experienced executives from the major U.S.- and European-headquartered oil and oil service companies to fashion this financial makeover. He also transformed Yukos technically and operationally by recruiting engineers to raise production and profitability at the company's major oil fields. From Schlumberger, he hired Michel Soublin, a Frenchman, as an interim chief financial officer. He recruited Joe Mach, also from Schlumberger, to oversee oil production. Mach brought in Stephen Chesebro, who had been chief executive of Tenneco Energy, a U.S.-based, $4 billion company. When it came time to recruit a permanent C.F.O. Chesebro telephoned an old colleague, Bruce Misamore, then forty-nine, who had spent his career as a senior financial executive in the American oil business, including seventeen years at Marathon Oil.

"Look," Chesebro told Misamore, "I just got a call from a buddy of mine who is working for a large Russian oil company. They're looking for an American C.F.O. My wife told me not to call you, but I just thought, 'Is this something that you might be interested in?'"[15]

Misamore had semiretired; he was financially secure and comfortable, busy with his family, but he was relatively young. He decided to look into Yukos; it sounded like an adventure.

Misamore knew enough about Russia to know that you didn't want to get yourself in the wrong situation there, lest you end up raked by automatic weapons fire one morning. He read in depth about an earlier dispute that had taken place between Mikhail Khodorkovsky and the American investor Kenneth Dart. From his initial research, he could tell that Khodorkovsky was "somebody who had the bad boy image for a while." But he could detect as well that the young Russian was sincere about transforming Yukos into a normal corporation, with proper audits, governance structure, and accounting. Khodorkovsky had already hired a former auditor from BP and he had retained the accounting firm Price-waterhouseCoopers to implement generally accepted accounting principles. He had also retained the consulting firm McKinsey & Company and, with its advice, had "totally revamped the structure of the organization," Misamore recalled.[16] Khodorkovsky had also recruited international directors to his board and had a goal to win a listing for Yukos on an international stock exchange. Misamore decided to take a chance; he accepted the position of chief financial officer of Yukos and moved to Moscow in 2001.

Every Monday afternoon Misamore met with Khodorkovsky in the Russian's office in the old Soviet-era Menatep bank building on Ulansky Pereulok, fronting Moscow's Garden Ring. Khodorkovsky usually wore jeans, a turtleneck, and a sport coat. Misamore came to regard the young billionaire as a visionary and a natural leader. He lacked some management skills, perhaps, but he possessed a powerful intuition about Russia's present and future, Misamore believed.

In late 2001, just as the Bush-Putin energy negotiations were launched, Khodorkovsky told Misamore that he wanted to explore the sale of some of his and his partners' shares in Yukos—up to 25 percent—to a foreign oil company. Given the immense amount of oil reserves controlled by Yukos, this would be a great prize in equity oil holdings for whatever corporation bought in. For Yukos, it would diversify ownership and provide opportunities to explore and develop oil reserves outside Russia.

Misamore was enthusiastic. "One of the reasons I came here, Mikhail, was because I wanted to have a big impact on the future of Russia, through the example of Yukos," he told Khodorkovsky. "This guy Putin—all the reforms, and his support for reforms—everything seems to be going very well."

"Be very careful," Khodorkovsky had replied, as Misamore recalled it. "What is going on behind the scenes—and nobody can see it—is that Putin is stacking the Kremlin with former K.G.B.-ers and former Russian

military types. This is very, very dangerous because Putin is basically consolidating his power through these groups of ultraconservative Russians.
Once he gets that power established in the Kremlin, it is going to be
worse." That was why they had to move diligently; Khodorkovsky believed
they could succeed in selling a piece of Yukos, but they would have to
watch the Kremlin carefully. Although he did not discuss it at length with
Misamore at their weekly meetings, Khodorkovsky seemed to have his
own strategy to counter Putin: He was spending money to build ties to
members of Russia's parliament, the Duma, to create a political bulwark.
Khodorkovsky never explained his strategy fully, and perhaps he was too
opportunistic and improvisational to have a careful one, but it appeared
that he envisioned two prongs of activity to outflank Putin's cabal: a deal
with a Western oil corporation to infuse Yukos with cash and political
links abroad, and a network of allies in Russian politics.[17]

On an afternoon in early 2002, several large black armored cars pulled
up outside John Browne's home in Cambridge, England, and "numerous burly bodyguards" emerged, as Browne recalled it, to scan the horizon.
When the bodyguards were satisfied, Mikhail Khodorkovsky stepped
out. He went inside to join Browne over lunch. Khodorkovsky struck the
BP chairman at first as being an unassuming man, but as he spoke about
Yukos and its place in Russian politics, Browne grew nervous.

Khodorkovsky said he was prepared to sell a quarter of his company,
plus one share, a percentage that conferred certain rights to the buyer
under Russian law. Browne said he did not feel that was enough of a stake
in Yukos to justify an investment. "You can have twenty-five percent, no
more, and no control," Khodorkovsky answered, as Browne recalled it. "If
you come along with me, you will be taken care of."

Khodorkovsky went on to talk about "getting people elected to the
Duma, about how he could make sure oil companies did not pay much
tax, and about how he had many influential people under his control," as
Browne put it later. The BP chairman felt that Khodorkovsky "seemed too
powerful" and that there was "something untoward about his approach."

After the Cambridge lunch, Browne broke off negotiations.[18] Browne's account, written later, smacked of some of the benefits of hindsight. But it was true that he now turned away from Yukos as a target for BP's deeper push into Russian oil. When Browne eventually found another partner in Moscow, it would have consequences for Khodorkovsky and Lee Raymond, just as his fist-mover merger with Amoco had pushed Exxon and Mobil together a few years before.

Bruce Misamore served with Rex Tillerson of ExxonMobil on the U.S.-Russia Business Council's board of directors and respected him; Tillerson, in turn, had gotten to know Khodorkovsky. It was obvious that ExxonMobil had become a leading partner in the Bush administration's Russia oil initiative, which might bring the Kremlin to support a sale of Yukos shares.

Misamore was aware, however, of the rumors about Khodorkovsky's political ambitions in Russia and of his rivalry with Putin—that he might even be planning to challenge Putin or his successor for the Russian presidency. It would be difficult to negotiate with ExxonMobil if Khodorkovsky intended to move openly into political competition with Putin.

One Monday afternoon, Misamore asked his boss about that possibility. He thought Khodorkovsky would be a great president of Russia, he said, but what would it mean for Yukos?

Khodorkovsky said he could never run for president, for two reasons. He was perceived as an oligarch who had stolen his fortune, and while that was not true or fair, he continued, it was a perception so widely distributed within Russia that it was not realistic to think that he or any 1990s-era billionaire could be popularly elected. Second, Khodorkovsky said, one of his parents was Jewish. "There is no way with one Jewish parent I could ever get elected president." Still, he said, "I am going to be very politically active. I have to be. This is my country."[19] He sounded noble, but his ambition was clearly dangerous, since challenging Putin politically did not look like an easy matter, and it clearly violated the gentlemen's agreement that generally prevailed between the Kremlin and the oligarchs.

Khodorkovsky backed the Open Russia Foundation with Yukos funds to support civic and philanthropic projects; less publicly, he continued to contribute to the coffers of individual members of the Duma. Soon Khodorkovsky's lobbyists were seen working the floor of the Duma, relaying instructions to their allies at voting time.[20]

H is audacity extended to Washington. Khodorkovsky donated $1 million to the Library of Congress in 2002. Separately, he worked Republican circles and won an invitation to table one at the National Prayer Breakfast, which President Bush attended.[21] As in Russia, he advertised himself to the Bush administration as a transformational figure. He traveled to Washington with a PowerPoint lecture outlining how Yukos could improve America's security as an oil importer.

His presentation was entitled "Russia, the Persian Gulf, and World Oil Supply." Its thrust was that private oil companies in Russia, led by Yukos, could build a new network of pipelines to export more Russian oil to the United States and China, reducing dependency on unstable oil exporters in the Middle East. One of the slides forecasted that Russian exports of crude oil could more than double to 6 million barrels per day by 2010, out of total production of more than 9 million barrels per day. (The forecast proved to be accurate.)

Khodorkovsky believed that Russia should build new export systems. Lukoil and other privatized Russian companies had already drawn up plans for a new pipeline that would run north to the Russian port of Murmansk, just to the east of northern Finland, on the Barents Sea. Murmansk was free of ice year-round, and while the seas in the region could turn rough, they were navigable by the largest oil tankers.

From Murmansk, "oil can be transported from Russia to the U.S. by a shorter route than the one used for transporting crude from the Middle East," one of Khodorkovsky's slides declared. Russia could eventually supply 10 percent or more of American oil imports and serve Europe reliably, too. "The geopolitical imperative for today is to diversify energy sources and concentrate on the most reliable ones," the presentation declared.[22] Given the liquid, interconnected nature of global oil markets, it was not obvious why a Murmansk route would specifically benefit the United States over the long run, other than, perhaps, by providing an additional hedge against supply disruptions elsewhere. But the notion that direct imports by the United States from Russia would bolster geostrategic cooperation took hold at the highest levels of both governments, even among some officials who understood the fungible character of global oil markets. And Khodorkovsky emerged as the most active proponent of this vision in Russia's private sector.

Khodorkovsky and Misamore believed their Murmansk vision enjoyed the full support of both the Putin and the Bush administrations. Sandy Vershbow, the American ambassdor in Moscow, met regularly with Misamore and encouraged him. Privately, Vershbow was ambivalent. He called Khodorkovsky "sometimes defensive and even threatening" after a private meeting in Moscow in the fall of 2002, where they argued about what sort of contracts should govern future tie-ups between American and Russian oil companies.[23] Yet Vershbow, Vice President Cheney, and others in the Bush administration recognized that Yukos, under Khodorkovsky, was one of a handful of Russian oil companies that might be in a position to make a transformational deal.

On trips to Washington, Misamore met with senior officials on Bush's National Security Council. "They were quite happy to have the diversity of supply," he recalled, with a rising role for Russia, so as to "move away from the Persian Gulf dependency. The economics made a lot of sense because to the extent that you didn't have higher shipping cost, that could be translated through in the costs of crude oil to the U.S. There was a great deal of support."[24]

It seemed increasingly uncertain, however, where Vladimir Putin really stood on the proposed American oil partnership. Putin thought about the security of energy supply much more as a Risk game player than did the free-market idealists in the Bush administration. He saw oil and gas pipelines as physical valves that he could open or shut as he sought to reward or punish other countries, at least within the former Soviet Union. Russian gas companies and ministries routinely squeezed supplies to customers in Ukraine if they were unhappy about regional finances. Surely Putin was complicit in those squeeze plays.

When he met Putin, Commerce Secretary Don Evans continually tried to impart what he considered to be basic concepts of free global oil markets, that the markets were interdependent, not something that should be conceived of as susceptible to the manipulation of physical supply. Evans was trying to make a subtle argument: U.S.-Russian cooperation would deepen because of oil partnership, but the supplier-customer relationship would be indirect, subordinate to the integrated global market. Putin, on the other hand, seemed focused on ties that would bind Russia and the United States together in a mercantile arrangement. He wanted to explore deals that would connect Russian supplies directly to the United States through long-term contracts and investments. He was interested, for example, in building liquefied natural gas facilities that would bind American importers directly to Russian supply. That was the kind of supplier position Russia enjoyed in Ukraine—Moscow controlled a valve that it could open or close.

"Mr. President," Evans told Putin, "what you need to do is to quit thinking so much" about direct exports to the United States. The way to build cooperation was through joint corporate deals, investment, and shared participation in the global industry. "We are trying to open up supplies and get more out there in the global market. That's the big deal. Just get it into the market. It will find its way to the highest price."[25]

Khodorkovsky retained UBS, the Swiss investment bank, to explore how to handle negotiations for the sale of a 25 percent stake in Yukos.[26] Lee Raymond was interested in talking. He met with Khodorkovsky late in 2002 at a Bush administration–sponsored energy summit in Houston. Raymond told his colleagues that he found Khodorkovsky to be confident and knowledgeable. The Russian tycoon was probably getting himself too mixed up in politics, but it might be that he had little choice about that, given the intermingling of business and politics in his country.

"I'll never sell so that you have a majority stake," Khodorkovsky told Raymond, marking out the same position he had outlined for Browne.

Raymond said that was certainly Khodorkovsky's prerogative, but in that case, there could never be a deal between Yukos and ExxonMobil. "I would never take a minority stake in Yukos if there wasn't a clear way to become the majority owner," Raymond told him.

ExxonMobil's Russia watchers assessed that Khodorkovsky was increasingly anxious to pull cash out of Yukos. This gave them leverage, they calculated. Rex Tillerson interacted frequently with one of Khodorkovsky's partners, Yuri Golubev, who began to formulate a way to strike a bargain: ExxonMobil would buy a 30 percent stake in Yukos, with the understanding that it would later go to the Kremlin to win permission for the sale of a majority stake to the American corporation. Whether this was a realistic reading of how Russian politics and oil deals worked was highly debatable, but it reflected the ExxonMobil way abroad: control, contracts, and irrevocable authorities granted at the highest levels.[27]

Raymond pressed Khodorkovsky about whether he thought Putin would really approve of ExxonMobil taking a majority stake in Yukos. Khodorkovsky said he thought it could be done. Golubev, in his conversations with Tillerson, was more cautious about whether Putin would grant permission. It wasn't clear whether Khodorkovsky was sincere or gaming them as part of his multifaceted efforts to sell a stake in Yukos. Raymond and Tillerson weren't sure they could trust Putin, even if he said yes.

On June 14, 2003, Raymond and Tillerson flew to Saint Petersburg to meet with Khodorkovsky and others, and to participate in an energy conference in Moscow. At the conference, Khodorkovsky presented a lecture entitled "The Future Strategic Global Role of the Russian Oil Industry" in which he seemed to speak directly to Lee Raymond's worries about whether a deal with Yukos would be secure. "The rules of the game are being established," one of his slides declared. "We can now say that the Russian tax system as a whole has indeed stabilized and the interest that our Western colleagues now have in the stability of this system—since they are now playing on a level playing field, the same rules of the game as we are—give us confidence that this stability will last for quite a long time."[28]

Raymond talked again with Khodorkovsky. They both flew to Beaver Creek, Colorado, for the annual off-the-record World Forum staged by the

American Enterprise Institute, the conservative Washington think tank where Raymond served on the board of directors and where Lynne Cheney, the vice president's wife, had long worked on public policy matters. The Beaver Creek conference convened on June 19. Vice President Cheney flew out to attend a dinner of about a dozen people where Khodorkovsky was also present.

Raymond and the Yukos chairman held long discussions on the sidelines of the conference about ExxonMobil's proposed purchase. The vision conceived around President Bush's fire pit almost two years before seemed at last within reach.

The biggest sticking point for ExxonMobil remained whether Putin would give permission for the corporation eventually to take a majority stake in Yukos. That would be a major break with Russian precedent, a signal of a new era. Khodorkovsky and his team promised Raymond and Tillerson that they would lobby the Russian government. But they could not deliver the Kremlin's permission up front. ExxonMobil should buy a minority stake first, and win Putin's permission later.

"I'm never going to do that," Raymond told the Yukos team. "I would need to have assurance from the government at the beginning."

"This is not the right time to talk to the Russian government," Khodorkovsky said.

Khodorkovsky was fencing through the negotiation—and talking simultaneously with David J. O'Reilly, the Irish-born chief executive of Chevron. It was far from clear who was gaming whom.[29]

The Kremlin's clan of "siloviki," or security men, surrounded Vladimir Putin. The siloviki were a mysterious network of former K.G.B. and military led informally by former Interior ministry officers and a former military interpreter named Igor Sechin, Putin's deputy chief of staff. The siloviki formed Putin's base at the Kremlin, and yet the president also continued to speak and sometimes act in public as if he intended to transform Russia into a European democracy. "Putin thinks you can square the circle between capitalism and authoritarianism," one of Khodorkovsky's senior advisers at Yukos told an American visitor on July 1, 2003. "But you can't." The factionalism among ex-K.G.B. and other security men at the heart of Russian business and politics distorted decision making.

Yukos had moved into a new high-rise that spring; Misamore and his colleagues now worked in a cool, spartan, modern setting. The only way forward, Khodorkovsky had concluded, was to opt for "an American model" for Yukos, the colleague recounted in his fluorescent-lit office that July afternoon. True, there were those inside Putin's Kremlin who would resist. "It's a very real conflict," the Yukos official said. "It's not public yet."[30] Other oligarchs of Khodorkovsky's ilk—Vladimir Guzinsky and Boris Berezovsky, among them—had already been served with arrest warrants, the latter's earlier in 2003. The next morning, Russian authorities arrested Platon Lebedev, the chairman of Menatep and a major Yukos investor, on charges of defrauding Russia in a 1994 privatization deal. Khodorkovsky, too, was summoned for questioning, but released. The Yukos chief was on notice, but he plunged ahead.

The American embassy in Moscow cabled Washington to report that the detentions of Yukos leaders were a serious development, but one that would likely blow over: "Most analysts interpret the [government of Russia's] actions as a warning to Khodorkovsky to reduce his high-profile involvement in politics, which has included significant contributions to political parties and speaking publicly of the need to ensure that the upcoming elections are successful in producing a Duma that will pursue the reforms he favors. . . . Most analysts believe Yukos and the Kremlin will step back and quietly resolve their differences, at least in the short term."[31]

The Bush administration's energy diplomacy barreled ahead as well. In September 2003, Don Evans led a delegation of American energy executives to tour Russia in connection with yet another U.S.-Russian energy policy summit. Putin traveled to New York and celebrated the opening of a gas station in Manhattan owned by the Russian firm Lukoil, to demonstrate that Russian companies were investing in American markets, too.

Richard Grasso, then the chief executive of the New York Stock Exchange, invited about twenty American business leaders to meet with the Russian president while he was in New York, inside the exchange's ornate headquarters at 11 Wall Street. Lee Raymond flew in.

After the general roundtable session with the American executives, Raymond and Putin met separately in private. The ExxonMobil negotiating team had decided that it would be best to find out directly from Putin whether the Russian president would be prepared to allow ExxonMobil to eventually acquire a majority stake in Yukos.

Raymond had told Khodorkovsky that he intended to speak directly with Putin about this; Khodorkovsky had discouraged him. The ExxonMobil team interpreted Khodorkovsky's warning as only a negotiating tactic, designed to maintain Yukos's leverage as the exclusive source of communication with the Russian government about the proposed deal.

Seated in a stock exchange conference room, Raymond told Putin about the negotiations with Yukos. He explained that if ExxonMobil were to make an investment, it would do so only if there was an agreement in advance that the American corporation could eventually take majority control.

ExxonMobil didn't necessarily need to own all of Yukos, Raymond continued; if Russia wanted enough local ownership so that the company could be listed on a Russian stock exchange that would be okay. But ExxonMobil required a pathway to at least 51 percent ownership.

"You can basically decide how you want the other forty-nine percent," Raymond told Putin, according to an account of the meeting later briefed to ExxonMobil executives. "Do you want the government to own it? Do you want it to be listed on an exchange? But before I get started, I need to have an understanding of our ability to get to fifty-one percent."

It turned into a lengthy conversation. Putin talked expansively about the choices he faced in building oil pipelines to China, to feed that economy's thirst for energy. He put a piece of paper on the table, sketched a map on it, and started drawing lines showing possible pipeline routes. He talked about whether a pipeline should cross to China above or below the Aral Sea. They also talked about coal—it turned out that Putin had studied coal as a graduate student.

As to ExxonMobil's proposition, Putin asked Raymond, "If you have fifty-one percent, that means if I want to have Yukos do something, I'm going to have to come and talk to you?"

"Yeah, that's not so awful," Raymond answered. "That's true in a lot of places in the world."

"I'm not prepared to answer that today," Putin said.

"I'm not asking you to answer that today," Raymond told him. "You need to talk to your people."[32]

Khodorkovsky had also kept up his talks with Chevron, alongside those with ExxonMobil. The negotiations involved price and shareholding percentages, among other issues. As Bruce Misamore understood the terms, the discussions with Chevron involved some cross-ownership, whereby Yukos might acquire an interest in Chevron entities. With ExxonMobil, the terms under discussion were more one-sided, with ExxonMobil proposing straight up to buy an interest in Yukos.

Khodorkovsky asked Misamore which of the two American companies, Chevron or ExxonMobil, he would recommend as a partner. Misamore said that he felt Yukos's style of operations "was far more analogous to Chevron." It was more of a "laid-back culture." Also, if they took on Chevron as a partner it would be "more of a mutual learning concept." With ExxonMobil, by contrast, it "was going to be much more of a 'We know what we are doing, we are going to tell you how to do it' type of an approach."[33]

After the meeting with Putin in New York, according to a former senior ExxonMobil executive, "Raymond spoke quite optimistically about what he thought was going to happen."[34]

Lee Raymond's remarks about what Russia would have to do to satisfy ExxonMobil may have grated on Putin, however. "The report that we got back later was that Putin perceived him as just totally arrogant and

far too aggressive," Misamore recalled. "And he just really was totally
turned off by Lee Raymond—this big U.S. industrialist coming, and his
arrogance, and telling the president of a country how things are going to
be, almost. . . . Putin just was totally turned off by the guy—that was the
report we got."[35]

On September 1, 2003, BP announced a partnership with a Russian
firm to jointly hold oil assets as a new entity called TNK-BP. The deal was
complicated, but it effectively transformed TNK-BP into Russia's third-
largest oil company, with a London-based private corporation as a major
shareholder.

From Washington, Leonard Coburn, who was at the Department of
Energy monitoring the Raymond-Putin talks, assessed that "Putin was a
little scared" about what an ExxonMobil purchase of Yukos "would mean
for him."[36] Here was an American-headquartered oil giant obviously tied
to the Bush administration proposing to follow BP into a strategic Russian
industry—the primary source of Russia's national wealth.

K hodorkovsky's private chartered jet pulled into a fueling terminal at
the airport in Novosibirsk, in Siberia, in the early hours of October 25,
2003. The Yukos chairman was en route to inspect an oil field; he planned
to gas up his plane and take off again. Masked agents in camouflage dress
from the F.S.B., the successor to the K.G.B., stormed aboard in the darkness, their guns drawn. They grabbed Khodorkovsky and placed him under
arrest. They flew him to Moscow, where prosecutors charged him with
six counts of personal income tax evasion, overseeing corporate tax evasion, document forgery, theft, and other crimes.

The Prosecutor General's Office announced that Khodorkovsky's alleged crimes had cost Russia at least $1 billion in lost revenue. Khodor-
kovsky's spokesman at Yukos called the accusations "absurd" and said the
"brute force" used to arrest the chairman had been "humiliating for the
whole Russia law enforcement system" in the eyes of the world.[37]

The U.S. embassy in Moscow judged that Khodorkovsy's arrest "almost certainly must have been done with Putin's implicit or explicit approval," and it showed "that the authorities may want not only to humble

Khodorkovsky but to destroy him and even drive him out of the country." Vershbow urged the White House to take action.

"The timing of the latest investigations . . . amid rampant speculation of an imminent deal with ExxonMobil or ChevronTexaco, and immediately following Putin's U.S. visit—does not appear coincidental," the ambassador wrote in late October. "Khodorkovsky has refused to back down from the start, and for a while thought that he had beaten back his persecutors. . . . He was wrong."[38]

Less than eight weeks after their meeting at the New York Stock Exchange, Vladimir Putin had given Lee Raymond his answer. Why did Putin authorize Khodorkovsky's arrest? The latter's maneuvering to buy allies in the Duma in advance of parliamentary elections scheduled for December 2003 was probably the biggest factor. "There was clear information that Yukos supported candidates who could have formed a real, sizable faction," Milov recalled. "Putin is a person who is very influenced by these threats." The TNK-BP merger announcement on September 1, followed almost immediately by Raymond's discussion with Putin at the New York Stock Exchange, in which he sought a path to majority control, may also have inflamed Khodorkovsky's rivals at the Kremlin. Khodorkovsky was negotiating with Chevron, too, the siloviki knew. "I saw these notes saying, 'We might be losing our oil industry to foreigners in a couple of months completely,'" Milov said. "It was a kind of scare like that. This factor was involved. I wouldn't say it was the ultimate trigger, because this is a very complex story. . . . It was a competition for influence in the country, for control over the country."[39]

Raymond spoke placidly in public about Khodorkovsky's downfall. "Everyone ought to take a deep breath," he said after the arrest. "Rome wasn't built in a day. ExxonMobil wasn't built in a day. This is a long-term industry." He conceded that ExxonMobil had been interested in Yukos and had engaged in talks, but as to why it had fallen apart, "There are some things there I'm not privy to, in terms of the Putin-Khodorkovsky relationship. You know, I've got enough problems."[40]

In private, with colleagues and friends, Raymond could be more reflective, and even a little guilt stricken. He wondered if the advanced state of his talks with Yukos might have prompted or influenced Putin's move against Khodorkovsky, he told friends and colleagues. The reality almost certainly was that Putin and Khodorkovsky were on a path of irreconcilable conflict, no matter what. For ExxonMobil, the arrest placed a punctuation mark on two bold but costly failures: The corporation's search for shoot-the-moon purchases of oil and gas reserves in Russia and Saudi Arabia had now produced back-to-back zeros. There were those who blamed Raymond's truculence and ExxonMobil's general arrogance for contributing to the strikeouts, but even if Raymond were Prince Charming and his corporation played well with others, that would not alter the fact that Saudi Arabia had no trove of nonassociated gas reserves to sell and Russia's government had no stable plan to secure foreign investment in its oil fields. If even one of these two plays had panned out, ExxonMobil's reserve replacement challenges might have lessened in the decade ahead. Now the corporation would have to continue to scrap, and its reliance on places such as tiny Qatar and unstable West Africa would not ease anytime soon.

After Khodorkovsky's arrest, Raymond telephoned Cheney and asked for a meeting. He had kept the Bush administration out of his negotiations during the summer of 2003. He now told the vice president's office he didn't want anything from the administration, but he felt he owed them an explanation about what had happened, from ExxonMobil's perspective. Cheney suggested they meet away from the White House, at the vice president's official residence on the grounds of the U.S. Naval Observatory, on Massachusetts Avenue. He and Raymond spoke for about ninety minutes. Raymond recounted the history of the failed deal; he and the vice president exchanged assessments.

When Bush administration officials contacted the Kremlin to raise concern about Khodorkovsky's detention, Putin said that the rule of law in Russia had to run its course—wasn't that what the United States said it favored?

On January 29, 2004, a Russian commission denied licenses to ExxonMobil and Chevron for drilling in offshore blocks around Sakhalin, blocks that they had leased in 1993. Secretary of State Colin Powell met with his Russian counterpart, Sergey Lavrov, and handed over a letter of protest on behalf of the U.S. oil companies. The Bush administration was still fighting for the companies' prospects in Russia, but its campaign looked increasingly like a rearguard action, fought while in retreat.

"ExxonMobil and ChevronTexaco have invested approximately $60 million in exploration activities," Powell pleaded. He continued:

> The Russian government's failure to issue a license to ExxonMobil and ChevronTexaco would hurt the climate for U.S. and other foreign investment in Russia's energy sector and cast a shadow over Russia's reputation for fulfilling its commitments. It would also raise serious questions about Russia's commitment to our bilateral energy partnership. . . . It has been two years since our two presidents launched a strategic energy relationship. Since that time, we have not seen the concrete progress in the foreign investment climate for energy that our partnership was intended to promote.[41]

As the months passed and Putin's authoritarian retrenchment spread from media to oil deals to the direct suppression of democratic opposition, Bush and all of his advisers realized, sheepishly, that they had "drunk the Kool-Aid a little," as Don Evans told his colleagues. Global oil prices rose; Russia's government profited and felt less pressure to change. The Bush team concluded that as soon as Putin realized that rising global oil prices meant he did not need American or European capital to finance improvements in the oil sector, he reverted to autocracy.

After Evans left Bush's cabinet, in 2005, his telephone rang at his office in Midland, Texas, where he had returned. German Gref, his former counterpart in the Commercial Energy Dialogue, during the years of optimism, told him that Vladimir Putin would like him to fly to Moscow for a visit. Evans agreed.

He found Putin alone in his office, except for his interpreter.

"I would like you to be chairman of Rosneft," Putin said. Rosneft was the state-owned oil company Lee Raymond had examined and rejected on the grounds that it was a political labyrinth. Igor Sechin, the former leader of the Kremlin siloviki with which Mikhail Khodorkovsky had tangled, now served as an influential figure at the company. He was a beefy man with short, cropped hair.

Evans said that he was flattered and that he would think about it; he flew back to Midland. Putin called Bush to tell him about the job offer he had made.

This was the Putin they had all underestimated in 2001—the K.G.B. man whose idea of how to build an oil partnership with the United States was to provide a lucrative job to one of the American president's best friends, at the head of a Russian oil company heavily influenced by the Kremlin. Putin's offer also suggested ambivalence; he wanted both control and international credibility.

Evans thought about the offer for a few days, but never spoke to Bush about it. Some friends told him he was crazy to even think about it; others advised that he give it serious consideration. Evans told his friends that he did think a more globally integrated Russian energy industry could spur economic growth worldwide. He was also mindful of appearances. Gerhard Schroeder, the former chancellor of Germany, had embarrassed himself and his country by accepting a position on the board of Gazprom, the Russian gas giant, days after he left political office; he created the appearance that he and Germany were being paid off by Putin.

After a short period of reflection, Evans decided that working for Rosneft was not right for him. He telephoned Sechin, thanked him, but said he would have to decline.

Later, John Snow, Bush's second Treasury secretary, found himself in a meeting with Putin where the subject of the job offer to Evans came up. Putin marveled at Evans's refusal.

"You know," he told Snow, "if he had taken that, you could have cut your C.I.A. budget in half!"[42]

Putin misunderstood the American system as much as American analysts misunderstood him. Russian oil companies cut their deals from a position that was clearly subordinate to the state. In Putin's worldview, the recruitment of a Bush friend like Evans to Rosneft made perfect sense. The converse proposition—the idea that ExxonMobil would recruit a Putin consigliere to its senior-most executive ranks in Irving, in order to solidify U.S.-Russian relations—was highly unlikely. ExxonMobil had never been an arm of the Bush administration's Russia reset after 2001, events had demonstrated; it was a private global empire that would choose to align with Bush, or not, as its enduring interests required.

Thirteen

"Assisted Regime Change"

Theresa Whelan joined the Defense Intelligence Agency out of college in 1987. She served as a junior analyst of Africa as the cold war ended. During the George H. W. Bush administration, Whelan came to the attention of what was known to insiders as O.S.D.-Policy, a mixed civilian and uniformed staff that reported to the under secretary of defense for policy, and through that officeholder, to the secretary of defense. Whelan moved to O.S.D.-Policy's Africa desk and served there through the tumult of the early 1990s—the withdrawal of American troops from Somalia and the genocide in Rwanda. Later she worked on Balkans issues during the Kosovo conflict. When she returned as office director of the Pentagon's Africa policy unit in 2001, she was a seasoned manager of the Defense Department's overseas programs to train foreign militaries, to support international peacekeepers, to patrol ocean waters, and to covertly attack terrorists. A year after the September 11 attacks, George W. Bush promoted her again, naming her as the deputy assistant secretary of defense (or "Das-D," in Washington's vernacular) in charge of the Pentagon's Africa policy.

On November 19, 2003, a warm and rainy day in the capital, Whelan rode after work across the Potomac to a hotel conference room to deliver a speech. The occasion was the annual meeting of the International Peace Operations Association, a trade association of private security companies that had determined that "peace operations" was a better branding strategy than "corporate mercenaries." Many of the executives in the audience had an interest in whether Pentagon policy might encourage more contracts for private security firms, particularly in regions like Africa, given that America's uniformed military was increasingly overtaxed in Afghanistan and Iraq. Whelan spoke about the limitations of relying on contractors for military missions, but also about some of the advantages that private security firms offered in training armies in poor countries, such as the fact that corporate trainers could stay in the targeted nation for years at a time, building local expertise and long-lasting relationships.

She answered some questions after her formal remarks, received a round of applause, and then, "as often happens at those kinds of things," she recalled, about fifty people gathered around her "shoving business cards in my face, chitchatting." One man with a distinctly British accent caught her attention. He introduced himself as Greg Wales. He said he was an independent "security consultant" who worked with oil companies in West Africa, around the Gulf of Guinea. They talked about the region; Wales seemed knowledgeable.

"I'm going to be in town," Wales said. "Would you be interested in sitting down and talking more?"

"Sure," Whelan said. "Fine." She met regularly with security firms that worked with American oil companies in Africa, or their consultants. It helped her keep up with details about politics and violence in countries where American intelligence and diplomatic reporting could be very limited.[1]

Not too long afterward, Wales made an appointment to visit Whelan in her Pentagon office. She did not research his background. If she had, it might not have helped much; Wales was an elusive figure. He was an accountant by profession who had collaborated during the 1990s with private security and mercenary corporations active in diamond-rich regions of Africa.

During that autumn of 2003, Wales was involved, as it happened, in a conspiracy organized by British and South African military veterans to overthrow the government of Equatorial Guinea. ExxonMobil Corporation was the largest oil company invested there. The conspirators intended to replace the current president, Teodoro Obiang Nguema, with whose government ExxonMobil had signed its contracts, with an exiled opposition leader, Severo Moto, who lived in Spain, the former colonial power in the country. Moto had gone so far as to sign his own contract with the mercenaries, guaranteeing cash payments and future security contracts to be paid for by the country's oil wealth, if the coup plan succeeded.

Greg Wales had spent much of 2003 planning for the operation. He had written a number of strategy documents. One of them, entitled "Assisted Regime Change," had emphasized that it would be important to persuade all of the American oil companies active in Equatorial Guinea that their investments and profits would be protected after a change of government. "Foreign investors are to be reassured," Wales wrote.[2]

Wales was in charge of the coup's political strategy, particularly in the United States. He traveled to Washington and telephoned the D.C. offices of major oil companies doing business in Equatorial Guinea—as well as other companies that might be interested in moving in. He was vague, but he suggested that political changes might soon be in the offing in Equatorial Guinea.

He arrived alone for his meeting at the Pentagon. Whelan invited a Defense Department aide to sit in. As Whelan recalled it, the discussion touched generally on the rising importance of oil production in West Africa and the security challenges that seemed to come alongside. They talked about Angola, a major oil producer recovering from a long civil war, as well as the insurgency- and crime-wracked oil-producing regions of the Niger Delta, in Nigeria—"just the whole enchilada," Whelan recalled.[3]

They also addressed Equatorial Guinea. "He made a comment that he thought that Equatorial Guinea was probably not the most stable place in the world," she remembered. Wales described himself as being in the security and air transport business in Africa. "He said that his colleagues were reporting to him that there were a lot of well-heeled Equato-

Guineans . . . that were sort of prepaying or prescheduling flights out of the country in the event of some sort of problem. . . . So that was his one piece of intelligence—if you want to call it that—that the leadership in Equatorial Guinea was nervous, and it appeared that family members were making their plans for a getaway, just in case."

That the elite in Equatorial Guinea might be nervous and prebooking flights as a hedge against a coup did not strike her as particularly noteworthy. The country was like a small, unprotected, oil-endowed bank sitting in a bad neighborhood, just waiting to be robbed. During the 1970s, Frederick Forsyth had written his novel *The Dogs of War*, about a mercenary-led coup in a small African country, while staying at a hotel above Malabo's harbor; Forsyth himself later became entangled in a stillborn coup attempt against Equatorial Guinea's dictator. Coup plots seemed to blow through the country like its tropical storms. Leaders of Obiang's fractured and repressed political opposition were regularly accused by the dictator of fomenting his overthrow. Periodically, the president arrested his own relatives for plotting a move on his palace. In 2002, Obiang had detained and tried dozens of alleged plotters on treason charges. The poor standards of the trial were one reason why the State Department had so far refused Obiang's pleading for a license to hire M.P.R.I., the Virginia-based security firm, to train his police and military. The United States had earlier approved an M.P.R.I. license for maritime defense on the grounds that the Equatorial Guinean coast guard posed little threat to the country's citizens and might defend offshore oil platforms owned by ExxonMobil, Marathon, and Hess. Yet the Bush administration judged that training Obiang's land forces to shore up his dictatorship could not be justified until the government improved its human rights performance.

Theresa Whelan interpreted Greg Wales's purpose in visiting her as typical of the hustling, networking world of profit-making consultants in the global security field. "He wanted to make sure that he could say that he had a contact, that he had access to the Pentagon for some purpose in the future, whatever that might be. But I didn't think that he had a particular agenda, other than making it clear to us that he was a potentially valuable interlocutor," Whelan concluded. "It was the equivalent of a sales call or a marketing meeting. For our trouble in listening to him, we would get another perspective on the ebb and flow of the political situation in the region."

After Wales departed, Whelan made no record of the conversation and distributed no memos or e-mails to Pentagon colleagues, she said. It did not occur to her to do so because the entire encounter had been "pretty unremarkable."[4]

It is not known how Wales assessed the conversation, but it is clear that there was nothing routine about his thinking about the Pentagon's possible role in his coup plan. His fellow conspirators had the impression that winter, based in part on what Wales told them after traveling to Washington, that the United States might actively support their efforts to seize power—or at least that the Bush administration might not object. It is not clear why such an impression might have developed among the coup planners. Contemporary memos show that Wales and his colleagues knew they faced a risk of alienating the United States, particularly if major oil companies such as ExxonMobil concluded that the change of regime would jeopardize their investments. Threatening American oil interests in Equatorial Guinea, Wales wrote, might be "what gets the Marines coming in."[5]

Simon Mann served as Greg Wales's principal military partner in the plot under way that summer to overthrow Obiang. He was a lean and poised man in late middle age, a descendant of English cricket captains, who had grown into a formidably successful adventurer. After graduation from Eton, the elite preparatory school, he joined the Special Air Service, Britain's principal special forces unit. Later he founded Executive Outcomes and Sandline International, part of a network of corporations that provided mercenary military services to African governments in exchange for diamond mining and other business concessions. During the 1990s, Executive Outcomes won contracts with Angola, to battle a formerly anti-Communist rebel movement, and with Sierra Leone, to keep antigovernment rebel marauders away from that West African country's diamond mines. Mann grew wealthy, bought an estate in the English countryside, another in London, and married, apparently intending to settle down. He and his younger wife started a family, but in 2003, when he turned fifty-one, Mann became restless. Eli Calil, a Nigerian-born Lebanese oil trader with extensive contacts in Africa, and Severo Moto, the Equato-Guinean opposition leader in Madrid, recruited him early that year, Mann later said in a Malabo courtroom. "I agreed to do this for the money, yes, but also because I believed it was right," he said.[6]

That spring, Mann learned, he and others involved in the coup planning said later, that Spanish prime minister José María Aznar supported the plan. Aznar, a former Franco supporter who had become a successful conservative politician, oversaw an aggressive foreign policy. On March 17, 2003, on the eve of the invasion of Iraq, he appeared with George W. Bush, British prime minister Tony Blair, and Portugese prime minister José Durão Barroso in the Azores to provide a show of solidarity for the American-led plan to overthrow Saddam Hussein. "We are committed on a day-to-day fight against new threats, such as terrorism, weapons of mass destruction, and tyrannic regimes that do not comply with international law," Aznar declared. "They threaten all of us, and we must all act, consequently."[7]

"The Spanish PM has met Severo Moto three times," Mann later wrote, and he indicated that Spain would take concrete steps to support Moto's installation in power in Malabo if the mercenaries' coup plan succeeded. "He has, I am told, informed SM that as soon as he is established in EG he will send 3000 Guardia Civil. I have been repeatedly told that the Spanish Govt will support the return of SM immediately and strongly. They will, however, deny that they are aware of any operations of this sort." As Mann and his colleagues approached their launch date for the coup, Spain dispatched warships bearing marines to the Gulf of Guinea; they attempted to dock in Malabo, but were denied permission. The Spanish foreign minister was quoted as describing the naval deployment as a "mission of cooperation."[8]

Greg Wales's travels to Washington—his awkward approach to The-

resa Whelan at the Pentagon and his unsolicited telephoning of oil companies' lobbying offices—suggested a strain of amateurism in the plot at
odds with Mann's record of corporate security success during the 1990s.
The plan had several prongs. An undercover team led by a South African
named Nick du Toit infiltrated Malabo in late 2003 in the guise of businessmen. They set up a firm called Triple Option and claimed to be interested in fishing. In fact they carried out reconnaissance and prepared to
aid the coup team when it landed at the airport. The main external raiding party would be made up of veteran soldiers from South Africa and
Angola. They would streak into Malabo from southern Africa in a transport plane, seize the airport, and then roll toward the presidential palace
to capture or kill Obiang. As Du Toit later described it, "The advance
group and the arriving mercenaries would meet at Malabo's airport and
then drive to Obiang's palace, kidnap the president, and then systematically kill all other [Equato-Guinean] ministers. Obiang would be exiled
to Spain and a plane would arrive carrying Severo Moto and his supporters to form a new government." Du Toit would be paid either $1 million
or $5 million—there are documents reporting both numbers—if the coup
succeeded.[9]

Mark Thatcher, the son of Margaret Thatcher, the cold war–era conservative prime minister of Britain, lent the group funds, although he said
later that he had been misled about the purpose of his loan and had no
idea it was to be used in support of a coup d'état. Soldiers clued in on the
plan in South Africa and Equatorial Guinea spoke so loosely about their
plans that the British government picked up coup rumors simply by monitoring Africa's radio services. When Obiang visited South Africa in December 2003, his hosts passed intelligence that he should be watchful.
Angola's government passed him warnings as well. The intelligence was
accurate; the coup makers had moved toward a strike that winter.

Obiang traveled to the United States in February on one of his regular
private visits—apart from his occasional medical treatment, he seemed
to enjoy spending time in America—and he checked into the Four Seasons
hotel in Washington, a modern brown-brick building on the southeast-

ern edge of cobblestoned Georgetown. On a gray Monday afternoon, five executives arrived from Riggs, Obiang's principal bank, whose branch across from the White House had first attracted the president's attention as a place where he might store his oil wealth while deepening ties to American corridors of power. Equatorial Guinea's deposits now totaled about $750 million, by far the largest of any of the bank's clients.

Auditors from the regulatory Office of the Comptroller of the Currency were crawling all over Riggs that winter, however. A probe had been sparked by other accounts held at Riggs for Saudi Arabia's embassy in Washington and for the former Chilean dictator Augusto Pinochet. The regulators were now asking questions about Equatorial Guinea's accounts. They had recently raised concerns about offshore transactions in the accounts controlled by Obiang. Riggs and its management could face fines or worse if they did not come up with convincing answers. At the Four Seasons, the bank's delegation announced that it was particularly focused on several relatively modest wire transfers—totaling less than $1 million—to offshore companies that appeared to be linked to Simon Kareri, the Riggs account executive who serviced Obiang.

Obiang waved them off. The transactions were authorized to support the economic development of his country, he said vaguely. He refused to be drawn on specifics. When the bankers pressed, Obiang sent one of his sons, Gabriel, along with several aides, to go off and review the matter.

The two delegations bundled into cars and rolled over to Pennsylvania Avenue. Inside the columned Riggs branch, Gabriel looked over the records and explained that some of the transactions involving Kareri had been authorized, but at lesser amounts than had actually been transferred to the offshore accounts. This raised the possibility that someone, possibly Kareri, had been skimming money. The Riggs executives asked what the offshore companies receiving funds actually did. Gabriel was vague.

The bankers warned him that if he could not provide specific, verifiable descriptions of what the money sent offshore had been used for, Riggs might have to end its entire relationship with Equatorial Guinea, notwithstanding the great financial pain it would cause the bank. In the post–

September 11 world it was unacceptable for American banks to host accounts making international wire transfers to unknown front companies, they explained. Nobody was suggesting that Equatorial Guinea was financing terrorists, but the rules were inviolable.

Gabriel declined to explain. The transfers were "authorized by the government to pay for services," he said.

That afternoon, Riggs's risk management committee met to terminate the bank's relationship with Obiang. The bank's executives announced that $40 million in Obiang's accounts—an amount equal to the balance of his outstanding loans—would be frozen, pending resolution of the debts. The rest of his funds—about $700 million altogether—would be released in the form of cashier's checks, which authorized individuals, including the president himself, would be free to pick up at the branch across from the White House, so they could hand carry the checks to another bank of their choice for deposit.[10]

The Dodson family sold airplanes and parts from hangars and warehouses outside of Kansas City. One way they acquired inventory was by monitoring sales of surplus U.S. government airplanes by the General Services Administration, the agency responsible for federal buildings and property. In 2003, Dodson Aviation, Inc., purchased a 727-100 jet that had been put up for sale on a Web site called GSAAuctions.gov. They spruced the plane up and listed it for resale. About six months passed before an English firm, Logo Logistics, contacted them about a purchase. "Normal business guys in suits" turned up in Kansas to inspect the aircraft, as J. R. Dodson recalled it. They had foreign accents.[11]

The Dodsons handled sales through an escrow firm in Oklahoma City—typically, the buyers transferred cash into an escrow account, and when the conditions of the sale were met, the escrow firm released the funds to the Dodsons. The deal for the 727 closed on March 3, 2004; the Dodsons did not know the origins of the cash, only that it had arrived to the satisfaction of the escrow agent.[12]

Colleagues of Simon Mann arrived in Kansas and flew the plane to

South Africa. They landed at a small airport outside of Pretoria, where Mann and about five dozen mercenaries boarded in darkness.

From Spain, Severo Moto flew to the Canary Islands. Greg Wales and other conspirators joined him there at the Steigenberger hotel. On March 7, they boarded a leased Beechcraft King Air and lifted off for Africa, intending to rendezvous in the air with Mann's American-purchased jet. In Malabo, Du Toit's local recruits prepositioned cars and other vehicles at the airport, so the arriving armed mercenaries would have transport into town. Moto and Wales would circle so as to land in Malabo an hour after the coup plotters arrived and seized the capital. Moto had even written a speech in advance promising to transform Equatorial Guinea into the "star of Africa."[13]

En route, Mann and his crew flew to Harare, the capital of Zimbabwe, where he intended to pick up weapons and fuel before flying on toward Equatorial Guinea.

They never left the Harare tarmac. Zimbabwe police stormed the plane and arrested all of its passengers; they had been tipped off by South African intelligence. Mann managed to get a phone call through to Du Toit in Malabo to let him know that "problems had arisen." Mann and his mercenaries were taken to prison and, according to Mann, beaten and tortured into making confessions.[14]

In Malabo, Obiang's security forces arrested Du Toit and many others who had worked with him. They paraded the South African before diplomats and television cameras. He confessed that his purpose was to "carry out a coup against the Obiang regime" and to bring in Severo Moto from Spain "as the country's new leader."[15]

"The terrorists who have been arrested will go through a fair trial," Obiang declared to his people. If convicted, however, "because Equatorial Guinea has not abolished the death penalty, we won't forgive them. If we have to kill them, we will kill them." He urged his countrymen to watch out for other conspiracies, to "eliminate these terrorists. . . . Whoever presents themselves as a mercenary, there will be no need to let the President know. They must be liquidated—they must be killed because they are devils."[16]

Obiang might be accustomed to coup plots, but this one was enough to make any insecure, oil-endowed dictator's head spin. Its external tentacles ran around the world. Spain seemed to be involved—Obiang required little proof to conclude that Madrid was out to get him, but here the evidence looked substantial. There was circumstantial evidence to suggest that the United States might also have been covertly involved—the American origins of the plane carrying the mercenaries was one suggestive indicator, and the common statements of Bush and Aznar about ridding the world of dictators suggested the potential for secret collusion between them. Yet, hadn't he, Obiang, been generous again and again to ExxonMobil and the other American oil corporations gorging on Equatorial Guinea's oil? Weren't his contract terms among the most generous to American oil firms in Africa, or indeed the world?

Simon Kareri, the Riggs account executive, who would soon be indicted for his dealings with Equatorial Guinea's deposits and wire transfers, told Obiang that he suspected the Bush administration had been involved, according to an associate of Kareri's. The coup attempt explained all the pressure Obiang had faced over his Washington bank accounts, Kareri argued. How else could the events at the bank that winter be explained? The closure of Obiang's accounts just as the plot was moving toward execution suggested that the Bush administration had created conditions in which the Malabo regime would lose control over Obiang's $750 million in deposits just as Moto seized power, Kareri speculated. Was it a coincidence that Obiang had been told to move his money and that fungible cashier's checks were issued just as the coup attempt was being prepared?[17]

At least one of Obiang's Washington advisers believed that the Bush administration must have been involved, but the adviser could not turn up proof. Obiang was not entirely sure what to believe, but he could not in the end bring himself to conclude that the Americans had joined Spain in the conspiracy to oust him. About Spain's culpability, he had no doubt. Bush, Blair, and Aznar "discussed the need to get rid of dictators," he reflected later. "When you talk about something like dictators, you have to do an analysis: Which governments are dictators and which are not? Aznar took advantage of this . . . to advance the concept of bringing down the 'dictatorship' of Equatorial Guinea." Obiang doubted that the Bush administration knew about the plot in advance because "the American companies are the ones with the primary investments here." Spain was "jealous of American success here. . . . The mercenaries and Spanish companies were going to take over. For that reason, I can't say the Americans were involved. They would lose business to the Spanish and the British."[18]

The Bush administration's attitude toward Obiang's government nonetheless mystified him. More than $5 billion of investments by American oil companies were at risk in his country. The Mann coup made clear just how diverse, creative, and determined were the potential jackals circling tiny Equatorial Guinea, waiting for a chance to snatch its riches. Yet Obiang had been asking for security assistance from the United States, to protect the wealth of its oil corporations, and all he had been given was an M.P.R.I. license to train a coast guard. What good would a coast guard do if mercenaries or a neighboring military invaded Malabo and voided ExxonMobil's contracts? The oil-endowed autocracies of Saudi Arabia, Kuwait, and the United Arab Emirates had poor human rights records and hardly a whiff of democracy, yet they were treated in Washington as important strategic partners and received billions of dollars' worth of sophisticated defense systems—jet fighters, missile interceptors, the works. Why not Equatorial Guinea?

Obiang paid a handful of lobbyists to represent him in Washington. They advised him about political reforms and image management, but they had not resolved the basic problem, as he saw it, that he lacked sufficient access to the Bush administration. The oil companies operating in Equatorial Guinea told Obiang that he needed to upgrade his Washington presence. They could support his cause, but they could not conduct his lobbying for him. If ExxonMobil's Washington office, or those of Hess and Marathon, "oiled" Obiang's efforts to win favor from the Bush administration by becoming too directly involved in Malabo's rehabilitation, it would only discredit Equatorial Guinea further. Obiang reached out to two of the most successful lobbying firms in Bush's Washington: Barbour Griffith & Rogers and Cassidy & Associates.

Richard Burt, a former *New York Times* reporter who had served as the American ambassador to West Germany during the Reagan administration, helped to manage the Obiang account under contract for Barbour Griffith. At Cassidy, one of Obiang's aides called Amos Hochstein, a young former Capitol Hill aide. Intrigued, Hochstein used Google to research Equatorial Guinea; the search returns were not particularly encouraging.

He traveled to New York to meet with Obiang's prime minister, Miguel Borico, at The Pierre hotel. The president was in a mood for fresh thinking, the prime minister reported. Obiang had been "embarking on an American strategy," as his oil wealth grew, to protect and align himself with the world's most formidable oil-consuming superpower, and yet "he wasn't getting anywhere," one of Obiang's advisers recalled.

"You're in deep trouble," the Cassidy lobbyist told them. Hochstein was a liberal Democrat. He was uncomfortable with the account, but his firm had decided to go forward. "I'm not going to lobby for you. What I can do is help you understand what you need to do to change your relationship with the United States government. I'll try to be the translator between you and the American government." Cassidy accepted Obiang as its client for a retainer of more than $1 million a year, a handsome sum in the Washington lobbying arena.[19]

The question of whether the Bush administration had winked in advance at the Mann-led coup plot lingered, sowing distrust and uncertainty in the U.S.-Equato-Guinean oil partnership on which ExxonMobil depended. At an African counterterrorism conference for regional intelligence leaders held in Libya around this time, Obiang's director of internal security approached Mel Gamble, the Africa division chief of the C.I.A., and accused him outright of sponsoring the Simon Mann–led coup plot against Obiang.

"It was a U.S. aircraft," Obiang's spy chief pointed out.

"Look, you can buy a lot of things in the United States," Gamble answered. He denied that the United States had any involvement. "You can buy weapons from the U.S., too," he said, but that didn't mean that the Bush administration was involved or even aware.

Senior intelligence officers from Angola and Algeria overheard Gamble's pleading. They joined the discussion and backed their American colleague: Just because African coup plotters bought equipment in the United States did not mean that the Bush administration knew what was going on. They knew this from their own experiences, they affirmed.

The Central Intelligence Agency had no station in Equatorial Guinea at the time of the Mann-led coup attempt. The agency covered the country out of a base in Lagos, Nigeria; operations officers there might make one or two trips to Malabo each year, to make contacts and survey the political landscape. Reporting on economic issues such as oil production had been cut back during the late 1990s. After September 11, terrorism became the C.I.A.'s overriding focus, and by 2003, the Africa division was doing all it could to stave off the transfer of its personnel to Iraq and Afghanistan. Even at the Pentagon, which was developing new military-to-military and counterterrorism contacts in Africa's oil-producing regions, Equatorial Guinea "just wasn't up there" as a priority for American intelligence collection, said Theresa Whelan. "We had no access."[20]

Was it credible to think that neither the Central Intelligence Agency nor the Pentagon knew about the coup in advance? South Africa's intelligence service plainly did, and Britain's picked up advance word as well. Either the United States was so obsessed with terrorism and Iraq that it did not have the capacity to pick up not-so-top-secret reporting by allies, or it did know, and successfully buried its awareness after the coup failed.

The evidence from the plotters' testimony, although tainted by the circumstances of their various incarcerations, does suggest that Prime Minister Aznar was involved, or at least was informed in advance. (Aznar, through his office, has denied that allegation.) If Aznar was involved, it is at least conceivable that he warned the Bush White House in advance about what to expect in Malabo, using a narrow, closely held intelligence or White House channel that did not reach the wider American security bureaucracy. Even that theoretical possibility seems doubtful, however. Tipping Bush in advance would risk soliciting the White House's objection to the coup, not least because of American corporate oil investments in the targeted country.

The coup attempt did catalyze the Bush administration to move closer to Teodoro Obiang Nguema. The exchange between the C.I.A.'s Gamble and his Equato-Guinean counterpart at the Libyan intelligence conference suggested the flavor of the administration's dilemma. Obviously, Obiang now had reason to doubt the United States—the Riggs scandal and the purchase of the coup plane from Kansas would have made anyone in his position nervous. The policy of cautious ambivalence that the administration had pursued toward Obiang's regime, notwithstanding the enormous investments in the country by American-headquartered corporations, now had to be reexamined. Among other things, if the Bush administration did not reach out quickly to Obiang, he might assume the worst about the events in March and reassess his commitment to Exxon-Mobil, Marathon, and Hess.

On June 18, 2004, the administration delivered to Obiang what he had been seeking since Bush's inauguration—a high-level meeting to discuss "in depth," as Secretary of State Colin Powell put it, the enduring ties between the United States and Equatorial Guinea. Obiang arrived at 2 p.m. at the State Department in Foggy Bottom. He ascended a special elevator and sat down with Powell in the secretary's outer office. The White House's senior director for African affairs, Cindy Courville, joined the meeting, along with several other senior American diplomats. Obiang was thrilled: "I feel as if I am meeting with President Bush himself," he declared.[21]

Obiang tried to explain to Powell and his aides the "complicated" situation his country faced because of its rough neighborhood and its distinctive status as the only Spanish-speaking country in Africa. He said Equatorial Guinea had been very poor in the past, but now enjoyed the benefits of being an oil producer. He wanted to move toward democracy, he explained. He also wanted the Bush administration's help in "protecting U.S. investments in oil and natural gas" in his country. He pointed out that he had applied years earlier for a license to pay American trainers to increase the skills of Equatorial Guinea's national police and armed forces, but so far he had been refused. He asked Powell to reconsider. He wanted to "modernize" his "military and security forces." He wanted M.P.R.I., an American company, to take this mission on. Powell agreed to review the request.

The secretary mentioned that he had "heard about" Equatorial Guinea's "recent banking problems."

Obiang replied that the Riggs matter was "not very clear" to him. He explained the history of Equatorial Guinea's deposits and dealings at the Washington bank. He had made these banking arrangements in part because "the U.S. oil companies stipulated they preferred to deposit oil payments into a U.S. bank." Obiang said he had to personally approve every payment from the Equato-Guinean treasury, and therefore, he "did not understand the allegations that oil payments went to him personally."

The logic of the president's explanation was not obvious. All Powell could think of to say was that the United States "supported" Equatorial Guinea in its "efforts to resolve these banking problems." The secretary added that he hoped Obiang would use his country's oil "windfall" to help his people, and that he would "get it right" and not repeat the mistakes of other African oil producers. Powell added, "We are here to be a friend and not to preach or lecture."

Amos Hochstein and his colleagues at the Cassidy lobbying firm saw the failed coup and the meeting with Powell as the opportunity their client needed to break out of Washington's punishment box reserved for notorious African dictators. The American government coddled dictators in the Arab world, but authoritarians in Africa enjoyed less margin for error. The attitude of career foreign service officers at the State Department had long been, "This country doesn't matter," recalled one of Obiang's advisers. "It's a bunch of crooks running a dictatorship in the middle of nowhere."

Hochstein approached the White House. He proposed that the National Security Council come up with a "road map" of governance changes that, if implemented by Obiang, could create conditions for an expansion of American security assistance to the Malabo regime. Cindy Courville, at the National Security Council, was assigned to work on Cassidy's proposal. The Bush aides laid out steps Equatorial Guinea would have to take to win American favor: prisoner releases, followed by substantial public investments in health care and education.

The State Department prepared new "talking points" for meetings with Obiang's ministers and representatives to outline concrete steps Equatorial Guinea could take on "human rights benchmarks" in order to win quick approval for the M.P.R.I. license to train the country's still-notorious police and military. "As you know, this license was previously granted on a very limited basis, due largely to human rights concerns," talking points written seven months after the coup attempt noted. "While we continue to have concerns, we recognize and applaud you for your leadership in making advances in this area. We are prepared to work with you and your Government to reach agreement on the conditions under which we could approve the license in its entirety." The license would be "subject to a series of criteria and conditions" involving Equatorial Guinea's human rights performance. State had already "discussed" the proposal with M.P.R.I., which had accepted it.[22]

Separately, and even more quietly, the Bush administration encouraged Obiang to develop a commercial and security partnership with Israel, whose military and internal security specialists had developed a global business out of selling advisory, training, intelligence collection, electronic surveillance, and arms supplies to small, weak regimes in difficult places.

As he processed these offers and developments in the coup attempt's aftermath, Obiang "made an intellectual decision that the U.S. was not involved," an adviser recalled. "He needed to believe that." He was prepared by late 2004 to entrust his security to the United States and to Israel.[23]

Cassidy scheduled a meeting between Obiang and Mel Gamble of the C.I.A. Obiang's evolving plan was to have M.P.R.I. and Israeli forces work with his military and national police, but he also wanted to propose that the C.I.A. train Equatorial Guinea's intelligence service to help protect against future coup attempts that might endanger American oil companies. The Bush administration, for its part, set aside its qualms: The administration approved the expanded M.P.R.I. license to provide internal security training to Obiang's regime.

Gamble met Obiang at The Pierre hotel in New York. Amos Hoch-stein, the Cassidy lobbyist, and an interpreter joined the meeting.

"Yes, I think we can do this," Gamble told Obiang, referring to the plans to strengthen Obiang's security forces. "But you've got two issues. One is human rights. Two is, how are you going to pay for this?"

Obiang answered that he understood the C.I.A. had enough money in its budget to pay for these sorts of intelligence training programs.

Gamble was not about to recommend to his superiors at Langley that the C.I.A. divert budget funds from its global campaign against Al Qaeda to shore up the oil assets of a small-time dictator, even if ExxonMobil and other American companies might benefit. Gamble spoke some Spanish, but he turned now to the interpreter. "Tell the president that we're just a small service with a small budget compared to the money that he has in his bank account."

"I'm not telling him that," the interpreter said.

"Look, tell him, or I'll say it to him in my broken Spanish."

The interpreter started hesitantly. "My boss wants to say . . ."

Obiang laughed. "Okay." He understood, he said.

"Mr. President," Gamble said, "we're in business."[24]

The next time coup makers cast eyes on Equatorial Guinea and the property of ExxonMobil, Marathon, and Hess—as they would, soon enough—they would reckon with new defenses. Other authoritarian lead-ers might have grown frustrated with all the human rights talk in Wash-ington and moved on to a security and oil partnership with China or France by now. Obiang, however, believed that he would be more secure with the United States than with any other global power. Cassidy helped keep him tethered to Washington; Amos Hochstein resigned from the account, but others at the firm continued to work the White House and the State Department to deepen security and economic ties as much as possible. Instinctively, with the aid of his checkbook, and in spite of his prolonged resistance to American ideas about how he should organize his government and protect the rights of Equato-Guinean citizens, Obiang had found American assistance to change his regime—for the purpose of reinforcing it.

A ndrew Swiger had been rising rapidly through ExxonMobil's management ranks when, soon after the coup attempt, he was selected by Lee Raymond and the Management Committee in Irving to explain the corporation's dealings with Obiang to the United States Senate. Democratic staffers on the Senate's Permanent Subcommittee on Investigations had been looking into Riggs for more than a year; they had scheduled a hearing to review and publicize their findings.

Swiger had taken charge of Africa operations shortly after the Mobil merger. There was almost nothing about ExxonMobil's relationship with the Obiang regime that the corporation wished to discuss in public, other than its charitable campaign to fight malaria. Asked occasionally about its ties to a government with such a poor human rights record, ExxonMobil spokesmen repetitively and briefly stated that the company followed the law and condemned human rights violations "wherever they may occur." The Riggs bank scandal had exposed many of the details of ExxonMobil's financial ties to Obiang—its rental of land from regime officials, its investments in businesses controlled by Obiang relatives, and its funding of scholarships for elite Equato-Guinean children selected by the president. The Justice Department opened an inquiry into whether the corporation might be operating in violation of the Foreign Corrupt Practices Act (F.C.P.A.). Now Swiger would have to explain and defend ExxonMobil's decisions.

Early on the morning of July 15, 2004, he arrived on the third floor of the Dirksen Senate Office Building on Constitution Avenue, just to the north of the Capitol dome. Norm Coleman, a Republican from Minnesota who chaired the investigations subcommittee, gaveled the hearing to order at 9:06 a.m.

Swiger read a prepared statement about ExxonMobil's work in Equatorial Guinea. He explained why the corporation's lawyers and executives had concluded after painstaking reviews that its investments in real estate and businesses in Malabo that were controlled by Obiang and his close relatives were legal under the Foreign Corrupt Practices Act and proper as a matter of corporate responsibility. Essentially, the corporation's de-

fense was that it was exempt from some of the normal F.C.P.A. requirements because in Equatorial Guinea, there was no market for local services
other than that provided by the Obiang family. This was indeed an established defense to F.C.P.A. allegations in such circumstances. Equatorial
Guinea "has no law limiting or even defining conflict of interest. Most
ministers continue to moonlight and conduct businesses that often con-
flate their public and private interests," a State Department cable from
Malabo reported. "There are no arm's-length transactions here."[25]

"ExxonMobil is committed to being a good corporate citizen wherever we operate worldwide," Swiger read out to the Senate committee.
"We maintain the highest ethical standards, comply with all applicable
laws and regulations and respect local and national cultures. These principles and practices apply to our operations in Equatorial Guinea. . . .

"The practical realities of doing business in developing countries
are challenging. Equatorial Guinea, like many developing nations, has
a limited number of local businesses and a small population of educated
citizens. . . . In such countries it is sometimes necessary to do business
with a government official or a close relative of a government official. But
it is still expected that we do business ethically and comply with all U.S.
and local laws."

Carl Levin, the committee's ranking Democrat, took up the questioning. He criticized ExxonMobil for failing to cooperate with the committee's investigators. Amerada Hess and Marathon had been fully cooperative
with the Senate committee's probe, but ExxonMobil had stonewalled,
he said.

Levin then asked Swiger whether improving the social and governance conditions in Equatorial Guinea was a condition of ExxonMobil's
decision to do business there.

"It is not, Senator," Swiger said.

"Does it trouble you that you have a business partner like this
dictator?"

"Business arrangements we have entered into have been entirely commercial, have been at market-based rates, arm's length transactions, fully
recorded on our books," Swiger answered robotically. Then he seemed to
improvise a little: "They are a function of completing the work that we're

there to do, which is to develop the country's petroleum resources, and through that and our work in the community, make Equatorial Guinea a better place."

"Make it what?"

"A better place."[26]

Fourteen

"Informed Influentials"

James Rouse, the U.S. Army veteran who ran ExxonMobil's Washington office, retired in 2004. Lee Raymond appointed Dan Nelson, previously the lead country manager in Saudi Arabia, as his successor. Nelson stood six feet eight inches tall. With his silver hair, broad shoulders, and Naval Academy–bred deportment, he seemed to embody the popular image of an oil industry lobbyist; among other things, he looked like someone who might be coming or going from a steakhouse. In fact, The Prime Rib on K Street, downstairs from the ExxonMobil office, was one of his favorite haunts. Through his background as a U.S. Marine infantry officer, Nelson had credibility with the war-saturated Bush administration, although in private, he could be skeptical about Bush's military activism abroad.

One of Nelson's closest friends was Chuck Hagel, the Republican senator from Nebraska, Nelson's home state, and a fellow military veteran. Hagel was a leading opponent in Congress of the Kyoto Protocol and other prescriptions to control greenhouse gas emissions; he also was an increasingly outspoken critic of President Bush's foreign policy. Hagel's outlook was not easy to categorize, but in general, he saw himself as a

skeptical realist about the ability of the United States to coerce and transform other nations, and he was put off by the belligerence of the Bush
administration. Nelson increasingly shared Hagel's views. The ExxonMobil chief lobbyist characterized himself to colleagues in Washington as
fiscally and economically conservative, but a realist in foreign policy and
a libertarian on social issues such as gay marriage. Increasingly, Hagel,
Nelson, and other Republican realists in town worked on Lee Raymond
to rethink his associations with the more outspoken, militarily activist
sections of the Republican Party, those shorthanded as the "neoconserva-
tives," such as some of the scholars and advocates at the American Enterprise Institute, a free-market think tank. Raymond was in the running to
become A.E.I.'s outside chairman, but Nelson warned him that while the
institute had plenty of economists with whom Raymond would agree, its
foreign policy thinkers had become doctrinaire and were too activist to
be aligned with ExxonMobil's worldview.

Nelson built connections to Democrats as well. He and his wife
bought a $2 million town house on Leroy Place in Washington's historic
Kalorama area. Their neighbors happened to include Phillip and Melanne
Verveer; the latter was a longtime confidante of Hillary Clinton's. The
Verveers got to know Nelson and persuaded him to encourage ExxonMobil to support a program called Vital Voices, designed to empower women
in developing countries.

Nelson had no particular experience in lobbying. He had what ExxonMobil valued more: an insider's knowledge of the oil industry, as
well as business and political credibility, particularly in the eyes of Republicans.

"It's time to do things differently," Raymond told his K Street lobbyists
around the time that Nelson arrived in Washington. He didn't specify
what he meant. Raymond had no complaints about the departing James
Rouse, but the change in leadership offered a chance to become more
active, more visible—not so much to lobby on specific legislation, but to
try to educate Washington more successfully about ExxonMobil. Nelson
expanded the number of outside lobbyists under contract with the corporation, building a network of about twenty former senators, congressmen,

Capitol Hill chiefs of staff, and regulatory specialists to support the in-house K Street team.

Energy policy debate in Washington tended toward all-or-nothing pronouncements that were divorced from technical and economic reality—hydrogen would be the next big energy source, or ethanol, or wind. Raymond retained his long-held biases against federal subsidies for alternative energy, but he had learned through bitter experience that it was easiest to make his case by talking about the energy industry's global structure and the embedded place of oil, coal, and gas. One of Raymond's goals as Bush's second term began was to launch an education campaign about fossil fuels in Washington.

On April 13, 2005, Raymond arrived at the White House with Dan Nelson. They passed through security at the entrance to the West Wing and crossed the carpeted hallways lined with photographs of the president to meet Allan Hubbard, the National Economic Council's director and a close friend of Bush's. Hubbard had attended graduate school at Harvard with the president, during Bush's carefree period.

The president increasingly harbored doubts about America's dependency on oil imports. Global oil prices had been rising steadily since 2004, from about $25 per barrel to above $40 per barrel. Rising demand from China and India, the Iraq War, and instability in Nigeria were among the reasons. Higher oil prices had sent retail gasoline prices in the United States soaring, touching off a wave of popular anger and threatening the pace of the country's recovery from the 2001 recession. At the White House and the National Security Council, midlevel aides met continually to discuss policies and diplomatic strategies that might ease oil prices. The president seemed restless about the subject. He remained skeptical to agnostic about climate change. Bush also understood that global oil markets were liquid and interdependent and that "energy independence" was at best a complicated goal for the United States, if it was realistic at all. Nonetheless, he seemed increasingly focused on the costs the United States paid in security and in its economy for its reliance on volatile, expensive imported oil.

Bush thought out loud with his advisers about ways the United States might change the pattern of its relationship with the Middle East. He displayed excitement and curiosity about nascent hydrogen technologies that might revolutionize automobiles and eliminate oil as a source of transportation fuels. The president had hardly turned against the oil industry—he remained an ardent supporter of expanded domestic drilling, for example—but he was asking questions in private about whether and how it might be possible to find a technological breakthrough that would end America's dependency on oil imports within a single generation.[1]

Bush's friend and adviser Al Hubbard became the vessel of the president's ambivalence. He was a principal liaison for ExxonMobil and other oil lobbyists, and they had trouble figuring out where Hubbard was coming from on their issues. It almost seemed as if George W. Bush "felt like he needed to do something that disassociated him with the traditional oil and gas" corporations and yet, simultaneously, the president "was always very supportive" of ExxonMobil and the industry, recalled one executive involved.

In the face of the creeping White House doubts, ExxonMobil applied its standard medicine: PowerPoint education, laden with forecasting data. The meeting with the president's leading economic adviser would be just one in a series, part of a sustained campaign to impress ExxonMobil's energy policy analysis on decision makers.

With Dan Nelson seated beside him, Lee Raymond told Hubbard that ExxonMobil had recently completed a detailed analysis of the world's energy economy, looking out at the next twenty-five years. The forecast made clear, Raymond said, that much of the popular debate about transformational alternative energy sources was misinformed—it was laced with unrealistic fantasies about the pace at which the world's energy economy would or could change. Oil and gas were here to stay, ExxonMobil's economists and planners had concluded; fossil fuels would be central to global economics and security until 2030 and beyond. Raymond sought to brief this forecast to as many staff in the Bush administration and Congress who would listen. Raymond and Nelson offered to bring one of the ExxonMobil forecast's authors, Scott Nauman, to Washington to present the findings in detail to White House policymakers. Hubbard agreed; he asked Vice President Cheney's energy aide F. Chase Hutto III to make the arrangements. The next day, Hutto fired off e-mails to schedule ExxonMobil briefings for White House aides, environmental policymakers, and officials at the National Security Council.[2]

In Washington and elsewhere that spring, ExxonMobil advanced a carefully designed, research-tested campaign to persuade political and media elites that while the oil industry should not necessarily be loved, it should be understood as inevitable. The "Conceptual Target" for this education and communications campaign, according to a 2005 ExxonMobil public affairs document, would be "Informed Influentials."

These were people who "seek to be informed and pride themselves on being able to handle complex issues." They would come from "all walks of life," such as business, government, and the media, and they would be "aware of, and concerned about, the current debate and issues surrounding the world energy resources/use as well as climate change." The ideal audience would be "open-minded," as well as "information hungry" and "socially responsible." The characteristics of the elites ExxonMobil sought to educate were derived in part from statistical modeling that Ken Cohen's public affairs department had commissioned in the United States and Europe, to understand in greater depth the corporation's reputation among opinion leaders. That model had allowed Cohen and his colleagues to forecast how elites would react to particular statements that ExxonMobil might make or actions it might take. The research found, among other things, that it would be beneficial for the corporation to brief elites about the findings of its in-house analysts' long-term forecasts about the global energy economy.

The purpose of the campaign would be to "grow understanding and respect for [ExxonMobil's] position [about] the tough energy challenges the world faces."[3]

The archives of ExxonMobil's Corporate Strategic Planning department contained twenty-year forecasts of energy demand and oil prices from as long ago as the 1940s. Economists, analysts, and executives presented the projections to the Management Committee each year. In 2000, as he oversaw the first forecasts generated by the combined planning departments of Exxon and Mobil, Lee Raymond had asked the analysts, "What did you say about 2000 in 1980?"

Raymond's subordinates "immediately thought that what I was trying to do was criticize them," he recalled. That was, in fact, the typical impression he made.

"No, no, no," he assured them. "What I'm trying to understand is, what did we miss? What things didn't we see right?"[4]

It turned out that in 1980, Exxon's forecasters had been half right and half wrong about the future. They had correctly predicted, within 1 percent, the total amount of energy the world would consume in 2000—a remarkable feat. They had been wildly off, however, in forecasting oil prices; the price trends they had predicted, following the spikes and upheavals of the 1970s, had been much too high. Analyzing this failure, Raymond and his colleagues reached two conclusions. One was that they had badly underestimated the pace at which technological improvements within their industry would make it easier over time to find new deposits of oil, increasing global supply and tamping down prices. The second was that geopolitical disruptions played such an important role in the price of oil that normal forecasting based on supply and demand equilibrium was not realistic to pursue.

Raymond decided to stop asking for price forecasts as part of Exxon-Mobil's long-term planning process. For one thing, the forecasts were so chronically inaccurate that they provided a built-in excuse for any manager whose project failed to meet financial expectations; the manager could just blame the economists for their inaccurate price predictions. "We cannot forecast the price of oil in the short term—so how do you run the business?" Raymond asked his colleagues. The answer, he said, was to manage on a "steady-as-you-go basis and try to make sure the fundamentals are right." Rather than forecasting price, Raymond decided to concentrate instead on predicting volumes—the amount of oil and other energy sources global consumers would demand over time, and also the amount of available supply.[5]

The internal forecasts typically looked out two decades, although some went longer. They complemented the extended cycles of ExxonMobil's capital investments—up to fifty years, in the case of some oil and gas extraction projects, and up to a century, in the case of the longevity of its American refineries. Around 2004, the corporation's forecasters began to shift their baseline target date to 2030. The work they completed seemed compelling enough to form the basis for the education campaign aimed at Informed Influentials. By the time of Raymond's visit to the White House, the corporation had ordered up a glossy book filled with colorful charts entitled, "The Outlook for Energy: A View to 2030."

The forecast opened with a comprehensive picture of the present. In 2005, the world's 6.4 billion people consumed about 245 million barrels per day of "oil equivalent" energy—that is, actual barrels of oil and other liquids (84 million of those) and the equivalent of 150 million barrels per day of other sources of energy, such as natural gas, coal, hydropower, nuclear power, biomass, wind, and solar power. To calculate how this portrait might change by 2030, the corporation's analysts first adopted the World Bank's prediction that the world's population would grow to 8 billion. They then examined, one by one, the economic growth prospects for about one hundred different countries and regions worldwide. Historically, ExxonMobil's analysts believed, the pace of a country's economic growth typically explained about two thirds of its changes in energy consumption; population changes explained only about one third. Economic activity, in other words, not the number of people, would be the most important factor in future energy demand. When they added up all of their individual country predictions, ExxonMobil's analysts concluded that the world's economy would grow on average by about 3 percent per year until 2030.[6]

In each country they also examined what types of energy were likely to be consumed—how much transportation fuel for cars and trucks, and how much energy for generating electric power. They assumed, based on the historical experiences of the United States and Europe, that as poor people around the world grew richer, they would buy more and more cars. They calculated that national populations would feel sated in their automobile consumption only when they reached about eight hundred cars per one thousand people, a rate of ownership that America and the European Union were approaching. The ExxonMobil forecasters made additional assumptions about the rate at which hybrid cars were likely to be adopted, the rate at which office buildings and refrigerators would become more energy efficient, the rate at which wind farms and nuclear power plants would be built, and the rate at which governments around the world would impose taxes on carbon-based fuels or caps on greenhouse gas emissions.

They concluded that worldwide energy demand would grow by about 35 percent overall by 2030 and that demand for oil and gas liquids would rise by about 22 percent, to 108 million barrels per day. Far from a green or clean energy future, they foresaw that energy-poor countries would burn fossil fuels increasingly as they industrialized. Flat or declining oil consumption in the United States and Europe, due in part to more efficient hybrid cars, would be more than offset by gasoline consumption in Asia's fast-growing economies, particularly in China, where ExxonMobil's forecasters expected that the number of cars and light trucks in service would grow from about 12 million in 2005 to about 110 million in 2030. (By comparison, there were about 220 million vehicles in the European Union in 2005, and about 240 million in the United States.)[7]

The transportation sector—cars, pickup trucks, heavy trucks, airplanes, ships, and trains—was the most important factor in the global market for liquid oil. Three quarters of the roughly 20 million barrels of oil the United States consumed each day was as transportation fuel; the rest went to industrial uses, such as the manufacture of plastics. Virtually no oil went to generate electricity—coal, natural gas, hydroelectric, and nuclear energy provided the main sources of electric power generation. It drove the analysts and forecasters in ExxonMobil's Strategic Planning department in Irving crazy when they heard radio talk-show hosts and politicians advocate that the United States should quickly build more windmills to free itself from dependency on oil imports from the Middle East; unless all-electric cars and vehicles spread very rapidly in the United States, windmill construction, whatever its pace, would have little impact on the amount of foreign oil the United States consumed.

Cohen's public affairs colleagues digested the 2030 analysis into a series of PowerPoint slides and texts. After 2004, the forecast became the predominant topic of speeches and briefings delivered by ExxonMobil executives and managers around the United States and in Europe. ExxonMobil systematically scheduled private briefings with policymakers, background sessions at think tanks, talks at universities and colleges, presentations to Wall Street analysts, and speeches at economic clubs and chambers of commerce. The rollout had all the automated, charmless tone of other O.I.M.S.-influenced campaigns by the corporation—a tsunami of color-coded pie charts, bar graphs, and global maps, read out unemotionally by executives wearing dark suits. By placing ExxonMobil's presentations, speeches, and lobbying briefs in a dense vernacular of statistics and economic forecasting, the 2030 campaign sought to reposition the corporation by eschewing political and ideological arguments that often provoked instant and emotional resistance from opponents. Instead, the corporation would let the facts, as ExxonMobil's analysts conceived them, speak for them. "Realistic" and "reality check" became two of Lee Raymond's favorite phrases as he presented and analyzed the 2030 forecast in public appearances.

"I note that Raymond is no longer seeking to gainsay the science behind climate change," Andrew Warren, director of the Association for the Conservation of Energy in Great Britain, wrote in frustration after sitting through one of the chief executive's presentations in London, early in 2005. "Instead he simply predicts an endless rise in the demand for the fossil fuels his company sells, and maintains that there is nothing that can be done to alter that."[8]

This was, in crude summary, the judgment ExxonMobil sought to infuse through its elite-targeted education campaign. Hardly anyone outside of the industry truly grasped the gargantuan scale of global energy production. Titanic changes in the patterns of energy use over decades would be required to create even modest changes in fuel consumption patterns. ExxonMobil's analysts did not downplay alternative energy's prospects. They projected that solar, wind, and other rising alternative sources would grow very rapidly until 2030—by more than 10 percent per year. Yet, because of the concomitant increases in worldwide economic activity and population, at the end of the forecast period wind and solar would still make up only about 2 percent of total supply.

Growth in oil consumption was inevitable, ExxonMobil's analysts held, because the movement of large numbers of poor people into wealthier lifestyles was also inevitable, particularly in Asia. Did anyone seriously expect middle-class Chinese or Indians to fashion their lifestyles and buy cars any differently from how Japanese, Koreans, Germans, or Californians had done? The oil industry's growth patterns would shift toward Asia, but the industry's expansion and profitability seemed assured.

ExxonMobil's 2030 exercise suggested, by implication, the distinctive role that climate policy would play in oil's medium-term future. The essence of the forecast's message was that the development of the global economy and population ensured that oil production would rise. By mid-century, some breakthrough in battery technology or solar panel arrays might reduce the costs of those energy sources so radically that they could compete economically with oil and coal in free markets, but ExxonMobil's in-house scientists did not believe such a breakthrough was conceivable before 2030. Until then, there was only one unexpected development, one "black swan" intervention that could shift the curve of rising global oil demand: a decision by governments to limit greenhouse gas emissions by heavily taxing or capping the use of carbon-based fuels.

The ExxonMobil forecast numbers suggested that to make an impact on oil demand, the world's governments would have to reach a unified conclusion that climate change presented an emergency on the scale of the Second World War—a threat so profound and disruptive as to require massive national investments and taxes designed to change the global energy mix. European governments had come closest to attempting such a policy, and ExxonMobil's forecasters had figured Europe's carbon pricing policies and alternative energy subsidies into the 2030 numbers. To reshape the global oil industry, however, the governments of China, India, the United States, and many other countries would have to adopt similar or even more aggressive carbon taxing policies. ExxonMobil's planners concluded that this was highly unlikely, if not all but impossible; they predicted, therefore, that CO_2 emissions would rise by an additional 30 percent worldwide between 2005 and 2030.

The corporation's forecasters assumed, essentially, that the world's governments would lack the political will to tax fossil fuels heavily enough to force any big shift away from oil. The issue here was not whether the world had the technologies to forswear oil; it was whether governments, panicked about climate change, would intervene to change price incentives to favor clean energy, knowing that such an intervention might curtail overall economic growth, at least for a time. In August 2004, the Princeton University scientists Robert Socolow and Stephen Pacala published an influential article in *Science* that declared, optimistically, "Humanity already possesses the fundamental scientific, technical, and industrial know-how to solve the carbon and climate problem for the next half-century. A portfolio of technologies now exists to meet the world's energy needs over the next 50 years and limit atmospheric CO_2 to a trajectory that avoids a doubling of the preindustrial concentration. Every element in this portfolio has passed beyond the laboratory bench and demonstration project; many are already implemented somewhere at full industrial scale." However, Socolow estimated that the technologies he and his coauthor had in mind—solar, wind, and nuclear power, among them—would require a carbon tax of about $100 per ton to be economically competitive fast enough to stabilize emissions before midcentury.[9] For world governments to enact such a tax, or set equivalent caps on greenhouse gas emissions, they would have to be galvanized by deep fears about a warming world.

Raymond continued to fund advocacy groups that promoted skepticism of mainstream climate science; he considered such funding just another example of the corporation's possessing the courage of its convictions when others lacked them. ExxonMobil traded spots from year to year with Walmart as the largest corporation in the United States, by revenue, and its reach and influence continued to exceed that of many of the world's midsize governments. Its employees, retirees, shareholders, and customers numbered in the millions. Lee Raymond did not believe, however, that ExxonMobil's scale required it to act as some sort of consensus-building institution on matters of public policy.

The criticism he received for funding anti-Kyoto groups was exaggerated, Raymond told a reporter. "The facts are you don't have to spend a lot of money to aggravate the proponents" of greenhouse gas limitations. "We think we have a responsibility. If we think people are about to make some bad policy decisions that are going to have a big impact for a long period of time, somebody's got to say something."[10]

William Freudenburg's work as a sociologist at the University of Wisconsin touched upon environmentalism, law, and society. He had earned his doctoral degree at Yale University and had published over the years in academic journals on subjects such as risk assessment. "A funny thing happened to me one day when I picked up the telephone," he recalled in an essay published in *Sociological Forum* in March 2005. "I learned something new about the mechanisms of corporate influence in science."[11]

As ExxonMobil appealed the punitive damages verdict imposed against the corporation by Alaskan jurors in the *Exxon Valdez* oil spill case, it funded a complex, quiet campaign to bolster its prospects. The effort unfolded in tandem with Ken Cohen's 2030 forecast campaign and the corporation's residual attempts to seed doubts about climate science. Freudenburg's experience was distinctive in part because it offered a rare, contemporaneously documented account of the strategic analysis that undergirded ExxonMobil's most subtle forms of campaigning to shape policy and ideas.

One of the corporation's executives telephoned Freudenburg to explore whether he might accept funding to develop an article about the impact of punitive damage awards on American society. "Naturally, we have a range of expert witnesses and so forth, but we find that it's also helpful to have people working on articles that come out in academic publications," the executive explained. "We've often worked with economists, for example. A lot of them feel that punitive damage awards are very inefficient, compared to other approaches such as regulation. . . . That's a perspective we're quite comfortable in supporting. But we're exploring whether we might want to work with professors in publishing things from a few other perspectives, too.

"Basically, what we're exploring is whether it's feasible to get something published in a respectable academic journal, talking about what punitive damage awards do to society, or how they're not really a very good approach," the ExxonMobil executive continued. "Then, in our appeal, we can cite the article, and note that professor so-and-so has said in this academic journal, preferably a quite prestigious one, that punitive awards don't make much sense. . . ."

Freudenburg scribbled notes; he decided that the details of corporate influence strategy he was absorbing might ultimately be more interesting than the commissioned consulting work ExxonMobil had in mind. He decided to string out the offer, not to undertake it, but to study its purpose.

His handler continued: "Or maybe it could be something along the lines of how difficult it is to prevent these kinds of things [accidents like the *Valdez* wreck] under any circumstances. It's a little like the *Challenger*. . . . The people involved weren't really all that venal."

A few days later, Freudenburg spoke again with his ExxonMobil contact. He asked questions about how the corporation constructed its influence campaigns. He found that the ExxonMobil executive assigned to him "doesn't come off at all like an ogre." He was always careful to stress the corporation's "interest in a rational approach."

Freudenburg asked how publication of an essay in an obscure academic journal that hardly anyone read could be of any help to a corporation as large and well resourced as ExxonMobil. The executive admitted that such work "wouldn't do much good" with trial juries, who tended to reach their verdicts on a "nonfactual" basis. Once a case was appealed to panels of judges, however, the prospects to shape their thinking improved. ExxonMobil would submit a copy of the academic journal article with its legal briefs. "The judges themselves don't usually read them, but often their clerks will read them . . . and quite a few of the clerks, nowadays, are pretty open to these kinds of arguments. . . . Quite a few of them now come out of a law and economics program or something like that. . . .

"It's possible to offer small amounts of support to academics who already show some tendency to express views that [ExxonMobil] finds congenial," the executive said. In addition, "You can sponsor workshops and so forth, but that gets tricky. For one thing, once you get to that point, you pretty much have to invite both sides."[12]

Eventually, ExxonMobil submitted findings from this academic work to the United States Supreme Court to support its challenge to punitive damages arising from the *Exxon Valdez* spill. The decision ultimately went the corporation's way and made important new law favorable to American businesses. David Souter, the Supreme Court justice appointed by President George H. W. Bush, joined in the majority's opinion. In a footnote, however, Souter mentioned the social science evidence submitted by ExxonMobil. "Because this research was funded in part by Exxon, we decline to rely on it," he wrote dryly.

Lee Raymond turned sixty-seven years old in August 2005. He had spent almost forty-two of those years as an employee of Exxon and had served as ExxonMobil's chairman and chief executive for a dozen years, a long run at the top by the timelines of corporate America; he was an informal dean of his oil industry class. When he was at headquarters in Irving, Raymond often ate lunch in the subdued formality of the alcohol-free Rockefeller Room with his most senior lieutenants, including, the two who were competing to replace him, Ed Galante and Rex Tillerson. When he traveled out of Texas, he flew on the Challenger Global Express designated as One Hundred Alpha and he lingered in Hawaii, Augusta, and Pebble Beach to play golf and relax. He and Charlene, his wife, were building a new home in Palm Springs, California, and they would soon acquire another home near Phoenix. The Raymonds remained highly private. A luxurious, subdued retirement now awaited them—if the corporation's board of directors could persuade Raymond to take it up.[13]

Some on the board felt they had struggled since 2001 to persuade Raymond that he had to take succession and retirement seriously. Raymond felt he had done so; he had set up a contest between Tillerson and Galante over the top job, a competition that Raymond told his board was entirely genuine—a close call. "It will be a few years before you are able to figure out how good they really are," he had said when he first appointed the pair.

By 2005, fearful of Raymond's stalling, the board "communicated softly" with him that the directors felt that it was time to make a definitive move. The message they delivered was, "Gee, Lee, you're now sixty-six." Raymond understood their worry that he might become, as he quipped privately, "the next Sandy Weill," referring to the banker who had hung on at Citigroup until he was seventy-three, finally retiring as chairman just two years before the bank nearly collapsed from imprudent bets on the American mortgage market. He assured them that he was ready to go. "Believe me, I will never be the new Sandy Weill."

He understood that he was plugging up career movement down the executive ranks, he told the board. He was less certain, he said, about which of the two finalists for his job the board should endorse.

"Why don't you just tell us who ought to do this?" James R. Houghton, a director who served as chairman at the international glass and ceramics maker Corning, Inc., asked at one board meeting.

"I'm not sure that's my job," Raymond answered. "It's the board's job."

"But you're the only guy who really knows them."

"I accept that point, but I am interested in any perspectives any of you have—and you have an obligation."[14]

It seemed clear to some within ExxonMobil and on the board that for all of Raymond's achievements in financial management and corporate strategy, his Dick Cheney–like bluntness had become a liability. Worldwide scientific and engineering talent recruitment and retention as well as lobbying strategy in Europe and the prospect of a post-Cheney Washington all argued for a leader of ExxonMobil, after Raymond, who could maintain the same level of financial and operational discipline, but project a gentler, quieter, more modern and inclusive voice.

It was not obvious which of the two finalists would be the better communicator. Galante had grown up in Queens, New York, and on the South Shore of Long Island. As he rose, he managed Exxon's massive Baton Rouge refinery and later served as Raymond's executive assistant in the years after the *Exxon Valdez* disaster—"the most thankless job in the world," as a former Exxon executive put it, in part because as the chief operations officer on Raymond's staff, Galante had to decide when to wake up the boss with news in the middle of the night. "Sometimes it's not much fun to wake Lee up at four a.m.," the former executive noted. As the succession contest solidified, Raymond appointed Galante to run the corporation's worldwide downstream portfolio—massive refineries from China to the Middle East to the American South. Some visitors found Galante to be affable, outgoing, and comfortable in comparison with his rival, Tillerson. A *Fortune* reporter, Nelson D. Schwartz, went so far as to offer ExxonMobil directors advice in print as they approached their decision: "If either of the candidates to succeed Raymond can address the company's tattered image . . . it's Galante."

Tillerson drawled unabashedly. He was a lifelong Texan bred in its small towns, and as his wealth accumulated, he bought a ranch outside of Dallas. His office in the third-floor executive suite in Irving contained a "Frederic Remington-style sculpture of a horse," which struck Schwartz as part of a style that "might be called masculine not-so-*moderne*." It was true that Tillerson was not one to toss around French terms. Yet he did seem comfortable in his own skin, relaxed, willing to hear different views. He was credible in industry circles but more accessible and less defensive than Raymond when addressing skeptical audiences.[15]

How important were communication skills, anyway? The board was choosing a leader at an operations-focused, highly profitable corporation in an innately unpopular industry, not a game-show host. Raymond knew that his board worried chronically about ExxonMobil's public image. Some of the corporation's outside directors had come to ExxonMobil from liberal university campuses or industries such as retail sales or telecommunications, where a corporation's public reputation was fundamental to the ability to attract customers. Raymond didn't see ExxonMobil or the oil industry as comparable.

"The facts are, with the exception of the service stations, everything we produce has no interface with the public," Raymond told his directors at a meeting in Japan, as the internal evaluations of Tillerson and Galante neared their end. "Crude oil, natural gas, chemical products—the public doesn't know where it comes from. The only interface we have is the service stations."[16]

The retail gasoline stations so ubiquitously visible and so familiar to Americans returned notoriously low profit margins to all large oil companies. If the goal of ExxonMobil was to have a better public reputation, Raymond continued, maybe it should consider getting out of the retail business altogether and become a lower profile, highly profitable industrial company, with visibility more comparable with a company like Dupont. The public and politicians became inflamed when ExxonMobil reported its gargantuan quarterly profits in part because many people thought those profits were extracted from their wallets at the retail gas pumps where they stopped to fill up twice a week. When they drove to and from work or to the grocery store, ExxonMobil signs blared at them from street corners, reminding them of the corporation's presence—and of the rising price of gasoline. Rather than choosing a new chief executive whose job would emphasize the rehabilitation of ExxonMobil before that hostile public, why not just retreat from view?

This was not as radical an idea as it might have sounded, but it was not going to be the basis for the board's succession decision.

After the succession contest was established, at the annual board meeting each October, when Raymond asked all ExxonMobil executives to leave and then spoke to the directors about Galante and Tillerson, he always framed his report by saying that he could offer his views about the men's strengths and weaknesses relative to each other. He did not know, however, how strong a leader either of them would prove to be in an "absolute" sense, tested over many years and compared with other Fortune 500 chief executives. Raymond told the board that the most important quality his successor would require was toughness—the ability to stand up to governments, pressure groups, environmentalists, and special pleaders of all types. His advice was a projection of how he saw himself.

Galante had supporters on the board until the final decision was made. Gradually, however—one board road trip or meeting retreat after another—the weight of opinion gathered around Tillerson. Raymond finally recommended Tillerson directly. He told colleagues that he felt he owed whoever followed him a firm endorsement, so that he would not leave any lingering doubts in the minds of directors who had deliberated over the decision.

ExxonMobil ended the succession contest publicly in 2004 by appointing Tillerson as president and the sole number two. In the summer of 2005, the corporation confirmed that Raymond would retire at the end of the year and that Tillerson would follow him as chairman and chief executive.

I n the last year of Lee Raymond's leadership, ExxonMobil earned a net profit of $36.1 billion, more money than any corporation had ever made in history. That broke the previous record of $25.3 billion, set by Exxon-Mobil the year before. Even if the profits made during the late 1950s and 1960s by such postwar corporate giants as General Motors, Ford, International Business Machines, and General Electric were adjusted for inflation, none could match the size of ExxonMobil's 2005 profit. During the span of Raymond's tenure, from 1993 to 2005, ExxonMobil's market capitalization—the total value of the corporation's shares in the stock market—rose from $80 billion to $360 billion. The company also paid out $68 billion in dividends during that time. It was difficult for an oil corporation of ExxonMobil's size and experience to fail, particularly after oil prices began to rise in 2004. Yet ExxonMobil's performance reflected in substantial part Raymond's relentless focus on cost and efficiency.[17] Raymond's record on behalf of the corporation's shareholders was by now less well known than his record as a self-appointed climate scientist. As part of his transition to retirement, Raymond was in the running to become the chairman of the John F. Kennedy Center for the Performing Arts in Washington, D.C., a prestigious and visible position, but environmentalists in the Kennedy family blocked him. On Wall Street and within the industry, however, Raymond commanded considerable respect. Competitors such as Royal Dutch Shell hosted farewell dinners for him, where he was feted as one of the most accomplished leaders in oil industry history and perhaps the most effective in the United States since John D. Rockefeller himself.

Whose profits were they? Under the law, of course, they belonged to ExxonMobil's shareholders, to be managed for the shareholders' benefit by the corporation's board of directors, subject to the rule of law. In political terms, however, oil profits were distinct. They arose from the sale of energy products, particularly gasoline, that the American public had no practical choice but to purchase. Some energy industry profits—those made from the sale of electric power to homes and businesses—were capped and regulated in the United States by state utility commissions whose mission expressly included protection of the public interest. It was in some respects an accident of American political history—as well as an expression of the enduring power of the largest oil corporations—that electric energy was treated as a public entitlement subject to close regulatory scrutiny, while gasoline was not. Even setting aside all ideological arguments about the costs and benefits of free versus regulated capitalism, the incentives ExxonMobil and its peers followed—Wall Street signals, competitive signals, and obligations under the law to maximize shareholder value—had practical consequences for working- and middle-class families. As *Petroleum Intelligence Weekly* put it, "What many of the companies have in common is a reluctance to sacrifice high financial returns for stronger output growth." There were surely many efficiencies in this system, but one of its problems proved to be poor long-term performance and underinvestment by the big companies in oil exploration and production, which contributed to tighter supply and more volatile prices that occasionally socked American consumer budgets unexpectedly.

Unarguably, the margin for error in the global oil supply system was shrinking. Just a few weeks after ExxonMobil announced Lee Raymond's prospective retirement, Hurricane Katrina gathered force over the Bahamas, crossed into the Gulf of Mexico, and came ashore near New Orleans; the storm claimed more than eighteen hundred lives and caused about $80 billion in property damage. A month later, Hurricane Rita smashed into Texas and caused about $11 billion in damage. America's five largest oil refineries lay in the paths of the two storms. Chaos and shutdowns in the gasoline supply chain caused retail gas prices in the United States, which had been rising steadily during the previous year, to spike suddenly toward three dollars per gallon. In general, gasoline prices rose and fell in tandem with global prices for crude oil, but occasionally, as in this case, bad weather or strikes or other local disruptions could cause a spike upward. Within a few weeks, senators and congressmen responded to outraged phone calls and e-mails from angry, financially strapped constituents by introducing legislation to prevent ExxonMobil and other large oil corporations from reaping windfall profits from popular misery. It did not help ExxonMobil's public relations position that it announced on October 25 record third-quarter profits of just under $10 billion.

Lee Raymond flew to Washington on an ExxonMobil jet and arrived around 9 a.m. on November 9 at the Dirksen Senate Office Building on Constitution Avenue. There were few Washington rituals that more aggravated Ken Cohen and his colleagues in Irving's public affairs department than congressional hearings called for the purpose of theatrically interrogating Raymond and other oil industry chiefs about rising retail gasoline prices. Faced with the complex problem of the oil industry's role in America's economy and environment, about which it was not prepared to act seriously, the U.S. Senate could be relied upon to hold inflammatory and partisan hearings. Ted Stevens of Alaska gaveled a joint hearing to order as Raymond took a seat beside David O'Reilly, his counterpart at Chevron, as well as senior executives from Shell and BP. "Energy Prices and Profits" was the title Stevens had selected for the day's questioning.[18]

The eighty-two-year-old senator immediately fell into argument with Barbara Boxer, a liberal Democrat from California, about whether it was necessary to have Raymond and the other witnesses stand before the cameras, raise their hands, and swear to tell the truth, as tobacco industry executives had been forced to do at a 1994 congressional hearing, before they testified falsely that cigarette smoking was not addictive. Stevens refused; he said it was not necessary. "I remind the witnesses as well as the members of these committees, federal law makes it a crime to provide false testimony," he declared.

"Did your company or any representatives in your companies participate in Vice President Cheney's energy task force in 2001, the meeting?" Senator Frank Lautenberg of New Jersey asked Raymond.

He answered in a single word: "No."

Lautenberg moved on; all of the other executives at the witness table issued similar denials. His question, with its reference to "the meeting," was in some respects ambiguous, but Raymond's answer could be defended as truthful only in the most technical, lawyerly sense. He had met one on one with Cheney to discuss the energy task force's broad mission,

ExxonMobil's Washington office had been in contact with the White House during that review and the parallel review of climate policy, and Raymond had spoken with Energy secretary Spencer Abraham a few weeks before the task force finished its work, in what Abraham later called a "telephonic meet and greet."[19] By commonsense definition, these were forms of "participation." It would have been easy enough for Raymond to construct a truthful but self-protecting explanation about his energy policy contacts in Washington, but seven weeks from retirement, he evidently could not be bothered. Nor would he ever be held accountable for his testimony.

With apparent weariness, Raymond addressed question after question from both Republican and Democratic senators about the nature of global oil markets, how prices were set, and what ExxonMobil might do to control them. Afterward, Senate staff composed dozens of questions and submitted them to Raymond. The ExxonMobil chairman spent some of his last hours in power at the corporation signing off on answers to the same fundamental questions about oil, science, and American power that he had been attempting to control for more than a decade.

"The National Oceanic and Atmospheric Administration has projected that the country and the Gulf of Mexico have entered a cyclical period of twenty–thirty years during which the Gulf and coastal areas are likely to experience a greater frequency of hurricanes and higher odds of those hurricanes making landfall in the U.S.," Jeff Bingaman of New Mexico began. "What preparations has your company made to deal with a greater hurricane frequency?"

"Whether there will be a greater hurricane intensity or frequency in the future remains unclear," Raymond answered. "Evaluating the future frequency and impact of weather events is an imprecise and uncertain area of science."

"What is the relationship between the price of oil that Americans are paying and the profits you are making?" asked New Mexico's senator Pete Domenici.

"In fact, the vast majority (approximately 70 percent) of ExxonMobil sales and profits are made outside of the United States," Raymond replied.

"Because oil is a globally traded commodity, the absolute level of crude oil price, established on a global basis, is a key factor impacting American consumer costs and energy industry earnings."

"Do you believe that Americans are dangerously dependent on oil and its refined products?"

"No. The emergence of abundant, affordable energy over a century ago provided a key foundation for the tremendous gains in living standards and quality of life achieved in the United States and throughout the world. . . . We do not view the projections for increases in production from the Middle East as a significant concern."[20]

Lee Raymond leveraged his friendship with Vice President Cheney one last time. ExxonMobil's upstream division was negotiating with the Abu Dhabi National Oil Company over a stake in a 50-billion-barrel complex oil field called Upper Zakum. The government of the United Arab Emirates was willing to sell a 28 percent interest in the field in exchange for technology and engineering work that would enhance production and profitability. The terms offered by the U.A.E. were tough—"The government takes something like 99 percent" of revenue, Frank Kemnetz, ExxonMobil's regional president, remarked. Yet it was an immense prize, one of the largest undeveloped fields available in the world, hosted by a small, friendly emirate that possessed just less than 10 percent of the world's oil and the fifth largest reserves of natural gas. The U.A.E. depended upon American military protection for its very existence, yet American oil companies had managed to secure only 13 percent of the foreign participation available to international majors; European and British firms had 60 percent. The Upper Zakum sale would provide the corporation that bought in with a substantial boost to its booked oil reserves. Initially, ExxonMobil, BP, Chevron, Shell, Total, and a Japanese company submitted bids. The Supreme Petroleum Council narrowed the field to ExxonMobil, BP, and Shell, and the negotiation ultimately came down to a competition between ExxonMobil and Shell. A wide gulf of perceptions emerged between how the U.S. embassy in Abu Dhabi saw its efforts on ExxonMobil's behalf and how the corporation saw them.

The embassy "has a long tradition of advocating on behalf of U.S. oil companies for government contracts and tenders," the post reported to Washington. "Our close and continuing relationships with the powerful elite in the U.A.E. has no doubt led to increased U.S. exports, selection of U.S. firms for various contracts and tenders, and positive resolution for U.S. firms in commercial disputes." Yet in the clinch on the Zakum talks, ExxonMobil believed that State was not doing enough to pry the deal loose from Shell.[21]

Finally, Dan Nelson persuaded Raymond to place a call to Cheney. "What in the hell is with this country?" was the thrust of Raymond's message. The largest corporation headquartered in the United States by profits, a locus of American employment and shareholder wealth, could not persuade State to intervene aggressively in a prospective overseas deal even when the only competitor was a non-American firm? The vice president's office later reported back that even they had been unable to persuade State diplomats to lobby hard for ExxonMobil—an early indicator, perhaps, of Cheney's declining stock during Bush's second term, or else a confirmation of Raymond's long-standing hypothesis about State's general uselessness and antipathy toward American oil companies. Ultimately, ExxonMobil's representatives were told, Vice President Cheney had picked up the phone and called contacts in the U.A.E. government himself. ExxonMobil won the exclusive right to negotiate for the project, pushing Shell aside. Raymond flew to the emirate early in October 2005 and met U.A.E. president Khalifa bin Zayed "in order to allow Abu Dhabi to raise any major outstanding issues" in the deal. The issues that remained were "mostly about money" and Zayed appeared sanguine. And now that ExxonMobil was on track, it wanted the Bush administration to back off: When Energy secretary Samuel Bodman arrived the following month, the embassy briefed him: "ExxonMobil would prefer that we do not carry a strong, specific advocacy message on its behalf for the Upper Zakum bid, citing the sensitive nature of the negotiations and the timing." ExxonMobil could be as maddening a partner for State diplomats as for its peers in the oil industry; the corporation wanted what it wanted, and it was not easy to please. ExxonMobil soon finalized a twenty-year contract to raise production in Upper Zakum; a U.A.E. official involved in the talks em-

phasized that "Exxon's technical proposal was the deciding factor" and
that given the geological complexity of the oil field, Abu Dhabi was "more
interested in know-how" than money. The corporation's "capabilities to
increase oil recovery and efficiently build production capacity were key
considerations" in its success in winning the deal, ExxonMobil reported
publicly. The corporation's executives often claimed that they did not
require favors from the U.S. government, did not take direction from the
White House, and preferred global independence. The reality was more
complex. The corporation had a direct line to Cheney and negotiated with
State and Abu Dhabi as its interests dictated.[22]

Raymond retired on January 1, 2006. Between the day he started
work at Exxon in 1963 and the end of his career, he had seen whipsawing
change in the global energy industry: the nationalizations and price shocks
of the early 1970s, the price collapse of the 1980s, the cold war's end and
the opening of new oil frontiers in Africa and Central Asia, two oil-fueled
wars in Iraq, the emergence of global warming as a threat, and the market-
upending growth of China and India as oil importers. Since 1993, Raymond had steered the corporation through these events with his eye
firmly fixed on ExxonMobil's profits. "This is perhaps the single biggest
and most powerful legacy of Lee Raymond—raw profitability," wrote Paul
Sankey and Adam Sieminski of Deutsche Bank, in an assessment timed to
Raymond's retirement. "The current level of cash flow being generated by
the company is unprecedented by historic standards."[23]

ExxonMobil's return on capital employed, the metric Raymond had
long promoted as the best indicator of an oil company's performance,
came in at 31 percent during his last year, a jaw-dropping number and the
best in the corporation's peer group that year. The gap between Exxon-
Mobil and its competitors in this self-assigned category reflected in part
the superior performance of its chemical business and its downstream
refinery division, where Raymond and his colleagues had driven annual
profits to $8 billion, a fourfold rise in four years. Raymond had risen
within Exxon mainly as a downstream performer and he left behind the
oil industry's "strongest refining and petrochemical businesses—bar none,"
as the Wall Street analyst Mark Gilman put it.[24]

For all of these stellar financial accomplishments, yellow warning lights were blinking about ExxonMobil's future. Annual oil and gas production remained flat, no higher than it was at the time of the Mobil merger, despite repeated promises from Raymond and other executives that production would rise. Upstream oil and gas production generated industry-leading profits because of ExxonMobil's discipline in project investment and operations, but strategically, in Gilman's view, Raymond "did not position the company properly in the upstream business," where most of the industry's profits and potential lay. The Mobil acquisition was a triumph, Gilman believed, but afterward, "they milked the developed inventory that had been previously established—there was little left for his successor to draw on." The rise of state-owned oil companies meant that in the long run, access to new oil and gas properties would require cooperative partnerships with myriad governments and foreign rival companies, but Raymond had exacerbated Exxon's historical "organizational arrogance," and so they were "not, in my view, the favored partners."

Gilman was something of a contrarian about ExxonMobil's performance; many other Wall Street analysts praised the corporation while offering few caveats. But in Gilman's analysis, the failure to find new oil was also inextricably tied to ExxonMobil's relentless drive for superior profits. The very discipline that ExxonMobil bragged about to Wall Street analysts at the annual presentation—its insistence that it would release cash to invest in new oil projects only if the returns would be 15 percent or more annually—had imprisoned the company in a cycle of flat or declining production and reserve replacement struggles. "The basic problem is that their threshold returns are too high—way too high—so they end up with gobs of excess cash" without a sound strategy for long-term production growth. This was not necessarily a problem limited to ExxonMobil: "As big oil announces record profits," wrote the analyst Amy Myers Jaffe, "you have to ask yourself: Can anyone out there find oil anymore? What happened to those wildcatters of yesteryear? Do they only search for stock dividend plays now, not promising extensions to geologic structures?"[25] The corporation claimed that 2005 was the twelfth consecutive year in which it had found enough proved reserves of oil and gas to replace the amount pumped and sold, but this was true only if an investor accepted ExxonMobil's self-generated rules for reserve counting and ignored the rules issued by the S.E.C.

Doubters like Gilman and Jaffe might question long-term strategy, but ExxonMobil's reputation as the best steward of shareholder capital delivered a premium share price, in comparison with its oil industry peers. UBS Warburg estimated toward the end of Raymond's run that ExxonMobil's stock, because of market expectations about its superior future performance, enjoyed a 7.3 percent price premium in comparison to BP's and even more in comparison to Royal Dutch Shell's. The corporation's net income in 2005 was greater than the combined profits of the next five largest publicly traded American corporations, as ranked by revenue. To be sure, sheer size and rising oil prices caused by factors outside of ExxonMobil's control were largely responsible, but Raymond had forged the management systems that put the corporation into position to reap the rewards.

After the 1999 merger, ExxonMobil had jockeyed with Microsoft and General Electric for the status of largest American company by total stock market value, but as Raymond retired, ExxonMobil moved into the top spot. Walmart earned more revenue in some years, but as a retailer, its profit margins were relatively thin. ExxonMobil generated more profit than any other company in the world. The corporation had 83,700 employees and 2.5 million individual shareholders.[26] If that community of interest was defined as ExxonMobil's "population," it was about the same size as tiny, oil-rich Kuwait's, yet the latter's national income in 2005 was only about $100 billion, or less than a third of ExxonMobil's revenue. ExxonMobil citizenship, then, particularly for senior managers and executives with lucrative restricted stock packages, had become highly rewarding, even if it involved constraints on personal freedom that a Kuwaiti subject might also recognize. Even the corporation's lower-paid employees enjoyed wages, retirement security, and income growth unavailable to the vast majority of comparable American workers.

After he formally retired, as part of a $1 million per year transition consultancy, Raymond embarked with his wife, Charlene, on an ExxonMobil Challenger jet from Dallas Love Field to Paris. For almost a month, the Raymonds circled the world on a farewell tour, touching down in Norway, London, and Singapore before they cleared customs and reentered the United States in Hawaii. They stayed a few days on the Big Island. Raymond's journey from Watertown, South Dakota, was over, and his family was now wealthy beyond imagination. Following formulas for executive retirements the corporation had previously established, and taking into account the compensation he had deferred over the course of his forty-year career, ExxonMobil's board of directors awarded Lee Raymond pension benefits as a lump sum of $98 million, restricted shares worth $183 million, stock options with a potential value of $70 million, the $1 million consultancy, and reimbursement of his country club fees. Altogether, his retirement package was worth just under $400 million.[27]

PART TWO

THE RISK CYCLE

Fifteen

"On My Honor"

Ken Cohen's public affairs team in Irving crafted a formal transition plan to shape outside perceptions of ExxonMobil's change at the top. The essential message was continuity. "Clearly the financial model was working," an executive involved said later. "You do not change a winning game." The new Management Committee sought to reassure Wall Street that the corporation's scale of profit making would continue, and that Rex Tillerson was not arriving with grand new ideas about how the business should be restructured. When he met analysts in March 2006 for the first time as chief executive, Tillerson recited the doctrine of 2030 and pledged that ExxonMobil's strategy "remains unchanged."

Tillerson sought nonetheless to reset the corporation's communications with its opponents and the public. "It's true that I wanted to change the tone of ExxonMobil and of our industry," he said later. He felt the corporation was "misunderstood" and that he "owed it to the industry" to try to reduce some of the friction ExxonMobil generated. "The industry was so good to me, [but it] was given a bum rap," and it was in his power to correct some of that image problem, he believed. "Let me assure you we never set out for the company to be public enemy number one," he said. At issue was more than just feel-good public relations strategy. By choosing Tillerson over Ed Galante, the board and Lee Raymond had handed the corporation back to an upstream executive with a presumed gift for forging international partnerships with oil and gas owners from Moscow to Luanda, at a time when a question hanging over the company and the industry was its long-term ability to replace reserves. Raymond had demanded that the world deal with ExxonMobil on its own terms—and had often succeeded. But the unilateralism by which he steered his private empire mirrored an American hubristic age that was fading by 2006, and the oil and gas properties on which ExxonMobil profits mainly relied as Tillerson took charge had mostly been acquired during the 1990s or earlier. To win new oil and gas access in what the author Fareed Zakaria would soon call a "post-American world," Tillerson might benefit from a collaborative touch. At the same time, Cohen's department worked as much as possible to depersonalize the leadership transition. They certainly did not want to promote some sort of cult of corporate leadership centered on the new man in charge.[1]

It would not have been an easy cult to construct in any event. Tillerson had grown up mainly in small towns in Texas and Oklahoma, the son of a midlevel, modestly compensated professional organizer at the Boy Scouts of America. His mother was a devout but independent-minded Christian who volunteered as a social worker on ambulance runs.[2] From these influences Tillerson grew up to identify himself as a lifelong Eagle Scout. He frequently cited the Scout Oath and Scout Law in corporate speeches. (The Oath: "On my honor I will do my best to do my duty to God and my country and to obey the Scout Law; to help other people at all times; to keep myself physically strong, mentally awake, and morally straight." The Law: "A Scout is trustworthy, loyal, helpful, friendly, courteous, kind, obedient, cheerful, thrifty, brave, clean, and reverent.") Boy Scout language soon found its way into ExxonMobil promotional materials, to describe the corporation's values and ambitions.

Apart from a divorce and remarriage during his early career, there was little evidence of complexity in Tillerson's life. He was fifty-three when he succeeded Raymond: Exxon was the only company for which he had worked after college. He did not possess the fierce intellectual independence (or the sometimes gratuitous meanness) of Lee Raymond. If Tillerson were to match Raymond's financial achievements as ExxonMobil's leader, however, he would have to learn to think for himself. If he were to successfully manage adversaries that ranged from environmentalist campaigners to sub-Saharan insurgents, he would also have to grapple with the world as it really was, not as the Boy Scouts wished it would become.

Tillerson's parents both grew up in Wichita Falls (motto: "The City That Faith Built"), a North Texas plains town that continually reckoned with tornadoes and had a population of just more than one hundred thousand in 1952, when Rex was born. On the day of his birth, the *Wichita Daily Times* happened to publish its annual oil edition, which contained about two dozen stories with headlines such as "U.S. Oil Production Reaches New Record." (A public service ad in the edition observed, "There Is No Security in Foreign Oil for the Defense of Our Own Borders.") The city had been an oil boomtown in the 1920s and 1930s, after the discovery of the giant Electra oil field. By the fifties and sixties, its oil revenue was declining, and the town was struggling to diversify. The Tillersons lived in a modest one-story house in a working-class neighborhood called Faith Village that had been erected for returning veterans. When Rex was six, his father, Bobby Joe, who had been a bakery salesman, took a job that would change the family's destiny: He became assistant district executive for the Boy Scouts of America's Wichita Falls Council. The family moved several times during Rex's childhood before Bobby Joe took a job in the Sam Houston Council, one of the country's largest, and settled in Huntsville, Texas.[3]

The focus of Rex Tillerson's young life was scouting, and he diligently pursued the public service and other tasks required to earn his Eagle Scout rank. In 1970, he left Huntsville to enroll at the University of Texas at Austin. He was by his own account a "slightly above average" civil engineering student. He evaded the city's blossoming music counterculture and served instead as a bass drummer in the Longhorn marching band. He joined Kappa Kappa Psi, which was less a fraternity than a service organization within the university band. It nonetheless had an initiation ritual in which new recruits would be "taken on a ride." The euphemism described a rule-constrained kidnapping regimen. During a prescribed window of time, fraternity pledges could be seized unexpectedly, often in the middle of the night, bundled into a car or truck, driven into the countryside, often stripped of all clothing, and abandoned. A mercy rule required the kidnappers to provide each victim with at least a dime to make a phone call. Another rule required that victims be kidnapped in pairs, so no one would be abandoned alone.

Faith in free enterprise also influenced Tillerson; he later listed his favorite book as *Atlas Shrugged*, Ayn Rand's 1957 philosophical novel that became a touchstone for diverse conservatives, libertarians, and advocates for unfettered capitalism.[4] Exxon recruited him when he was a senior at U.T. He had offers from two corporations and turned down a higher salary to work for the Standard Oil successor. His intuition that he would fit in proved correct. Tillerson worked initially in south Texas and passed successfully through the corporation's vetting for potential managers and leaders; after that, he began the typical eighteen-month rotations and reassignments that Exxon employed to groom future executives.

About two years out of college, Tillerson married a Huntsville High band mate, Jamie Lee Henry; they soon had twin boys. As Tillerson's career took off, however, his marriage fell apart. By 1983, Jamie Lee had returned to Huntsville with the twins, and Rex had married a divorced mother of one, Renda House. In 1988, Rex and Renda had their own child, another son. They shared a passion for horses and settled eventually in Argyle, Texas, away from the Dallas social scene and closer to Fort Worth.

Renda was an effusive cowgirl, a barrel racer at rodeos. She continued to race competitively well into middle age. The Tillersons became leaders and intimate members of the Texas barrel-racing community. They also became active in Fort Worth's world of cutting horses and attended auctions to bid tens of thousands of dollars for seasoned competitors. They bought a ranch in Kamay, about a dozen miles south of Wichita Falls. They called it Bar RR Ranches (for Rex and Renda). As Tillerson rose into ExxonMobil's Management Committee and enjoyed more and more of the wealth and privilege that came with success at the corporation, they also purchased a $2.5 million lake house in the Hill Country, outside of Austin. Texas was his world; nothing but the demands of business travel could persuade him to leave. Renda served on the board of the National Cowgirl Museum, and the couple gave generously to Republican office seekers in the state, often to middle-of-the-road conservatives such as Kay Bailey Hutchison, as well as to Rex's alma mater in Austin.[5]

Increasingly, Tillerson gave over much of his extracurricular charity work to the Boy Scouts of America. He served on the boards of the Dallas chapter and the national organization. Tillerson had never lived outside the United States—even when working closely on the Russian operations, he had been based in Houston—and he had never served on the board of a global corporation besides ExxonMobil, as Raymond had done at J.P. Morgan. If his emphasis on scouting seemed parochial to some within ExxonMobil, Tillerson seemed to regard it as a constructive, universal moral and management system. Within ExxonMobil, he implemented a program of medals or coins for outstanding performance that employees understood to be modeled on the Boy Scouts' merit badge program. Employees with exceptional ability or motivation could meet certain criteria and obtain a full collection of ExxonMobil medals, which were minted in the style of the coins handed out to troops by military commanders. There was a medal available for team leadership, another for teamwork, one for safety performance, and one for technical excellence, which was particularly difficult to obtain. Even managers who laughed and scoffed ironically at Tillerson's merit badge–inspired regime competed to complete their collection, either because they were naturally competitive and couldn't help themselves or because they thought it would enhance their career prospects.[6]

Tillerson asked Ken Cohen to remain in his role overseeing all public policy and outside communication strategies. The decision to keep Cohen in place spoke to the consensus at the top of the company that ExxonMobil had no profound public affairs crisis to solve, only a need to change the tone of its messaging. Tillerson did recognize that ExxonMobil had backed itself into a corner on the climate question and that the arrival of a new chief executive presented an opportunity to chart some sort of new direction—at a minimum in the way the corporation characterized climate change, and perhaps in its lobbying and advocacy positions as well. Yet he viscerally resisted concessions to the corporation's campaigning opponents. The message he heard from environmentalists as he came to power at ExxonMobil was, he said, "Get in line. You're outta line right now, get in line."[7] Instinctively, he refused. Privately, he ordered a review of the issues.

During 2006, he formed a small interdisciplinary group at headquarters, of which Ken Cohen was a member, to review ExxonMobil's climate policy, think tank funding, and public affairs strategy. The group included economists, scientists, and corporate planners. They intensively analyzed alternatives to the climate policy positions ExxonMobil had articulated and supported under Lee Raymond. One manager who participated saw the exercise as an effort by Tillerson to carefully reset the corporation's profile on climate positions so that it would be more sustainable and less exposed. The review, conducted in secret and rolled out only to selected executives upon completion, included a case-by-case evaluation of think tanks and advocacy organizations funded by ExxonMobil and active on climate issues. The committee's work was deliberate; it would be late 2006 before it produced firm conclusions about how to gradually unveil a modified position on both the validity of climate science and the implications of that science for public policy. "We just looked at it," said another participant. "We took each option and said, 'Okay,' and analyzed each one." ExxonMobil was well known in the oil industry for passing on an oil field auction if the timeline did not align with its by-the-book evaluation process; the same was true for public policy questions.[8]

The Tillerson committee's assignment was tricky because it involved legal risks. Greenpeace, the Union of Concerned Scientists, and other campaigners had accused ExxonMobil of replaying the science-manipulating techniques of tobacco companies. The comparison itself served as a warning: To achieve their goals, environmental groups might, as antitobacco activists had done, file class-action tort litigation, accusing ExxonMobil of fraudulent efforts to suppress greenhouse gas regulations. It was hard to imagine how ExxonMobil could ever be as badly affected by lawsuits over global warming as tobacco companies had been by suits over smoking's dangers, but Cohen and other corporate lawyers could not afford to be complacent. (The tobacco companies had initially regarded their own class-action suits as an immaterial nuisance and had been proven wrong.) One of the challenges facing Tillerson's interdisciplinary climate team, then, was to reposition ExxonMobil's arguments about warming to more fully account for consensus scientific opinion, without admitting that any of the corporation's previous positions had been mistaken, for that might open a door to lawsuits.

This was doubly difficult because climate scientists themselves issued stronger and stronger warnings after 2005—these included scientists funded by ExxonMobil. The scientists who refined and perfected the Global System Model at the Massachusetts Institute of Technology, for example, reported steadily more dire forecasts when compared with those Lee Raymond had earlier cited to support his arguments about such forecasting. In 2003, the Global System had issued a median prediction of a 2.4-degree-Centigrade rise in global temperatures by 2100; a few years later, that number had more than doubled.[9]

If possible liability lawsuits had not been a factor in the corporation's choices, ExxonMobil might have crafted a straightforward climate communications plan for Tillerson: We have been following the science closely all along; that science is now more alarming than it used to be; and so we have adjusted our thinking and our policies. In any event, Cohen's committee crafted a more convoluted plan whose core message sounded something like: We were always right, but we were misunderstood.

The political climate in which they considered the dilemma Lee Raymond had bequeathed them was changing even faster than the weather. Early in 2006, *An Inconvenient Truth* debuted at the Sundance Film Festival. The documentary highlighted Al Gore's lectures about the dangers of climate change; it would earn a record $50 million at the box office and eventually win an Academy Award. In ExxonMobil's K Street office, the corporation's lobbyists screened the film a half dozen times, scribbling notes and fashioning talking points about how to attack Gore's arguments. They gamely went forth on Capitol Hill to do so, yet increasingly they felt like front-trench soldiers battling on in a losing war that required new eyesight from the generals at the top, in Irving. And as the Bush administration's attitude toward climate policy wobbled and Gore's advocacy swept through popular culture, and particularly through opinion-shaping elites on the two coasts, the sense among the planners and strategists at ExxonMobil's Irving headquarters was "Uncle. We get it. We won't capitulate, but we will reconsider."[10]

Cohen had dispatched a public affairs colleague named Lauren Kerr to Washington to work on climate issues. By the time of the ExxonMobil leadership transition, she had grown into a significant force on K Street, working from an office next to the Washington chief, Dan Nelson. Kerr spoke out publicly to defend ExxonMobil's climate policies, but she also managed a plan to reposition the corporation as a patron of serious, credible scientific and technological research in the field. She shepherded the large, continuing ExxonMobil donations to M.I.T. and also to Stanford University, to support research into breakthrough alternative energy technologies—programs ExxonMobil thereafter cited as evidence that it was not anti-science. Kerr fed advice, policy research, and political analysis into Irving's climate policy review process.[11]

The committee was restrained not only by the questions about legal liability, but also by a desire among Tillerson and other senior executives to remain loyal to Lee Raymond's legacy. Tillerson had served as Raymond's executive assistant early in his career; he and Raymond had worked closely on the Management Committee in the midst of the most intense climate policy controversies; and it was Raymond, after all, who had chosen Tillerson as his successor. Raymond had tried, as he departed, to deliver to his successor an ExxonMobil board of directors that was as united in its support for Tillerson as possible, despite the fact that at least a handful of directors had favored Ed Galante for the top job. Initially, at least, a smooth leadership transition and mutual loyalty between Tillerson and his new aides, and Raymond and his old-school loyalists, seemed possible.

Even a Boy Scout–loyal protégé will distance himself from a mentor once in power—the story is at least as old as Shakespeare—and in doing so he may generate resentments, even conflict. The more Tillerson made clear that he intended to change the tone of ExxonMobil's communications, the more he implicitly criticized Raymond's approach to climate lobbying and to leadership, in particular. Tillerson also relentlessly spoke of his desire to perpetuate the management culture of discipline and exactitude that Raymond had built, even if he sounded at times as if he was overcompensating. "In terms of showing my predecessor respect . . . he doesn't need anybody's endorsement," Tillerson said later. "He has my great respect. His accomplishments will never be equaled again."

And yet the uncomfortable truth was that Tillerson and Raymond disagreed: The former believed that ExxonMobil had a communications problem on climate, whereas Raymond, now under contract to consult for the corporation and moving toward full retirement, did not. In Raymond's view, as he made clear through his residual loyalists inside the corporation, the question was whether ExxonMobil had the courage of its convictions. The history of Standard Oil, in Raymond's reading, was one of standing firm and taking the heat when necessary.[12]

Tillerson's own views about climate science were not greatly different from Lee Raymond's. Tillerson held a bachelor's degree in civil engineering; Raymond held a doctorate in chemical sciences. Tillerson did not claim or wish to project the same sort of independent scientific expertise that Raymond had offered about climate science. Tillerson remained relatively quiet on scientific questions. "During Lee's reign, Rex never expressed any concern whatsoever" about ExxonMobil's policy positions, recalled an executive on the board of directors. "He was fully on board. . . . Lee would say, 'The scientists on the other side are wrong.' Rex would say, 'It's more complicated than most people understand.'"

As the 2006 midterm elections approached, an era of Republican hegemony in Washington faded. War, corruption cases, and episodes of sexual misconduct by several Republican officeholders set conditions for a Democratic surge in November. As Tillerson settled into authority and Raymond departed toward the end of the year into full retirement, Ken Cohen and his Washington colleagues refined their public policy strategies so they could respond not only to the proselytizing of Al Gore, but also to the larger challenge of inconvenient Democrats.

Under Lee Raymond, ExxonMobil had aligned itself with the Republican Party to a greater extent than many other large oil and industrial corporations. Ken Cohen chaired the ExxonMobil Corporation Political Action Committee, reporting to the chairman's office. Cohen made decisions about the P.A.C.'s political donations only after holding internal hearings with senior executives from Washington and the major business divisions in Houston and Fairfax. The P.A.C. distributed about $700,000 during each two-year election cycle; during the 2000 and 2004 cycles, only 5 percent of those contributions went to Democrats. That was the lowest percentage of any of the largest oil corporations active in American politics. The P.A.C.'s rate of contribution to Democrats crept up slightly as the party's tide seemed to be lifting during the 2006 cycle, but it remained in single digits. (BP did not make political donations in the United States under John Browne; Chevron and Shell's American subsidiary generally gave between a fifth and a quarter of their contributions to Democrats. Even Conoco, the most explicitly conservative oil company after ExxonMobil, gave about 15 percent of its P.A.C. money to Democrats.) Ken Cohen had dispatched Walt Buchholtz of the Washington office—the issues manager who advised anti-Kyoto groups such as The Heartland Institute—to work as a volunteer at the 2004 Republican Convention in New York, alongside other lobbyists. Raymond had also approved a six-figure donation to fund George W. Bush's second inaugural festivities. The ExxonMobil chief maintained a few friendships with industry-friendly Democrats, such as John Dingell of Michigan, but the party knew where ExxonMobil stood as a corporation in partisan competition. "There is no question there is a new phase of scrutiny for Exxon. . . . They have a self-righteousness that sooner or later will catch up with them," said New York senator Charles Schumer, a longtime nemesis of the K Street office. Wisconsin's Democratic senator Herb Kohl, a career business executive whose family founded an eponymous retail chain, declared that ExxonMobil reminded him "of the tobacco industry."[13]

"We need a conversation with Democrats," Dan Nelson told his colleagues in the ExxonMobil office on K Street as the 2006 midterms neared.

That meant redirecting more ExxonMobil P.A.C. donations to Democratic candidates. The problem was that ExxonMobil made almost all of its decisions on the basis of mathematical analysis—and under the corporation's internal, unpublicized political ranking system, the Key Vote System, Democrats looked hopeless. The system was linked to the issues management binder that Cohen kept—the key public policy issues on which ExxonMobil had formulated positions. Analysts identified legislative votes in Congress that were related to the issues list in Cohen's binder. Congresspeople and senators were then each rated on the basis of the votes they cast on these issues—much as liberal and conservative advocacy groups publicly rated members on votes tied to ideological litmus tests. No Democrat in Congress scored above a 50 percent rating under the ExxonMobil Key Vote System; John Dingell, an industry-friendly committee chairman, did no better than 30 percent. Even rock-solid Republicans from midwestern farm states scored poorly because they supported subsidies for corn-based ethanol, which ExxonMobil opposed.

Overall, in the view of its internal critics, the system failed to distinguish between truly key votes and routine ideological votes in Congress, when congresspeople and senators cast votes because, effectively, they had no political choice but to appease local voters or ride the party line. ExxonMobil's rigid adherence to the Key Vote numbers had helped drive P.A.C. giving toward the very safest Republicans—a fact that had seemed to bother Lee Raymond not at all, but which had deprived ExxonMobil of ties to Democrats who might be sympathetic to at least some of its lobbying priorities. Cohen and Tillerson eventually agreed, after discussions within the P.A.C. committee during 2006, that they would have to take a wider view.

The Washington office was a generally congenial place to work. Anniversaries marking longevity as an ExxonMobil employee—ten years, fifteen, twenty, twenty-five—were celebrated with cakes and huzzahs, as were birthdays. There was some ideological diversity in the office—Susan Carter, a Democrat, lobbied the Bush administration on ExxonMobil's behalf, and Lorie Jackson worked to make global health and women's issues a priority, influenced by the multiple Hillary Clinton connections at the office. On the whole, however, this was a lobbying shop constructed by and for Republicans. Nelson's congeniality went only so far; to insist that his lobbyists become good time managers, he would lock them out of meetings if they turned up late. Inside the Washington offices of Chevron and British Petroleum, there was an explicitly international atmosphere; Chevron's Nigeria lobbying specialist was a Nigerian, for example. At ExxonMobil's office on K Street, there was less diversity and more of a military flavor.

The ExxonMobil lifers who lobbied Capitol Hill—Jeanne Mitchell in the House of Representatives and Buford Lewis in the Senate—worked alongside "issue" managers with subject specialties such as tax policy or chemical industry regulation. The largest contingent of lobbyists remained the foreign policy specialists working with Robert Haines, who spent most of their time maintaining relations at foreign embassies and at Foggy Bottom. The only Democrats the K Street team knew well on Capitol Hill tended to be those from southern or western states who voted like moderate Republicans on energy and tax issues. What a House of Representatives ruled by Nancy Pelosi would mean for ExxonMobil was not a question the corporation's Washington office could answer with great sophistication.[14]

Ken Cohen worked closely in Irving with ExxonMobil's chief in-house Democrat, Theresa Fariello. She had joined the corporation from the Clinton administration's Department of Energy in 2001 and managed worldwide public policy issues at headquarters, reporting to Cohen and acting as a liaison with the K Street office. Fariello was a single woman, traditional in some of her views, liberal in others, and she worked comfortably in ExxonMobil's corporate culture. She, too, was a committed supporter of Hillary Clinton's and worked actively to aid her candidacies when Clinton ran for the United States Senate and later plotted a presidential bid. Fariello had attended George Washington University as an undergraduate and then earned law degrees at Georgetown and George Mason. For more than a decade, she worked in the Washington office of Occidental Petroleum during a period of controversies that arose from the corporation's work in war-torn Colombia. Fariello arranged a few consultancies and retainer contracts with lobbyists in Washington closely connected to Democrats—she chose outside lobbyists she knew and trusted, such as

David Leiter, who had worked as chief of staff to Senator John Kerry during the 1990s and then worked on alternative energy technologies at the Energy Department during the second Clinton term. Intelligence from Fariello's Democratic network fed into the climate policy review and the preparations for a Democratic-controlled House after November.[15] Dan Nelson also brought on Louis Finkel, who had worked for years for a moderate Democrat representative from Tennessee, Bart Gordon, then the ranking member on the House Committee on Science and Technology.[16]

On November 7, 2006, Democratic candidates swept to victory nationwide, and the party took control of the House with a thirty-one-seat gain. A few weeks later, Ken Cohen flew to Washington to deepen his scrutiny of what this might mean for ExxonMobil.

He drove west on Interstate 66 on a chilly, dry afternoon in early December to Warrenton, Virginia, in the foothills of the Blue Ridge Mountains. A few miles along Country Road 605 he turned through split rail fencing into the secluded grounds of the Airlie Center, a large farm that had been converted into a self-styled "island of thought" a half century earlier, and which served as a retreat for government and business leaders. Cohen had scheduled a three-day Opinion Leader Dialogue with environmental and human rights activists. The program would be an opportunity to test out ExxonMobil's emerging climate policy and to engage more deeply with nonprofit activists. Checking in, too, at the elegant reception desk were Sherri Stuewer, who had succeeded Frank Sprow as ExxonMobil's vice president in charge of climate and environmental policy; Jamie Spellings, a vice president for corporate planning; David Kingston, a vice president for the downstream, or refining, side of the corporation's business; and Mark Sikkel, a vice president on the upstream side with responsibility for Asia and the Middle East.

As part of ExxonMobil's broader campaign to engage and persuade Informed Influentials, Cohen's department had developed the dialogues as a hybrid institution—part private retreat, part focus group, and part lobbying briefing where ExxonMobil could roll out its 2030 PowerPoint slides for environmental leaders, human rights researchers, journalists, and think tank analysts, as well as test some of its advocacy positions. The dialogues were designed and managed in such a way as to suggest that ExxonMobil considered its Irving and Houston executives on the one hand, and influential Democratic-leaning nonprofit leaders in Washington on the other, to be members of slightly different species who would require a safe, controlled setting in which to assess each other peaceably.

That evening the ExxonMobil executives mingled awkwardly with their fourteen invited guests—two senior energy-policy analysts from the Brookings Institution, a human rights activist at Freedom House, climate specialists, business ethics professors, socially responsible investors, and religious activists. Most of the guests were very liberal, but the group included at least one conservative Christian leader. They shared concerns about ExxonMobil's record of corporate citizenship. During the cocktail hour, one of the guests, who worked at an environmental nonprofit, chatted casually with Cohen about her most recent project, and she mentioned the brutal hours she was putting in to get it finished. "He was shocked," she recalled. "He said that he thought people who worked in environmental groups in Washington had a cushy life."[17]

Another participant recalled thinking of his corporate hosts: "These were clearly thoughtful, smart, articulate people—they just lived in a totally different world than we live in." The *New York Times* had just published a story about Lee Raymond's $398 million retirement package; in response to the incredulous asides of their guests, the ExxonMobil executives labored to explain the difference between pensions and stock options and restricted stock, in an effort to suggest that the package was not as rich as it might appear. "You know you can't win on that message, right?" the participant thought as he listened. "You're talking to people who can't even take the Acela to New York." The Acela was a fast, expensive intercity train between Boston and Washington; to conserve funds, some nonprofit groups ordered their employees to take slower, cheaper trains—or the bus.

The next morning they assembled in a conference room around a table arranged as a hollow square. The agenda included two "dialogue sessions" on climate change and a third on corporate transparency and human rights. As ever, the ExxonMobil team ran through the PowerPoint slides laying out the corporation's forecasts of oil-and-gas-dominated energy demand and sources until 2030.

Cohen shared some of his internal polling about the corporation's reputation. In one survey ExxonMobil had received 47 percent approval for overall corporate citizenship and 24 percent for environmental stewardship. Environmental issues remained a challenge. In countries such as China, the ExxonMobil executives acknowledged, environmental regulation was being taken more and more seriously. As middle classes grew around the world, so would environmental concerns, they knew.

On climate change, Cohen and Stuewer flashed PowerPoint slides outlining draft language of a new formulation of ExxonMobil's position. "They were really dancing around the question of certainty" about the risks of global warming and the evidence that man-made activity contributed, recalled Leslie Lowe, one of the participants.[18]

Lowe introduced the metaphor of having insurance against fire: Why *not* work against man-made contributions to climate change, even if there remained uncertainty about every last detail of cause and effect?

Yes, the ExxonMobil side responded, but you don't spend all of your money in life on insurance. You calculate how large and valuable an asset you are trying to insure, and how big a risk you face. Climate was like everything else ExxonMobil did: It was a matter of risk management, Cohen emphasized.

The participants talked about imposing a price on carbon, through gasoline taxes or other formulas. "If you tax gasoline, people will be hurt," Cohen said. Even if you tax gasoline and then rebate the money to middle-class and working households, commuters would just be forced to take the rebates and "go out and buy gas with it," the ExxonMobil executives argued.

The nonprofit leaders asked Cohen about the funding he had provided to groups such as the Competitive Enterprise Institute and The Heartland Institute that had so stridently attacked the validity of mainstream climate science. Cohen told them that as part of ExxonMobil's review of its options on climate policy, the corporation had decided to pull funding from the most controversial groups. The disclosure was the beginning of a quiet campaign to clarify that ExxonMobil had altered some of its public policy funding—without quite admitting that what it had done earlier was wrong or misguided. The more strident groups were a distraction, Cohen indicated; they were focused heavily on the validity of climate science, whereas ExxonMobil now wanted to leave that subject to focus the debate on research and policy choices.

The participants on both sides spoke gently; they were trying "to be ever so polite," Lowe recalled. That night at dinner, she found herself sitting with Dave Kingston, the downstream vice president. In an unthreatening tone, she asked, "Look, you're a science-based organization. How can you not accept the science that is basically confirmed by most mainstream thinkers?"

Kingston talked about the inherent uncertainties in weather modeling and forecasting.

She listened patiently, then asked, "What are you going to say to your grandkids when they say, 'Grandpa, why did you fuck up the planet?'"

The ExxonMobil executive just chuckled.[19]

It proved difficult for each side not to simply ratify its assumptions about the other and move on. "They did get defensive, but they didn't lose their temper," a participant recalled. "It was sort of, 'You don't understand.' I felt like someone was sort of patting me on the head, 'You poor thing.'" Some of the nonprofit leaders felt disrespected, manipulated, and sensed little but condescension and closed mindedness from Cohen's team over the course of the Airlie Dialogue; others left more impressed by the corporation's subtlety and sophistication than they had been when they went in. One guest in the latter group marveled, nonetheless, about how the ExxonMobil executives "were so wedded to their oil and gas image, identity." It seemed perplexing to him that such intelligent, research-driven executives could come up with a communications and positioning strategy that seemed so obviously to be self-limiting, if not an outright loser. Perhaps they let the research tell them what they wanted to hear.

On January 11, 2007, ExxonMobil forwarded an e-mail from Ken

Cohen to the Airlie participants. The Irving climate policy committee's work had now yielded firm decisions by Tillerson to support a new communications campaign to try to clarify and redefine ExxonMobil's position without creating legal jeopardy.

"ExxonMobil's position on climate change continues to be misunderstood by some individuals and groups," Cohen began. As to the new stance: Climate change presented risks, Cohen wrote, despite the scientific uncertainties, and so it would be "prudent to develop and implement strategies that address the risks, keeping in mind the central importance of energy to the economies of the world. This includes putting policies in place that start us on a path to reduce emissions . . . among other important world priorities, such as economic development, poverty eradication, and public health." Tillerson was prepared to say publicly, "We know our climate is changing, the average temperature of the earth is rising, and greenhouse gas emissions are increasing." That went further than Lee Raymond had ever gone, yet Tillerson would not go so far as to accept a causal link between rising greenhouse gas emissions and rising temperatures— the fundamental finding of climate scientists about global warming.[20]

In 2007, the Intergovernmental Panel on Climate Change reported that it was now an "unequivocal" fact that the earth's surface temperature was rising and that there was "very high confidence" that human activity was a factor. Amid an emerging scientific consensus of such a firm character, Rex Tillerson's evasive silence and small concessions could hardly be counted as a decisive turn. Subtly, however—so subtly that it was very difficult for outsiders to detect the change—the corporation slid during 2007 into a new position.

Even after the dialogue at the Airlie Center, it was as if Cohen and Tillerson felt they needed to keep auditioning and refining the nuances of their advocacy language in public, adjusting to each audience's reactions, like theater producers developing a risky new musical in secondary cities. In essence, however, ExxonMobil would henceforth decline to offer a formal opinion about whether the burning of fossil fuels contributed to global warming (a neutrality that protected its legal defenses), but the corporation acknowledged, for the first time, that it would be sound public policy, nonetheless, to limit man-made greenhouse gas emissions to at least some extent, because of the potential risk that the worst climate change forecasts might prove to be correct.

It remained palpably painful to extract this last admission from Ken Cohen's lips, however.

"Are you saying you now accept that human intervention is the main source of global warming?" a climate researcher asked during one of Cohen's early 2007 appearances aimed at publicizing ExxonMobil's new lobbying language.

"There is no question that we understand the physics of the warming caused by CO_2 and we welcome the discussion of what the in-depth link is. . . . We are involved in this discussion."

Still, he went on, having learned by now to interrogate himself before others did: "Should we be on a path to do something about anthropogenic emissions? The answer is yes."[21]

Sixteen

"Chad Can Live Without Oil"

Scores of oil platforms had spread across the red clay ravines and subtropical bush of southern Chad by 2006. There were by now 368 wells in all, scattered across just over twelve thousand square miles. Each rig pointed at the sky like a thirty-foot rocket prepared for launch. In the dry seasons heat baked the ground to an amber shade and blown dust caked the equipment. When the rains came, low clouds hung in wisps around the metallic caps topping the derricks. At the main production facility at Kome, known within ExxonMobil as Kome-5, some of the Africans who worked behind the tall fences illuminated by safety lights, and who found their lives constrained by ExxonMobil's extensive rule making, facetiously referred to the compound as Guantánamo. The joke's suggestion of exclusion captured a larger truth about ExxonMobil's operations in Chad. Since it was not feasible for the corporation to devise solutions to all of the problems of a very poor country, the logic of ExxonMobil's systems management argued for engineering Chad and its pathologies out of the equation to the greatest extent possible. ExxonMobil's one thousand expatriate workers and managers rotated in and out of the country on twenty-eight-day tours without ever interacting with the world outside their compound chain-link fences, which were nine feet high and topped with a barbed-wire anticlimbing barrier. In addition to Kome-5, there were 9 ExxonMobil camps across southern Chad, as well as 450 oil production areas and 50 construction sites. The corporation's expatriate workers and managers flew on commercial or chartered flights into N'djamena's international airport, transferred without leaving the grounds to an ExxonMobil shuttle plane, flew to the corporation's private airport at Kome, and then checked into barracks or bungalows within the fenced perimeter. For the next four weeks they could neither drink alcohol nor leave the facility. The corporation's roughly five thousand Chadian employees followed the same rules. Those who worked in Kome-5 checked through the employee security gate for twenty-eight-day shifts and could not leave but for approved emergencies, unless they were prepared to give up their lucrative salaries. It was like being entombed, one of the Africans remarked.[1]

Inside the Kome-5 complex, ExxonMobil employees enjoyed twenty-four-hour electricity, high-speed Internet service, satellite television programming, refrigerated and packaged food, and as many other amenities of middle-class Houston as it was possible to supply by airlift at reasonable cost. Immediately outside the fence stood a shantytown known locally as Quartier S'Attend, "the Quarter of Waiting." It had sprung up as a settlement for hopeful Chadian laborers when the oil field construction work began. After the initial building slowed, the settlement became an iniquitous way station for Chadians entering or leaving "Guantánamo." Its stalls were stocked with bottles of the potent millet home brew "billi-billi," or the even more potent "Argue." Prostitutes from Cameroon staffed the back rooms. In the bush beyond the quartier there were small villages where the homes were constructed from kilned brick and thatched straw, with tall, conical roofs shaped like a farm wife's sun hat. Oxen, pigs, donkeys, horses, and shirtless children wandered through sandy lanes. Here there was little reliable electricity.

"Public relations efforts have fallen short of the locals' expectations and a growing resentment is building towards the oil project and its workers," noted an embassy cable written about six years after the World Bank, ExxonMobil, its corporate partners, and the government of Chad had agreed to a radical plan to ensure that the country's oil profits were not stolen or wasted.[2]

The oil region presented a booming industrial economy behind the ExxonMobil fence but little sign of sustainable progress outside. Rather than engineering social and economic change to improve the conditions of most ordinary Chadians, the bank experiment had created conditions under which ExxonMobil could produce and sell oil the way it did everywhere else, in a setting where that might otherwise be impossible. Without the bank, the corporation could not have easily profited from its reserves in Chad, but whether the bank succeeded with its broader development goals was fading as a material concern for ExxonMobil. Chad's geology had proved challenging, project costs had ballooned over budget, and oil production remained lower than originally forecast. But rising global oil prices cured these business ills. During 2005, the corporation's oil sales from Chad reached the point where it had recouped the costs of construction and investment; the project was now profitable and even susceptible to Chadian taxes at a 60 percent rate, over and above a 12.5 percent royalty ExxonMobil had paid on gross oil sales to Déby's regime from the beginning.

The security of ExxonMobil's operation and investment remained a worry, however. ExxonMobil and its partners recognized "that the oil fields in southern Chad are of strategic interest to any group seeking to overthrow the regime," the embassy noted to Washington. As in Indonesia, ExxonMobil protected its oil field perimeter with a forward defense. In southern Chad there were very few armed rebels; President Idriss Déby had brutally wiped out an incipient guerrilla movement as ExxonMobil prepared to construct its oil wells. By 2006, the groups that most threatened President Déby operated from Sudan, many hundreds of miles away. Still, the desert could be traversed quickly: Rather than sit behind its fences and wait for some unexpected rebel group wearing bandannas and wraparound sunglasses to arrive in pickup trucks toting rocket-propelled grenades, ExxonMobil employed about twenty-five hundred unarmed private security guards in southern Chad and equipped them with white sport utility vehicles and portable radios. The Jeeps conspicuously patrolled the roads around Kome and bumped through villages, watching for outsiders. With ExxonMobil's encouragement, camouflaged soldiers from the formal Chadian gendarmerie shouldered automatic rifles and maintained checkpoints and encampments along the roads to and from the oil area, prepared to seal off the region in the event of trouble.[3]

American intelligence officers picked up and passed to ExxonMobil reporting that "suspected extremists" had surveilled the oil fields; afterward, the security officer at the U.S. embassy was invited by the corporation to inspect the facility to assess its vulnerabilities. He found it wanting. There were holes in the fences and the Chadian paramilitaries protecting the region were found "either in their T-shirts and flip-flops playing cards and eating or sleeping in their tents." The facility was "totally unprepared for any level of terrorist attack. . . . The Gendarmes often go months without pay. . . . What is provided [by Chad's government] lacks numbers, motivation, discipline and training. Their morale is poor and their equipment is not good." Locals complained that the police preyed upon them. For its part, after its travails in Indonesia, ExxonMobil was doubly sensitive about allowing local forces guarding its facilities to shoot aggressively at attacking guerrillas. The corporation's Chadian executives told the U.S. security officer that they were more likely to shut down and evacuate Kome than defend it, if Chad failed to do the job. In the worst case, since they had already made their money back, they just might "sell what remains to the Chinese." To defend ExxonMobil, the assessing officer suggested that the United States "should consider providing to the local police and Gendarmes . . . basic to advanced firearms training as well as instruction in police and patrol tactics to counter any terrorist threat."[4]

A small leadership and analysis team from ExxonMobil Global Security supported the local guard force from the capital, N'djamena. The team worked from a compound of bungalows and offices that made up the corporation's country headquarters on the Avenue Charles de Gaulle. They were former police and retired military men. They built an intelligence-collection operation separate from and in many ways better sourced than that run by the small Central Intelligence Agency station in the nearby American embassy. ExxonMobil enjoyed privileged access to Déby and his military and intelligence advisers because, in 2006, the corporation and its partners would transfer to Déby some $774 million, whereas U.S. government aid, all totaled, including the counterterrorism assistance provided by the C.I.A., was in the neighborhood of 1 percent of that amount. That equation had changed little since the World Bank–sponsored oil project began; Déby's aides hardly required calculators to understand where their interests lay.

Statisticians at ExxonMobil's upstream division headquarters in Houston helpfully prepared PowerPoint slides for use by the U.S. embassy—slides that emphasized the outsize benefits of the corporation's activities in Chad for the United States: one thousand American jobs per year during the construction phase of the Chad oil project; two hundred expatriate jobs for Americans in the country generating about $70 million in total revenues; a projected 24 million barrels of direct oil exports to American refineries annually; and more than $1 billion in profits returned to American shareholders over about six years. American diplomats in N'djamena transmitted statistics, fed to them by ExxonMobil's country manager, in cable after cable to Washington. In N'djamena, ExxonMobil's security officers used their influence and inside access to Déby's security regime to school themselves like political science scholars in the region's alphabet soup of liberation armies and quasicriminal gangs, and they monitored reporting inside Déby's palace for emerging threats, particularly any mutiny that might reach the south, target ExxonMobil directly, or otherwise call for an employee evacuation plan to be activated. The Global Security team continually planned, mapped, and exercised evacuation routes into Cameroon by road and air.[5]

The salaries earned by the corporation's local guards poured into a corrupt, inflation-wracked regional economy. Before oil the south was a region of marginal subsistence farming. Sorghum and cotton were staple crops. The rains were erratic, and the roads to market could be impassable when wet. The construction and employment boom that followed ExxonMobil's arrival flooded the area with cash. Farming families that might previously have earned $125 in a year received as much as $7,500 in one-time payments for land rights. Food prices spiked; the cost of a sack of staple millet doubled. World Bank development specialists encouraged schemes of micro lending to local households in the hope that this temporary influx of capital could be converted to self-sustaining entrepre-

neurship, but too many borrowers absconded or blew their funds on
motorbikes and billi-billi. Labor contractors connected to Déby's administration brokered construction jobs in the oil fields and skimmed off what
a Chadian court later determined to be $7.5 million in wages, money that
never reached the workers who actually built the derricks. (About half
that amount was later paid out to four thousand workers in a settlement;
the stolen money was never traced.)

At times, depending on the ambassador, the U.S. embassy in N'djamena
provided optimistic assessments to Washington. ExxonMobil is "regarded
as a model company, particularly in respect to continuing good community relations and environmental consciousness," one cable reported.
The corporation has "upheld American standards in terms of worker
entitlements to wages, safety and health." But ExxonMobil's disciplined
systems and investments did not resolve the country's weak governance,
corruption, poverty, or unmet aspirations. "The influx of oil revenue has
created little improvement, frustrating heightened expectations," a confidential U.S. State Department-led fact-finding team reported in early
2006. "Many Chadians told the interagency team they were 'better off
before the oil.'"[6]

Armed with advice from consultants in poverty reduction and corporate responsibility, ExxonMobil allocated modest sums to local projects in
the south intended to demonstrate Irving's commitment to global citizenship. Among other things, the corporation constructed a water storage
tower in a village about a twenty minutes' drive on rutted roads from the
Kome fences. In a decision that seems unlikely to have crossed Lee Raymond's desk during the latter part of his tenure, ExxonMobil's regional
managers installed solar panels to generate the electricity needed to pump
well water into the newly constructed village tank. Thieves ripped the
panels out, however, and soon all that remained was the storage tower and
the metal rails that had once held the solar panels in place. Theft at ExxonMobil's properties was rampant; the corporation lost $500,000 worth
of equipment in 2006 alone. Kome-5's security doors were a favorite target; locals made beds from them. Popular disenchantment increased when
local officials imposed what was regarded as an "Esso-imposed 6 p.m. cur-
few" in response to the looting.[7]

During the hopeful days when the World Bank and Exxon had planned the Chad project as a pioneering experiment in nation building and social engineering, Exxon erected a health clinic in Kome. When the Chadian government also requested a nurse, medicine, and equipment, the corporation refused—it locked up the building until Chad hired its own health care workers to staff it.

ExxonMobil's managers in Chad took justifiable pride in the $16 million in annual wages, training, education, and exposure to global norms in health and education that the corporation provided to the Chadians in its direct employ. The knock-on benefits of these improved lives would be substantial, if hardly enough to right Chad. Yet ExxonMobil considered the construction of deeper social and physical infrastructure in the country to lie outside of its responsibilities. That was why the corporation had so purposefully recruited the World Bank into the high-risk Chad oil project in the first place. When ministers in Chad's government or local human rights activists begged ExxonMobil to build a health clinic or lay a road, the corporation typically demurred, explaining that oil production was its core competency and that it intended to follow the letter of its contract. As oil production grew and Idriss Déby did not become a better president, and Chad's social and health indicators failed to improve significantly, ExxonMobil's executives privately blamed the World Bank. The bank had simply not done what it promised to do when it endorsed and funded the 2000 plan to manage oil revenue for the greater good of Chadians, ExxonMobil's managers argued. There was some truth in their complaints. ExxonMobil operated on time and under budget in a way that a sprawling, multinational bureaucracy such as the World Bank never could. The oil company's criticism assumed, however, that the experimental governance goals embraced by all of the project partners—including ExxonMobil—had been realistic in the first place. By 2006, this no longer seemed a defensible claim.[8] Irving prided itself on its realism. The most honest assessment was that the two main parties to the Chadian oil project had always been ExxonMobil and its partners and shareholders on the other side, and Déby and his kleptocratic clansmen on the other.

In Doba, the flat market town nearest to the oil fields, a revenue-sharing scheme supported by the bank endowed the local government with tens of millions of dollars. The political chiefs did build a school, but generally favored grandiose construction projects over health and education services. They spent $4.4 million on a soccer stadium in Doba that could seat twenty thousand, notwithstanding the lack of a professional soccer league of any significance. School, road, and hospital building projects went forward more encouragingly, but when they were finished, the buildings stood empty for lack of nurses, doctors, and teachers. Local construction budgets and expenditures were grossly inflated by the skyrocketing costs of materials and the large number of skimming hands involved. Unmowed grass soon bent like wheat across the soccer stadium's field. Young mothers in the area still died in childbirth at rates comparable with the era before oil. "It's not easy to hide the sun with your hands; the money is there in large quantities," remarked Boukinebe Garka, a member of a national commission meant to help supervise oil revenue. "But it is not being spent very well."[9]

As Idriss Déby's government grew wealthier, the president also became more vulnerable to mutiny. Like his acquaintance Teodoro Obiang in Equatorial Guinea, Déby increasingly looked like a nervous dragon sitting on a pile of treasure, waiting to be assaulted by coup makers. In 2003, Déby had pushed for alterations to the constitution that would allow him to become president for life, and he succeeded in abolishing term limits by 2005, a betrayal of Chad's shaky democratic parties that had further isolated Déby politically. He had come to power by force years earlier, backed by Sudan's secret police, and so it did not require a paranoid imagination to think that he might only go out by a similar method.

Sudan, to the east, was engulfed by violence in the province of Darfur, which bordered Chad. Marauding Arab militias loyal to President Omar Al-Bashir waged a scorched-earth campaign through Darfur's villages and towns after 2003. At least 300,000 people died and about 2.5 million fled their homes. About 240,000 Sudanese civilians crossed the border into Chad and crowded into camps, where they were formally recognized by the United Nations as international refugees. At first, Déby tried to play a balancing role in the Darfur crisis, mediating between his former patron, President Bashir, and his newer patrons in Washington. By 2005, however, Déby perceived that the Darfur conflict might threaten his grip on power. Many of the non-Arab rebels battling Sudan's government belonged to Déby's own Zaghawa tribe. Bashir had been urging Déby to bring these rebels under control—to prove, in effect, that he was a friend of Sudan's. Suppressing the rebels was beyond Déby's means, but in any event, he needed the Zaghawa networks for his own security. In 2005, he aligned himself with Khalil Ibrahim, the powerful Islamist leader of a Darfur rebel faction. Once he did that, Déby made plain to Khartoum that he had changed sides, defaulting on his historical debts to President Bashir.[10]

In reply, Sudan's security services offered vengeful support to Mahamat Nouri, the head of a Chadian rebel group, the Front Uni Pour le Changement, or the "United Front for Change." Déby picked up intelligence that a Nouri-led rebel invasion from Sudan's territory, for the purpose of removing him from power, could come at any time.

Déby's defenses to the east and around the capital were weak. His Ministry of Defense paid out about 70,000 salaries, but only about 20,000 of those "soldiers" possessed uniforms and occasionally turned up at their jobs, while only about 4,000 were armed, trained, and prepared for combat. Moreover, if Déby ever ran out of the cash he required to pay his tens of thousands of ghost military salaries, he would invite mutiny—this threat was a constant drain on his cash flow. Mercenary Algerian, Ukrainian, and Mexican pilots on rotating contracts flew the ramshackle planes and helicopters in the president's tiny air force.[11] The professional French-manned garrison and air force training mission at N'djamena's international airport offered a much more convincing defensive deployment—but only if, in the heat of a crisis, the French government decided that it was in its interest to have its soldiers shoot at Déby's enemies, a highly uncertain prospect.

Such was Déby's predicament late in 2005: He feared an imminent rebel invasion aimed at overthrowing his regime, but he lacked adequate means to defend his palace. The absurdity, from his perspective, was that at the very moment this threat loomed, he was becoming richer than many of his neighboring dictators—or "authoritarian leaders," as Washington sometimes preferred. Under the 1988 oil production contract, as world oil prices rose and ExxonMobil recouped the costs of its initial oil field investments faster than expected, the revenue flowing to Déby also soared; it now exceeded all previous projections. His government took in $300 million in 2005.[12] Yet Déby could not spend his own money freely because of his good governance compact with the World Bank. He was banned from buying as many guns, desert vehicles, and attack aircraft as he felt he needed to defeat Bashir's rebel proxies. Déby had gone along with the surrender of some of Chad's sovereignty to the World Bank back in 2000 because it was necessary to get the country's oil flowing. He was not going to surrender his office to a foreign invader to preserve the plausibility of some Westerner's utopian ambition. "It was very hard to explain to Chadian opinion that we must buy weapons," recalled Mahamat Hissène, a minister in Déby's cabinet. "So we did not develop real power to respond to attacks coming from outside. . . . Unfortunately, in front of us, we don't have politicians, we have technicians. The World Bank and I.M.F. sent us technicians who don't care about the security of the country."[13]

Late in 2005, Déby at last announced that he would introduce amendments to Chadian law that would break his government's bonds with the World Bank. By doing so he threatened to upend the nation-building compact and international financial rules that had prevailed, however raggedly, long enough to reward ExxonMobil for enduring high political risk in Chad. Just as the corporation reached breakeven and its local profits began to gush, Déby's desperation for money and guns threatened to bring the whole project down.

R on Royal, an engineer with many years at the corporation, managed ExxonMobil's country office in N'djamena. In addition to his security team, a handful of expatriate and Chadian public affairs advisers—lobbyists and analysts—advised him. Royal reported in turn to the Africa division in Houston. Until the crisis that erupted that autumn, Royal's job had been merely grinding and thankless. Idriss Déby paid little attention to the details of his country's oil contracts or operations; one American embassy cable described him as "terribly ill-advised and grossly uninformed" about oil.[14] As a result, Déby's minions—cabinet ministers, labor union allies, governors in the south, and other assorted rent seekers—felt free to harass ExxonMobil at every turn, seeking money to skim or for other advantages. In addition to the chronic thefts from ExxonMobil facilities, Royal had to respond to wildcat strikes, court cases arising from labor disputes, public protests, complaints by environmentalists, delays and threatened fees imposed by immigration officials who issued visas, customs delays, and other challenges. Chad's politicians and labor leaders might be poor and some of them might be unsophisticated, but they had been schooled in obduracy and provocation by French colonialists, which made them formidable. A typical matter in Ron Royal's in-box was the announcement, during 2005, by the Chadian civil aviation authorities that because ExxonMobil had overlooked a certain legal provision, the corporation would have to immediately start paying the country's struggling state-owned airline $150,000 a month in royalties for the right to avoid having ExxonMobil workers fly on the Chadian airline's planes. Chadian officials also were deeply suspicious about the relatively low prices the country's sour, heavy oil attracted on world markets. Some of their worries seemed irrational—it was as much in ExxonMobil's interest as in Chad's to receive the highest possible per-barrel price. Yet the wide gap—more than $20 per barrel—between the price of Chad's oil and the prices for more attractive, benchmark blends of oil that were published in world newspapers raised suspicions with Déby and his aides. ExxonMobil flew some of Déby's oil advisers to Fairfax, Virginia, and London to show them how ExxonMobil bought and sold oil on the world market, and how it negotiated to win the best possible prices for Chadian barrels from shipment to shipment. Even this show-and-tell had limited impact: Afterward, one of the Chadian delegates, Abdelkarim Abakar, remarked to an American diplomat that the "visit solidified the question of why Doba crude was priced so low when the price of oil in the international markets was priced at such a high level," and he argued that if ExxonMobil "was truly willing to communicate openly" about the issue, "it would have organized this visit a long time ago."

Ron Royal and his team met regularly with American diplomats in N'djamena, including Marc Wall, the ambassador, who had succeeded the part-time novelist Chris Goldthwait. Wall was a career diplomat, a thin, professional, silver-haired man with extensive experience in troubled countries. Normally, Royal did not ask the embassy to lobby on Exxon-Mobil's behalf with Déby or his aides; ExxonMobil took care of its labor, security, and government lobbying hassles on its own. That changed during the first weeks of 2006. The crisis Ron Royal and ExxonMobil faced that winter was instigated from Washington. Its principal architect was the World Bank's new leader: Paul Wolfowitz.

In his previous position as deputy secretary of defense, Wolfowitz had been an intellectual leader and a passionate defender of the Bush administration's decision to invade Iraq and overthrow Saddam Hussein. In 2005, as the war descended into chaos, and as its costs in blood and treasure to the United States rose precipitously, Wolfowitz left the Pentagon. He arrived at the World Bank in a complex political and psychological position: He was deeply unpopular in many parts of the world, unrepentant about Iraq, and yet he seemed eager to demonstrate his commitment to poverty alleviation and liberal development goals. Presented in his first months on the job with the conundrum of Idriss Déby's defiance of World Bank principles of good governance, Paul Wolfowitz decided to make an example of him.

When he reflected publicly on Africa, Wolfowitz grouped its countries into three broad categories: About a third of the continent's nation-states seemed to be hopeless basket cases, including Sudan, Somalia, and Zimbabwe. About another third, endowed with oil and other natural resources, were doing better, but were struggling with the resource curse—Angola, Equatorial Guinea, and Chad were examples. Another third were becoming successful, registering positive economic growth rates for ten years or more—Botswana, for example, and also previously failed states such as Mozambique and Rwanda. Wolfowitz felt the World Bank had an opportunity to promote a more nuanced picture of Africa's economy. The continent's negative image "wasn't helped by large rock concerts that talk about what a miserable, failing place it is," he remarked. He did not blame Bono or U2 for this image problem; he blamed dictators like

Zimbabwe's Robert Mugabe. Chad's ruler was nowhere near as bad as Zimbabwe's, but no good would come from appeasing Déby on the matter of his arms purchases, Wolfowitz concluded.

Chad has a sovereign right to decide how to spend its oil money, Déby argued to Wolfowitz over the telephone on the night of January 5, 2006.[15]

Their conflict had reached a climax. Wolfowitz was threatening to freeze the money ExxonMobil and its partners deposited in Chad's Citigroup accounts in London—an action he could take by invoking the bank's rights under its pipeline lending agreements. Even though the amount the bank had lent to Chad was relatively small—less than the annual oil revenue Déby now received—under the loan terms, the bank could seize Déby's oil funds.

Wolfowitz told Déby that he accepted that Chad was sovereign. However, Chad's sovereign government had entered into certain contractual commitments with the World Bank to spend oil revenue on the health, education, and welfare of Chad's people. It was in the interests of Chad's sovereign government, Wolfowitz argued, to show the world that it was an honorable party to the agreements it had made.

They talked through French translators that night for two hours. Neither of them budged. The next day, Wolfowitz announced that the World Bank would withhold new loans to Chad; a freeze on its bank account would follow if Déby still refused to compromise.

Wolfowitz tipped Déby into a rage. His aides told Ron Royal at Exxon-Mobil that if Exxon went along with the World Bank any longer, Chad might order the corporation to shut down all oil production.

Royal met with Ambassador Wall at the U.S. embassy. Closure of Chad's fields "would be catastrophic," he said, and would likely set off a chain of loan defaults. ExxonMobil wanted a cooling-off period and had asked the World Bank for time to negotiate a compromise, but the bank had responded with a letter to Irving, declining ExxonMobil's request.

Royal revealed to Wall for the first time that ExxonMobil was on the verge of profitability in Chad, and that, therefore, the corporation "was now in an income tax-paying position." Given rising world oil prices, ExxonMobil and its partners might soon hand over to Déby's regime between $80 million and $200 million in initial tax payments—enough, perhaps, for Chad to pay off the World Bank altogether and extricate itself from its commitments to social investments.[16]

Royal met with Déby and warned him that "dismantling oil operations and forcing the World Bank to leave" would "jeopardize the reputation of the country and the possibility of foreign investment." He also told Déby for the first time that Chad would soon receive a tax windfall—if the president allowed ExxonMobil to keep pumping oil. Déby was surprised—he asked for specifics, but remained noncommittal about any compromise. Royal contacted Wall again and told him that he feared a "melt-down" and a final collapse of the entire Chadian oil project—six years and several billion dollars of investment, so far.[17] Royal proposed various plans by which ExxonMobil might win support from the Bush administration to defy Wolfowitz's hard line. Under Royal's plans, ExxonMobil would pay Déby royalties while international talks proceeded. The oil company would put some of this cash directly into Chad's treasury, defying and undermining Wolfowitz's freeze.

Ambassador Wall took up the case. He met with Déby and found the Chadian president "taken aback" by Wolfowitz's "decisive actions." Wall perceived that the United States had many interests in Chad besides the World Bank's development goals: the economic benefits and jobs associated with oil production, counterterrorism, the care of refugees from Darfur, and the need for Chad's cooperation in bringing the Darfur conflict to an end. The ambassador tried to promote the idea that it would be in the interest of both Déby and the World Bank to reach a settlement.[18]

By now ExxonMobil had made its own choice clear: It was more interested in the survival of Chad's oil production than it was in the World Bank's experiment in nation building. If Déby found a way to pay back his bank loans, and also stuck to the letter of his oil production contract, ExxonMobil would stay with him, according to State Department cables and ExxonMobil managers involved. The corporation wanted to keep its options open: "Esso is seeking to stress its neutral position vis à vis the dispute between the [World Bank] and the [government of Chad], as it is not a signatory to the agreement," Wall reported to Washington. ExxonMobil described its general approach to troubled African countries where it produced oil by emphasizing that the corporation was merely "a guest . . . and as a guest we've got to show respect. . . . It's not up to us to go into a sovereign country and tell them how they ought to be governing their people." That was an Orwellian defense in this case, because the Chad oil project had been made possible for ExxonMobil in the first place precisely because the corporation had supported the World Bank's plan to control the uses of Chad's oil funds. Yet by declining to sign the final bank agreement, ExxonMobil had positioned itself so that it was no longer accountable—as the bank's deal with Déby fell apart, the corporation stood aside. If anything, the corporation was subtly encouraging Déby to defy Wolfowitz. "We like the format we had," Andre Madec, an Exxon-Mobil global community relations executive said. But he refused to criticize Déby for his decision to balk.[19]

Robert Zoellick, the Bush administration's deputy secretary of state, telephoned Wolfowitz and talked with him about the violence in Darfur and the gathering rebel attacks on Chad, sponsored by Sudan's notorious intelligence service. Wolfowitz said he felt he could still work out a compromise with Déby.

He was wrong; Déby refused to accept the bank's new proposals, which were designed to maintain social spending but allow some more defense spending.

In pickup trucks and sport-utility vehicles, toting automatic rifles, the self-declared soldiers of the Front Uni Pour le Changement struck N'djamena on April 13. Gunfire resounded in the capital but Déby's loose-knit defenders proved just stalwart enough to chase the rebels back toward Darfur. Chad's rebels had been thwarted, but only temporarily. As a World Bank analysis put it, "The government brought the situation under control through the course of the day, but the situation has remained tense. . . . The tension is likely aggravated by the new petroleum resources, which have raised the stakes associated with power, and by the paucity of tangible results associated with oil revenues to date."[20]

Déby was again furious. He organized a "popular" rally of his supporters in the streets of N'djamena. He declared that if the world would not back him, he would defy the world: He threatened to expel all two hundred thousand refugees from Darfur and shut down all oil production in Chad by the following Tuesday, if the World Bank did not immediately meet his demands. "You have just been eyewitnesses to the attacks by Sundanese mercenaries," Déby's prime minister, Pascal Yoadimnadji, declared in a communiqué issued to all of N'djamena's ambassadors. "We regret to state that the International Community closes its eyes to the inimical behavior of the Government of Khartoum. . . . This is a particularly laughable situation. The oil is Chadian. Its exploitation must first of all profit the Chadian people."

ExxonMobil drew down to six core staff in N'djamena, including Ron Royal, who stayed on; others of his staff withdrew from the rebel raid to the relative safety of the oil fields in the south.

Royal met Déby's prime minister on the evening of April 14. He suggested openly that Chad get out of its social investment obligations to the World Bank by paying off its loans. At that point, ExxonMobil "would be free to pay royalties directly" to Déby's regime, bypassing Wolfowitz's strictures. The prime minister asked if ExxonMobil might be willing to lend Chad the money to pull off this maneuver.

Déby asked Ambassador Marc Wall to visit him the next day. Wall had been overseeing evacuations by American Peace Corps volunteers and other aid workers spooked by the rebel attack. The ambassador drove with his deputy, Lucy Tamlyn, to one of Déby's private residences in the capital, safely secluded from the presidential palace, a rebel target.

Wall found the president in the company of his new wife, Hinda Déby Itno. They spoke in French and took their places; Tamlyn took notes.

Wall acknowledged Déby's successful defense of the capital. "Washington is deeply troubled by the current turn of events in Chad," he said sympathetically.

Déby said he had sent a letter to President Bush "asking for understanding of Chad's predicament." Even now, he continued, Sudan's government had unleashed a new convoy of sixty rebel trucks filled with armed men toward Chad's eastern city of Abeche. His military problems were far from over. "I have spoken repeatedly to the international community, but the international community has failed to respond. A small country such as Chad cannot at the same time face an armed invasion as well as shelter refugees."

Half of the rebels captured after the raid on N'djamena were Sudanese nationals, Déby said. The United States had publicly condemned any efforts to seize power in Chad by force. His implication was clear: Why then would the United States stand by and allow such an invasion to succeed, given that ExxonMobil was here and that Déby was cooperating on counterterrorism and Darfur?

Wall asked about Déby's threat to shut down the ExxonMobil oil consortium on Tuesday. Déby had declared he would close the pipeline if ExxonMobil did not start making its payments directly to Chad's government, rather than through the London accounts controlled by the World Bank.

"This is not a topic for discussion," Déby answered unequivocally. "It's our money. The money belongs to the Chadian people." He added that he needed the ExxonMobil royalties to pay his troops.

Wall said that if Déby shut down oil production, he would deprive Chad's government of future payments.

"Chad can live without oil," Déby said.[21]

He agreed, however, to wait until the end of April before issuing his order to shut down production. The president's bargaining was transparent. He was angry, yes, but he was also seeking leverage with the Bush administration.

"We are not cowboys," Déby added. But the World Bank "has pushed our back to the wall."

His oil threat galvanized the Bush administration's attention. Don Yamamoto, a State Department envoy, flew into N'djamena on April 24. Yamamoto was a principal deputy assistant secretary of state, barely higher ranking than Ambassador Wall, but he was nonetheless the most senior American official to visit Chad in years. He carried with him a letter from Secretary of State Condoleezza Rice.

Wall rode with Yamamoto to the presidential palace. Spruced up with the help of oil revenue, the palace had marble floors, clean carpets, Greek columns, and painted murals that told of Chad's strength in colorful allegory. Chadian special forces soldiers in U.S.-supplied camouflage and desert head scarves protected doorways and leaned against the walls. In Déby's reception room a presidential portrait graced one wall; there were ornate white leather chairs with oversize, thronelike armrests.

Tamlyn and two other embassy diplomats completed the American delegation; Chad's foreign minister, the country's internal security director, and two note takers flanked President Déby.

"Chad currently has the full attention of the United States," Yamamoto began. He thanked Déby for agreeing to postpone the shutdown of the oil pipeline. He mentioned the letter he was carrying from Rice. The envoy unfolded a French translation and read it aloud. Its essence was that the secretary understood Déby's situation was a very difficult one and that the United States remained committed to a successful partnership with him.

Déby said that he would like to convey his gratitude to Secretary Rice. Chad wanted "a good relationship with the United States." As to his troubles with the World Bank, "We've been looking for a resolution for a year and a half."

"I know the challenge Chad faces in maintaining stability," Ambassador Wall assured him. "Our suggestions are all designed to find a way forward for a more stable future for Chad."[22]

They struck an agreement in principle on April 26. The upshot was that Déby would have greater freedom to spend money on his military and ExxonMobil could keep pumping oil.

Ron Royal told Ambassador Wall and the visiting envoy, Yamamoto, that he and ExxonMobil headquarters in Texas were "extremely appreciative" of the Bush administration's efforts.

The World Bank project now lay exposed as a failed experiment. The bank's presence in the oil deal had ensured that Déby allocated somewhat more funds to domestic development than he likely would have otherwise, and it probably created more space for Chadian opposition parties and civil society than Déby would have otherwise allowed. Several thousand Chadian families in the south benefited from education and incomes by working inside the ExxonMobil compound. Otherwise the project had not achieved its goals: It did not create a template for international management of resource wealth in poor countries; it did not prevent Déby from diverting funds to cronies and defense spending; it did not reduce corruption; it did not create political or social stability; and it had not yet improved Chad's abysmal poverty indicators. In 2000, when the project was approved by the Clinton administration, Chad ranked 167th out of the 174 nations assessed by the United Nations Human Development Index, a table of quality of life indicators, and the U.N. estimated that Chad's average life expectancy was forty-seven years. In 2006, after six years of reform experimentation and several years of oil revenue, Chad ranked 171st out of 177 nations assessed, and its average life expectancy was forty-four years.[23] The oil project did, however, allow ExxonMobil and its partners to generate several billions of dollars of top-line revenue, to forge a path to profitability just as world oil prices spiked, and to embed themselves with Déby's regime, positioning the corporation for additional oil deals beyond those covered by the original bargain. Nobody was held accountable for the experiment's failures; the project's successes belonged to Déby, ExxonMobil, and its consortium partners.

U nlike the World Bank's representatives, ExxonMobil's rotating country managers in Chad managed President Déby's expectations successfully. "They were the first to help us," said Déby's adviser Mahamat Hissène. "Every time we had problems with them, they expected to sit down and discuss it. We understand they have more experience than us. We understand they are here to make benefits. We think our benefits are linked."[24]

Déby's goal after his confrontation with Wolfowitz was cash in hand; he needed funds right away to pay more military salaries, and he wanted to buy Russian-made Sukhoi attack aircraft and helicopters that could blast Sudanese rebels in pickup trucks from the air. ExxonMobil agreed to accelerate the timing of payments Chad was due under its original contract, without altering the basic revenue-sharing terms. Déby might be borrowing against his own future—precisely the opposite of the World Bank's goal for the country—but from ExxonMobil's perspective, that was his sovereign decision as a party to their contract.[25]

Chevron and PETRONAS, the project's minority partners, interpreted their tax obligations to Chad differently from ExxonMobil. As Déby received large payments from ExxonMobil in 2006, and as he sorted out a temporary understanding with the World Bank, he was surprised to learn that Chevron and PETRONAS believed they owed him nothing at the moment, because they had yet to recoup certain costs and investments. Chevron cited a mysterious agreement dating to 2000 that the corporation claimed relieved it of certain tax burdens; Déby and his aides claimed to have never heard of the document. Déby did not take the news of Chevron's defiance calmly.

On August 17, 2006, Ron Royal invited Ambassador Wall to a new office building ExxonMobil had opened in the capital. He noted that Déby had recently restored diplomatic relations with the People's Republic of China and that Chinese oil executives were lurking around the capital. The ExxonMobil manager said he feared the American corporation's presence in the country might be threatened now by Chinese competition. The Chevron tax dispute seemed a symptom of rising troubles and could have knock-on effects on ExxonMobil's production and sale of Chadian oil. Déby's regime had indicated, for example, that it might soon try to reopen the convention under which ExxonMobil operated.

Given the "highly tense environment," Royal told Wall, ExxonMobil would seek a meeting between President Déby and Rex Tillerson.

Déby soon announced that he was throwing Chevron and PETRONAS out of Chad. "A revolution has begun," Déby announced to a government-organized crowd of a thousand people in the capital. "We are only receiving the crumbs that are called royalties. . . . This is a flagrant injustice."

This would be a limited sort of revolution, however, he explained. "One company has not failed in its obligations," the president said. "I'm speaking of ExxonMobil, with whom we will continue to work."[26]

Two days later, a young United States senator from Illinois arrived on an American military charter at N'djamena's airport. Barack Obama, in office less than two years, was on his way back to the United States from travel to Kenya, Somalia, and Chad's eastern refugee camps. Obama had not been in national office when the World Bank's experimental project in Chad was born. His interest in the country derived from his interest in the Darfur crisis. Déby's ambassador in Washington had met with Obama before he traveled and had recommended to Déby that he make time to meet the senator personally when Obama passed through the capital. They scheduled a thirty-minute courtesy call at the airport.

Obama took his seat and thanked Déby for his cooperation with the United States on Darfur and counterterrorism.

Déby in turn thanked Obama for visiting Chad and for his interest in Darfur. The crisis had profoundly affected Chad, he said. Cross-border raids into his country from Sudan continued. The April 13 raid on his capital had been an effort by Sudan "to destabilize the country, bring in a regime favorable to Khartoum, and inflict harm on Sudanese refugees in Chad."

They talked in some detail about negotiations under way to settle or at least stabilize the conflict in Darfur. Déby rarely let a meeting with an American official go by without emphasizing Chad's needs and shopping lists. He told Obama that while their countries enjoyed cooperation on counterterrorism, "Chadians still required equipment, and had submitted requests in the past year to U.S. authorities."

"If a request was submitted, then the Pentagon would be reviewing it," Obama assured him.

Déby then raised the subject of his threats to throw out Chevron and his problems with the oil companies.

"I am trying to ensure that Chad benefits from oil production," Déby told Obama. "The Chadian people cannot benefit from the country's oil as long as Chevron and PETRONAS refuse to pay the taxes they owe. They claim they have a legal basis for not paying income taxes," but the agreement they signed was not approved by the National Assembly and did not have the force of law.

By deciding to confront the oil companies, Déby continued, he was seeking only to reduce the "economic inequality" between the companies and Chad.

"I can't speak for the United States or Chevron," Obama replied. Still, he continued, "two principles need to be considered: that the Chadian people should benefit from the country's natural resources, and that contracts need to be observed."

Obama said that Chad "could benefit from foreign investment, but if the rules of the country's business environment changed, foreign investors would be more hesitant to enter Chad's economy." He said he hoped the dispute could be resolved and that Chad "would develop a business environment where contracts were respected."[27]

Déby was accustomed to this sort of lobbying by now. No matter the party membership or political ideology of American visitors, when it came to oil, diplomats and politicians always seemed to emphasize their belief in the rule of law and the sanctity of contracts.

The two men spoke much longer than scheduled. After about ninety minutes they ended their meeting and went outside to hold a press conference, to speak mainly about Darfur. Afterward, Obama flew back to Washington. At the time, Déby's advisers gave little thought to their president's encounter with the junior senator from Illinois. "Nobody in Chad really understands the situation in the United States," recalled Hissène. As for Obama, "Nobody was betting on him at the time."

Déby flew to Paris to meet with Chevron's chief executive, David O'Reilly. Their talks stalled, but Déby expressed an interest in meeting with one of Chevron's international negotiating consultants, Andrew Young, the former U.S. ambassador to the United Nations. Young, on behalf of a consulting firm called GoodWorks International, flew into N'djamena. With his help, Chevron negotiated an agreement that would clarify its tax obligations and those of PETRONAS. As part of the deal, the oil companies agreed to make a onetime payment in 2006. The oil companies paid President Déby's government a single lump sum of $281 million. A Chevron negotiator privately told the American embassy that the settlement was "not the worst, but not the best." Chad's oil would continue to flow; Chevron and ExxonMobil would continue to sell it.[28]

Seventeen

"I Pray for Exxon"

During the late 1960s, Exxon Corporation erected a gas station near the three-pronged corner of Jarrettsville Pike, Paper Mill Road, and Sweet Air Road in Baltimore County, Maryland. The corner was nestled in thick woods, rolling hills, horse farms, and muddy streams that joined into the Gunpowder River and ran to the Chesapeake Bay. Modest aluminum-sided brick ramblers with long driveways and multiacre lots dotted the region. In the early days after the Exxon gas station opened, new homeowners in the neighborhood topped up their wide-finned Impalas or their 8-cylinder muscle cars while commuting to jobs at restaurants or insurance offices or the industrial sections around Baltimore Harbor. In 1984, Exxon shut its first station in the neighborhood and opened a new one nearby, in the midst of the three-way intersection: Jacksonville Exxon, station number 2-8077, as it was known in the corporation's vast system of retail gasoline manufacturing and distribution. Suburban sprawl encroached as the years passed, and later, new subdivisions of brick McMansions with granite countertops and chef's appliances sprang up in the woods. Doctors and city executives refurbished old flagstone farms and transformed them into elegant country estates each worth

a million dollars or more. The small ramblers from the 1960s seemed
dwarfed by the larger new residences, but all of the area's homes rose
steadily in value as the great American housing bubble inflated. By 2006,
Jacksonville Exxon served a growing and economically diverse community
in northern Baltimore County: professionals, small-business owners, retirees, and middle-class commuters. All along, for more than three decades,
a single family, the Stortos, had operated the Exxon-branded stations and
auto repair facilities in Jacksonville; day-to-day management had eventually passed down to a daughter, Andrea Loiero.[1]

Altogether, ExxonMobil sold about 14 billion gallons of gasoline to
American drivers each year. Andrea Loiero reported to a downstream
division of the corporation in Fairfax, Virginia, on the site of the old Mobil
headquarters, which oversaw this retail system. There were almost 29,000
ExxonMobil-affiliated gas stations worldwide, about 14,000 of them in
the United States. Market researchers conducted public opinion surveys
after the merger and discovered that consumers valued and felt loyal to
each of the Exxon and Mobil brands, and so they concluded that there
was no reason to change any of the names, or to create a combined ExxonMobil brand. More than 8,000 of the gas stations carrying one or the
other name were owned and operated by independent distributors who
paid ExxonMobil for the right to use the brands and who agreed to abide
by the strict rules in franchise contracts. Another 1,000 or so stations were
referred to within the corporation as Heritage Mobil stations, branded as
Mobil and owned directly by the company. Some were operated entirely
by ExxonMobil employees; others were owned by ExxonMobil but operated by an independent dealer under contract. There were also about
2,200 Heritage Exxon stations similarly organized. Jacksonville Exxon was
a Heritage station owned by the corporation but managed under contract
by the Storto family. It had operated this way since it had opened.

Running a gas station had become steadily more complicated since
the 1960s. The typical retail snack and grocery shop under a red Exxon
roof now generated as much as or more profit than gasoline sales did.
Managing the retail business required expertise in credit cards, customer
reward programs, and packaged food supply. Technology and regulation

had at the same time transformed the gas station's physical plant into an intricate system of electronic monitoring systems, interconnected pumping systems, computerized inventory managers, alarms, and console boards. The blinking monitors set up behind the thick safety glass where the cashiers and station managers worked allowed ExxonMobil corporate managers to see from a distance, for example, when a particular dealer like the Stortos needed more gasoline, so that deliveries could be scheduled efficiently. The gas station business had become infused with new technical jargon: What customers referred to casually as the gas pump was now known within ExxonMobil as the M.P.D., or multi-product dispenser. Its digital systems might allow a driver to use a single handle to pump multiple grades of gasoline. A modern gas station's electronics required continual supervision, to ensure that the systems were operating properly and that gasoline sales were being captured and credited correctly.

The scene at Jacksonville Exxon on the brisk winter morning of January 12, 2006, reflected this new complexity. On one side of the station tarmac that day, a contractor had arrived to fix a submersible sump pump that pulled the gasoline out of the ground and delivered it to the multiproduct dispensers. The contractor was drilling holes in the asphalt. As this work proceeded, around 9:00 a.m., an ExxonMobil tanker truck also arrived to refill the station's 12,000-gallon underground storage tank.

As the tanker driver directed gasoline down a thick hose into a storage vat, alarms rang out suddenly—they signaled that gasoline was leaking somewhere in the station's system. All of the multi-product dispenser islands at Jacksonville Exxon shut down automatically, cutting off befuddled customers in midsale.

The tanker-truck driver came inside and spoke to the cashier. "I think I overfilled the regular tank," he said, referring to the station's underground storage vat. Spilled gasoline from the tanker hose had set off the station's gasoline leak alarm system, he suspected.[2]

At every stage of its operations—from oil wells in Africa to filling stations in America—ExxonMobil relied on outside contractors to perform much of its technical work. Halliburton and Schlumberger constructed oil and gas wells for ExxonMobil around the world. Companies specializing in offshore oil production leased their ships and crews to the corporation to drill wells in deep ocean water. Similar business practices had become the norm in ExxonMobil's retail gasoline division. Contractors, not corporate employees, serviced ExxonMobil station managers under fixed-price agreements: They mowed lawns, painted walls, and they also installed and repaired electronic and gasoline storage systems.

When a leak alarm sounded at any Exxon station in the mid-Atlantic region, it automatically alerted a call center in Greensboro, North Carolina, operated by an ExxonMobil contractor called Gilbarco Veeder-Root. The contractor's technicians in turn telephoned another independent company in Connecticut called I.P.T., which was responsible for dispatching maintenance specialists to Exxon stations. That January morning, in response to the ringing alarm, I.P.T. telephoned Alger Electric, which had a subcontract in the Baltimore area. An Alger truck turned up at the Jacksonville station within two hours of the alarm's first bell to diagnose and fix the problem, so that the Stortos could begin selling gasoline again.

Alger's technicians had learned through experience that leak detector alarms at Exxon stations usually did not go off because there was an actual gasoline leak. The detectors were sensitive devices that could be triggered by any number of causes—a paper jam inside the office, a faulty electrical component, or simply because the station was running out of gasoline. "A very big majority" of times that Alger was called to Exxon stations to inspect a leak alarm, it turned out that the alarm had been set off by something other than leaking gasoline, David Schanberger, an Alger manager, said later.

On January 12, the Alger technician first checked for evidence that the gasoline delivery driver had overfilled the storage tank as he had reported to the cashier. There was no evidence of such a spill, however. Then he ran troubleshooting tests on other station equipment. He concluded that a motor in the pumping system was at fault; he unplugged the motor from the leak detector wires, replaced it with a new motor, reconnected it to the leak detector, reset the alarm, and departed. Jacksonville Exxon was back in business and Alger Electric, the repair contractor, was "under the impression . . . that everything was working properly."[3]

Russell Bowen had worked for Exxon and then ExxonMobil for thirty-seven years. As a territory manager he worked from his Maryland home and looked after scores of gas stations in his home state, Delaware, the District of Columbia, and northern Virginia. Bowen lived just eleven miles from Jacksonville Exxon. He had known the Storto family for many years. On February 16, 2006—about six weeks after the morning incident with the ringing leak alarm—he was driving back from a corporate meeting in Fairfax when his cell phone rang. It was Andrea Loiero, the Jacksonville Exxon's manager.

"I've got a problem," she said. "I'm missing some gasoline."

Bowen asked what she meant. How much was she missing?

About 24,000 gallons, she said. That was a lot—double the capacity of an underground storage tank at the station—and not easy to misplace. Bowen figured that the gasoline was not actually missing physically, but that the problem was probably a faulty meter or a glitch in an inventory computer program. Still, he thought they should be cautious. "Shut everything down," he told Andrea. "I'll come on over there."[4]

Bowen had been around the retail gas station business long enough to remember how things were done before all the computers came in. "Back in the day," as he put it, each gasoline pump had a meter on it called a "totalizer" that kept track of how much gas was dispensed to customers. At the end of each day, the station manager would take a reading off each mechanical pump totalizer and check it against cash register receipts. To complete the inventory check, the manager would grab a dipper stick, go outside, drop it into the underground storage tanks, and measure the level of gasoline, to make sure the level conformed with the totalizer readings and the register receipts. A station manager would expect to lose a few gallons here and there because of small spills around the pump and the like, but otherwise, he would expect the daily totals to be aligned.

It was dark when Russell Bowen arrived at Jacksonville. He first checked the electronic multi-product dispenser totalizer readings; the meters were computerized now but they still counted up the number of gallons of gas pumped that day. Bowen first wanted to be sure that the measuring device was working properly; he bought a gallon of gasoline, pumped it into his own car, and then rechecked the totalizer to see if the sale had registered. It had.

Inside, he found Andrea Loiero in a nervous state. Bowen asked to see her daily inventory records. He saw that Jacksonville Exxon had been posting a "negative variance"—missing gasoline—on the scale of hundreds of gallons each day since early January. Between January 13 and January 31 alone, 14,501 gallons appeared to be missing. Why had it taken her six weeks to notice that so much gasoline seemed to be missing? Had she been failing to undertake her required daily electronic inventory reconciliations? It was just like "doing your checkbook," Bowen said later. ExxonMobil rules required station managers to report to the corporation if they experienced significant losses of any grade of gasoline. Loiero seemed agitated and confused about what her daily inventory records showed—how the math worked.

"Did any leak alarms go off?" Bowen asked her.

"No."

The more they talked, the more Andrea Loiero seemed to be "bouncing around," thinking out loud about the work that had been done by the Alger Electric contractor back in early January.[5] Had that caused the gasoline to go missing somehow?

Bowen called to ensure that another contractor was on his way to Jacksonville to start running new tests on the station's equipment. He told Andrea to calm down, think through the events of the last six weeks, and write out a chronology that could help them diagnose what had happened.

He returned in the morning and found a new contractor had also arrived with metal tanks full of helium gas. The technician drilled quarter-inch holes in the concrete near the multi-product dispenser to inject gas into the ground, to search for evidence of leaks in the lines or tank walls, through which gasoline might have escaped. About an hour later, he called Bowen: There was a single fluid leak near the station's big underground storage tanks. It was about the size of the holes that the contractor—who happened also to be from Alger Electric—had been drilling during his sump pump repair project back in January.

Exxon Spill in Jacksonville, Maryland

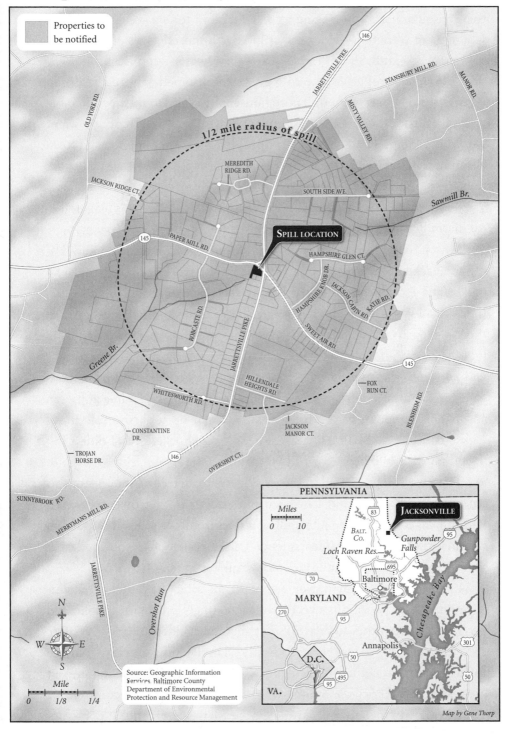

Properties to be notified

1/2 mile radius of spill

OLD YORK RD.
JARRETTSVILLE PIKE
146
STANSBURY MILL RD.
MISTY VALLEY RD.
MANOR RD.
JACKSON RIDGE CT.
MEREDITH RIDGE RD.
SOUTH SIDE AVE.
Sawmill Br.
145
PAPER MILL RD.
SPILL LOCATION
HAMPSHIRE GLEN CT.
HAMPSHIRE KNOB DR.
JACKSON CABIN RD.
KATIE RD.
ROBCASTE RD.
JARRETTSVILLE PIKE
SWEET AIR RD.
145
Greene Br.
FOX RUN CT.
WHITESWORTH RD.
HILLENDALE HEIGHTS RD.
BLENHEIM RD.
CONSTANTINE DR.
JACKSON MANOR CT.
TROJAN HORSE DR.
146
OVERSHOT CT.
SUNNYBROOK RD.
MERRYMANS MILL RD.
JARRETTSVILLE PIKE
Overshot Run

N
W   E
S

Mile
0   1/8   1/4

Source: Geographic Information Services, Baltimore County Department of Environmental Protection and Resource Management

PENNSYLVANIA

Miles
0   10

83
JACKSONVILLE
BALT. Co.
Gunpowder Falls
95
Loch Raven Res.
695
70
Baltimore
MARYLAND
270
95
95
50
Annapolis
301
D.C.
50
495
95
VA.
Chesapeake Bay

Map by Gene Thorp

The news was catastrophic: In all probability, about 24,000 gallons of toxic gasoline had been leaking into the ground for six weeks in an area where there were houses located within five hundred yards of the Exxon station. Worse still, those homes drew their water supplies from wells drilled on their properties; the county system of piped and treated drinking water did not serve them. Those household wells would be vulnerable to contamination from the gasoline in ways that piped county water would not. Since January, scores of local families and children had been consuming their well water, bathing in it, and cooking with it, unaware that they might be imbibing and dousing themselves with diluted gasoline. How could this have happened? Why didn't the station's leak detector alarm bells ring as they were supposed to do?

Bowen relayed the findings up ExxonMobil's chain of command. His report reached Steven Polkey in Fairfax. Polkey was an Englishman who had joined Exxon out of a British university, rose through the corporation's retail gasoline businesses in the United Kingdom and Europe, and then moved to the United States in 2004 to take a senior position in ExxonMobil's Safety, Health, and Environment department—"She," as it was known to acronym-savvy corporate insiders. Polkey was now responsible for all of the environmental issues involving spilled gasoline at Exxon and Mobil stations in the United States. When he took the call about Jacksonville, he said later, "I was stunned, I was shocked. . . . I didn't believe that we could have lost 25,000 gallons."[6]

There was no precedent in the retail gasoline division for a leak of this scale—especially one that had been allowed to unfold over six weeks without detection. Worse, the spill had occurred at a gas station known within ExxonMobil as a Consequence I site because it was located near houses that relied on aquifer wells for drinking water.

ExxonMobil promoted itself as one of the safest large industrial corporations in the world; its executives increasingly scoffed in public and private at competitors such as BP that seemed accident prone. By 2006, ExxonMobil's record was certainly better than the industry norm, but the truth was that, nonetheless, accidents with serious environmental impact or in which workers were injured were regular events at the corporation. They involved pipeline spills, trouble at refineries, accidents at construction sites, and losses of inventory of dangerous chemicals. Since the *Exxon Valdez* accident, under Frank Sprow, the daredevil adventurer and dangerous game hunter, Safety, Health, and Environment had seen worker fatality rates, in particular, fall to the point where if one did occur, it seemed a shocking anomaly inside the corporation. Yet such a massive global enterprise that daily moved toxic materials from beneath the earth to customers at 29,000 retail gas stations, plus large refineries and chemical plants, could hardly expect zero accidents. On the retail gasoline side, ground leaks of as much as 1,000 gallons of gasoline occurred periodically at American stations or storage facilities. ExxonMobil's Public Affairs and Safety, Health, and Environment departments had developed standardized playbooks to respond to such events. The protocols included prepackaged talking points for communicating with alarmed members of the public. The leak at Jacksonville was exceptionally large, but it was exactly the sort of accident that ExxonMobil's playbook was intended to address. When Steven Polkey hung up after receiving the news about Jacksonville, he set ExxonMobil's spill response plans into motion.

Public Affairs faxed Andrea Loiero talking points to use if homeowners around the station or journalists turned up asking questions about all the commotion and activity now taking place at the Jacksonville station, which had been closed to customers as contractors drilled and dug to determine the extent and flow of the leaked gasoline. "We're investigating," she was to tell neighboring homeowners. "We'll provide an update."

Outside Jacksonville Exxon, a sign soon appeared that explained the station's closure as well as the presence of so many mysterious trucks and workers drilling in the ground: "Please excuse our appearance. We're working to serve you better. Fueling facility is temporarily closed for upgrade."[7]

Transportation fuel—the production, refining, and distribution of gasoline, diesel, jet fuel, and the like—is the second-largest segment of the worldwide energy economy, and the fastest growing. Power generation—the production of electricity—is the largest. The fuel economy's worldwide growth is mainly a function of rising incomes in previously car-deprived poor countries. In Europe and the United States, however, by 2006, gasoline consumption had reached a plateau and possibly peaked forever.

The retail gasoline stations had always been an unglamorous stepchild division within ExxonMobil. If the value of the land on which corporation-owned stations sat was factored in, the division's profit margins were embarrassing by ExxonMobil standards, particularly compared with upstream oil and gas production or chemical manufacturing. Since the 1980s, like other large international oil corporations, Exxon had been steadily divesting itself of retail stations: Its total of 29,000 Exxon and Mobil stations worldwide in the early days of Rex Tillerson's reign as chief executive represented less than half of the 62,000 stations Exxon had operated under its brand alone three decades before.

The environmental and legal aspects of the gas business looked particularly unfavorable. In the euphoric postwar automobile age, neither citizens nor government officials in the United States paid much attention to gasoline's toxic properties or to the consequences of so much sloshing and spilling of gasoline on and beneath the ground. Only after the rise of environmentalism and the birth of the Environmental Protection Agency in 1970 did federal regulators begin to look seriously at carcinogenic effects of gasoline exposure and to tighten the loose, expedient storage and cleanup practices of retail gasoline sellers. Like other oil corporations that had previously pumped gas without much reason to think about environmental impact, Exxon discovered, as the laws tightened after 1970, that it would henceforth be responsible for hundreds of "remediation sites"— that is, sites where gasoline had leached into the ground at some point in the pre-E.P.A. era and where it now had to be located and scrubbed out as best as possible, whatever the cost.

By 2006, ExxonMobil managed four thousand environmental remediation sites around the United States.[8] Of those located at gas stations, many involved "historic spills," as they were called, that dated to the pre-environmentalism era. The origins and extent of these old leaks were often unknown—all that could be said was that gasoline had somehow gotten into the ground, contaminating the soil and any water that might lie beneath.

To prevent the recurrence of such leaks, the E.P.A. issued regulations in 1998 requiring gas station operators to upgrade their storage tanks, improve the tank hulls, and install more spill buckets and other protections. It also required station operators to install leak detection systems that were better than the old system of physical inventory reconciliation.

The purpose was to protect people from health damage caused by exposure to gasoline. So far as federal scientists could determine, there were two elements in gasoline that might be damaging, if a person suffered sufficient exposure: benzene and methyl tertiary butyl ether, or MTBE.

Benzene is an aromatic hydrocarbon compound long known to cause cancer; it has been formally designated as a "known human carcinogen" by the U.S. Department of Health and Human Services. Benzene was widely used as a gasoline additive in the 1950s. Its use was discontinued, but as the government moved to replace lead in gasoline, to attack air pollution, benzene made a limited comeback as an additive. Regulators limited the amount that could be blended into gasoline, however—no more than 1 percent, precisely because of fears that spilled gas might accidentally leach into groundwater accessed by household drinking wells.

MTBE, the second dangerous element in gasoline, was developed in laboratories to raise gasoline octane ratings; after 1990, government policy encouraged its use to enhance the amount of oxygen emitted when cars burned gasoline, to reduce urban air pollution caused by tailpipe emissions. Nobody had studied MTBE's health effects, however. Later, based on laboratory tests involving rats, the E.P.A. concluded that MTBE was a "potential human carcinogen at high doses." That tentative finding led to fast policy reversals by state and federal regulators, who ordered plans to reduce and eventually eliminate MTBE from gasoline. The E.P.A. finding about MTBE's potential health effects also stimulated massive numbers of lawsuits against oil companies by cities, towns, businesses, and individuals who claimed to have been affected by historical gasoline spills where MTBE had been present in the fuel. ExxonMobil found itself a defendant in hundreds of these cases after 2001. PACER, the computerized system containing records of lawsuits in the American federal court system, contained dozens of listings of civil cases where ExxonMobil stood accused of negligence for allowing MTBE to leach into groundwater because of gasoline spills—even though it had been encouraged by the government to put MTBE into its gasoline in the first place. The corporation's law department managed these suits as a kind of high-cost division of legal operations, seeking to minimize ExxonMobil's financial exposure. ExxonMobil's Washington lobbyists pushed unsuccessfully for Congress to enact laws that would exempt oil corporations from liability, on the grounds that the government had encouraged MTBE's use. Separately, the corporation accepted that the additive should be phased out: "ExxonMobil recognizes that MTBE use in gasoline has caused concern with some customers," one of its lobbyists, D. L. Clarke, wrote to a state air pollution regulator in 2003, "and we support phase down of MTBE use in a manner consistent with maintaining reliable and affordable gasoline supplies."[9]

On the other side of the issue stood a network of plaintiffs' lawyers who saw MTBE as an opportunity to sue oil companies and win lucrative verdicts. By the time of the Jacksonville Exxon leak, American plaintiffs' lawyers who previously had represented victims of tobacco marketing, asbestos exposure, or faulty medical devices traded information and scanned for news of new gasoline leaks and spills. For the Baltimore area plaintiffs' bar—ambulance chasers, to their critics—it would have been difficult to imagine more enticing news than that which circulated in the last weeks of February around northern Baltimore County: that 24,000 gallons of MTBE- and benzene-laced gasoline had spilled in an area of homes dependent on groundwater wells, and that the world's largest and least popular publicly traded oil corporation directly owned the gas station responsible for the leak.

In this way, the irresistible force known as Stephen Snyder came to meet the immovable object branded as ExxonMobil.

Stephen Snyder grew up in a modest row house in West Baltimore. His father and uncles owned clothing stores. In high school, Snyder recalled, he rarely did so well as to earn a B. As an undergraduate at the University of Maryland, he at last began to study, and at the University of Baltimore School of Law, he excelled. He had the gifts of a natural salesperson and worked his way through school selling magazines—he was so successful that he soon was earning more than his father, even before he entered law. At twenty-four, he set up an independent legal practice devoted to "contingency-fee" cases, in which he generally sued corporations on behalf of individuals and got paid only if he won damages or settled for cash. "I don't think you could hire me for an hourly rate, no matter what," he explained later. "If I win, I have to have some skin in the game, a piece of the action." He won his first million-dollar medical malpractice verdict in the 1980s and kept going. United Cable settled a racial discrimination case with him in 1990 for $106 million. The accounting firm Ernst & Young settled over a business bankruptcy matter for $185 million. He won a jury verdict against a bank for $276 million. A contingency attorney such as Snyder generally took about a third of such verdicts as his fee.[10]

By the time of the Jacksonville Exxon gasoline leak, Stephen Snyder had reached his late fifties. His silver hair was receding from his forehead; he wore his hair cropped. He was not a tall man, but he was broad-shouldered and powerfully built. He had more wealth than even most successful lawyers could imagine. He had fathered five children by two marriages, and two of his sons had followed his footsteps and joined his law firm. And yet Snyder remained deeply restless, driven, and insecure. "How did I do?" he would eagerly ask anyone within earshot after a court appearance. "I just wish he'd take a deep breath and relax," his second wife, Julie, said. "It's never enough."[11]

Snyder displayed his wealth conspicuously: a diamond-studded Rolex watch; a gold chain with "Steve" encrusted in diamonds; an alligator-skin briefcase; expensive tailored suits. His office wall displayed a framed check written to his firm for $70 million. He almost lost a New Jersey trial when jurors mistook his Rolls-Royce in the parking lot for that of his client. He defied conventional thinking about how lawyers should comport themselves: He flashed his wealth inside the courtroom because he believed jurors would lean his way if they believed he was rich and successful. He wept and shouted at witnesses. He ignored judges when they ruled him out of line.[12]

Some members of the corporate bar dismissed Snyder as "more show man than lawyer, a flashy cynic who manipulates unsophisticated jurors by twisting the facts," the *Baltimore Sun* put it. Even within his own tort or plaintiffs' law community, he remained emphatically and annoyingly in second place in the city of his birth. Peter G. Angelos, another street-smart University of Baltimore law school graduate, had earned an immense fortune in contingency-fee asbestos and tobacco cases and had used his winnings to purchase the Baltimore Orioles baseball team. Whereas Snyder's greatest verdicts exceeded $100 million, Angelos had gotten rich from billion-dollar tobacco and asbestos cases.

Snyder was desperate to catch up, to land his own white whale: Jacksonville Exxon seemed to have that potential, or so Snyder concluded as he solicited clients soon after news of the leak became public. He had not been tracking MTBE litigation nationwide, but soon educated himself. The attraction of the Jacksonville case had little to do with its complex environmental aspects. Snyder and his colleagues were drawn instead to the fact that the station had put up a misleading sign during the first day or two after the spill and then ExxonMobil had given talking points to the station manager that she found to be "lies." Those were the sorts of facts that could turn a jury's emotions against a giant corporation.

Snyder found himself in a race with Angelos once again. His rival's firm signed up as clients property owners around the Jacksonville station. Lawyers for the two competing firms prowled the same neighborhoods, seeking to recruit as many homeowners affected by the leak as possible. On Robcaste Road, Steve Tizard and two of his neighbors decided to interview the firms contending for their business. They met Angelos's team and more than a dozen other firms. When it was Snyder's turn, he arrived with his entire law firm, even as he declared he was not sure he wanted to take the case. It was "a show of force," Tizard recalled. "I was getting sold every second. He was just so arrogant and nasty." Eighty-nine families within the general vicinity of the station, including Tizard's, eventually agreed to go with Snyder.[13]

Exxon executives quickly removed the case from the Safety, Health, and Environment department and handed it over to the law department, putting it into the operations queue with the other MTBE cases the corporation faced. By now ExxonMobil's in-house legal strategists had a playbook for such cases. In accidents like Jacksonville's, the corporation had learned that there was usually no point fighting the basic question of legal responsibility; instead, the goal of its defense strategy was to avoid punitive damages. In the *Exxon Valdez* case, Lee Raymond had refused to bend by paying punitive damages, and his stubborn determination eventually made new and favorable law for corporations at the United States Supreme Court. (The Court held that formulas under which actual damages found at trial might be multiplied to determine punitive damages could be constitutional, as long as the multiplier was relatively low, such as one or two times the actual damages.) "The strategic call we made," Lee Raymond recalled, "was that the punitive damage issue is moving our way. . . . So we are just going to hang in there. That was the strategy for twenty years. And we just called it right. And had good lawyers." Exxon-Mobil was hardly going to depart from these principles while defending itself in local trials over spilled gasoline. As to actual or compensatory damages—payments to homeowners for the actual losses they incurred because of the Jacksonville Exxon's leaking gasoline—Exxon's representatives told residents that in principle the corporation was willing to pay for declining property values and proven medical claims, including documented emotional distress. With Stephen Snyder's clients, however, settlement negotiations failed. Snyder's firm felt ExxonMobil's lawyers were trying to skim off the clients in his group with the strongest cases, and settle with those, while leaving Snyder with the weaker cases at trial. He urged his clients to hang together, and they did.

Because the case involved MTBE claims, it was initially assigned to the federal court system, to be consolidated with all of the other MTBE cases accumulating around the country. Snyder wanted to try the case before a state jury, on his home court, where he knew the rhythms and rules best. State court juries tended to award punitive damages more readily than federal juries. It took some maneuvering, but Snyder eventually won a decision removing the case to the Maryland courts in exchange for his agreement to drop claims specifically related to the health effects of MTBE.

ExxonMobil honed its defense strategy as the trial date approached.

Gasoline prices in the United States were rising, and oil companies were more unpopular than ever. The corporation could expect hostility from at least some jurors. Therefore, it would try to win sympathy from the jury by forthrightly admitting that it was at fault; it would apologize to the plaintiffs and the jurors; and it would invite the jury to determine what actual damages local homeowners deserved. At the same time, Exxon-Mobil's lawyers would defend adamantly against the claim that it owed punitive damages. Before the Jacksonville trial opened, ExxonMobil paid $4 million to the Maryland Department of the Environment and accepted responsibility for the spill. The corporation spent, according to its representatives, another $38 million on cleanup efforts in the neighborhood around Jacksonville Exxon—it dug into the groundwater, installed test wells to monitor for the presence of benzene or MTBE, and used chemical and other treatments to clean and eliminate gasoline residues from the aquifer.

To win his billion dollars, or at least something close to it, Stephen Snyder would have to persuade the jury that the Jacksonville gasoline leak was more than just an accident. He would have to show that ExxonMobil had acted maliciously, fraudulently, or with gross negligence, a standard that might amount to a finding of "willful blindness." He had to show that greed and corporate cover-ups lay behind the Jacksonville leak—and therefore, ExxonMobil should be punished or deterred with an award of heavy punitive damages, beyond the actual losses of the homeowners, in order to send a signal to the corporation's executives and to other companies in the oil industry. Snyder figured that if he won about $150 million in actual damages, and if the jury was outraged enough by Exxon's actions, he might win a multiplier for punitive damages that could push the total verdict toward $1 billion. If he did that well, he hoped to withstand appellate scrutiny or at least force Exxon into a high settlement.

Snyder subpoenaed hundreds of thousands of pages of documents and e-mails from ExxonMobil's retail gasoline and safety divisions. As he and his partners painstakingly read through them before trial, they found what they felt was a winnable fraud case that could produce a billion-dollar jury verdict. Snyder decided to turn the trial into a story about the alarm bell that hadn't rung at the Jacksonville station after the gasoline leak began to flow on January 12.

The story involved a leak detector system called the EECO 3000. It was one of two different electronic alarm systems the corporation used at its stations nationwide—and of the two systems ExxonMobil employed to comply with federal regulations, it was the more problematic. Internal documents showed that the EECO 3000 was highly sensitive and prone to false alarms.

ExxonMobil had decided to replace the EECO 3000 before the Jacksonville leak occurred, but it had not moved quickly to do so. Exxon said the devices were safe, just harder than they should be to operate, and therefore the pace of replacement was just a routine business matter. The company that originally manufactured the system had been sold; in 2004, the successor company informed ExxonMobil that it would no longer support the leak detector with spare parts. Budgetary constraints and corporate planning timelines meant the EECO 3000 changeover was proceeding gradually.

Snyder concluded that the totality of evidence added up to fraud. His argument was that to enhance its gargantuan profits, ExxonMobil had avoided coming to terms with the EECO 3000's fatal flaws; it had failed to act promptly to replace the system at stations near homes that relied upon groundwater wells; and the corporation had sought to hide evidence of the system's troubles. It was perhaps not as obvious a jury-ready story of corporate neglect and greed as the case of the *Exxon Valdez* captain with a documented alcohol problem, but given the unpopularity of oil corporations and of ExxonMobil in particular, it might be good enough to bring home a fraud verdict from a Baltimore County jury.

"It was a lemon," Snyder said of the EECO 3000. "They knew it. It is a dark secret. It was the skeleton in Exxon's closet."[14]

On an autumn morning, Snyder and dozens of his clients filed into Courtroom 2 on the third floor of the Baltimore County Courthouse, a massive prisonlike concrete box in suburban Towson, Maryland. Judge

Maurice W. Baldwin Jr., a senior visiting judge assigned to the case from nearby Harford County, entered the courtroom and settled on his raised bench. On the wall to his left hung oil portraits of robed judges. Wood paneling, plush carpeting, and upholstered blue vinyl chairs contributed to a heavy, sleep-inducing aesthetic. They all might as well get comfortable; Stephen Snyder intended to speak at length about the cause he had now shouldered.

"Members of the jury, this is a gas leak that should not have happened," Snyder declared in his opening statement, pacing before the jury box. "It is a leak that took place because Exxon made a corporate decision to disregard the health and the welfare of the citizens. This is a company that decided that profits are much more important than safety."

Snyder warned the jurors that the trial would take months, and he urged them to pay close attention to the details. "It is not a contest between the lawyers—who wears the flashiest suit or jewelry. I will win that contest."

"Stipulate," came the deadpan response from the ExxonMobil defense table.

There sat James F. Sanders, a trial lawyer from Nashville, Tennessee. Sanders had participated in ExxonMobil's trial defense in the *Valdez* case more than a decade earlier. He was one of the trial attorneys ExxonMobil relied on in its most risky, sensitive jury cases. Sanders had tested over years the best ways to reach jurors who might be naturally skeptical about the motives of a giant oil corporation.

Among other things, as the Jacksonville Exxon trial unfolded, Sanders would avoid badgering witnesses or arguing vehemently with Stephen Snyder, no matter how provocative or outrageous Snyder's behavior or accusations became. To build an emotional connection with jurors on behalf of an unpopular corporation, Sanders believed he had to come across as entirely reasonable, calm, humble, and interested only in a modicum of fairness on behalf of his client. His southern accent and soft voice reinforced his demeanor. Let Snyder bluster and thunder; Sanders would slip in behind him and speak calmly of common sense.

ExxonMobil's alleged greed lay at the heart of Snyder's accusations, but he had to calibrate his charge. "No one is saying in this case that Exxon intentionally allowed 26,000 gallons to go into the ground," he explained to the jury. "Exxon *did* knowingly allow unreliable and defective equipment that they knew was a lemon—they knew for seven years and they did nothing about it because they didn't care about residents and the environment. All they cared about was profits."[15]

M r. Snyder's time with you was quite a performance," Sanders replied when his turn arrived. "And I will tell you from the very beginning that it is not my intention to try to match the performance. Indeed, I'm not going to perform at all. I'm not going to try to match the jewelry or his suits. . . .

"The most important thing that I have to say to you is the first thing that I'm going to say to you: And that is, we are sorry. We are sorry for the leak. We are sorry that the leak went on for over 30 days without being discovered. We are sorry at the magnitude of this leak and the spill into the community. . . . We apologize. We apologize to the Plaintiffs in this room. We apologize to the Plaintiffs not in the room. We apologize to you. We apologize to the community. We apologize to the State of Maryland. . . .

"Now, we do not—do not—accept liability under some of these theories you heard about" from Snyder, he went on. "We do not accept liability for fraud, we do not accept liability for any intentional misconduct, and we do not accept liability for anything that says we did anything intentionally or with malice. We don't accept that. But we do accept liability to pay for the harm that you find was actually caused to the people who were actually harmed."[16]

The fraud charge centered on the EECO 3000 leak detector would be the "battleground," as Snyder put it later, of what became a five-month trial. Day after day, Snyder presented ExxonMobil internal documents and cross-examined corporate witnesses in an effort to prove that Exxon-Mobil managers and executives knew the EECO 3000 was dangerously unreliable because it gave off so many false alarms, and that ExxonMobil accepted this flawed leak detector because it did not want to spend the money necessary to replace all of them at once—not even in Consequence

I areas such as north Baltimore County, where a gasoline leak could infect household wells.

"You would agree, sir, that you all at Exxon had the economic where-withal to replace the [alarm system] in one day if you wanted to do it across the country? You had the money to do it?" Snyder asked John Greco, an ExxonMobil manager in charge of gas station construction around the United States.

"I can't speak to that, no."

"Is it not a fact, sir," Snyder demanded, "that this was the system that continuously alarmed, 99 percent of the time, for reasons other than a leak, and it just wasn't trusted by you all at Exxon?"

"I would disagree with that statement."

"And is it not a fact, sir, that you all at Exxon knew, you knew that the leak detector had alarmed [at Jacksonville]—you knew it at Exxon, and you all ignored it?"

"That's incorrect."

Exxon's defense turned on its assertion that however many false alarms the EECO 3000 might emit, the detector still found leaks accurately. It had done so in Jacksonville on the January morning in question, James Sanders told the courtroom. The contractor fixing the station's submersible pump on January 12 had unknowingly drilled a hole in one of the gas lines; the alarm had sounded as it was supposed to do; and a second contractor had arrived to resolve the leak issue.

This second contractor thought he had fixed a false alarm problem by replacing a motor in the leak detector system, but in fact, he had missed the real trouble, the gasoline leak, and then, compounding his error, in resetting the alarm he had inadvertently calibrated the EECO 3000 improperly, so that it would no longer sound as gasoline spilled into the ground in the days to come. That might not have mattered so much if Andrea Loiero had conducted her daily inventory checks properly and noticed the missing and leaking gasoline within a day or two, but she, too, had failed. Like many industrial accidents involving complex systems and human beings, the Jacksonville spill had arisen from small errors compounding one upon the other, ExxonMobil argued. But there was no fraud: The leak detector had sounded its alarm; it was the human beings involved who failed to diagnose the alarm correctly. Therefore, there was no gross corporate negligence involved.

Snyder called some of his clients, the local homeowners, to take the stand to speak about their emotional experiences—their anxiety about not knowing if the water they had been drinking might leave them with cancer in later years, and the distress of lost wealth as news of their contaminated property spread and home values fell. Some of the witnesses wept. Some spoke of their fears for their children and grandchildren. Almost all of them expressed anger about the inflexibility and arrogance they said they experienced when they dealt with ExxonMobil public affairs and legal officials after the accident.

Ricci DePasquale, the owner of a local pizza parlor, told the jurors about his children: "At one time the youngest gulped down a little bit of the water in the bathtub and asked his mother if he was going to die because he drank contaminated water. No child should ever have to say something like that."

Snyder bore in on Exxon's strategy of apology and appeasement. "In this case, you have heard Mr. Sanders apologize on behalf of ExxonMobil. . . . How do you believe that Exxon has handled this entire situation?"

"Exxon has handled this for Exxon, not for the people of the community of Jacksonville," DePasquale answered. "They have taken care of themselves, not us. . . . My neighbors shouldn't be up here spilling their heart out. They should have been taken care of. . . . We should never have to come to this and go through this."

"As an affected resident in this community, do you accept Exxon's apology?"

"I pray for Exxon; I don't accept their apology."[17]

"My small law firm from Baltimore County took on the world's largest corporation," Snyder declared when closing arguments finally arrived. Speaking of his homeowner clients, he continued, "I was sort of floored by what I saw in this courtroom. . . . I heard people's hearts pouring out. People breaking down on the witness stand, and I can be pretty tough, but there were times when I had to hold back the tears."

At ExxonMobil, "they do not have their priorities in order. Hopefully, you'll correct that. . . . You heard their stories. They cried out to you. This is their avenue for change. This is their opportunity for justice."

James Sanders spoke gently. He appealed to the jurors' sense of responsibility, after they had invested so many days and hours in their roles as citizen-judges: "Here we are," he said, "a big Texas corporation, international, profits that you have read about, all this other stuff—how do you treat us the same way that you treat people sitting out here in the audience? Man, that's tough. But you have to. That's what is really hard about this. You have to treat us fairly as you have been treating me during this trial." If they felt sympathy for him as ExxonMobil's lawyer, if they liked and trusted him, they should apply that trust to their verdict and reject the fraud charge.

ExxonMobil's vulnerability in the trial—its potential billion-dollar problem—lay with Snyder's emotive performance, his efforts to pull the jury into a change-the-world mind-set, from where they could unleash their pent-up anger at Big Oil. Sanders therefore aligned himself with those in the jury who might admire Snyder's passion—and then he turned this sympathy for his opponent around.

"I like Mr. Snyder," Sanders said. "I am fond of him. I respect his abilities, which are considerable. He is quite a character. . . . He is an absolute handful, but you have to love him. But I don't agree with how he mangles the facts. . . . In my wildest imagination, I would never have been able to come up with some of the theories that he has come up with in this case. . . . It is ingenious. It is brilliant. It is wrong, but it is brilliant."[18]

The jurors filed into Courtroom 2 after twelve days of deliberations. They announced their verdict: $150 million in actual damages and zero dollars in punitive damages. The verdict for actual damages was high: The jury awarded all of Snyder's clients 100 percent of the appraised value of their homes, even though some of them had sold their homes for hundreds of thousands of dollars and none of the homes was appraised as worthless. Of the $150 million, $71 million was for emotional distress,

$61 million for property loss, and the rest for the costs of future medical monitoring.

Snyder was still looking for his first billion-dollar case. The jury's decision showed that proving "intentional malice is an extremely uphill climb," he explained.

James Sanders had told the jurors repeatedly during the trial that ExxonMobil would pay whatever they thought was fair by way of actual damages. Nonetheless, once the verdict was in, ExxonMobil rejected the jury's decision and declared it would appeal. This was ultimately Rex Tillerson's decision; he followed the legal strategies and policies bequeathed to him by Lee Raymond, and before Raymond, by corporate lawyers dating back to Standard Oil's defiance of antitrust reformers. "Compensatory damages should not be so high as to essentially be punitive instead of truly compensating for actual harm caused by the spill," the corporation said in a statement.[19] Judge Baldwin upheld the verdict on initial review, but ExxonMobil said it would appeal again. It would be years before the families around Jacksonville Exxon would see a dollar from the corporation, if ever. "Don't mess with Texas" remained the ExxonMobil law department's ethos, and the corporation's strategists believed that if they made exceptions for one set of accident or tort victims, they would only be challenged and exploited by others—whether in Baltimore County or Aceh, Indonesia.

Eighteen

"We Will Need Witnesses"

The Supply Chain Building in Cluster I of ExxonMobil's gas fields in Aceh, Indonesia, stood on cleared land ringed by palm trees that rustled in ocean winds. The facility contained a processing plant for cement and chemicals used for cleaning pipes in the Arun gas fields. Grasses and rubbery, purple-tinted shrubs encroached on the chain-link fences and partially obscured from view a dozen or so beige storage tanks; not even ExxonMobil's disciplined tree-trimming systems could keep Aceh's littoral forests at bay. Farmers from nearby Mee village often wandered along the fence line, as did their cows and goats.

By the spring of 2007, villagers no longer had to endure the occasional crack of a rifle shot or the risk of arbitrary detention by Indonesian army soldiers or the Acehnese separatist rebels known as the G.A.M. On August 15, 2005, in Helsinki, Finland, Hasan di Tiro and his comrades in the rebel leadership had signed a peace accord with the government of Indonesia. Afterward, G.A.M. disarmed its cadres and reorganized to enter peaceful democratic politics. The Indonesian army, or T.N.I., withdrew most of its troops from Aceh. All that remained of the war were its ghosts, the missing and unaccounted dead.

Years earlier, ExxonMobil had benefited from the Bush administration's intervention against G.A.M., when the administration quietly urged the guerrillas to stop targeting the corporation. The violence had subsided, but ExxonMobil's reliance on the administration for protection persisted. The venue had now shifted from Aceh's jungle to Washington courtrooms. At issue was whether America's largest private corporation bore any responsibility for the Aceh conflict's legacy of death, injury, and torture.

The peace deal between G.A.M. and Indonesia's weak, nascent democratic government had produced no serious investigations of the war's abuses. About a thousand young Acehnese men remained missing. For the most part their relatives accepted that they were dead, but many ached to bury their sons or husbands on family compounds. Occasionally farmers and villagers stumbled over human remains in the places where the T.N.I. had previously established interrogation centers, such as along the perimeter of ExxonMobil's gas fields. In December 2005, for example, some villagers living near the corporation's Cluster II gas fields found human bones and decomposed clothing in a sack buried in the ground. The sack contained a wallet with an identity card inside: The victim, a young G.A.M. volunteer named Kaharuddin, had been arrested at his home in 2003. His widow reburied her husband's remains in their tree-shaded yard. When new bones were discovered in this way, the word spread, and families of the missing traveled to the informal excavation sites, seeking evidence that might bring their own searches to a close.[1]

During the third week of June 2007, an Acehnese civilian discovered bones inside a safety tank within the ExxonMobil facility at Cluster I, near Mee. Exactly how the discovery was made and how the searcher got inside the fence was never made clear. A survivor of T.N.I. detention had publicly identified the Supply Chain Building as a place where soldiers in camouflage uniforms had interrogated him and his brother; the brother was among Aceh's disappeared. By June 20, news of the bone find had circulated. Crowds gathered at ExxonMobil's chain-link gate and demanded to be let inside. The corporation's unarmed guards, some of them Acehnese, yielded. The villagers spread the excavated bones on a warehouse floor and took photographs. The ExxonMobil security guards established a police line to protect the evidence. The crowd was excited and agitated.

Some of them noticed a mound of rubble and dirt outside that looked as if it might be a makeshift grave. They demanded that the ExxonMobil guards dig up the mounds, but the guards refused. They instead summoned the police; a deputy commissioner arrived and announced to the villagers that they must control themselves. He said they would need a forensic team as well as permits from ExxonMobil before they could dig. The crowd of villagers and farmers decamped to the North Aceh police station and obtained an audience with the police chief, who promised to investigate. "We will need witnesses because what we will dig up [on ExxonMobil property] is someone else's house," he said.[2]

Later the police suggested that the villagers had been mistaken and that what they had found in the storage tanks were the remains of animals. Photographs of the bone fragments, to an amateur eye, were ambiguous; there was nothing as obvious as either a human's skull or a cow's jawbone. In any event, the Indonesian army had no intention of allowing a serious investigation to proceed. General Bambang Darmono, who had commanded Indonesian military operations in Aceh between 2002 and the final peace agreement, explained to an interviewer: "If we keep digging up the past, the problem in Aceh will never be resolved. Everyone violated human rights laws. It wasn't only the Indonesian armed forces, [but] the cops, the government, the G.A.M. . . . Do we want to keep talking about that? If we do, we will always talk about violence and there will be no peace. But if we want to have peace, we have to bury everything."[3]

After the Bush administration intimidated G.A.M. into ceasing its direct attacks on ExxonMobil's gas fields in Aceh in 2001, which allowed the corporation's operations in the province to resume, the war had turned even more brutal.[4] President Megawati's military-aligned government in Jakarta imposed martial law. To destroy G.A.M., she deployed about 22,000 T.N.I. soldiers, 12,000 paramilitary police, and authorized the recruitment of 10,000 militiamen. She curtailed civil liberties and banned or restricted international aid and human rights groups in Aceh. Shielded from scrutiny, the Indonesian military detained, tortured, and executed hundreds of young Acehnese men after 2001.[5]

Baharuddin, a young farmer whose family lands lay in the vicinity of ExxonMobil's gas fields, recounted a story that was emblematic of widespread testimony about the abuses. One afternoon in December 2003, Baharuddin walked, with a friend named Noordin, to an old mosque elevated on stilts and overlooking rice paddies near the Pipeline Road. He intended to have his hair cut at one of the informal stalls beside the mosque. T.N.I. soldiers on foot patrol walked by. They approached and asked, "What are you doing here? Are you a member of G.A.M.?"

"No, I'm just a villager," Baharuddin answered.

"You're lying. You just came down from the mountains." They hit him. Then they force-marched him and his friend across the rice fields to an interrogation center in the battalion commander's post, a little farther down the Pipeline Road. Interrogators showed Baharuddin the wallet of a man named Ismail and accused him of extorting money from him. "Just confess," one of his interviewers, an Acehnese, said gently. "You'll be fine. If not, they'll beat you up."

"I'd rather be beaten—I didn't do anything," he answered.

About eight soldiers came upstairs. They carried two iron bars, a razor blade, salt, and the delicacy known as star fruit. First they drew bloody lines on his chest with the blade. Then they sliced open his ears. They sprinkled salt and fruit juice on the wounds. Baharuddin thought, he recalled, "If I confess, it won't mean the beating will stop." The T.N.I. were notorious among local villagers for continuing with their punishments no matter what their detainees said. He worried, too, he said later, that someone else might get into trouble because of any confession he made. The soldiers cut off sections of his ears. They beat him in shifts. They forced him to lie down, put a wooden board on his back, and then walked over him two at a time.

After about five days, they gave him a cigarette to smoke and invited him to speak with the battalion commander. The commander leafed through a thick book of G.A.M. suspects. "You have an hour" to confess, he announced.

"Why don't you shoot me now," Baharuddin recalled answering.

"Okay, dig a hole," the commander finally said. They went outside and dug a trench between the interrogation post and a nearby electric pylon.

"You can invoke your God and tell him to release you," the commander told him. Baharuddin did so.

Rather than shoot him, however, the soldiers took him back inside, tied him up, and locked him upstairs again.

A soldier named Reza, who was being punished, was imprisoned alongside him that evening, perhaps as a ruse. "This is your last night," he advised, as Baharuddin recalled. "If you confess, they'll release you." Baharuddin doubted this. Reza fell asleep, and Baharuddin found the strength to untie himself. He crawled slowly down the stairs in the early hours of the morning, slipped outside, crossed over to the Pipeline Road, and ran. He hid in a small brick factory as the sun rose, stopped a passerby, and sent a messenger to his family. They brought him to a unit of G.A.M. guerrillas. He enlisted in the rebellion immediately. His mutilated ears provided him credibility with his peers.[6]

His story was one among hundreds from 2003 and 2004. Throughout, ExxonMobil operated the Arun gas fields and the liquefied natural gas conversion factory without interruption. In Jakarta, Bill Cummings, who ran the corporation's public affairs office under country manager Ron Wilson, recruited some local specialists in community relations and dispatched them to Aceh to try to win enough favor from the local population to avoid a repetition of 2001, when G.A.M. had attacked ExxonMobil repeatedly because of its alliance with the T.N.I. The corporation's operations in the midst of the conflict were precarious, but G.A.M. had pulled back for the most part from striking at ExxonMobil or its employees. The corporation routinely engaged in community relations efforts where it extracted oil and gas; in Aceh, a creative engagement could also be protective.

ExxonMobil's campaign in Aceh was prescribed in Texas, however; it had to follow the worldwide corporate manual approved in Irving. Under the corporation's Operations Integrity Management System, or O.I.M.S., the comprehensive instruction book and guidelines adapted after the *Exxon Valdez* wreck, the same community relations strategies employed in Colorado had to be employed in the conflict zone. To handle complaints, for example, the public affairs department set up a voice mail number where Acehnese citizens could leave messages, although relatively few of them had phones at the time. Then the ExxonMobil managers were to take down the messages and respond within a set period of time prescribed by the manual. Corporate managers also attended events around the province and handed out ExxonMobil hats to Acehnese villagers.

ExxonMobil's "idea of community development was, 'You give me a proposal and we'll assess it,'" one former employee familiar with the campaign recalled. "The people [Acehnese] would say, 'Give us ten cows for community development.' Exxon would assess it and say, 'Well, that fits with our approach,' and they would give the cows, and the next day the cows would be roast beef and there would be a big party in the village."[7]

Cows not eaten immediately wandered around with "ExxonMobil" stenciled in red on their backsides. "It was just, 'Give, give give.'" Exxon-Mobil "went into this Father Christmas mode," the employee recalled. "They built volleyball courts—just slabs of concrete in paddy fields. You had so many volleyball courts around Aceh. They were just slabs of concrete, but G.A.M. imposed a tax on all construction, so G.A.M. liked this."

ExxonMobil opened a health clinic in Aceh that sometimes treated as many as three or four hundred patients a day, according to the former employee involved. The doctors prescribed Valium to reduce the stress of farmers and villagers living amid the violence. The stress could be particularly intense for Acehnese employees of the oil corporation. G.A.M.'s intelligence officers knew of the local employees' salaries, where they lived, and who their family members were. Rebel extortionists would telephone and say, "We're going to tax you thirty million rupiahs," or about $3,300. If an employee received such a call, he or she could negotiate the price, but would have to pay—or face death. "The local employees were really out on a limb," the person involved in the outreach campaign recalled. "G.A.M. was everywhere." ExxonMobil's expatriate managers "never quite understood what was going on."[8]

As the war calmed, *John Doe I et al. v. ExxonMobil Corporation et al.*, the lawsuit filed in Washington, D.C., on June 19, 2001, by Terry Collingsworth, on behalf of eleven anonymous alleged victims of the T.N.I.'s violence around the gas fields—seven John Does and four Jane Does (some of whom were representatives of their dead spouses)—seemed to some of Aceh's human rights activists to offer the only plausible means by which the ghosts of the province's torture rooms and civil violence might be exorcised. Through the lawsuit's fact-finding, some measure of transparency and accountability about Aceh's past might be established, they believed. The democratic government of Indonesia had neither the means nor the will to overcome the resistance to investigations by T.N.I. leaders; that would require the distance and relatively neutral setting of an American courtroom. From the beginning of the *Doe* case, however, Exxon-Mobil used the deep sensitivity within Indonesia about the subject of Aceh's human rights abuses as a bulwark in its own legal defense strategy.

If the case ever came to trial, ExxonMobil's lawyers made clear in early court hearings, the corporation's defense would pursue the argument that Indonesia's generals, not ExxonMobil's executives, were responsible for any actionable violence and, therefore, any financial liability.

To lead its defense, ExxonMobil retained Martin J. Weinstein, a former federal prosecutor specializing in corruption and overseas bribery cases, who was a partner in the Washington office of Willkie Farr & Gallagher. Among other previous assignments, Weinstein had investigated Pete Rose's gambling activity on behalf of the commissioner of Major League Baseball. He was a wealthy, forceful K Street litigator. Weinstein was an active political contributor, primarily to Republican candidates.

In *Doe*, his strategy was to have the lawsuit thrown out before its factual allegations about human rights violations in Aceh could be tried. Weinstein asserted that the case should be set aside or dismissed because it would interfere with the Bush administration's efforts to enlist Indonesia as an ally against Al Qaeda after the September 11 attacks. Indonesia had, in fact, been a locus of radical Islamist groups that had trained and aligned themselves with Osama Bin Laden. Weinstein's approach asked, in effect, which was more important: the appeasement of the Indonesian military, to induce cooperation against Al Qaeda, or the promotion of universal human rights, through the normal functioning of American civil law?

The *Doe* case fell by random assignment to Judge Louis F. Oberdorfer of the United States District Court in Washington. The judge was a former clerk to Supreme Court Justice Hugo Black; he had served in Robert F. Kennedy's Justice Department and had been appointed to the bench by President Jimmy Carter. By the time the Aceh matter landed on his desk, however, Oberdorfer was eighty-two years old. He occasionally appeared at hearings wearing an oxygen mask, which unnerved some of the lawyers who argued before him.

It was established American law that domestic courts should not lightly interfere with a president's foreign policy. Federal judges regularly asked for advice or comment from the State Department when a civil case like *Doe* might affect American interests abroad. Martin Weinstein asked Judge Oberdorfer to make such a query of State. Even apart from Exxon-Mobil's deep connections to the Bush administration, and the administration's earlier intervention against G.A.M., Weinstein had ample reason to think that Secretary of State Colin Powell and his principal legal adviser, William H. Taft IV, the great-grandson of the former Republican president, would lean ExxonMobil's way. State's lawyers often looked skeptically at the novel use of the eighteenth-century Alien Tort Claims Act to seek civil damages in American courts on behalf of overseas abuse victims. "It's another one of those pesky ATS [Alien Tort] cases keyed to corporate liability," State Department lawyer David P. Stewart wrote to colleagues when Oberdorfer's request for advice landed at Foggy Bottom.[9]

In fact, Taft and his colleagues were torn. They wanted to discourage Alien Tort cases. They wanted to support the White House's plan to build a new counterterrorism partnership with Indonesia. But they did not want to send a message to the T.N.I. that would lead its generals to believe they enjoyed a free hand in Aceh or on human rights issues. Taft sent Oberdorfer a six-page letter that tried to express this balance. The letter aligned the Bush administration with ExxonMobil's arguments, but also attempted to pressure Indonesia's military to improve its human rights performance. "Adjudication of this lawsuit at this time would in fact risk a potentially serious adverse impact on significant interests of the United States, including interests related directly to the on-going struggle against international terrorism," he wrote. "It may also diminish our ability to work with the Government of Indonesia on a variety of important programs, including efforts to promote human rights." If the lawsuit went forward, Indonesia might reduce or end cooperation with the Bush administration on "issues of substantial importance to the United States."[10]

It was a victory for the oil company, but not a total one: Taft had used wiggle words, more so than in some other State objections to cases of this type. "At this time" was a phrase Taft had employed more than once, for example. Oberdorfer and his clerks seemed to take the hint. ExxonMobil urged the judge to dismiss the *Doe* case outright, but rather than ruling promptly, he delayed. He also ordered the corporation to preserve all the documents and evidence that might be relevant if the case did go forward.

M ore than one hundred thousand Acehnese died in about twenty minutes on December 26, 2004, when ocean waves unleashed by an earthquake crashed onto northern Sumatra. The water buried scores of villages in mud and carried ships several miles into the jungle. The tsunami decimated both sides of Aceh's civil war; its toll exceeded by many orders of magnitude the carnage from man-made violence around the gas fields. The shock of the disaster, as well as the massive international relief and development effort that followed, set conditions during 2005 for breakthrough peace talks between the government and G.A.M.

On May 24, Judge Oberdorfer summoned the *Doe* lawyers back to his courtroom.

"It says in here," the judge said, referring to the text of Taft's original letter, that 'adjudication of this lawsuit at this time would in fact risk a potentially serious adverse impact on significant interests of the United States.' Now, that was at that time, which was July 2002. At this time, does the same thing apply?"

A lawyer from the United States Attorney's office, assigned to monitor the proceedings, stood to report that he had recently "touched base" with the Bush administration's State Department and was told that the letter "remains their view at the present time."

"Your Honor," Martin Weinstein said, "this case involves alleged conduct in which the government of Indonesia and the military of Indonesia, in the midst of fighting a civil war on its own soil, has taken certain actions, and whether or not they are proper in the midst of civil war. The conduct of ExxonMobil as a contractor . . . is inextricably intertwined with the conduct of the Indonesian military."

It had been almost four years since Terry Collingsworth had traveled through Aceh's war zone to interview victims of the T.N.I., and he had yet to even receive a decision on jurisdictional issues in the case. Judge Oberdorfer had just turned eighty-six.

Collingsworth, who was a full-faced man with blue eyes and thick reddish eyebrows, which he sometimes cocked in a rakish manner, now rose to report that two of his John Doe plaintiffs had been murdered during the years that the case had been on hold.

"We sent somebody after the storm who was Acehnese and was able to move around," he said. Nine of the eleven plaintiffs had survived the tsunami. "Two of our plaintiffs have died, but unfortunately it was not due to natural causes. . . . They were killed by the military, we believe, that were operating on behalf of ExxonMobil."[11]

Oberdorfer searched for a way to begin taking testimony from Exxon-Mobil executives, to move toward trial. He probed the lawyers for a plan that would sidestep the State Department's concerns about Indonesia's sensitivities. The judge soon ruled that the case should go ahead. He threw out the controversial Alien Tort Claims Act elements but kept the lawsuit alive as a tort case—the international equivalent of a slip-and-fall lawsuit. The Acehnese victims had standing to sue ExxonMobil in a D.C. court, Oberdorfer concluded, because ExxonMobil had a substantial office on K Street, from which managers such as Robert Haines had overseen the corporation's security operations in Indonesia.

Collingsworth still feared for his clients, he said.

"But you're not ascribing that to any action taken by Mr. Weinstein's client, are you, the fact that they were killed?" Oberdorfer asked. "I mean, has there been a disclosure" of the real identities of the John Does and Jane Does in the case?

"Yes, there has. . . ."

"To make it very clear," Martin Weinstein answered, "this case has been around for five years. We have precisely stayed clear of this issue until now" because ExxonMobil feared it would be blamed for any harm that came to the Doe plaintiffs, no matter the cause. "So to be quite clear, we want to see these people happy and healthy and in whatever current state they are because even without having their names, we're going to be blamed for what happened to them."

Collingsworth explained that he was not accusing ExxonMobil of murdering his clients—certainly not directly. He went on, however: "It is not our theory of the case, and Mr. Weinstein knows this, that [Exxon-Mobil employees or lawyers] went to Indonesia and ordered our clients to be harmed. . . . Our theory of the case is primarily one of [Exxon-Mobil] having set up a project in such a way with a security force that is dangerous, that we're entitled to know, 'Did you anticipate that danger? What did you do from the United States to anticipate and ameliorate that danger?'"[12]

Collingsworth crystallized why ExxonMobil was really in the dock: that it had failed to anticipate the consequences of its operations in Aceh, and had failed to move actively to protect civilians as best it might. The corporation might not have directed any of the violence, but if its leaders had exercised sufficient care and activism, torture and killings might have been avoided. ExxonMobil's presence in Aceh fueled conflict; therefore, the corporation had a duty to act preventively. Collingsworth's accusation demanded, in effect, that ExxonMobil consider the repercussions of its presence anywhere, especially in weak, poor societies prone to conflict, where oil and gas windfalls provided ample motive to anyone with a gun.

ExxonMobil appealed Oberdorfer's decision to proceed; more delays followed. As *Doe* languished, the corporation moved to strengthen its protections against future lawsuits of its genre. Lee Raymond agreed, as he prepared to step down, to have ExxonMobil finally join BP and other oil and mining companies in the Voluntary Principles on Security and Human Rights regime, which Raymond had initially rejected in 2000 and 2001. The Bush administration had since endorsed the system. It committed corporate signatories to communicate their human rights standards to local security forces; authorize the use of force around their properties only in proportion to threats; and vet local militaries serving them for known rights abusers. Mike Farmer, the longtime head of ExxonMobil Global Security, adapted these rules and expectations into a new human rights and security manifesto within the O.I.M.S. operating system. He rolled the human rights regime out to ExxonMobil affiliates worldwide as Rex Tillerson took office.

Tillerson's embrace of the compact, like his review of climate policy, marked another turn from ExxonMobil's stubbornness during the first Bush term. The new human rights system required audits and reviews of whether local security forces protecting ExxonMobil oil and gas fields were in compliance with the standards defined by the Voluntary Principles' steering committee. ExxonMobil assigned an Irving-based executive, Kevin Murphy, to join that leadership committee. The corporation was true to its self-image: It might not readily join squishy liberal-minded regimes like the Voluntary Principles, but if it did so, it would go all out. Longtime critics of ExxonMobil in the human rights community marveled as the corporation's new approach unfolded under the O.I.M.S. spur: "It's like implementing human rights through a police state," Arvind Ganesan of Human Rights Watch quipped.[13]

Although it was not foolproof, the Voluntary Principles would shield ExxonMobil from human rights cases like *Doe* in the future by helping the corporation to argue to a jury or judge that it had taken all the steps it reasonably could to protect civilians in its areas of operations. This might be self-interested, but the compact had been constructed to appeal to such corporate instincts. The practical effect would be to bring Exxon-Mobil more closely into line with the corporate responsibility practices of other large multinationals—to socialize Exxon in an era of global norms increasingly influenced by civil society. This was the emerging pattern of the Tillerson era: ExxonMobil's new regime learned its lessons late, never admitted the corporation had been wrong in the past, but it shuffled nonetheless in new policy directions, seeking a sustainable form of corporate normalcy and legitimacy. All that remained now was to dispose of the potential legal liabilities from the past. Burying the *Doe* litigation by delay, it seemed to Terry Collingsworth and his colleagues, was a core element of ExxonMobil's strategy.

Collingsworth had brought on Agnieszka Fryszman as a cocounsel to handle courtroom arguments and litigation strategy for the John and Jane Doe plaintiffs. She was a Brown University and Georgetown Law graduate who specialized in human rights, antitrust, and class-action cases on behalf of small businesses, women, and abuse victims. After September 11, she had provided pro bono representation to families of victims of the attacks as well as detainees at Guantánamo. She was accustomed to well-funded corporate opponents, but she found the ExxonMobil legal team at Willkie Farr, led by Weinstein, to be aggressive beyond the norm. "They wanted to grind us into the dust—it was scorched-earth," she said later. "They have way more resources and they use them. They file motions to reconsider and they file simultaneously in all possible courts. They know we're just two people."[14] The paper blizzard from Willkie Farr veered so far out of control that at one point a magistrate working with Judge Oberdorfer ordered all lawyers in the case to sign an affidavit affirming that they had read Rule 12—the principal guidelines on federal legal procedure, which prohibited the gratuitous use of filings to harass an opponent—before they could file another motion.

Because of ExxonMobil's appeals, the *Doe* case wound its way toward the United States Supreme Court. ExxonMobil retained the prominent Supreme Court advocate Walter Dellinger and the well-known white-collar defense lawyer Theodore V. Wells Jr. to represent the oil corporation. The Supreme Court asked the Bush administration for an opinion about whether it should take on the appeal. The issue fell to the administration's solicitor general, Paul Clement. His office invited the two sets of lawyers to separate informal meetings at the Justice Department.

Collingsworth had by now moved into a new private firm. One of his partners, Bill Scherer, was a Florida Republican who had helped President Bush during the 2000 vote recount. Scherer believed in the human rights agenda behind the Aceh case and sought out a meeting at the White House to put in a word for his partner's cause. He met in Washington with one of Karl Rove's successors on the political side of the West Wing. His White House interlocutor heard him out and then told him, straightforwardly, "That's up to Dick Cheney."[15]

In a reception area at the solicitor general's office at the Justice Department, on Pennsylvania Avenue, Agnieszka Fryszman and her team waited for their session with Clement. The intimidating figures of Dellinger and Wells emerged—they had gone first and had just completed their presentations on ExxonMobil's behalf. "They are grinning, laughing," one of the participants recalled. "They think their meeting has gone really well. They're high-fiving each other, popping champagne bottles—not literally, but you can tell they're going out to celebrate."

Fryszman's team found their meeting to be a tough grind. They were interrogated about their anonymous clients, particularly about whether there was really enough evidence to withstand scrutiny if the case went forward. The human rights lawyers were impressed by the quality of the government interrogators, but they had no confidence about how the decision would come out. They went out afterward and knocked back shots, trying to keep their spirits up.[16]

They soon had reason to be joyous: Clement decided against Exxon-Mobil and told the Supreme Court that there was not enough at stake in American foreign policy to justify extraordinary action by the high court. Time had changed the Bush administration's thinking about the balance of American interests in Indonesia: During the long delays in the case, not only had the Aceh war been settled, but Indonesia had evolved toward stable democratic politics and was enjoying rapid economic growth. The country could now afford scrutiny of its past human rights problems; such a trial might even strengthen its democracy.

Judge Oberdorfer suffered a stroke shortly after Clement's decision. He prepared to retire. As he recovered from his illness, he composed an opinion ordering the *Doe* case to trial, one of the last written decisions of his career. He came out on the side of the Does. "Plaintiffs have provided sufficient evidence, at this stage, for their allegations of serious abuse," the judge concluded.[17]

ExxonMobil appealed again. Its lawyers had outlasted Oberdorfer. A new judge might see the issues differently. In Aceh, among the original eleven Does, two more of the plaintiffs died as the appeals went on, leaving widows and estates to pursue their claims.

Nineteen

"The Cash Waterfall"

n Hugo Chavez's Venezuela, ExxonMobil's strategy for attracting good-will in the midst of political contention relied on the powers of art. The corporation staged a salon exhibition in Caracas at regular intervals, with prizes for the best work. It held the event at the National Art Gallery on the Plaza de los Museos. A serene, neoclassical building housed the gallery on grounds that contained a manicured interior courtyard, weep-ing willows, and a small pond. Tim Cutt, an American who served as the president of ExxonMobil's Venezuela operations after 2005, presided over the exhibitions with the careful decorum of a museum curator who must please donors and patrons of impossible quirkiness and diversity. The tone Cutt and his colleagues sought to convey at the events was, *The art speaks for itself; it brings us all together*. In Venezuela's eroding democracy, how-ever, that wish proved increasingly difficult to fulfill.

The trouble began as the 2006 presidential election approached. The long struggle between President Hugo Chavez and his opponents intensi-fied. A son of schoolteachers, Chavez had enrolled in a military academy as a young man, played baseball, wrote poems, fought in counterinsur-gency campaigns, and rose to the rank of lieutenant colonel. During the

1980s, he became involved in leftist movements seeking to challenge Venezuela's business and landed elites for power; with his red beret and fiery rhetoric, he emerged as a populist leader. He was jailed for his role in a 1992 coup attempt and later won the presidency by promising to restructure Venezuela's corrupt, inequitable economy for the benefit of the poor. Once in office, he acted with increasing ruthlessness to consolidate power. Democratic, civic, business, and military forces opposed him.

Venezuelan artists seized on the annual ExxonMobil exhibition at the National Art Gallery as a forum for dissent: They submitted to the contest only paintings oiled in black, with the map coordinates of Venezuela, from west to east, lightly etched on the dark canvases. ExxonMobil's Caracas executives decided they had little choice but to mount the stark paintings all around the gallery.

The artists demanded the microphone at the exhibition's opening. The corporation's mortified local public affairs team negotiated an agreement to let them speak briefly; Tim Cutt yielded the floor. The artists explained one after the other that their paintings depicted the future of Venezuela under Hugo Chavez: It would be bleak, they said, in case anyone missed the symbolism of their canvases. They denounced the president and his encroachments on civic freedoms. ExxonMobil's executives nodded politely.

This was the sort of awkwardness that had persuaded the corporation's engineer executives to steer clear of politics in the countries where they worked, and to concentrate as narrowly as possible on those issues that enabled oil and gas production. Now ExxonMobil had opened the door to art-as-politics and there was little the Caracas office could do to reverse course, its local executives believed. If they canceled the next contest, they would signal fear or, worse, that ExxonMobil was somehow choosing sides in Venezuela's polarized polity. It was not clear whether Chavez or his opposition would prevail across the long arc of years by which ExxonMobil measured geopolitics: As recently as 2002, the president had barely survived an uprising and coup attempt.

Once the ExxonMobil art wars were launched, the Chavez regime acted decisively: It dispatched its own cadres to the next corporate salon. The Chavez loyalists took the floor and delivered pointed speeches against

Yankee imperialism. There was nothing the ExxonMobil executives could or would do to stop them; the National Art Gallery was state owned, and these were Venezuelan government representatives. Cutt and his aides reported the incident to their supervisors at the ExxonMobil upstream division in Houston. Their shared conclusion, said a former executive involved, was that "this is going to be harder and harder to handle."[1]

ExxonMobil had multiple interests in Venezuela: downtream filling stations, some oil production ventures, and supply agreements that directed Venezuelan oil to a large refinery in Chalmette, Louisiana, which ExxonMobil owned jointly with Venezuela's state-owned oil company. The web of contracts, financing agreements, and supply linkages meant that confrontation with Venezuela's leader could prove unusually messy and costly. The dependency was mutual: Venezuela's oil was unusually sour, not suitable for most refineries worldwide, whereas the Chalmette facility was tailor-made to handle it profitably. Moreover, as the U.S. embassy noted succinctly, "Chavez's priority is regime survival."[2] He needed oil royalties, profits, and taxes—the principal source of revenue for the Venezuelan treasury—to pay for expanded social spending born of his self-styled Bolivarian revolution. After the failed 2002 coup attempt, Chavez signaled in speeches and rambling television interviews that he intended to take greater control of Venezuela's oil industry, to challenge what he described as the dominance of foreign profiteers. Yet some of ExxonMobil's executives calculated that the president could not afford to spook international oil corporations precipitously or to trigger even more capital flight from Venezuela than was already taking place in response to the president's policies.

As Chavez gathered power and increasingly employed the populist rhetoric of resource nationalism to stir his followers, ExxonMobil responded initially with appeasement. During 2004, Chavez came under intense pressure from the democratic opposition, a wave of resistance that culminated in the scheduling of a referendum in August of that year to determine whether he could continue in office. Part of the charge against him was that he was jeopardizing Venezuela's economy through his rash, irrational, corrupted populism. Chavez pressured the international oil companies in Venezuela to sign and publicize agreements with his regime on the eve of the election. Televised signing ceremonies would signal confidence in his presidency by bastions of global capitalism. ExxonMobil had been talking with successive Venezuelan governments for nine years about a multibillion-dollar petrochemical investment that would supply plastics to Latin America's burgeoning economies. It had never been able to close even a preliminary deal. Now Chavez volunteered to sign an initial understanding—if ExxonMobil would agree to a televised signing ceremony three days before the referendum vote, effectively handing Chavez the American corporation's endorsement.

The Bush administration sought to contain Chavez and hoped democratic forces would overthrow him peacefully. The referendum was a critical moment. The administration tried not to undermine Chavez's opposition by embracing them openly, but there was no question which side of the vote Bush was on. ExxonMobil's executives understood perfectly that the administration would prefer them to take no steps that would strengthen Chavez on the eve of a critical vote about his legitimacy and tenure in office. On the other hand, here was a long-sought investment opportunity finally on offer. The corporation capitulated to Chavez. It agreed to a televised signing ceremony just three days before the election.

Charles Shapiro, Bush's ambassador in Caracas, asked an ExxonMobil executive why the corporation would accept such a clearly supportive contract signing on the eve of an "event that has convulsed Venezuela's political life." The executive replied that ExxonMobil "could not think of any other issues to raise" with Chavez's government in the contract negotiation, and the Venezuelans had been "pressuring the company [to] sign immediately." Lee Raymond, then ExxonMobil's chief executive, telephoned the Bush White House to tell them of his decision. His deputies in Venezuela signaled that Raymond might be willing to meet Chavez personally if the petrochemical talks advanced far enough. Amid fraud allegations, Chavez won the vote.[3]

The peace ExxonMobil purchased that summer did not last. Politics, not the maximization of profit, drove Chavez's thinking about Venezuela's oil industry. Reasserting Venezuelan control over the country's oil was for Chavez an irresistible opportunity. As his assaults on the contract terms enjoyed by the international oil majors intensified, ExxonMobil's executives and lawyers decided that this time, they would not give in. They also decided to maximize Chavez's pain. They concocted a legal ambush—carried out, in its final act, on a wintry Friday afternoon in the Manhattan offices of a prestigious law firm—to seize by stealth more than $300 million in cash from the fiscally strapped, debt-laden Venezuelan regime. It would be one of the largest asset seizures ever attempted by an American oil corporation. The Bush administration struggled to punish Chavez for his anti-American policies in a way that measurably pinched him. ExxonMobil, when it finally aligned with the administration's perspective, developed a practical scheme. The corporation's motivations were pecuniary—the interests of its private empire, not the policies of President Bush, provided the cause. The plan involved a mechanism of modern global finance known to its participants as "the cash waterfall."

E xxon had been thrown out of Venezuela once before, in 1975, when the country's elected government embraced the global trend of oil nationalizations. (Venezuela's government claimed its intervention was not a full expropriation, but Exxon insisted that it was, and the American government backed Exxon up as the company fought for compensation.) To ease Exxon's departure, Venezuela cut some unpublicized sidebar deals, an American official and a former Exxon executive said later. The state-run oil giant, Petróleos de Venezuela, known by its acronym, P.D.V.S.A. (pronounced as "peh-de-vay-suh"), allowed Exxon to purchase Venezuelan oil at discounted rates, for refining and onward sale.

Nationalization had proven to be disastrous for the Venezuelan oil industry the first time around. P.D.V.S.A. had many outstanding engineers and executives educated at international universities, but under political control the state-run company could not acquire the capital and technology required to maintain oil production. Venezuela's oil output fell by more than half, from 3.7 million barrels per day in the mid-1970s to just 1.8 million barrels per day by the mid-1990s.

Falling global oil prices also played a role in the industry's collapse, in part because much of Venezuela's oil was "extra-heavy," meaning it was laden with sulfur, acid, salt, and heavy-metal contaminants. (An industry-wide system developed by the American Petroleum Institute designated oil as "light," "medium," "heavy," or "extra-heavy," on a scale that, among other things, compared the density of a particular batch of oil with the density of water; extra-heavy oil was denser than water. The A.P.I. scale used numerical degrees to describe grades of oil, ascending from heaviest to lightest. Extra-heavy oil was less than 10 degrees, whereas light or "sweet" crude, the most suitable for refining, could be as high as 48 degrees. Oil rated higher than 31 was labeled "light.") Heavy oil required extra production steps to prepare it for sale. Among other problems, the oil did not flow smoothly in its natural form. The costs of these additional processes meant that most heavy and extra-heavy oil could be extracted profitably only when global oil prices were high. P.D.V.S.A. lacked the technologies to produce Venezuela's reserves economically as prices fluctuated at low levels during the 1980s and 1990s.

Venezuela reversed its attitude toward outside corporate oil investment as the cold war's end spurred privatizations worldwide. The country's oil-dependent economy had long stagnated, and rates of poverty had risen, in line with the grim forecasts of resource curse theorists. To generate more revenue for development, the government opened talks with international oil companies about deals that would allow foreign ownership of Venezuelan crude again, through joint ventures with P.D.V.S.A.

Some of the deals involved the country's rich, underdeveloped vein of extra-heavy oil, in the Orinoco River Basin, which lay in Venezuela's wet coastlands to the east, near Guyana. The Orinoco River snaked thirteen hundred miles to the Caribbean through tropical palms and slash-and-burn agricultural fields. The United States Geological Survey estimated that the basin held between 380 billion and 650 billion barrels of recoverable oil, perhaps double Saudi Arabia's endowment. This vast reserve lay beneath the river basin's muddy soil, but the petroleum was unusually viscous and contaminated. American, Canadian, and European oil companies had begun to experiment by the late 1990s with new technologies that could efficiently "upgrade" heavy oil near wellheads and refine it to a lighter blend, suitable for international markets. Mobil was a

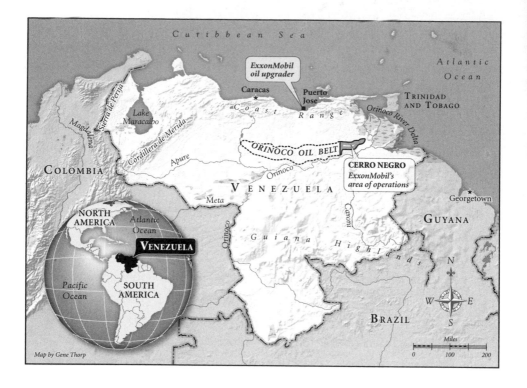

Map by Gene Thorp

leader in the field. Its executives opened talks with Venezuela's oil ministry about an Orinoco heavy-oil project in 1991 and finalized a contract
six years later. The deal was known as the Cerro Negro Association Agreement. (There were four such associations created to mine Orinoco's reserves. Total of France, Statoil of Norway, ConocoPhillips, Chevron, and
BP all participated, either as operators or as minority holders.)

Mobil's civil engineers cleared a muddy expanse in a valley surrounded
by thickly forested mountains and erected an "upgrader," a modest word
to describe a facility that, when completed, would resemble in visual di-
mension the vast refineries of northern New Jersey. Its pipes, flaring
smokestacks, and white oval storage tanks soon formed a gleaming, belch-
ing industrial park in the midst of Venezuela's rural poverty. Construction proceeded during the Exxon merger. After first oil flowed, senior
executives from Irving, including Rex Tillerson, proudly flew in by jet and
helicopter to inspect the achievement. "Exxon was extremely proud," re-

called a U.S. government official who toured the facility. "This was the new frontier—the first time anything on this scale had been tried. . . . It worked. It was totally new. It had been a big risk."[4]

Hugo Chavez read and espoused the usual Marxist-influenced texts, but he saw himself as a synthesizer of old and new political ideas. Chavez later said he was "gullible" and believed he might be able to construct a mixed capitalist and socialist system.[5] The ambiguous remarks he made about business and ideology during his early years in power led some international oil companies to think they might yet be able to hold on to the oil deals they had made during the opening of the 1990s. Gradually, however, Chavez moved more forcefully against his domestic opponents, and as oil prices gyrated and the economy deteriorated, he grew desperate for new sources of revenue.

The president purged P.D.V.S.A. of engineers and technocrats he regarded as hostile to his regime. He fired about twelve thousand employees—mostly executives and administrative staff—after the company participated in a national strike called to bring him down.[6] He replaced the ousted managers with political cadres who tacked Che Guevara posters on their office walls and whose knowledge of oil production and accounting was often limited or nonexistent. He looted P.D.V.S.A. for revenue—the company handed over about 70 percent of its gross revenue to the Chavez regime in 2006, including about $10 billion for social spending projects. In a year when most global oil companies posted record profits, P.D.V.S.A. lost an estimated $3.7 billion. To stay afloat, Chavez authorized mass borrowing—$4 billion from China, in exchange for special access to Venezuelan oil, and another $6 billion from international bond and financial markets. He cut oil production and supply deals with Syrian, Iranian, Indian, and Indonesian corporations. By the time of the ExxonMobil art wars, Chavez was running P.D.V.S.A. like a political Ponzi scheme: He overpromised to his impoverished Venezuelan followers, then milked the oil industry's revenue to pay for those promises as best he could.[7]

Chavez railed to his followers about the low royalty rates paid to Venezuela by ExxonMobil, Chevron, BP, Total, Statoil, and other international majors during the 1990s, when he had been in opposition. Exxon-Mobil's complex at Cerro Negro enjoyed a royalty rate of just 1 percent during the project's early years of production. That low rate (distinct from the corporate taxes the government collected, which were substantial) had been agreed upon by Venezuela to assure Mobil that it could recoup the investments in the upgrader complex before Venezuela took a larger share of revenue. After nine years, the royalty rate would rise to 16.67 percent, but that event still lay years away.

Chavez started to pressure the international oil companies over their royalty deal soon after he won his referendum, the victory ExxonMobil had aided. He threatened to unilaterally bring forward the 16.67 percent rate. The companies could not be sure whether or how quickly Chavez would act. The U.S. embassy cabled Washington that ExxonMobil, "presumably looking at potential risks around the world," had declared privately and repeatedly that it "takes sanctity of contract very seriously." Yet the corporation had invested $1.5 billion in its Orinoco operations and planned to spend at least $700 million more over the contract's thirty-five-year life. Would it really pull out of Venezuela a second time, and so early in the project's tenure, before profits had flowed amply?

Norm Coleman, a Republican senator from Minnesota, traveled to Caracas and met ExxonMobil executive Mark Ward. Coleman noted that the other international oil giants had decided not to protest Chavez's initial probes on the royalty issue.

"ExxonMobil perhaps has a different perspective on contract sanctity than other companies," Ward replied. "For ExxonMobil, the sanctity of contracts is paramount."

The mixed messages sent by different companies—some accommodating, others defiant—created difficulties, Coleman said.

Ward answered that the other companies operating in Venezuela had been "blackmailed" by Chavez. Their oil holdings in the country were in some cases more important to their global reserve reporting than was the case for ExxonMobil.

Such brave talk would soon be tested. Chavez drifted through 2006 and never forced the royalty issue. But as an election scheduled for December of that year approached, he went after the Orinoco deals in full bore. He had committed himself to massive social spending and he lacked financing options. "We're moving toward a socialist republic of Venezuela, and that requires a deep reform of our national constitution," he announced. "We're heading toward socialism, and nothing and no one can prevent it."[8]

Lee Raymond and then Rex Tillerson trotted out a standard ExxonMobil script when they spoke about anti-American, anticorporate resource nationalism in Venezuela, Russia, the Middle East, and elsewhere. Exxon-Mobil's executives had seen oil nationalization waves come and go over many decades, they asserted, and yet in the long run, most governments would see that their economic interests lay in partnering with private corporations. Before his retirement, Raymond had spoken of the particular problem of Hugo Chavez with a hint of condescension: "I worked in Venezuela a long time ago. . . . I guess my comment would be: 'Patience.'" Tillerson preferred the language of business realism, but his thrust was the same: Latin American governments enamored of resource nationalism should recognize that it was in their own interest "to find a way to invite and open up to foreign investment, because of the technologies and the know-how that's needed" to benefit fully from their oil and gas reserves.[9]

ExxonMobil's vocal stance about contracts had a pragmatic aspect; it was a form of bargaining by deterrence. The corporation operated in about two hundred countries and it had major oil production operations in several dozen. If it renegotiated contracts in one country, others would surely take notice and might exploit the opening.

By the time of the Hugo Chavez imbroglio, the thinking of Exxon-Mobil's senior executives about the sanctity of contracts had evolved beyond business strategy into a philosophy of global governance. The spread of international law regimes and trade treaties had given birth to global business arbitration forums at the World Bank and international chambers of commerce. In its contracts with national oil companies or foreign governments, the corporation inserted elaborate clauses guaranteeing Exxon-Mobil's rights to international arbitration before these bodies if the host country tried to alter contract terms, royalty rates, or taxation. Through these provisions, ExxonMobil evaded the conundrums of two hundred different systems of national property rights; it drew all of the host governments with which it contracted into a universal system of arbitration at the World Bank and the international chambers. The corporation's purpose, said the industry consultant, was to "approximate a global law," one defined not by national parliaments or the United Nations, but by the binding dispute resolution regime of ExxonMobil's worldwide contracts. ExxonMobil relied upon this system more than on the United States government. And the corporation's international competitors took a free ride on ExxonMobil's hard line—Chevron, BP, Shell, Total, and the rest benefited in general from the education and contract standards campaigns that ExxonMobil mounted with oil-owning governments, but the competitors retained flexibility. They could more easily make contract compromises when it suited them because they did not have such a prominent declaratory policy.[10]

When Vladimir Putin during 2006 demanded to renegotiate one of ExxonMobil's remaining contracts in Russia, President George W. Bush telephoned Rex Tillerson to discuss the affront, according to reports of the call that circulated among the corporation's managers. The Bush administration was by now thoroughly disabused of its romanticism about oil capitalism under Putin; its optimism had ended when Putin arrested Mikhail Khodorkovsky, the president of Yukos, with whom Lee Raymond had negotiated unsuccessfully during 2003. Three years later, Khodorkovsky remained in prison; he made impassioned speeches about democracy while appearing periodically in Russian courtrooms, confined to a cage, and he was emerging as an unlikely symbol of credible dissent. Bush said that his administration stood ready to dive into oil diplomacy to push back against Putin's attempts at renegotiation. Tillerson thanked the president, but afterward, through its Washington office, the company begged the Bush administration to stay away. The message ExxonMobil's K Street staff sent to the White House was, in essence, Putin is one of the less offensive heads of state we deal with; we'll do much better on our own.[11]

A t ExxonMobil's headquarters, Rosemarie Forsythe, the former National Security Council aide, still managed the political risk department. She reported to Tillerson's Management Committee, which reviewed dilemmas such as the ones in Russia and Venezuela by reference to color-coded, tab-divided binders Forsythe prepared. These divided the world's nations into three groups: democracies, authoritarian regimes, and transitional governments. The last were characterized by chronic instability; Venezuela was an emblematic case. More and more of the world's oil and gas lay in red-shaded transitional countries, as they were marked in the confidential ExxonMobil binders. This made the pursuit of a global, reliable system of contract enforcement all the more imperative, in the Management Committee's opinion.

It also made political forecasting and project planning excruciatingly difficult. ExxonMobil's corporate planners had mastered the art of long-term planning for variability in the cost of extracting oil, variability in rates of economic growth, and for the geological surprises that might arise after drilling began. But who could predict the political futures of Venezuela, Nigeria, Indonesia, Russia, Iraq, Iran, or Saudi Arabia over two decades or more? The best that could be hoped for, Rex Tillerson believed, was to "think about a range of outcomes in any given country" and try to position the corporation so that it could adjust to extreme events. In some countries where ExxonMobil invested in long-term projects, the "fundamentals," as an economist would put it, looked unsustainable—large, young populations; high unemployment; and authoritarian or dysfunctional systems of government that could not meet the needs of the population. The question in these countries was how long it would take before something exploded, and then, when it did, how the upheaval might affect the corporation's investments.[12]

The impact of Venezuela's turmoil, in particular, was not confined to its own borders. Instability in Caracas—as well as in Nigeria, Iraq, and Iran—contributed to steadily rising global oil prices after 2003. Benchmark per-barrel oil prices crossed $40 in 2004; $50 in 2005 and $60 in 2006. The average weekly price of a gallon of unleaded gasoline in the

United States topped three dollars for the first time in American history in September 2005; the price fell back some the following winter, but then climbed back to $3 in the summer of 2006. Adjusted for inflation, American gasoline prices reached historic highs after three decades of flat or declining trends.[13]

Soaring demand for oil from China, India, and other fast-growing emerging economies stoked the price rise. Between 2003 and 2007, China's oil consumption and net imports grew by about 50 percent. China's prospective thirst for oil as a transportation fuel, to power the cars of its burgeoning middle classes, created a psychology of scarcity.

Along with soaring demand came less provable claims that the world might be physically running out of oil. Matthew Simmons, a Houston-based oil industry consultant who specialized in financial matters and who was not a professional geologist, published an influential book in the summer of 2005 that argued, on the basis of his review of U.S. geological data about Saudi oil fields, that the kingdom had reached the peak of its capacity to pump oil and would soon enter a long decline.[14] Saudi Arabia was not only the world's largest oil producer; it was also the most important to international markets and prices. The kingdom exported the great majority of its production. Among the world's major producers, it could most easily raise and lower production volumes to respond to changes in global demand. If Saudi fields were tapped out, as Simmons claimed, the long era of low or relatively stable oil prices enjoyed by the world economy from the 1980s onward would be in jeopardy.

Saudi Arabia insisted that Simmons's forecasts were wildly off base, but its penchant for secrecy continued to stoke such reports. As oil prices rose, the kingdom launched an investment and construction program to raise its production capacity to 12 million barrels per day from about 10 million. But its project could not easily or quickly undo the psychology of scarcity that Simmons and other end-of-oil commentators generated after 2005. There could be little doubt in any event that Saudi Arabia's ability to increase or lower global oil prices by adjusting the amount it pumped from day to day would be diminished in the future. The world's surplus oil production capacity—that is, the amount of oil that could feasibly be pumped each day but was held back for market, economic, or political reasons—peaked in 1985.[15] After that, global oil supply and demand moved closer to equilibrium.

Rex Tillerson and ExxonMobil's Management Committee scoffed at the idea that the world was running out of oil. The corporation prepared PowerPoint slides to document that governments and industry analysts had badly underestimated the amount of oil in the earth throughout the twentieth century. Time and again, forecasters failed to anticipate how technological innovation would free up or "discover" oil previously thought to be unrecoverable, ExxonMobil executives argued in their slide shows. The corporation demonstrated that in 1925, the U.S. Geological Survey estimated the world's conventional oil reserves to be only 60 billion barrels. By 1950, mainstream estimates had risen to between 750 billion and 1.5 trillion barrels. By 1975, typical estimates were in the range of 2 trillion barrels. By 2000, they had grown to between 2.5 trillion and 3.5 trillion. These swelling numbers did not even account for extra-heavy oil and tar sands oil deposits in places such as Venezuela, Canada, and Russia—perhaps another 4 trillion barrels. Obviously, the actual amount of geological oil had not changed during these decades; all that had changed was the ability of engineers to locate it and pump it profitably. As Tillerson put it: "With new technology, we're always finding more oil. . . . We will achieve a peak, because it is a finite resource. But that time is well beyond where we are today." The problem in the global oil markets, Tillerson and his colleagues declared again and again—and the reason Americans so often gasped and sputtered about prices when they pulled into their local stations after 2005—had not to do with geology. It was a result of geopolitics.[16]

Rising prices did more to hurt the United States than just pinch its drivers' budgets. Higher prices made oil-exporting governments richer at the expense of importers such as America. BP's economists estimated that oil-exporting countries enjoyed a $3 trillion windfall between 2004 and 2007. That wealth provided radical and authoritarian governments such as those in Iran and Venezuela with extra muscle and room to maneuver—whether to purchase arms for proxy militias or to forge new compacts with thirsty importers such as China and India, alliances that might constrain American power. For its part, by 2007, the United States had become more dependent on foreign oil imports than ever before. This not only exacerbated its dependency on governments such as Venezuela's, it also put the country's prosperity at risk. During the 1980s and 1990s, spending on oil, measured as a percentage of U.S. gross domestic product, hovered under 2 percent; as prices soared after 2003, that spending rose to above 5 percent. History showed a strong correlation between such energy price spikes and the onset of recessions.[17]

The expropriations threats emanating from Venezuela contributed to those rising prices. Chavez also threatened ExxonMobil's share price. If the corporation lost its Venezuelan production and its booked oil reserves in that country, the corporation's publicly reported worldwide oil reserves would shrink. The annual challenge of reserve replacement had not lessened as Tillerson imprinted his leadership on ExxonMobil; if anything, the pressures were rising. Tillerson and other executives might console themselves that expropriations come and go, and that Exxon had left and returned to Venezuela before, yet they would be departing a country proximate to the United States with an oil endowment of enormous size and durability. The numbers from ExxonMobil's current Venezuelan operations were not large—well under 5 percent of total reserves and production—but with the reserve replacement equation so tight, all losses made a difference, and the shock of being expelled would probably knock down confidence in ExxonMobil shares, which would in turn depress the wealth of executives and employees. The question facing ExxonMobil early in 2007 was whether, for the sake of principle and long-term global strategy, Tillerson and the Management Committee were prepared, nonetheless, to walk.

On January 13, 2007, Hugo Chavez told the Venezuelan congress that he would enforce a law requiring that P.D.V.S.A. seize majority shares and become the sole operator of all oil projects in the Orinoco River basin. If the international companies currently in charge of those projects wished to stay on as minority owners, they could renegotiate terms. Chinese, Russian, Indian, Belarussian, Vietnamese, and Cuban operators would be entering the Orinoco basin, Chavez announced. "He who wants to stay on as our partner, we'll leave open the possibility to him," Chavez said. "He who doesn't want to stay on as a minority partner, hand over the field and good-bye." He added playfully, switching from Spanish to English, "Good-bye, good luck, and thank you very much."[18]

In fact, Chavez was prepared to negotiate. Into the spring of 2007, each of the oil majors found itself in maddening, opaque, shifting talks with Venezuela's oil technocrats. Fundamentally, they would have to accept a subordinate position to the Chavez regime for the first time and lower rates of return. Within that framework, however, Chavez was ready to deal.

Tillerson and Cutt took a two-track approach: They took all the steps necessary to leave Venezuela by the June deadline Chavez had announced, and simultaneously, they negotiated to stay.

BP was ExxonMobil's minority, nonoperating partner in the Cerro Negro project. The corporation's Venezuelan country manager, Joe Perez, admitted privately that "BP's greatest fear was that Exxon would pull out." At this point, he also conceded, BP "is basically hiding behind Exxon" and its tough-sounding bargaining position.[19]

Cutt and Tillerson had four options: Leave Venezuela and invoke their contractual right to international arbitration to recover their investments and lost earnings; sell their share in Cerro Negro to Venezuela; sell their holding to another company; or accept Chavez's terms and become subordinate to P.D.V.S.A. in a joint venture. ExxonMobil executives told the U.S. embassy that the chances they would capitulate this time were "close to zero." A sale on terms reasonably close to market price seemed optimal.

Under Venezuelan labor laws, if it planned to shut down by the end of June, ExxonMobil had to take public steps as early as March to prepare to lay off workers. Even if its employees found new jobs under Venezuelan management, their compensation and benefits would shrink. Tim Cutt rented out the movie theater near the ExxonMobil office—located in downtown Caracas in a mixed-use complex of offices and retail stores—for regular all-staff meetings, replete with popcorn, to keep the employ-

ees informed. He provided updates on the corporation's negotiations with
the Chavez regime. Carlos Rodriguez, ExxonMobil's Venezuelan-born,
U.S.-educated government affairs director, and Milton Chaves, another
Venezuelan who worked on government relations from Houston, sometimes joined or supported Cutt's presentations. From some of the employees, Cutt and his colleagues heard angry, even menacing complaints. Cutt
became so anxious about the loyalty of his own workforce that he installed a metal detector at the entrance of the executive suite.

The corporation's expatriate executives worried, too, that they might
be arrested suddenly in Caracas and perhaps made the objects of some
theatrical show trial concocted by Chavez. They kept cell phone and
emergency numbers for the petroleum attaché at the American embassy,
Shawn Flatt. Carlos Rodriguez had regular breakfast meetings with Flatt
to keep him up to date, but the corporation was wary about being identified with the American embassy, and so its Caracas managers minimized
the embassy liaisons to the most essential matters, such as planning for
evacuation if one was required because of violence or threats to American
employees.[20]

In Washington, Tillerson met with Venezuela's ambassador to the
United States, Bernardo Alvarez, on May 16. He told the envoy that ExxonMobil must have a confidentiality agreement with the Chavez regime
before it could negotiate in earnest. "We're looking for a win-win solution," Tillerson said, but he warned that the corporation "was willing to go
to arbitration if it had to do so."

Cutt confided to the embassy as the final deadline neared that Tillerson and the Management Committee at headquarters had "shown surprising flexibility in attempting to reach a deal" with Chavez. The chances
of giving in to Venezuela's demands were apparently not so close to zero
after all. For example, Cutt disclosed, they would "swallow hard" and give
up rights to international arbitration if all the other deal terms were sat-
isfactory.

On June 25, however, Cutt called again to declare that ExxonMobil
had given up. The two sides were "billions of dollars apart," he said.

For the second time in just over three decades, the largest private oil corporation in the world would withdraw from Venezuela, a country that might hold the world's largest reserves, or was at least second to Saudi Arabia, where ExxonMobil also had not a single barrel of bookable reserves.[21]

BP, Chevron, Total, ENI of Italy, Sinopec of China, and Statoil all negotiated compromises during the weeks that followed; they accepted new terms as minority owners, subordinate to the Chavez regime. Only ConocoPhillips joined ExxonMobil in refusal and departed.

The decision marked one of the first tests of Tillerson's willingness to endure economic losses for the sake of policy and principle. The corporation's local government affairs team gathered and drove over to the twin P.D.V.S.A. towers in Caracas's cluster of downtown skyscrapers. As Carlos Rodriguez took notes, Tim Cutt announced to the vice minister of energy that ExxonMobil would be pulling out of the Cerro Negro project and that it intended to file claims against Venezuela in international courts of arbitration to recover damages. It was an uncomfortable meeting that ended quickly.[22]

Cutt and the engineers at Cerro Negro prepared to turn over the pride of ExxonMobil's Latin American operations to Chavez's political cadres in April. It was a painful endeavor, not only because the local executives would be walking away from a complex they regarded as state of the art, but also because scores of ExxonMobil's Venezuelan employees would lose their jobs in a country where unemployment stood at 9 percent.

ExxonMobil's lawyers and accounting analysts prepared for their legal campaign that summer by doing some mind-boggling math. Hobert E. Plunkett, a University of Alabama graduate who served as an asset enhancement manager at the corporation, later explained to a federal court how he calculated the damage Chavez had caused to ExxonMobil. The Cerro Negro project had twenty-eight years to go under the terms of the 1997 contract. The contract had a clause that laid out the formula under which ExxonMobil's damages should be figured. This was called the Threshold Cash Flow Formula. It allowed ExxonMobil to estimate how much money it would have made in Venezuela after operating expenses, royalties, and taxes if it had not been forced to relinquish owner-ship. Plunkett arrived at a round number: somewhat more than $11.9 billion.[23]

ExxonMobil did not immediately share its thinking about its damage claim with Chavez or his aides. Cutt and his Caracas colleagues concen-trated on winding down operations as smoothly as possible. When they visited Orinoco after handing control to P.D.V.S.A., they saw the scores of new employees sitting around in blue jeans without apparent respon-sibilities, and they saw that political propaganda had replaced their ubiq-uitous safety notices on some of the walls, but they kept their opinions to themselves. In public, on Wall Street, and elsewhere, Rex Tillerson described the breakup dispassionately: "Our situation in Venezuela is a pure and simple contract. The contract was disregarded."[24]

Cutt gathered the local staff at the movie complex near the Caracas office for a sort of farewell party—"ExxonMobil Idol," a talent show in-spired by *American Idol*. The corporation's senior managers, with Cutt fronting, took the stage in hats, dark glasses, and chains, with their under-wear hanging out of their pants, to perform a gangsta rap number. They chanted inside jokes into their microphones, to the roar of laughter and applause from the remaining staff.[25]

The show's theme accurately reflected ExxonMobil's mood about Hugo Chavez. The corporation's executives would not think to call the plan they had in mind "revenge"; they forswore emotion about business and legal decisions. When their plan was revealed for the first time in a New York federal courtroom a few months later, ExxonMobil's lawyers insisted that they had done nothing untoward or vengeful. It was a con-tract matter, they said, pure and simple.

The banking and legal system known as the cash waterfall was designed to control the flow of money generated by the sale of Cerro Negro crude oil. ExxonMobil and P.D.V.S.A. issued $600 million worth of bonds to international investors to finance construction of the massive upgrader complex in the Orinoco basin. Given the long record of political instabil-ity in Venezuela, nobody was likely to buy these bonds—at least not at an affordable interest rate—unless there were guarantees about repayment. The lawyers and investment bankers who organized the bond sale therefore constructed a Common Security Agreement to protect bond purchasers. This agreement established the cash waterfall, as it was termed by the participants, at the Bank of New York, headquartered in Manhattan. The waterfall was a web of restricted bank accounts through which revenue from the sale of Cerro Negro oil flowed in a prescribed manner. Receipts from oil sales went first to pay for the operations of the Orinoco project and second to pay interest to bondholders. Only then did leftover funds cascade into accounts for each of the main project partners, ExxonMobil and P.D.V.S.A.[26]

The Bank of New York's role was to ensure that all of these legal obligations were met before either ExxonMobil or P.D.V.S.A. took out the money that reached the bottom of the cash waterfall. The agreement included collateral and other guarantees to assure bondholders that the Bank of New York could enforce the system's rules.

The cash waterfall had flowed smoothly for almost a decade. Oil came out of the ground in the Orinoco basin; the upgrader lightened the oil and removed its contaminants; the oil flowed through pipelines to ships at a Caribbean port; and the ships delivered regular loads to Chalmette, Louisiana, for final refining into commercial products. When Chalmette confirmed receipt of a particular shipment, it released its payment to the Bank of New York, which in turn sent the money flowing down the waterfall accounts. The bank routed some cash back to Cerro Negro to cover operating expenses, it set some money aside to make monthly interest payments to the bondholders, and then it released the remainder—many tens of millions of dollars annually—to ExxonMobil and P.D.V.S.A.

There were still $538 million worth of bonds outstanding under the cash waterfall agreement in the spring of 2007, when Hugo Chavez abrogated the Orinoco oil agreements. Some of the bonds were due in 2009, others in 2020, and still others in 2028. The bondholders—investment banks, pension funds, mutual funds, hedge funds—became nervous when ExxonMobil signaled publicly that it might pull out of Venezuela. If the corporation no longer operated Cerro Negro, the cash waterfall system might not work so well, and at a minimum, the prices of project bonds would fall because of investor anxiety about the future. On April 27, 2007, Cerro Negro bondholders declared that because of the prospective actions of Hugo Chavez's government, Venezuela and ExxonMobil had legally defaulted on their joint obligations as issuers of the bonds. Under the cash waterfall system, the bondholders had the right to seize collateral if this default declaration was confirmed.

ExxonMobil had already handed over control of Cerro Negro to the Chavez government, but the corporation contacted the Chavez regime and offered to work closely with it on the bond problem. ExxonMobil was still on the hook for 50 percent of the bond issue; hundreds of millions of dollars were at stake.

J. R. Massey, ExxonMobil's vice president for operations in Canada, South America, and the United States, flew down to Caracas. Massey was a Texas A&I graduate who had worked at ExxonMobil for thirty-six years. He told his Venezuelan counterparts, as a lawyer for the Chavez regime recalled it, that "regardless of the differences which remained between the ExxonMobil companies and Venezuela" over the nationalization, "there was no reason not to cooperate in good faith to restructure the financing."[27]

On June 1, ExxonMobil and P.D.V.S.A. jointly retained the Wall Street investment bank of Lazard to advise them on how to make their nervous bondholders happy. Lazard concluded that the best way to clean up the mess would be for the government of Venezuela to buy back all the outstanding bonds through a cash tender offer and then borrow money on its own by other means. Chavez's aides at P.D.V.S.A. eventually accepted this advice. The Venezuelans even agreed to pay for Exxon-Mobil's share of the bonds.

Lazard initiated the complex tender process by which the Cerro Negro bondholders would first be given a chance to decide whether to accept the repurchase offer. If enough did so, they would later surrender their bonds and receive cash payments from the Chavez government. All of the parties retained high-end law firms in New York and Washington to handle the paperwork.

The offer succeeded and the full bond repurchase was scheduled to "close" on December 28, 2007, in a meeting similar to the document-signing sessions familiar to the sellers and buyers of residential property. Curtis, Mallet-Prevost, Colt & Mosle, Venezuela's New York law firm, agreed to host the closing in a conference room at its flagship office, an angular glass-walled skyscraper at 101 Park Avenue. For the lawyers and bankers involved, the year-end closing date meant they would have a disrupted holiday season, but a remunerative one, once their deal fees were distributed.

ExxonMobil and Venezuela were anxious for the deal to close, too. The cash waterfall had been stopped up through most of 2007. Money continued to flow into the Bank of New York accounts from oil sales, but until the disputes with debt holders were fully resolved, it could not go out. As the year wound on, more and more cash had accumulated in each of the stopped-up accounts of ExxonMobil and P.D.V.S.A. By December, ExxonMobil's account held $242 million and Venezuela's contained about $300 million. One purpose of the closing meeting at the Curtis law firm was to confirm that all the legal obligations to bondholders had at last been met so that these huge sums could be released, to be booked as corporate revenue before the year ended.

ExxonMobil's lawyers and finance specialists knew all about the $300 million building up in Venezuela's cash waterfall account because "one of the principal and fundamental economic objectives" of the bond repurchase, as a lawyer for Venezuela later put it, was to make sure that Exxon-Mobil got its own money out of the stopped-up accounts.[28]

On December 28, Venezuela's lawyers at Curtis assumed that everything was in order—both sides would be rewarded with a cash windfall when the paperwork was formally completed that afternoon.

All along, as the bond repurchase neared completion, ExxonMobil had been working secretly with a corporate repo man: Steven K. Davidson, a litigator at Steptoe & Johnson, a global law firm founded in Washington, D.C. Davidson was a practitioner in arbitration and corporate asset seizures who worked from Steptoe's Washington office on Connecticut

Avenue. By 2007, he had become one of the world's leading specialists in the art of seizing and liquidating assets on behalf of large, aggrieved companies. For Motorola, which had fallen into a dispute with a Turkish company over a $2 billion loan, Davidson had seized a yacht in Israel; private jets in Bermuda, France, and the United States; real estate in Britain, Germany, and the United States; and bank accounts in Switzerland and New York. There was a streak of ruthlessness in his work that made him a natural for ExxonMobil.

Over Christmas, Steptoe lawyers secretly prepared court documents to freeze the $300 million in Venezuela's Bank of New York cash waterfall account. They argued in the documents that the money was needed as security against future arbitration awards that might pay off ExxonMobil's outstanding claims against the Chavez regime.

On Thursday, December 27, the day before the closing, Steptoe lawyers contacted the federal court clerk in the Southern District of New York in Manhattan and asked for a hearing before the "emergency" judge on call. They drew Judge P. Kevin Castel. In Castel's courtroom on Pearl Street, the Steptoe attorneys handed up a prepackaged filing of affidavits, draft orders to seize the Venezuelan funds—and also a request that Castel place the entire matter under seal immediately, keeping it secret so that neither P.D.V.S.A. nor its lawyers at Curtis would know, as they signed the bond closing documents the following day, what ExxonMobil had in mind.

"In view of the urgency of the matter and the precise timing required," Steven Davidson wrote, he also requested that Castel designate two young female lawyers at Steptoe to serve the asset seizure papers so as to "avoid the delays that may occur if the U.S. Marshal is directed to serve the order."

J. R. Massey told Judge Castel that P.D.V.S.A.'s "sole asset in the United States is its share of the Project Accounts in New York, sometimes referred to as the 'cash waterfall' (approximately $300 million)." (P.D.V.S.A. also owned CITGO gas stations and refineries in the United States indirectly, through a subsidiary, but ExxonMobil did not seek to move against these assets.) It was "highly unlikely," Massey continued, that

ExxonMobil could force Hugo Chavez's government to make payments on any future arbitration claim "unless the cash waterfall is restrained."[29]

Castel signed off on all of ExxonMobil's requests and also agreed to keep his ruling secret. The next day, lawyers and executives from Exxon-Mobil joined the oblivious lawyers for P.D.V.S.A., Bank of New York, Lazard, and the bondholders in a Curtis conference room above Park Avenue. Robert Minyard, the treasurer of ExxonMobil's upstream division, attended with ExxonMobil's bond deal lawyers from the Los Angeles firm of Latham & Watkins. The oil corporation's delegation gave no hint that anything was out of the ordinary.

A large bond closing is an excruciatingly detailed event, carefully sequenced and choreographed on blinking computer screens, all watched over by nervous lawyers. Hundreds of millions of dollars previously deposited by Venezuela to pay for the bonds went out by wire that morning to each bondholder who had agreed to sell; when the sellers received their money, they sent back electronic confirmation messages. In the late morning and early afternoon, the confirmations blinked through one after another. Signatures were affixed, photocopies exchanged.

At about 2:10 p.m., Minyard took physical possession of the certificates of the repurchased bond shares and left the building. The Bank of New York's representative announced to those remaining in the conference room that it had received enough confirmations from bond sellers to declare the closing officially done. "I am prepared to release the collateral."[30]

That meant the money in the cash waterfall accounts could flow again. Bank of New York immediately initiated a wire transfer sending the $242 million lying in ExxonMobil's stopped-up account to the corporation, free and clear. ExxonMobil's lawyers soon confirmed that the wire had gone through. They had their money.

For Venezuela's account, the Bank of New York had a special Federal Reserve number to wire out the $300 million to the Chavez government. Before the bank's representative could push the final button, however, a new electronic message arrived—an "Order of Attachment" signed by Judge Castel, freezing the Venezuelan funds in place. "The amount to be

secured by this Order is Three Hundred Million Dollars," the order declared. The bank's lawyers told Venezuela's lawyers that there was nothing
they could do; the court had spoken, and the money would stay put.

Arbitration law pays well, but is not rich with emotional reward.
For the Steptoe attorneys the late-December Friday-afternoon seizure of
$300 million belonging to Hugo Chavez's government was like hitting a
walk-off home run in the bottom of the ninth before a full house at Yankee Stadium. It was the sort of thing the lawyers involved would put on
their résumés for years to come, as evidence of their litigating prowess.

They would have to celebrate in quiet, however. ExxonMobil did not
call public attention to what it had done, it did not offer any cowboy-
toned declarations, and it did not permit its outside attorneys to do so,
either. The corporation's executives did not see profit in crowing about
Hugo Chavez. Dictators came and went; nationalizations came and went.
They had their $300 million—the money was now frozen, awaiting final
rulings in the arbitrations to come.

ExxonMobil "has come to this Court . . . with unclean hands," Joseph
Pizzurro, the lead attorney for Venezuela at Curtis, wrote in an impassioned filing to a federal judge a few weeks afterward.

In its undisclosed appearance before Judge Castel, the corporation
had told the judge that it needed a freezing order in secret because it did
not wish to interfere with the orderly closing of the bond repurchase, but
"it never told the Court *why* it did not want that transfer to be interfered
with." The reason, Pizzurro continued, "was simple": ExxonMobil wanted
to make sure its $242 million—money provided to it by the government
of Venezuela, after months of cooperative negotiations—was moved out
of the cash waterfall accounts before it acted to seize its partner's money.
ExxonMobil's lawyers, as part of their cooperation with the Venezuela
side over the bond deal, had even signed an agreement late in 2007 affirming that there was "no provision of law . . . order [or] injunction" that
would "prohibit, conflict with or in any way prevent" the funds in the cash
waterfall accounts from being released once the bond repurchase was
completed.

"Obviously, if P.D.V.S.A. . . . had known that [ExxonMobil] was going to breach its obligations and representations," Venezuela would never have gone through with the bond deal. ExxonMobil "knew this full well, which was the reason its litigation counsel sought to keep the court file under seal." The whole charade, Pizzurro wrote, was little more than "a scheme calculated to conceal its misrepresentations, not only from P.D.V.S.A. . . . but from this Court."[31]

Steven Davidson, for ExxonMobil, said he couldn't understand why counsel for Venezuela seemed so upset. "They call it 'unclean hands,'" Davidson told Judge Deborah Batts, who had inherited the case from Castel, at a hearing on February 13. "They say that we did something that's somehow inappropriate with respect to the transaction that closed at the end of the year," Davidson said. "Now, Your Honor, we would, as a matter of fact and law—we would say that's simply incorrect. . . . There was nothing inappropriate done whatsoever with respect to that closing."

Pizzurro was the managing partner at Curtis, whose reputation with its client in Caracas had obviously been damaged by the December 28 fiasco. He had been schooled in Boston. He had a thick head of white hair and a boxer's pugnacious face, with a flattened nose. He struggled to persuade Batts that Hugo Chavez had been wronged. "It was sort of a chicken-and-egg situation," Pizzurro tried to explain when Batts called on him. The ambush beside the cash waterfall had been a complicated affair, he noted: "We wouldn't get the money until we paid them off, but once we paid them off, we could get the money. . . . Mobil was cooperating at all times. Indeed, they went and hired with us Lazard, to come up with and help with a restructuring of the financing, to create a solution" to ensure ExxonMobil got its own $242 million.

By sneaking into court on December 27 to obtain orders under seal, Pizzurro went on, ExxonMobil had acted in bad faith. "To pretend that that wasn't part of the plan and the scheme, to pretend that . . . it had nothing to do with why they needed to get this under seal, and to pretend that that has nothing to do with a breach . . . is simply not credible."

If Judge Batts was concerned about the ethics of ExxonMobil's executives or lawyers in this case, however, she gave no indication. Under the law, she said when Pizzurro had finished, a freeze order of this type was appropriate. All that was required, legally, to maintain the freeze on the account was "a probability or possibility" that Chavez's government might not make good if it eventually lost out to ExxonMobil, Judge Batts said. Therefore, "I'm confirming the attachment. The matter is adjourned." If Hugo Chavez wanted his money back, he would have to seek it in the court of ExxonMobil's choosing.[32]

Twenty

"Moonshine"

O n January 31, 2006, President George W. Bush delivered a State of the Union address in which he declared that the United States was, unfortunately, "addicted" to oil. A generation after President Jimmy Carter had declared America's oil dependency to be the "moral equivalent of war," and as casualties in Bush's Iraq War accumulated, the president laid out a timetable for greater American energy independence, driven above all by his exhaustion with radical suppliers such as Venezuela and Iran:

> Keeping America competitive requires affordable energy. And here we have a serious problem: America is addicted to oil, which is often imported from unstable parts of the world. The best way to break this addiction is through technology. . . . And we are on the threshold of incredible advances. So tonight I announce the Advanced Energy Initiative—a 22 percent increase in clean-energy research. . . .
>
> We must also change how we power our automobiles. We will increase our research in better batteries for hybrid and electric cars and in pollution-free cars that run on hydrogen.

We will also fund additional research in cutting-edge methods of producing ethanol, not just from corn but from wood chips and stalks or switch grass. Our goal is to make this new kind of ethanol practical and competitive within six years.

Breakthroughs on this and other new technologies will help us reach another great goal: to replace more than 75 percent of our oil imports from the Middle East by 2025. By applying the talent and technology of America, this country can dramatically improve our environment, move beyond a petroleum-based economy and make our dependence on Middle Eastern oil a thing of the past.[1]

For the president, a recovering alcohol abuser and the scion of a political family whose fortune came from the oil patch, to choose the metaphor of addiction for the country's oil economy was a striking and even radical rhetorical decision. His announcement of a national goal to free the United States from Middle Eastern oil imports was more familiar, but striking nonetheless. As with most political speeches, however, the gap between word and practical policy was vast. Since 2001, the Bush administration had regularly advertised its willingness to invest federal funds in alternative energy technology research as a means to deflect calls for more immediate policies to combat climate change, such as a national limit on total carbon dioxide emissions or a carbon tax. In that sense, Bush's new technology investment plans in 2006 were nothing new and were correctly seen by environmentalists as another effort to finesse the climate issue. Even so, there was no reason to doubt the sincerity of the president's wish that capitalism and technological innovation might quickly produce breakthroughs that would make America's oil dependency obsolete.

Like many Americans, Bush seemed to receive the September 11 attacks as a message that reliance on unstable oil-producing Muslim regimes such as those in Saudi Arabia, Iran, and Iraq was in some generalized or intuitive sense debilitating and unsustainable for the United States. America consumed roughly 20 million barrels of oil each day, of which up to 12 million were imported. Of the total, more than two thirds was used to fuel cars, trucks, and airplanes. To change oil import dependency would almost certainly require changing transport fuels. Bush called one of his first alternative energy technology research programs—an effort at the Department of Energy to develop hydrogen fuel cell technology as an alternative to the gasoline combustion engine—the FreedomCAR and Fuel Partnership. The implication of the "freedom" mantra seemed to be that if Americans could find a way to drive their open highways and shop in their sprawling suburbs without purchasing oil from the Middle East or Venezuela, the nation's liberties would be strengthened. Yet it was the policy of the United States not to ask American households to make sacrifices to achieve this outcome. Much of America's need to import oil resulted from the country's gas-guzzling habits—per capita and per highway mile driven, Americans in their S.U.V.s and six-cylinder sedans consumed much more gasoline than any other people on earth. This relative overconsumption was, for many conservatives, including ExxonMobil's Lee Raymond and Rex Tillerson, simply another manifestation of American freedom. A European or Asian authoritarian government might have attacked the problem of overconsumption of gasoline by taxing, planning, regulating, or otherwise mandating change. Yet Bush and Republican leaders in Congress made no serious effort to require Americans to drive more fuel-efficient cars. The president's only initiative was to pursue a technological leap away from oil dependency by investing in basic research. This placed his thinking firmly in the tradition of emotionally declared, inadequately engineered American energy policies dating back to Richard Nixon.[2]

ExxonMobil's executives regarded the president's romanticism about energy "independence" and alternative technologies as misguided. Describing the United States as addicted to oil was "an unfortunate choice of words, quite frankly," Tillerson said. "To say that you're addicted to oil and natural gas seems to me to say you're addicted to economic growth."[3] ExxonMobil executives privately traded theories of psychological analysis about Bush—perhaps the president felt compelled to distance himself from his family's history in the oil business or he feared his legacy as president would be "oiled" if he did not forcefully endorse alternative energy investments. On the big policy issues that mattered most to ExxonMobil—taxation, climate, domestic drilling—Bush had delivered, or had made a valiant effort in Congress. Yet the president also returned again and again to what senior ExxonMobil executives in Irving and Washington believed were fantasies about alternative energy breakthroughs that might lead to some Valhalla of energy independence.

"Cheney didn't have that problem," an ExxonMobil executive recalled. "Cheney was well-grounded. . . . 'Don't screw with the facts and let's deal with it' kind of thing. And Bush just was always afraid he was being associated with the oil industry." That analysis of the president was perhaps too easy. Bush genuinely wished to remake the world that had led to the September 11 attacks—radicalism stoked by Saudi clerics, generations of young Arab men coming of age in stultifying societies. It was obvious even to a free-market traditionalist like Bush that America did its share to enable such dysfunction by importing so much oil and assuring the wealth of anti-American exporters. There was no reason to doubt that Bush would be pleased if research investments in alternative energy that he directed were one day remembered as the catalyst that ended the oil age. Like presidents of both parties before him, however, he lacked the depth of conviction, the political coalitions, and the scientific vision to do more than toss relative pennies into a wishing fountain.

At the ExxonMobil K Street office, Dan Nelson and his colleagues worked the phones to track what might emerge in the energy policy sections of the annual State of the Union, but in the case of the "addicted to oil" speech in 2006, they failed to intervene successfully. Drafts of the president's address were very closely held. They learned that a push toward ethanol would receive heavy play, but they received no preview of the language repudiating oil and gas. Inside the White House that winter, economists and National Security Council senior directors who believed that global oil markets were liquid and interdependent—and who believed that "independence" from oil imports was an unnecessary and illusory goal—failed, too, to persuade the Bush speechwriters to remove the talk's reference to ending Middle East dependency. They argued about the liquidity of the oil market and whether a particular importing nation could determine which barrel of oil was "foreign" versus "domestic." But if one of the White House economists raised a question challenging the call for energy independence, the president's politically attuned speechwriters would respond, "The president has seen this language in ten previous drafts and obviously likes it—do you really want to get in his way?"

After the speech was delivered, Saudi officials and diplomats from other oil producers aligned with the United States, who felt they had been insulted, protested formally to the White House. "It upset the Saudis," the senior official recalled. "Démarches followed that speech."[4]

In Irving, Rex Tillerson sided with the Saudis. He believed, as did ExxonMobil's Management Committee, that "energy independence is not attainable, not any time in the foreseeable future," and that its pursuit was not desirable because it would set the United States on "misguided courses" that might raise the cost of energy in the American economy, destroy jobs, and disrupt trade alliances around the world.[5] As for Middle Eastern oil, in particular, it represented less than a quarter of American oil imports, and the percentage seemed likely to decline in the future, as African, Canadian, and other sources of supply grew. Most Middle Eastern oil went to Asia, and more would head east in the future.

What, specifically, would be the benefits of American energy independence? For each argument put forth by alternative energy and import-independence advocates, ExxonMobil's executives and lobbyists developed a response, which they delivered, particularly in the period that followed the "addicted to oil" declaration, in their own speeches at universities and economic forums, and on PowerPoint slides that ExxonMobil lobbyists handed out on Capitol Hill. These arguments addressed the main tenets of the "import-independence" advocates.

Energy independence would strengthen the American economy by reducing costly oil imports and thereby dramatically improving the country's trade and balance-of-payment deficits, about half of which were due to oil imports. But there was no proven or obvious correlation between a nation's trade balances and its per capita income or gross domestic product growth, ExxonMobil's economists argued. Japan, Singapore, Thailand, and other Asian economies relied more, proportionately, on imported energy than the United States, and yet their economies had thrived overall and had produced trade surpluses. Yes, reducing oil imports could meaningfully and helpfully reduce America's trade deficit, and thereby its total debts, but to blame oil for America's unbalanced economy and its struggles in global economic competition was specious.

Energy independence would reduce or eliminate the need for American military intervention and defense spending in the unstable Middle East. America's Middle East policies were constructed to defend Israel, check radical regimes such as Iran's and Saddam Hussein's Iraq, and keep global sea-lanes open for all commerce, to strengthen the world economy; in any event, U.S. defense spending as a percentage of national wealth was not badly out of line, by historical standards.

Energy independence would break the resource curse cycle by which American dollars financed hostile regimes in Venezuela, Iran, and elsewhere, and by which it encouraged the formation of corrupt authoritarian regimes in poor countries such as Chad, Nigeria, and Equatorial Guinea. The future of the world's energy economy lies in rising consumption in China, India, and other developing economies; their purchases would only replace America's. A technological revolution in which all oil consumption was rapidly displaced worldwide was unrealistic, given the pace of economic growth taking place in poorer countries and their thirst for energy.

Energy independence would free the United States from the threat of coercive supply disruption by hostile countries, such as occurred twice during the 1970s. ExxonMobil's own private war-gaming scenarios showed that such threats were not realistic, particularly given the security of supply to the United States from Canada, Mexico, and Africa; in any event, the strategic petroleum reserve had been constructed to manage any short-term dislocations, to buy time for more decisive military interventions, if needed.

Rapid adoption of clean energy would reduce the threat of climate change. Even if global warming threatened the United States—and ExxonMobil remained unconvinced about the certainty and magnitude of the threat—it did not warrant costly government intervention, which would impede economic growth and threaten jobs, particularly if the policy response did not include rising economies such as China and India, whose emissions were greater than those of the United States.

To Tillerson and other ExxonMobil executives, Bush's speech signaled how flawed and politicized American energy policy had become. If Tillerson and his Management Committee could write their own foreign and energy policy for the United States, it would involve, first, an acceptance of the interconnectedness of global oil markets—and an end to fantasies about national "independence" from those markets; and second, a recognition that carbon-based fuels would be central to the energy economy for decades, even if a significant tax on carbon was imposed eventually to address the risks of global warming. From those premises, ExxonMobil's executives would construct a deliberate, country-by-country strategy to maximize oil and gas supply through free-market competition and to enforce the sanctity of commercial contracts to support that effort. Such a strategy, if carried out with discipline and pragmatism, might ultimately provide the United States with adequate "energy security," as Tillerson put it, as opposed to "energy independence."[6] This was ExxonMobil's corporate foreign policy; it frustrated Tillerson that he could not persuade even the relatively oil-educated Bush administration to adopt it wholeheartedly, and worse, that the president, during his second term, worn down by Iraq, seemed to be drifting away from ExxonMobil's policy vision altogether.

ExxonMobil had amply proved that it could profit in the midst of weak American energy policies; indeed, the corporation's central role in the American energy economy was in many respects a function of Washington's inability or unwillingness to challenge assumptions about oil that had prevailed during most of the previous century. Confused and inconsistent American energy policy, then, was hardly a threat to the corporation; it was built into ExxonMobil's business model. The question that mattered more in Irving around the time of Bush's speech was whether the rapid, disruptive emergence of alternative energy technologies might upend the oil and gas industry. Could the pennies Bush tossed in the wishing fountain pay off unexpectedly?

Technological revolutions regularly overthrew incumbent industries and their corporate leaders; Moore's Law, which described how computing power was becoming ever greater and cheaper as the years passed, suggested that technology-driven upheavals in the American economy would likely occur more frequently than ever before. Since 2000, the Internet and related digital technology had radically disrupted the assumptions of retailers, publishers, and broadcasters in a very short time period. Was it conceivable that ExxonMobil might face a similar disruption from the sudden rise of a transformational alternative energy source?

Lee Raymond's view had been that it would be impossible to predict all technological change, but if ExxonMobil maintained a tightly disciplined process of planning and review, its executives might spot new technical trends early, as they started to make market impact. Tillerson proceeded from similar assumptions. Nothing in the oil industry happened overnight: Disciplined planning could catch innovation in its take-off phases.

The Management Committee reviewed the possibilities for technology surprise each year as part of its Corporate Strategic Planning exercise. At Irving, William Colton, a chemical engineer, finance specialist, and ExxonMobil lifer, led the reviews; he reported directly to Rex Tillerson. Colton was a sharp analyst with a slight smirk; he was not an economist, but his tours as an apprentice executive and later as treasurer of the company's upstream division had educated him about the corporation's financial model and its strategic challenges. Rosemarie Forsythe, the former National Security Council analyst who ran ExxonMobil's political risk department, participated in the reviews, along with a team of economists and business analysts. The corporation's 2030 "Outlook for Energy," which confidently predicted continuity in global oil and gas markets and foresaw only limited penetration by solar and wind technologies, reflected the main thrusts of the group's long-range forecasting.

What if they were wrong? Scores of competent economists predicted rising housing prices for decades ahead. Their inability to perceive a speculative bubble and to imagine a radical departure from historical trend lines would soon destroy some large banks and humble many others. ExxonMobil's own history of strategic prognostication could hardly provide its executives full confidence. Its planners had failed in the past to accurately forecast fundamental questions about the future of oil and gas prices. They had failed to predict some big changes in the American transportation economy, such as the popularity of S.U.V.s. Their basic analysis of the global energy mix had proven sound, although where it was a little off, it tended to suffer from intellectual conservatism—a failure to credit the possibility of more rapid change than conventional wisdom would typically credit. As part of their annual technological review, Tillerson and Colton decided to take a deeper look at some of the emerging new energy technologies just to be sure that ExxonMobil was not missing something that could be suddenly disruptive to the corporation's basic businesses.

They commissioned fresh assessments from ExxonMobil's own scientists, but also pushed the scientists to develop analysis from outside the corporation. The review produced proprietary, inch-thick white papers from outside scientists and technologists, papers designed to double-check or challenge in-house findings and assumptions.

By 2007, Irving's planners felt they had a solid grip on the state of solar and wind power technologies. While the corporation forecasted rapid growth in those industries, aided by government subsidies, ExxonMobil did not regard solar or wind as a meaningful threat to its business. For one thing, solar and wind systems generated electric power and had little direct impact on transportation fuels, the heart of the oil industry. Natural gas, ExxonMobil's principal contribution to power generation, promised to be price competitive with solar and wind for many years to come, and because of the relatively low carbon content in gas, that fuel would likely benefit from more intensive climate regulation, if such regulation emerged.

ExxonMobil now considered such regulation more likely than not: Although the corporation's Washington office still lobbied vigorously against a price on carbon, Strategic Planning in Irving now assumed, for purposes of its annual forecasting, that carbon taxation or limitations would be enacted in the United States in the future. There was inevitable tension between Tillerson's leadership group and Dan Nelson's Washington shop—in effect, the corporation's forecasters were betting against Nelson's ability to ward off climate bills. There was a risk that Irving's analysis would become self-fulfilling prophecy, if it led ExxonMobil to change its lobbying position on climate policy to appease rising Democrats. Nelson argued, in the manner of the former marine that he was, that capitulation would be premature and that ExxonMobil should stick to its guns.[7]

Only technology—not Washington policymakers—was likely to ambush ExxonMobil. The only area where a breakthrough might be deeply consequential would be in transportation—some leap that suddenly changed how most cars and trucks were powered. As of 2007, the most oft-discussed scenarios for such a revolution in transportation fueling involved the possible emergence of hydrogen fuel cells, biofuels derived from plants, and the growth of battery-powered vehicles. The team Colton helped to coordinate examined all of these possibilities in depth.

H ydrogen is the most abundant element in the universe, but it does not typically present itself in isolation; it promiscuously gloms onto other elements to form compounds such as water. Once separated, however, hydrogen can be burned to create energy while releasing relatively few harmful emissions. Hydrogen optimists have fantasized about its potential as an abundant, environmentally sound, affordable car and truck fuel since the 1960s, when the technology emerged from research to support the U.S. space program. After September 11, hydrogen advocates reached President Bush and prompted his decision to fund new research at the Department of Energy. However, a sober 2004 National Academy of Sciences study soon listed the "major hurdles" that stood in the way of a hydrogen miracle, noting, "The path will not be simple or straight-forward."[8]

ExxonMobil's corporate planning analysts had monitored hydrogen research for many years; they were well familiar with the element's industrial uses and difficulties, and they felt confident that the United States was not on the verge of any sudden, transformational turn into a hydrogen-based-energy economy. Sally Benson, the scientist who ran the ExxonMobil-funded research partnership at Stanford University, believed that "three or four near-miracles" were required before hydrogen could threaten gasoline, although she pledged her program to pursue them.[9] After Bush launched his FreedomCAR initiative, ExxonMobil appointed one of its transportation fuels specialists, Buford Lewis, to travel and speak worldwide about hydrogen's potential, but also, pointedly, about its limitations—Lewis traveled with carefully composed PowerPoint slides, derived from Irving's public policy issues binder. The slides acknowledged hydrogen's theoretical potential as a car and truck power source but laid out in detail the practical challenges. These involved the expense and risk of building a national hydrogen infrastructure, the problem that much dirty energy was often required to separate hydrogen from other compounds in order to burn it cleanly, and the difficulties of engineering viable fuel cells that could fit and operate inside vehicles economically.[10] Lewis was eventually recruited to serve as ExxonMobil's principal lobbyist in the United States Senate. He worked the Senate aisles alongside Dan Nelson by the time of Bush's "addicted to oil" speech.

B ush also pledged to redouble the federal government's investments in ethanol, a form of alcohol. Might that be the "black swan" fuel that upended ExxonMobil's oil and gasoline business? The chemical processes by which ethanol could be extracted from sugar and grain had been known to mankind for centuries—mainly because they produced an alcohol that made people drunk and happy, or at least temporarily distracted them from their miseries. Ethanol had also been burned as a fuel in the industrializing West since the early nineteenth century, but it was not as efficient as oil-derived fuels such as kerosene. It first emerged as a subject of possible federal regulation and mandates for use in the United States after the oil shocks of the 1970s. In that era's search for freedom from Middle Eastern oil, ethanol distilled from corn surfaced as a possible solution. The fuel's advocates—primarily in the agricultural Midwest—also promoted ethanol blends as a way to reduce air-polluting carbon monoxide emissions from burned gasoline. Initially, gasoline blenders added MTBE to their fuels to reduce carbon monoxide pollution. When concerns about MTBE's impact on human health surfaced, ethanol reemerged as a potentially safer alternative.

Ethanol refiners and corn growers around Chicago and other midwestern population centers organized themselves into a Washington advocacy group, the Renewable Fuels Association; hired lobbyists; and pushed Congress to adopt a mandatory national production level of ethanol—a forced march to freedom from oil. Ethanol mandates would also raise corn prices and enrich farmers. Like other farm subsidies, the push for ethanol in Washington was enabled by the constitutional makeup of the United States Senate, where sparsely inhabited farm states enjoyed disproportionate clout because each state elects two senators no matter its population base.

Many late-model American cars could burn fuels with up to 85 percent ethanol, mixed with gasoline, without being modified. The potential benefits of ethanol included the ability to grow corn for transportation fuel within the United States, substituting this for foreign oil, and the fact that ethanol blends produce fewer global warming emissions than pure gasoline. The drawbacks were substantial, however—primarily the lack of enough arable land, even in the vast United States, to produce the tons and tons of corn needed to manufacture ethanol in the volumes required to support American driving habits. Brazil, straddling the equator, produced ethanol efficiently from its abundant sugar crop, but the United States could not match that feat with corn. By directing so much corn production into ethanol manufacturing, the United States risked raising food prices at home and abroad. Nonetheless, in 2004, lobbied by farm-state politicians such as South Dakota's Tom Daschle, then the majority leader of the Senate, Congress passed a fifty-one-cent-per-gallon subsidy for ethanol. The next year, Congress enacted an Energy Policy Act that mandated an annual minimum use of ethanol, rising from 4 billion gallons in 2006 to 7.5 billion gallons in 2012. Lee Raymond refused to support the bill, given its webs of complexity and subsidy; he found it a "totally politically driven process." It requires about 1.5 gallons of ethanol to power a car as far as 1 gallon of gasoline would. Between this gap and the federal price subsidy, the true cost of ethanol ranged as high as six dollars per gallon in noncorn states by 2006, according to the calculations of Harvard University environmental studies professor Michael B. McElroy.[11] Two years later, Congress extended the ethanol mandate even more dramatically, setting a target of 36 billion gallons of annual biofuel use by 2022, or about 20 percent of total transportation fuel. That might not destroy ExxonMobil's gasoline-based business model, but it was a large number. Was it realistic? That was the fundamental question that the ExxonMobil team conducting internal technology reviews sought to answer.

Most of this hoped-for Congress-mandated biofuel would have to come from as-yet-unproven technological innovations that might allow "cellulosic" ethanol to be derived efficiently from dense plants—specially engineered stalks, wood chips, or switchgrass. The emergence of such a fuel might address the problem with existing ethanol technologies: that all the corn on all the land could not make enough ethanol to power 20 percent of American cars and trucks.

Fundamentally, ExxonMobil's executives doubted the transformational potential of ethanol; as with wind and solar, executives knew the technologies and chemistry well enough, from past work with them, to have earned their skepticism. Tillerson derisively referred to ethanol in public as "moonshine."[12]

ExxonMobil's downstream gasoline refining and retail division had produced and sold ethanol for years, and its technicians scoffed at the idea that it would be possible to produce anything like 36 billion gallons commercially by 2022. The corporation's strategic research partners at Stanford University noted, too, that all of the energy in all of the biomass on Earth is only six times greater than the total energy used worldwide; it would be hard, therefore, for plant-derived fuels to produce a dramatic change in energy production patterns, no matter how great the technological leaps forward. Still, plants did produce liquid fuels, which were convenient for transportation use, and humankind had already learned how to farm, so there was at least the potential for a transformational surprise.

In Irving, Strategic Planning turned to the corporation's Biomedical Sciences laboratories in Clinton, New Jersey, to head the review. Its scientists were asked to start with the most basic questions, familiar though some of them might seem: What is the science? What are the relevant technologies? Who are the leaders in the field? "We pulled together a pretty high-powered team to look at . . . literally all of the biofuel options," recalled Emil Jacobs, the vice president of research and development at

ExxonMobil's engineering division. They evaluated four factors: One was the ability to scale a fuel, that is, to produce enough volume to have a meaningful impact on American or global fuel markets. Second, they dug into specific technical challenges facing innovators in each fuel type. Third, they reviewed environmental impacts, such as land use, water use, carbon dioxide emissions, and other by-products. Finally, they analyzed cost and economic factors.[13]

They concluded that Congress's ethanol mandates were a fantasy. Globally, because of the efficiencies available to sugar-producing countries such as Brazil, ExxonMobil forecasted production of 40 or 50 billion gallons of biofuels annually within two decades; even so, that would amount to no more than about 3 percent of total oil production. In the United States, "the 35 billion gallons [the approximate federal mandate]—you can't do it—and there is no technology that can get you there" in the time frame ordered by the government, Ken Cohen, the corporation's public affairs chief, declared. Another Irving executive was equally blunt: "We just ignored it because we don't think it can be done." Only government subsidies and mandates made ethanol competitive in the United States. ExxonMobil refused to enter subsidized businesses because the subsidies might not prove to be durable, and also because any subsidy known to benefit ExxonMobil would almost certainly fall to political assault. Nonetheless, ExxonMobil was routinely accused of biofuel denial and obstructionism. Cohen replied, "It's strange for some to want ExxonMobil to get into businesses that would require a government subsidy."[14]

Battery technology interested ExxonMobil's corporate strategists most of all. If there was one emerging energy technology that seemed to have the practical potential to disrupt the oil industry's assumptions about the transportation economy, this was it. "I always put batteries in the category of game changers," an ExxonMobil executive involved in the strategic technology review recalled. The most important questions involved the potential for breakthroughs in the "energy intensity" of batteries, which referred to the ability of a car battery to store and release large amounts of power at a competitive cost.

ExxonMobil's corporate planners had already concluded by 2007 and 2008 that gasoline hybrid vehicles—those that combined existing battery technology with a traditional combustion engine—would be very success-ful in the marketplace, although the growth of sales of these vehicles, they believed, would be gradual and would not threaten the fundamentals of the oil business in the United States or abroad. A more disruptive scenario might involve the rapid adoption of all-electric or plug-in cars. That was a focus of ExxonMobil's white-paper research.

"In the case of batteries, we felt we had to bring in some outsiders," the executive involved recalled. "And they spent a considerable amount— I'm talking many, many months—looking at this. . . . 'Here's the technol-ogy, here's where they are, here's the likely breakthroughs or possible breakthroughs.'"[15]

ExxonMobil conducted its own in-house battery research, in its chem-icals division, and so it also employed scientists who were familiar with the core issues. The corporation also maintained research partnerships not only with Stanford and M.I.T., but with car manufacturers in Detroit and Japan; during the technology review, corporate planners reached out to all these sources to help study the technical questions about a prospective battery revolution.

The internal study's conclusion, the executive involved recalled, was that "there is just nothing there . . . no pathways we see to cracking that code." The main obstacle was "the cost of batteries as a storage device." Gasoline hybrids made "a lot of sense," particularly when you combined batteries with combustion engines that could increase their fuel effi-ciency in the years ahead, through improved engine designs. But the "tech-nology is just not there" for a step up to a radical new transportation economy dominated by all-electric vehicles, the study concluded.[16] There might be some early adoption of all-electric plug-ins, particularly if the sale of such cars was encouraged by government subsidies. Hybrid growth would probably contribute to an overall 20 percent decline in oil use for transportation in the United States within two decades, but that reduction would be more than offset by growth in car and truck consumption in China and other developing countries, and so the net effect to Exxon-Mobil, as a global oil producer, would be inconsequential. For the time being, the review found, ExxonMobil could rest easy: There would be adjustments ahead, but the gasoline and diesel industries were not about to be wiped out by a great leap forward in battery innovation.

Rex Tillerson believed that transformational technological change would upend the oil business and global energy economy eventually. Breakthrough batteries might be the pathway, or breakthrough biofuels, or cheaper, more efficient solar technology, or some combination of those technologies, or perhaps something unimagined in the present. Not anytime soon, however. For two decades and probably much longer, Tillerson's Management Committee concluded after its reviews were completed and digested in 2008, ExxonMobil could feel secure about its investments in oil and gas.

The corporation nonetheless remained vulnerable to unexpected, game-changing disruptions, its senior executives in Irving believed. In their view, however, the most likely strategic surprises would not involve technology. "Geopolitics, by far, is the largest uncertainty in the entire world of energy," one of the executives involved in the strategic technology review said afterward.[17] As the review wound down, events proved the point. By then, the most material and volatile facts in the global oil business arose not from alternative energy technology labs. They involved extortion rackets, kidnapping, and maritime piracy carried out by militia cult leaders in the swampy deltas of Nigeria. Here, too, ExxonMobil and the Bush administration struggled to align their interests and their influence.

Twenty-one

"Can't the C.I.A. and the Navy Solve This Problem?"

Around 9 p.m. on October 3, 2006, seven ExxonMobil contract workers—four Scotsmen, a Romanian, a Malaysian, and an Indonesian—sat drinking in Nancy's Bar, a bush-hut pub inside the oil corporation's walled compound in Eket, Nigeria, in Akwa Ibom State, on the eastern side of the Niger River Delta. The compound sat on a rise beside the wide, muddy Qua River in downtown Eket, a tumbledown town of market stalls, flophouses, water-streaked concrete buildings, and shacks with rusting corrugated metal roofs. Billboards, cell phone towers, palm trees, and church steeples protruded into the low skyline. Motorcycles and scooters poured out smoky exhaust as they swarmed like schools of fish through streets flanked by open drains. The stenciled names of small contracting businesses and places of worship (sometimes colocated) suggested the striving ambition of a faith-influenced oil hamlet: "Divine Hands Ventures," "Mount Zion Lighthouse," and "Success World."

Nancy, the pub owner, was upstairs cooking. She was the Nigerian wife of George McLean, one of the Scotsmen at the bar, a British army veteran who had been settled in Eket for a dozen years, employed by the ExxonMobil contractor Oceaneering International. The three other

Britons—Paul Smith, Sandy Cruden, and Graeme Buchan—were relative newcomers. They maintained cranes for Sparrows Offshore, which is based in Aberdeen, Scotland. About three hundred expatriates lived on the Eket compound, of whom about fifty were Americans. Their work supported one of the larger and more profitable oil and gas production operations in ExxonMobil's global portfolio. In recent months, the corporation had brought on line two large, floating offshore production platforms in Nigerian deep water. These platforms lifted ExxonMobil's production in Nigeria to a record 850,000 barrels per day, more than 15 percent of the corporation's worldwide total. High royalties and taxes limited profits but Nigerian crude was of high, sweet quality and commanded premium prices. The country's deep water beckoned with additional discoveries that might add to ExxonMobil's booked reserves. In a global oil system constrained by rising nationalism in Venezuela, Russia, and the Middle East, West Africa appeared to be a locus of abundant supplies available for the most part on free-market terms.

Automatic gunfire and pistol shots jolted the men at Nancy's. They heard shouts and cries at the ExxonMobil security gate. Eighteen young Nigerians burst into the bar. They wore head scarves and fired weapons in the air. They ordered African customers to lie down on the floor, and then they frog-marched the seven expatriates into the darkness. Two Nigerian security guards employed by ExxonMobil to protect the compound's perimeter lay dying outside. "Run!" the kidnappers ordered. The prisoners jogged for about five minutes to a bridge, where two large speedboats stood ready. The armed boys loaded the foreigners aboard, and the engines roared. They weaved and raced for about eight hours until they reached remote Delta swamplands.[1]

Within a day the kidnappers had issued their ransom demands to Sparrow Offshore and to the governor of Akwa Ibom State, a Christian politician named Victor Attah, who was a relatively effective governor by the Niger Delta region's abysmal standards. The Delta suffered from what the writer Chinua Achebe called a culture of "political godfatherism." The kidnappers demanded $10 million for the safe return of each of the four British ExxonMobil contract workers among the seven men abducted, or

$40 million in total. The largest kidnapping in the corporation's history was under way.

John Paul Chaplin oversaw ExxonMobil's operations in Nigeria. The corporation maintained offices in Abuja, the purpose-built capital in the center of the country, and Lagos, the commercial hub on the Atlantic coast, nearer the oil. Dozens of government affairs lobbyists, public relations specialists, security officers, and corporate intelligence collectors reported to Chaplin along with the usual array of engineers, lawyers, and labor supervisors. Especially during the first years of his tenure in the country, Chaplin had taken a somewhat optimistic view of Nigeria's oil potential. "It's like the Gulf of Mexico in the 1970s," he told American diplomats privately. Nigeria's gas reserves offshore could turn out to be the largest in the world, Chaplin thought. For all of its corruption and flaws, Nigeria's government had a decent record, to date, of honoring contracts with the big oil corporations that fed the government so much revenue; the government favored Western corporations and resisted China. Nor had Nigeria's rulers resorted to the sort of manipulated nationalism and populism roiling Venezuela. The centrality of Nigerian booked reserves and production to ExxonMobil's corporate performance reflected the broader rise of West Africa as a critical oil supplier to the United States after 2000. Nigeria was on track to soon pass Venezuela as America's fourth largest supplier of oil, after Canada, Mexico, and Saudi Arabia.[2] The corporation produced about the same amount of oil in 2006 from Nigeria, Chad, Equatorial Guinea, and Angola as from the United States and Canada combined.

All this provided ample reason, Chaplin thought, for ExxonMobil to try to continue to adapt to what was, admittedly, one of the world's roughest political and social environments among major producers. The corporation operated in joint venture with the Nigerian National Petroleum Corporation, and the N.N.P.C. was a mess, riddled with corruption and unable to keep up with necessary investments. When the government was not stealing outright, it operated at a hopelessly slow and inefficient pace;

bureaucratic approvals that might take six months elsewhere took twice as long in Nigeria. Civil and political unrest swept the country in waves. Even before the raid on Nancy's, Chaplin had found it increasingly difficult to persuade expatriate technical workers to come to Nigeria; if the problem of attracting talent worsened, it might threaten production.[3]

Unlike Royal Dutch Shell and Chevron, which ran many of their Nigerian oil wells onshore, in the midst of impoverished and politically disenfranchised Delta populations, most of ExxonMobil's operations took place in ocean waters eleven to seventy-five miles offshore. The corporation's onshore base in Akwa Ibom—a private airport and housing compound in Eket, and its nearby Qua Iboe Terminal on the Atlantic Ocean, where piped oil could be stored and loaded for transport—had not been greatly troubled during past phases of militancy and insurgency in the Niger Delta. Southern Nigeria's most restive antigovernment ethnic groups, such as the Ijaws, had little local presence. These factors fed a tendency toward complacency about security threats in Nigeria back at headquarters, in Irving. But during the months leading up to the Eket kidnapping, it had become clear that ExxonMobil was coming under serious and unprecedented threat.

The surge in Nigeria's importance to global oil markets seemed to inspire into action the political and criminal gangs in the Delta's oil-endowed swamps. Northern ethnic elites had long dominated politics and wealth hoarding in Nigeria; across decades, they had exploited the Delta's oil and left its people in poverty. The particular wave of violence that washed up at the ExxonMobil compound in Eket was only the latest manifestation of this conflict. In 2005, Nigeria's democratically elected president, the retired general Olusegun Obasanjo, sought to amend the constitution to extend his rule beyond two four-year terms. The power struggle that ensued among parliamentarians, Delta governors, and supporters of the president led to an upsurge of violence by armed gangs of thugs, students, and legitimately aggrieved insurgents—in Nigeria, it was never easy to separate criminals from political dissenters.

Obasanjo established the Niger Delta Development Commission to invest in the neglected south and address the deprivation that fed crime and militancy. ExxonMobil contributed more than $100 million annually to the budget as part of its operating agreements. Yet the commission failed to deliver; Chaplin grew discouraged. He complained that Exxon-Mobil and other oil majors could not "continue to be the only entities that address communities' needs because the companies simply are not equipped nor well suited to become quasi-governments." Chaplin felt that "now is the time for Nigerians to hold their government accountable."[4]

The country's incipient revolutionaries could be as nasty as the government kleptocrats they challenged, however. Delta gangs in early 2006 became newly audacious and ruthless. Kidnapping in the region had long been rampant, but in earlier eras, cases might be settled peaceably as kidnapper and victim sat together in a Port Harcourt bar, sipping beer and waiting for a final ransom to be determined—which they intended to divide between them. The region's kidnapping markets were highly evolved. Expatriate workers were assessed by kidnapping gangs on a "potential-for-payment scale," Royal Dutch Shell's local executive explained to a visiting U.S. senator, "with hostages from the United States or Western Europe garnering the highest ransoms and Russian, Indian and Asians the least." The oil majors tracked actual ransom settlements—Shell's matrix showed that the most recent ransoms were running at about $120,000. By 2006, however, raids and abductions led regularly to murder. Shadowy groups issued political demands, not just requests for ransom, and they spoke of revolution. Insurgent and pirate gangs deployed speedboats and raided corporate platforms offshore that they had never reached before. Once-orderly ransom marketplaces yielded to price uncertainty. This devolution unfolded very quickly in late 2005 and early 2006. The United States, Britain, France, and the Netherlands formed a consultative group, the Delta Working Group, based in their embassies in the capital of Abuja and in their consulates in the economic capital of Lagos, to evaluate the emerging crisis. "The question was," recalled John Campbell, then the United States ambassador to Nigeria, "have things fundamentally changed?"[5]

Campbell, an experienced career foreign service officer, believed they had. Insurgent groups in the Delta were issuing statements for the first time that threatened to shut down the country's oil industry. Their language increasingly attacked the Nigerian state and challenged its legitimacy to rule in the Delta. "If you put all this together, we are far beyond where we have been," Campbell argued to colleagues that early winter of 2006.

Nigerian military intelligence sources began to report specific threats against ExxonMobil in January. They said two groups, the Nigerian Ijaw Martyrs and the Ijaw Patriotic Front, had reportedly been handing out money to Akwa Ibom youths to join an attack on ExxonMobil's Qua Iboe Terminal. Chaplin's security team increased their alert status from Code Orange to Code Red, the highest possible level, at the targeted compound, but kept nearby Eket and other facilities at Orange. Threat upon threat followed.

On March 9, a militia group styling itself the Martyrs Brigade, which said it was acting on behalf of the Movement for the Emancipation of the Niger Delta, or M.E.N.D., threatened to carry out "massive attacks" on ExxonMobil's Nigerian affiliate unless the corporation paid new compensation to local communities, to make amends for a 1998 oil spill from one of its offshore pipelines. "ExxonMobil has continued to pay deaf ears [sic] to the pitiable plight of a now pained and severely exploited people," the brigade said in a written statement. Along with "all other nationalist and freedom-fighting units in the Niger Delta, we hereby declare a 21-day grace period for ExxonMobil to honor its obligation to compensate every community that was affected by that catastrophic spillage."[6]

An ExxonMobil security officer joined the Delta Working Group on March 14 and reported that a militant known as Comrade Owei was behind the protests and threats. ExxonMobil had in fact paid between $25 million and $30 million in restitution for the 1998 spill, but the corporation had turned away demands for compensation from communities that it "does not believe were materially affected by the spill," the officer said. The corporation felt that it was in a bind. It took the militant's ultimatum "very seriously," a second corporate official told the American embassy in Abuja, yet felt its options were limited. ExxonMobil depended primarily on a military command in the Delta, the Joint Task Force, and the notorious State Security Service for protection. Neither seemed prepared to confront the emerging threats. ExxonMobil reported that it had embarked on an advocacy campaign at all levels of Nigeria's government, arguing that Nigeria "cannot afford to allow M.E.N.D. and other militant groups to creep into Akwa Ibom" and threaten ExxonMobil as they already threatened Shell and Chevron in neighboring states. "Were this to occur, the crisis would encompass virtually all of Nigeria's oil producing coast." Chaplin and some of his colleagues feared a step-by-step escalation that might lead Nigeria's military-influenced government to unleash "a scorched earth policy, regardless of its impact on civilians," which would only make a bad situation worse for the companies.[7]

On April 29, a group of local youths, demanding entry-level jobs on offshore ExxonMobil platforms—cleaning, maintenance, and light construction jobs that typically went to foreign nationals recruited from India, the Philippines, or elsewhere—boarded speedboats and occupied a corporate barge twenty miles offshore. ExxonMobil's Nigerian managers asked an Eket labor commissioner, Chief Samingo Etukakban, to help persuade the young men to leave.

Etukakban was angry with ExxonMobil because in previous talks over jobs for Akwa Ibom youths, he felt that the corporation's public affairs officers had lied to him. Nonetheless, "we relented and went out to the barge" by corporate speedboat, he recalled. ExxonMobil expatriates were present, Etukakban said, trying to end the sit-in. A Nigerian navy warship soon turned up; ExxonMobil had summoned the navy for assistance. Nigerian officers and sailors—determined to prove that they could defend the property of international oil corporations—boarded the barge on May 1 and arrested everyone, including the mediators invited out by ExxonMobil. "We were detained in very terrible conditions—two people to one handcuff, lying on a concrete floor," Etukakban recalled. He spent twenty-three days in custody before a National Assembly member secured his release. The next day more youths with machetes forced their way into ExxonMobil's Qua Iboe Terminal and briefly held two expatriates hostage. The men escaped. Nigerian security forces fired on the demonstrators, killing two of them. ExxonMobil had been "lucky to operate in the most peaceful area of the Delta," said Chief Nduese Essien, the parliamentarian who freed Etukakban. But after the barge incident, by the summer of 2006, its local standing was deteriorating. ExxonMobil executives seemed unaware about how low its position had deteriorated among

some local politicians and their youthful supporters. The corporation repeatedly displayed a "lack of interest" in local issues and welfare, Essien
believed. "Everyone in the Delta is fidgety, and militants may be looking
for a pretext to expand their swath," the American consulate in Lagos
reported. ExxonMobil is "talking with local leaders" and yet, "even if we
keep the professional militants out, [ExxonMobil] will still have a difficult
time working with the local youth to resolve this situation."[8]

The abductions from Nancy's Bar signaled just how much had
changed. ExxonMobil had been warned.

The seven kidnapped ExxonMobil expatriate contract workers were
threatened by their kidnappers but not beaten during their first ten
days in captivity. They slept in a makeshift camp deep in the Delta's palm-
shrouded swamps, where muddy creeks and eddies snaked through thick,
humid foliage. The kidnappers seemed to be heavy drug users and often
preoccupied themselves by getting high. They fed their victims rice and
water, but the men felt hot during the day, cold at night, and wet per-
petually. They kept up their morale by talking about food and soccer.
Their captors opened talks by cell phone with Sparrows's chief executive
and with Victor Attah, the Akwa Ibom governor, a full-faced man who
espoused Christian principles, wore business suits, promoted grandiose
shopping mall and golf course developments, and owned a luxury home
in Lagos. Typically, governors were called in to mediate ransom agreements;
it was presumed across the Delta that leading politicians and their security
forces often took a piece of the action, although Attah himself had not
presided over a kidnapping industry in Akwa Ibom.

In Lagos, ExxonMobil security officers and counterparts from the affected contractor companies formed a crisis management cell and met
daily. ExxonMobil retained Controlled Risk Group, one of the major kidnapping management and security firms operating in the Delta. The consultants advised that it was important to "have only one channel of
communication between the kidnappers and the government." Typically,
kidnappers would use their victims' cell phones to reach out to family

members to negotiate, issue threats, and raise pressure on the employers. Controlled Risk contacted the families of the victims, passed along cell phone numbers that might be used in this way, and urged the family members not to answer. The Exxon crisis cell also urged Governor Attah to persuade the kidnappers to allow a delivery of humanitarian supplies to the hostages.

The British Foreign Office took a leading role. Washington involved itself as well. After September 11, the State Department set up an enhanced interagency crisis response team that could rapidly deploy to help governments respond to hostage takings, particularly those involving Americans. By 2006 the team had drilled for just the sort of crisis that ExxonMobil now faced. But the idea that American and British intelligence and security officers might parachute into Nigeria to sort out hostage crises made the Nigerian government "uneasy," as a State Department official involved put it. It did not thrill ExxonMobil, either. The corporation's security officers were at times reluctant to share information about kidnappings-in-progress with the American government, fearing that sensitive details might be released under the Freedom of Information Act or otherwise leak to the media, compromising negotiations. "Unless serious injury is imminent, companies prefer to negotiate without Embassy intervention unless intervention could be discreet," a cable to Washington from Abuja reported.

Ransom negotiations reached an impasse and the kidnappers panicked. They beat the four Scotsmen with sticks and slapped them around with machetes. They handed them cell phones and ordered them to tell their corporate bosses that they were "in danger of being shot."

One morning, the kidnappers beat Graeme Buchan again and then handed him a cell phone. One of the youths threatened him with a loaded gun and instructed him to report, falsely, that his fellow captive, Paul Smith, a father of two, had died of malaria—and that the others were at risk of imminent death as well. "I'm afraid the gun at my head might have uncovered a talent for acting I didn't know I had," Buchan said later.[9]

The kidnappers called Governor Attah to report that Paul Smith had died. Attah was furious, he recalled; the death of a British kidnapping victim snatched from ExxonMobil's fenced compound would devastate Akwa Ibom's reputation for business and development. "I do not talk to criminals," he snapped, as he recalled it. He hung up and ordered an aide to send a message to the kidnapper who had telephoned: "Tell him I hope he knows the cost of transporting a corpse from wherever it is back to the man's home country, because the man will want the body brought back to be buried." The governor hoped, he said later, that he might unnerve and rattle the kidnappers with this hard-line attitude.[10]

In Scotland, British police soon arrived at Paul Smith's home to convey the news of his death to his twenty-eight-year-old wife, Paula. Their elder son, Jordan, who was four years old, "was suspicious" of why the police had turned up, and so that night, Paula, devastated, decided to tell the boy the truth. "Daddy is just like the Lion King," she explained. "He's gone to heaven now and you won't see him again."[11]

The reported death did accelerate ransom negotiations. ExxonMobil maintained a firm public line against payments, but its declared policy could not constrain either its contracting corporations or the governor of Akwa Ibom. Attah recalled that he was besieged by calls from the American, British, Romanian, Malaysian, and Indonesian embassies—they pressed him so hard to resolve the kidnapping that he found it difficult to actually carry out the negotiations. In the end, he conceded, he authorized a ransom payment. It is not clear what advice ExxonMobil offered about this decision or whether it endorsed the payments or supplied funds. As to the kidnappers, Attah said, "They were some misguided boys from my state" who had "invited" and "escorted" elements of a more experienced, hard-core kidnapping gang from another Delta state to attack the Exxon-Mobil compound.[12]

State Department officials working with major American oil companies found by 2006 that they "diverged in our paths" on the Delta kidnapping issue, the State official recalled. "They would pay ransoms, and then we felt that was just actually contributing to the problem."[13] Hostage negotiating teams led by State's Diplomatic Security bureau did deploy to Nigeria, but then sat idle for lack of cooperation from the firms.

The kidnappers packed their hostages back into speedboats and drove them to a rendezvous point with officers of the State Security Service, or

S.S.S., the principal national Nigerian police and intelligence force. Assured of their payment, they freed their captives. ExxonMobil helicopters lifted the men to Lagos, where they at last boarded planes for home.

Paul Smith telephoned his wife, Paula, to explain that he was not dead. "He was completely calm," Paula recalled. "I was beside myself. All the family could hear me on the phone. . . . Everyone was jumping around all over the place." Once back in Scotland, Paul Smith issued a declaration: "I won't be going back to Nigeria."[14]

Influential scholarship documenting the resource curse emerged from the study of Venezuela's oil-induced woes, but Nigeria offered perhaps the most striking case study. Nigeria possessed a talented, well-educated elite; fertile land; and, of course, oil revenue. The country's earnings from oil and gas sales from the early 1970s to 2008 totaled about $400 billion. Yet nearly half a century after independence, the country's population languished perpetually near the bottom of the United Nations's human development index. Average Nigerian life expectancy remained only forty-six and one half years. Nine tenths of the population lived on two dollars a day or less. More than a third lacked sanitation and clean water, and the country's infant mortality rates remained among the world's highest. Such impoverished but less oil-burdened countries as Papua New Guinea and Zimbabwe ranked higher than Nigeria on the human development scale.[15]

Corruption, mismanagement, theft, and criminal violence were hallmarks of the government's performance. During the 1990s, the military dictator General Sani Abacha stole an estimated $4 billion of government funds, in addition to that taken by cabinet officials, state governors, and their affiliated youth gangs. International oil and construction companies conspired in these crimes or tolerated them with see-no-evil policies. Halliburton and its subsidiary, Kellogg Brown & Root, agreed early in 2009 to pay $579 million in fines to settle charges related to their participation in a joint venture that systematically bribed Nigerian officials across a decade to secure more than $6 billion in construction contracts; Albert "Jack" Stanley, the chairman of K.B.R., named to his position by Halliburton chief executive Dick Cheney about two years before Cheney departed for the White House, pleaded guilty to criminal charges after personally authorizing a $23 million payment to a Gibraltar consultant to win Nigerian contracts.[16]

In Abuja, "a tiny number of people have stolen a staggering amount of money," a Western diplomat there observed. The diplomat's work in liaison with Nigerian ministers routinely brought him into Abuja homes "that you would be embarrassed to build in Beverly Hills," mansions decorated with "ostentation that is just jaw-dropping." John Campbell, the American ambassador, referred to the capital's better neighborhoods as "an example of Las Vegas baroque." And this was what Nigeria's political overlords felt comfortable displaying in their home country, where fellow citizens could see it; they funneled much of the rest of their wealth abroad, into properties in London, New York, and Los Angeles.[17]

Poverty, disenfranchisement, and environmental degradation in the southern Niger Delta remained acute. Ken Saro-Wiwa led a nonviolent protest movement in the Delta to seek redress during the 1990s; Abacha arrested and executed him. Saro-Wiwa's idealism was exceptional in a resistance movement that increasingly migrated toward violence and crime. During elections in 2003, Delta political bosses armed youth gangs to compete for power; after the vote, the gangs moved into freelance rackets. They drew members, brand names, and cult practices from college campus fraternities: the Vikings, the Icelanders, the Outlaws, and their female counterparts, the Daughters of Jezebel, the Black Braziers, and the Viqueens.[18] They dealt drugs; siphoned oil from pipelines, or stole it in conspiracy with government officials or military officers; and they kidnapped Nigerians and foreigners for ransom. Unemployment ran high among the Niger Delta's young population; the criminal gangs were hiring, and if you could loll around the swamps, hold a gun, and occasionally take a few physical risks, you could have a paying job. As the gangs raised their political sights and economic ambition after 2006, their picaresque criminality—their head scarves, bandoliers, and speedboats; their bank robbery techniques, which included using massive charges of dynamite to blast away reinforced steel doors—seemed increasingly inspired by Hollywood.

This was the ethos from which the Movement for the Emancipation of the Niger Delta arose. M.E.N.D. became, after 2006, the dominant Delta insurgent brand. Central Intelligence Agency reporting from Nigeria during the period of ExxonMobil's Eket kidnapping episode described M.E.N.D. not as an organization with any true leader or hierarchy, but as "a label—at best an umbrella group or an umbrella label," as a consumer of the agency's reporting, who found the C.I.A.'s analysis credible, put it.[19] To avoid being targeted, M.E.N.D. lacked a central council that could declare who was an authorized commander and who was not. Its notional leader, Henry Okah, was an arms dealer who seemed to spend much of his time outside Nigeria; his supposed role as a *supremo* served as a convenience for a movement that was, in fact, made up of semiautonomous, extortionate gangs of varied strength and character. Consumers of M.E.N.D.'s press releases and Facebook videos might imagine a tight-knit band of swamp guerrillas fighting for justice against cold-blooded international oil corporations. There was some of that, but the private security analysts who advised ExxonMobil, Chevron, Shell, and other corporations on kidnappings and safety described M.E.N.D. more as a loose collection of armed young men, mainly from the Ijaw ethnic group, who used laptop computers to create an appearance of formidable coherence.[20]

M.E.N.D. activists or those using their brand name fought at times with Nigerian security services, but they also collaborated with the Nigerian navy in massive thefts of Delta oil from barges and pipelines—"bunkering," as it was known, a racket that independent analysts estimated generated between $4.5 billion and $6 billion in total thefts during 2008 alone.[21]

In Irving, the global political mapping exercise revised annually by Rosemarie Forsythe, ExxonMobil's chief political risk analyst, painted Nigeria as a bright red "transitional" country (as opposed to blue "democracies" and yellow "authoritarian" regimes), a category marked by internal instability. Forsythe had also developed maps showing where all the world's instances of piracy and similar crimes took place; Nigeria stood near the top of that chart, too. As Rex Tillerson settled into office and assessed the greatest global risks to ExxonMobil's oil and gas portfolio, Nigeria looked unstable; it was getting worse; it was increasingly influenced by pirates;

and yet its oil exports were central to the corporation's business model. Nigerian violence also stoked volatility in global oil prices and raised questions anew about America's energy security. ExxonMobil and the United States government, in alignment but each in its sovereign sphere, found themselves adapting after 2006, often in an atmosphere of confusion and argument, to the world that M.E.N.D. had created.

Tillerson was perhaps not ideally suited to assess Nigeria's moral swamps. His feel for political economies in poor countries was limited. Even during his rise within ExxonMobil's international divisions, he had never lived outside the United States. In any event, managing ExxonMobil's position in Nigeria in the post-Aceh era of the Voluntary Principles, heavy media scrutiny, and potential lawsuits would have been challenging even if Tillerson had been an anthropological expert.

In September 2005, on the cusp of taking power in Irving, Tillerson had flown into Abuja. President Obasanjo had been making noise about forcing Western oil companies in Nigeria to move beyond pumping crude and into the refining of gasoline and other products for local consumption. ExxonMobil had steered clear of Obasanjo because "he tended to pound tables" and make demands. Nigeria was about the last place in the world Lee Raymond wanted to spend time. Tillerson decided to engage, however. He met with the Nigerian president, flew down to Lagos, where he stayed in the corporation's Waterfront Guest House, traveled by helicopter to a few production sites, and departed. The corporation's message to American diplomats in the country was that they should "encourage deregulation" and work on "improving the investment climate."[22]

Tillerson remained hopeful—if not in a state of denial—about ExxonMobil's place in the hearts and minds of Akwa Ibom's population. After the Eket kidnapping, local insurgents took periodic potshots at ExxonMobil transport vans. Speedboat pirates menaced the corporation's offshore platforms. Still, Tillerson believed that ExxonMobil remained "largely . . . insulated" from the worst Delta violence and political dysfunction. In Akwa Ibom, Tillerson boasted, the "community in effect protects us when militants from outside . . . try to create problems. . . . We have good relations down there. That is because we made some good decisions at the beginning. And we look to that as a model."

The trouble the corporation endured as M.E.N.D. rose was "criminal in nature," Tillerson believed. He was "mindful of the security situation," but felt nonetheless that ExxonMobil had a winning formula for obtaining community allegiance in Akwa Ibom. This strategy was rooted, Tillerson thought, in firmness. He sought to imbue in locals the expectation that ExxonMobil knew "how you say no" and that the corporation "could not be intimidated and would act consistently."[23]

Exxon had inherited its operations in the Niger Delta from Mobil. The corporation's subsidiary, Mobil Exploration Nigeria, Inc., won its first license to explore for oil offshore of Akwa Ibom State in 1961; the first wells flowed later that decade. By the time of the merger, Mobil was on the way to becoming the second-largest international producer in Nigeria, after Royal Dutch Shell. Large volumes and the light, sweet quality of the oil made its Nigerian offshore properties exceptionally valuable.[24]

Mobil and then ExxonMobil recruited, paid, supplied, and managed sections of the Nigerian military and police assigned to protect the Eket compound, the roads that led from there to the Qua Iboe Terminal on the Atlantic, and the roads around Akwa Ibom's state capital of Uyo, a fume-choked city that housed the outsize development projects of Victor Attah. These included the shopping centers Mountain of Fire and Miracles Plaza and, after 2007, the equally ambitious, divinity-inflected construction projects of his successor, Godswill Akpabio. (Akpabio enjoyed a fortunate name for a career in politics in a faithful state; he handed out T-shirts to his youth gangs with slogans such as "Stop Social Vices" and "Support Godswill.")

The Mobil Police, as they were known locally, carried automatic rifles and wore black shirts emblazoned with a white arm patch that displayed the Mobil red Pegasus flying horse symbol first adopted in 1931 as a trademark by Mobil predecessor Standard Oil Company of New York. After the upsurge of violence in 2006, ExxonMobil's Chaplin reported, militants often stripped the Mobil Police of their weapons and "many officers have taken to removing their uniforms at the slightest hint of militant activity." The corporation's police established layered checkpoints, spaced every kilometer or so, on the major roads to and from ExxonMobil properties. "ExxonMobil: Take Ownership" declared the sign at the Qua Iboe Terminal entrance, surrounded by warnings posted by Mobil Police squadrons: "Military Zone," "No Stopping," and "No Waiting."[25] The Nigerian military deployed a mechanized battalion to reinforce security in the state, but most of the Nigerian government's support for the Mobil Police came from the S.S.S. ExxonMobil's Global Security unit in Nigeria appointed liaison officers to joint security task forces to coordinate convoy protection, perimeter security at ExxonMobil installations, and executive protection services. In addition to the Mobil Police, ExxonMobil hired and supervised an eight-hundred-man unarmed unit of the "supernumerary" or "spy" police in Akwa Ibom. The scope of their duties is unclear. The spy police carried corporate identity cards even while technically in the employ of their own government. At one stage, the supernumerary unit in Eket sued ExxonMobil for employment benefits. They argued that they were, in effect, corporate employees, not government police officers.[26]

Prior to the merger, Mobil had operated a successful program of community relations in Akwa Ibom, at least as local politicians perceived it. The corporation funded a soccer club, community buildings, water projects, and road building. Nigerian and expatriate Mobil executives curried favor with local political leaders. The corporation acted as "a neighbor and a brother," recalled Esseme Eyiboh, who represented Eket in the Nigerian House of Representatives. After the merger with Exxon, it became a "purely commercial drive." The corporation withdrew from a memorandum of understanding that Mobil had negotiated with local leaders and produced a new program, which they wanted "the community to accept . . . without making any inputs," said Nduese Essien, who negotiated with the corporation after the merger.[27]

With Mobil, Victor Attah recalled, political liaison was "a lot less mechanical," but with ExxonMobil, "it became a lot more rigid." He pleaded with the corporation to build a power plant, but its managers refused, declaring that such projects were "not their core area of business." Under ExxonMobil's rules, Nigerian politicians could not ride corporate airplanes unless it was strictly for oil business; special projects of the sort Mobil had accommodated before the merger were refused; the soccer club and local athletic programs were abandoned; and the corporation issued a new list of local projects it would support. "They have been operating on their diktats," said Essien.[28]

"You have to be willing to say, 'No, we aren't going to do it that way, we are going to do it this way; if we can't do it this way, we won't be here,'" Tillerson explained, speaking specifically about ExxonMobil's strategy in the Niger Delta. "This is the way my company has operated throughout the world throughout my entire career. We will walk away if we don't have an acceptable situation on the ground. That doesn't mean it's not tough, it doesn't mean we don't have problems. We manage it, but it can be done in a way that the local community benefits tremendously— and the Akwa Ibom state has benefited enormously. That is why we enjoy good relations."[29]

Tillerson's opinions echoed those of Governor Akpabio, who promoted a slogan, "Akwa Ibom *Ado Okay!*" or "Akwa Ibom Is Okay!" He sought to protect ExxonMobil. At a "gala night" to honor Chaplin, the governor declared, "Akwa Ibom cannot be safe for criminals; they will soon know that the state is not safe for kidnappers. Let oil companies and other firms know that the state is safe for them." In fact, Akpabio's supporters were engaged increasingly in a complex war with rival gangs, played out through tit-for-tat kidnappings. Nigerian-born ExxonMobil managers and employees, with their attractive salaries, were not immune. Governor Akpabio "has strong cult connections," said a U.S. official who tracked the governor's activities. "I'm told that many of the attacks on the roads . . . are being carried out by his militia—whether because he orders it or because they don't feel they are getting enough money is not clear." As Eyiboh put it: "We are an inch from insurgency."[30]

If the corporation enjoyed a measure of periodic stability in comparison with Shell and Chevron, it was hardly the result of its corporate strategy; it was because most of its oil production was offshore and therefore harder to steal or disrupt. Harder, but not impossible: M.E.N.D.-branded pirates were a determined lot.

In September 2006, President Bush signed National Security Presidential Directive 50, outlining American security strategy in Africa. The directive's stated objectives included building African capacity to govern and deliver social services, consolidating democracies on the continent, and bolstering fragile states. On November 15, 2006, about a month after the Eket kidnapping of ExxonMobil contract workers, Jendayi Frazer, the assistant secretary of state for African affairs, spoke at a maritime security conference in West Africa organized by the United States Navy. The conference was meant to rally regional governments into partnership with the Pentagon to improve maritime security in the Gulf of Guinea, as the Atlantic Ocean waters off Nigeria were known. "Achieving coastal security in the Gulf of Guinea is key to America's trade and investment opportunities in Africa, to our energy security, and to stem transnational threats," Frazer said. She continued: "Let us consider oil." If African governments protected oil commerce, they could prosper. But they required the goodwill of international oil giants. "If kidnapping of their workers and attacks on their facilities continue," those companies were unlikely to stay.[31]

African politicians, scarred by a century of resource-driven European colonialism, feared that the Bush administration viewed their oil as analogous to the oil of the Persian Gulf: as a vital American interest, one that might warrant military intervention, at least in extremis. Bush officials imagined themselves striking a more nuanced, postcolonial posture, one that emphasized encouraging African states to modernize and to rise from poverty. When an American official stood at a lectern flanked by U.S. Navy flags and spoke about oil security, however, the message was unavoidable: West Africa mattered to the United States in part because it possessed critical supplies of energy, and the American military stood ready to ensure oil flowed.

Would a U.S. military response to the Gulf of Guinea's struggles with piracy and insurgency serve ExxonMobil's interests? Before the M.E.N.D. uprising of early 2006, the major American oil corporations in Nigeria preferred to handle their own security problems in the region. Bunkering exacerbated corruption, militia violence, and inequality, but it was not necessarily a problem for ExxonMobil, because its contracts were written to absolve it from the costs of any thefts, and it was not necessarily a problem for global oil supply, because the stolen oil ultimately reached international markets. (As a practical matter, there was nothing to be done with stolen Nigerian crude but sell it.) Connie Newman, Jendayi Frazer's predecessor at State, recalled that in 2004 and 2005, as the trouble in the Delta first began to bubble, oil representatives seemed to have little interest in sharing intelligence or otherwise taking on the problem in partnership with the Bush administration. State officials who visited Nigeria flew over the Delta in Chevron or ExxonMobil helicopters, from which their guides would point out barges of the type routinely used in oil thefts, as if such larceny were part of the natural landscape. "You guys know about this bunkering—the militants don't have tankers," Newman argued when oil company liaisons visited her at Foggy Bottom. "I'm not saying you're doing it—but you know who's doing it, and you could share that information with us." But the companies demurred.[32]

The kidnapping and offshore piracy of 2006 started to alter their attitudes. "The cooperation of the oil companies turned one hundred eighty degrees," recalled a U.S. official in Nigeria at the time. The companies offered new levels of "coordination and information sharing."

The kidnapping surge was not the only new challenge to ExxonMobil. The corporation had begun to tow into Nigeria's deep water, one after another, massive offshore production vessels known as F.P.S.O.s, which stood for floating, production, storage, and offloading. These were oil production platforms in the form of enormous ships that hovered above oil fields, obviating the need to build pylons and platforms in such deep ocean water. The vessels were so huge and economically important, however, that they presented "a significant terrorist target," a U.S. government assessment concluded. The question facing Chaplin, ExxonMobil Global Security, and Tillerson was how to protect this investment. The corporation projected that by 2010 it would have one of the world's largest fleets of F.P.S.O.s floating off the Delta, some of them away from Akwa Ibom and in territory more accessible by Nigeria's most aggressive speedboat militants. Each vessel would produce 100,000 to 250,000 barrels a day of oil and other liquid products. Among other things, they were potentially combustible.

As trouble rose in Akwa Ibom, Chaplin had been reluctant to militarize ExxonMobil's response or to encourage Nigeria's government to do so. ExxonMobil initially decided against asking the Nigerian navy to protect its offshore fleet. Chaplin saw the navy as "amateurish with broken boats and no fuel," and some of its officers were probably involved with the militants in oil theft rings, as everyone in the Nigerian officer corps "retires with money." As for encouraging the Nigerian army to enter the Delta and attack the militants operating there, "The military was not an option that ExxonMobil hoped for," Chaplin said, because an incursion "would aggravate the problem by antagonizing local communities."

The corporation "has decided to go light on security out of concern that the presence of security would not function as a deterrent, but would be seen as a challenge to the militants to attack the facility," the Lagos consulate reported. "That a company would have to engage in these types of calculations for an investment of this magnitude demonstrates the extent to which the security environment for oil companies has descended."[33]

It got worse. M.E.N.D. and similar units attacked ExxonMobil supply boats as they moved through narrow channels and river ways. The corporation organized its supply ships into convoys for greater protection but Chaplin found the Nigerian forces were not willing or able to supply adequate security. The country's wealth depended upon coastal oil production but it lacked the basics of a coast guard.

In Washington, recalled a State Department official involved as the attacks worsened, "the oil companies kept telling us, 'Goddamnit, can't the C.I.A. and the navy solve this problem? We'll tell you where they [the militants] are. . . . Why can't you fix this swampy corner? It's a bunch of pirates. Why can't you just send the navy in there and fix this?'"[34]

The Pentagon had been reviewing that very question since at least 2004. At that time, Africa fell to the European Command, headquartered in Germany. That was an example of how tacked-on and neglected Africa policy often had been. September 11 had galvanized attention to the threat of Al Qaeda–inspired terrorists taking root in ungoverned spaces, of which Africa had many. With Taliban-inspired militias forming in northern Nigeria and M.E.N.D.-inspired oil insurgents and criminals rising in the south of the country, Nigeria "has the possibility of becoming the next Pakistan within twenty-five years," Johnnie Carson, then the Bush administration's national intelligence officer for Africa, noted.[35] The growing share of oil imported into the United States from West Africa by American companies, particularly from Nigeria and Angola, gradually attracted the Pentagon's attention. A proposal to form a distinct Africa Command at the U.S. Department of Defense, a command that would be hived off from the European Command, had surfaced within the Pentagon as far back as the late 1990s, but it was not until Delta militancy exploded after 2005 that the plan gained support from Defense Secretary Donald Rumsfeld. The command began initial operations in October 2007, headquartered at Stuttgart, Germany.

America's response when confronted with a problem such as Nigeria's security services was, as N.S.P.D. 50 outlined, to *build local capacity*. This was a mantra of postcolonial liberalism in the developing world, shaped by the belief that sovereign local governments should take the lead and not have Western solutions imposed upon them. When applied by Africa Command to the particular problem of piracy and kidnapping in the Gulf of Guinea, the philosophy produced plans to strengthen the capacity of the Nigerian navy, so the navy could control its own coastal waters and challenge seaborne M.E.N.D. gangs. The problem, noted an American official involved, was that "it was their admirals that were stealing the damn oil. And then they hired the M.E.N.D. to protect their theft, and they had to cut the M.E.N.D. in." Since the Nigerian navy collaborated with pirates, a corporate oil security analyst noted, did it really make sense to have the U.S. Navy train Nigerian navy officers in the most sophisticated techniques for, say, the storming and boarding of ships? Wouldn't that just ultimately create more skilled Nigerian pirates?[36]

The Pentagon's Africa policy office viewed the Niger Delta as "the perfect storm of political bosses mixing in with disgruntled populations, and Mafioso kind of enterprise that reached quite high in the Nigerian government," as a U.S. Defense Department official put it. The Pentagon's advice to Nigeria's government, nonetheless, was that "number one, they needed to improve their situational awareness" of what was happening from hour to hour on Delta swamp rivers and open Gulf of Guinea waters. Theresa Whelan's office shepherded the transfer to the Nigerian navy of excess U.S. Navy sixty-foot "buoy tender" vessels, for coastal patrolling. The Pentagon also spent $16 million to provide Nigeria with a "suite of sensors," as Whelan described them, tied into a command and control center installed in Lagos by the United States. The command center was designed to provide Nigerian navy officers with a real-time radar-enhanced picture of authorized and unauthorized sea traffic off the Delta coastline. The problem remained that Nigerian admirals did not actually wish to intervene in much of the unauthorized activity because it represented income to them.

From Europe, the U.S. Navy initiated the Africa Partnership Station, a program of periodic U.S. Navy patrols in the Gulf of Guinea, coupled with exercises and shore visits, that were designed to build up the Nigerian navy and the navies and coast guards of smaller neighboring countries. The program found traction in better-organized nations, such as Ghana, but its sponsors struggled in Abuja. The United States Navy had "never really come across an organization that behaves" like the Nigerian navy, said a U.S. official involved. American military officers "come in here and they see a navy with all the trappings, the ranks, the uniforms, and so on, and they think it's a real navy—poor, but earnest. But it's not that at all." It was not obvious what policies the Americans could bring to bear on a sister service that was mainly a criminal enterprise dressed up in epaulets. "It's hard to get used to the fact that Nigerian officials will lie to you straight up," the American official continued. "The chief of navy staff told us, 'There has been no incidence of piracy. You have been misinformed.'" In fact, American diplomatic and intelligence analysts documented nearly four hundred incidents of piracy in Nigerian waters between 2006 and 2009. ExxonMobil itself was struck in some seasons as often as three times per month. Arguably, the effect of American military assistance to the Nigerian navy had been to abet attacks on the property of America's largest oil corporation.[37]

Kidnapped, robbed, and suffering from steadily declining oil production volumes, as well as soaring maritime insurance rates, American and

British oil executives grew restless after 2006. Around the Horn of Africa, where piracy was about equally bad, statistically, the U.S. Navy led international coalitions to battle Somali pirates. Why not do the same in the Gulf of Guinea?

Nigeria's navy was down to two boats it could use to escort oil service ships to defend them from attacks. Nigerian oil output fell because of militant attacks and world oil prices rose toward record highs. Daily news stories of fresh M.E.N.D. attacks caused spot prices to gyrate wildly, pinching the world economy. Mark Ward, Chaplin's deputy, believed it was time to "help the [Nigerian government] rapidly increase its capacity to provide security" on the routes through the Bonny Channel used by supply ships servicing offshore production. "Any help" the United States could provide to Nigeria "for river training" would be "very good," Chaplin agreed. The corporation resisted Nigerian requests for special funds to buy more boats and equipment, but ExxonMobil did supply its own boats to the navy for use on rivers near corporate facilities and housing.[38]

In Washington, these suggestions generated brainstorming and war gaming about American options. "We were constantly beating off bad ideas" to provide training, supply equipment, and conduct joint exercises with Nigerian military units in the Delta, proposals that originated with the international oil firms, said a State Department official. The Pentagon and the economic bureau at State, which liaised with the large oil corporations and channeled many of these ideas, brought forward scenarios and war games that contemplated direct American military intervention in the Niger Delta, or high-profile exercises that might intimidate the kidnappers. "These were the guys in whose ears ExxonMobil was whispering," the State official recalled. "They wanted to persuade us, but it was never clear what they had in mind." M.E.N.D. challenged the viability of the Exxon way, and the corporation turned to the Pentagon. At Africa Command, Army colonels and navy captains rotated into Germany on short tours, the official continued, and they would declare, "'Let's have a Nigeria strategy.' They were constantly running war games in which Nigeria was the example. . . . 'What do we do if the marines have to seize the oil facilities?'"

A second American official in Nigeria recalled the prevalence, after 2006, of largely speculative "ideas that floated around" for deploying U.S. Marines in the Delta to address M.E.N.D. militancy and piracy.

The Pentagon proposed joint "riverine training exercises" to Nigeria, of the type suggested by ExxonMobil, but "they had no interest," a Defense Department official recalled. "There was talk about large force deployments in the Delta," the Defense official continued, but "we certainly never encouraged or validated that. . . . The Delta could easily suck up 100,000 [American] troops and you'd still not have it covered." The war games and troop proposals were more "just throwing out ideas," not action plans for an American military intervention, the official said. The very existence and repetitive recurrence of the Pentagon's tabletop exercises, however, struck some State officials as a discouraging example of the militarization of American foreign policy in Africa.[39]

Loose talk about riverine exercises and marine training packages made Nigerian commanders very nervous. The launch of Africa Command as a formal enterprise in 2008 (during its initial operations, it had remained subordinate to European Command) happened to coincide with another wave of violence in the Delta. Nigerian foreign minister Ojo Maduekwe pointedly complained of a "lack of conceptual clarity" about the Pentagon's intentions in the Gulf of Guinea. Beset by questions and criticisms from African capitals, Theresa Whelan felt compelled to declare publicly, "We have no intention of using Africa Command to try and control oil resources."[40]

What, then, was the true connection between the Pentagon's program to build up the Nigerian and other regional navies in the Gulf of Guinea, and the reality that about 25 percent of America's imported oil flowed through those waters?

"It is a fact that the United States government does not own any oil companies," Air Force major general Michael Snodgrass, the deputy commander of Africa Command, said at the Africa command's headquarters in Stuttgart. "And if the United States decided to take over any country because of its oil, who would then exploit the oil? It's up to the free market to do that."

What, then, is the American military's message to ExxonMobil or

Chevron if they point out that they are suffering attacks in the Niger Delta and offshore?

"Our message is nothing, unless the president of the United States directs us to go do something like that," Snodgrass answered. "It is not the mission of this command to provide the security. And we have no intention of going into an African nation and helping an industry, whatever the industry may be—be it the fishing industry, the oil industry, the textile industry, the fake African elephant industry—be protected within the confines of a sovereign nation. That's not our role. . . . So my response to those companies is, 'You need to work out the arrangement with that sovereign nation to your satisfaction. And if you can't, you might want to reconsider your investment.' We are not the guarantor of their security."[41]

N igerian piracy presented the first major test of ExxonMobil's decision to adopt the Voluntary Principles governing corporate conduct in defense, security, and human rights. After ExxonMobil's fiasco in Aceh, Irving did not want to be placed in a position where security guards had to shoot at Nigerian pirates. By implementing the Voluntary Principles after 2005, ExxonMobil had effectively adopted a different approach to security: passive defense, enhanced by surveillance and partnership with local Nigerian forces, however flawed they might be. That meant, as a practical matter, given the weakness of the Nigerian navy and the reluctance of the United States to intervene directly, that the safety of ExxonMobil's offshore platform workers and managers depended increasingly on a defense strategy that seemed inspired by the 2002 Jodie Foster movie, *Panic Room*, in which a New York divorcée and her daughter lock themselves in a sealed room in their apartment as burglars assault them unremittingly.

In general, the Voluntary Principles regime discouraged the direct use by private corporations of offensive tactics or military technology, such as the deployment of military radar on offshore oil facilities, which might make the platforms appear to be legitimate military targets. ExxonMobil Global Security did deploy patrol boats with unarmed observers in the Gulf of Guinea; the security boats motored out in forward sweeps, seeking to detect, as early in an assault as possible, armed attackers who might be

en route to ExxonMobil facilities. If pirates were seen approaching, defense protocols kicked in: broadcasted alerts, lockdowns, the disablement
of equipment, and retreat into interior safe rooms. Intelligence collection,
threat mapping, and surveillance were about as far as ExxonMobil was
now willing to go. The corporation's security officers used Google Earth
satellite photography to create graphic maps overlaid with the locations
and dates of recent attacks, and the sites of militant camps in Akwa Ibom.
They worked human sources and tried to understand who was who inside
the camps.[42]

ExxonMobil security officers called Africa Command in Germany as
attacks unfolded on the open ocean. "They call me a lot," a military officer
there said, "mostly to give me situational awareness.'Most of the time
we're not in a position to respond—that's part of what I tell them. It's not
unlike what the Coast Guard tells lots of people in the maritime industry
in the United States. . . . 'You own the first two hours,'" meaning that
self-defense would be required for at least that long.[43]

The oil corporations were "wary of what they share and how much"
with the Nigerian navy, a Pentagon official said, because the navy was so
transparently part of the crime problem. Even when ExxonMobil summoned the Nigerian navy in desperation, while under attack, said a third
American official, "they won't come. They have no will." ExxonMobil
therefore had little choice but to "go into lockdown . . . hunker down and
hope for the best."

This was not a conventional portrait of the powers of the largest publicly traded corporation in the most powerful military nation in the world.
It was, nonetheless, the reality ExxonMobil employees stationed offshore
of West Africa endured in an age of uncontrolled piracy, massive corruption, and the covenants of corporate responsibility won by international
human rights groups.

Rex Tillerson might boast that ExxonMobil did business on its own
terms around the world and walked away when conditions were unacceptable, but ExxonMobil determined that it could not afford to abandon
its booked reserves in Nigeria, even as M.E.N.D.'s provocations deepened.
After 2006, ExxonMobil relocated many of its Nigerian managers to a
secure headquarters building on Victoria Island in Lagos, and reinforced

its passive defense systems offshore. Periodically, after kidnappings and speedboat raids of particular virulence, the corporation evaluated whether the Delta's violence had crossed a threshold that might argue for the corporation's total withdrawal. None of the ExxonMobil reviews reached such a radical conclusion, however. "Where are they going to go?" asked an American official who worked with the company's managers. "They don't want these reserves off their balance sheets. . . . They need the reserves."[44]

Twenty-two

"A Person Would Have to Eat More Than 3,400 Rubber Ducks"

T he rising dependency of the United States on oil imports from the Gulf of Guinea did not only result from Americans' unusually high per capita guzzling of gasoline and diesel. About a quarter of American oil imports were taken up for industrial uses. The manufacture of commercial chemicals from oil and natural gas "feed stocks" accounted for a large proportion of this industrial use.

ExxonMobil Chemical, headquartered in Houston, made up the third of the corporation's major divisions, alongside upstream (oil and gas exploration and production) and downstream (refining and fuels marketing—the gasoline station companies). In 2007, the chemical division accounted for about 10 percent of ExxonMobil's record $40.6 billion in profits. Upstream dwarfed chemical, and the latter's executives often labored in the shadows of their oil brethren. Yet if ExxonMobil Chemical had been a stand-alone corporation that year, it would have been among the fifty most profitable companies in the United States. In many of its business lines the division quietly held the first or second market position in the world.[1]

The early-twenty-first-century politics of global oil and gas production turned on security, nationalism, climate change, and taxation. The politics of the chemical industry were distinct. They drew ExxonMobil's management and lobbying teams into legislative and regulatory debates about how best to manage the risks to human health posed by the use of industrial, agricultural, and household chemicals. Ken Cohen, the public affairs chief, had forged his career as an attorney in ExxonMobil Chemical. The communications and political strategies he developed for the entire corporation after 2000 reflected, in part, the science-debating, my-study-versus-your-study ethos of chemical regulation in Washington and Brussels, the headquarters of the European Union.

Since the 1970s, ExxonMobil Chemical and its brethren in the industry's principal U.S. lobbying arm, the American Chemistry Council, had won more regulatory battles in America than in Europe. As with climate change, chemical industry lobbyists feared and fought the migration of European regulatory philosophies across the Atlantic. And, after 2000, the regulatory issues that mattered most to ExxonMobil increasingly drew the corporation into a war of ideas about risk.

Some scientists and scholars advocated a "risk analysis" model to evaluate the dangers of chemicals. This relied on mathematical calculations about the probability that a certain use of chemicals might hurt people. The chemical industry favored this approach because it effectively placed the burden of scientific proof on those who wanted to stop a chemical's sale. Even where scientists might establish some risk to human health, that measure of risk could then be weighed against the benefits of the product. ExxonMobil and other opponents of greenhouse gas regulation had used risk analysis frameworks to strengthen their arguments in opposition to the Kyoto Protocol's goals. The corporation's allies in academia demanded that environmentalists "prove" cause and effect in climate-change science beyond a reasonable doubt, and even then, after 2006, when ExxonMobil finally conceded that human activity might be contributing to global warming, the corporation still resisted specific restrictions on carbon use on the grounds that the economic costs outweighed the environmental benefits. This was the burden of proof demanded by some advocates of risk analysis: First, prove the harm; then, if the harm is established, prove that the cost-benefit equation of proposed regulation is well balanced. ExxonMobil applied the same lobbying strategy to proposed regulation of its manufactured chemicals.

After 2000, a new idea arrived from Europe to challenge the assumptions of the risk analysis school: the precautionary principle. The idea can be traced to West German environmental regulations enacted during the 1970s on the basis of *Vorsorge*, or "precaution." Advocates of the precautionary principle argued that in cases where damage to society or people might be severe and irreversible, preventive action should be taken up front, even if there were important uncertainties about the relevant science. Although "it sounds like common sense . . . in fact, the precautionary principle poses a radical challenge to business as usual in a modern, capitalist, technological civilization," author Michael Pollan has noted. Under the principles of risk analysis, industry lobbyists could often overcome objections by environmentalists or food safety advocates "until someone finds the smoking gun," Pollan continued. The precautionary principle reversed the burden of proof and sought to address the problem of traditional regulation, namely "that long before the science does come in, the harm has already been done. And once a technology has entered the marketplace, the burden of bringing in that science typically falls on the public rather than on the companies selling it."[2]

To combat climate regulation, ExxonMobil had hired a pair of in-house astrophysicists to present scientific analysis. Yet ExxonMobil was not in the business of meteorological science. It did not operate satellites or sensing stations to measure glaciers or sea ice. The corporation's capacity and credibility as a participant in scientific argument about global weather proved, therefore, to be finite—ExxonMobil was obviously self-interested in pressing its arguments, yet its claims to expertise were at best limited.

As the corporation fought the rise of the precautionary principle in chemical industry regulation, however, its position was more favorable. ExxonMobil employed scores of chemists; it was on the front lines. At ExxonMobil Biomedical Sciences in New Jersey, a research-driven division of the company, the corporation had constructed laboratories that could conduct rat studies about the health effects of commercial chemicals. The lobbyists the corporation flew in to Washington to work on ExxonMobil Chemical's regulatory issues—Laura Keller, a senior issues adviser on chemical regulation, and Leslie Hushka, another registered lobbyist—were scientists who published in peer-reviewed journals; behind them stood dozens of other ExxonMobil scientists as well.

The ExxonMobil scientist-lobbyists engaged not only in debates about specific research and regulation, but also attended academic and regulatory conferences that reviewed the competing, overarching philosophies of risk management. ExxonMobil joined other corporations in funding the Harvard Center for Risk Analysis at Harvard's School of Public Health; the center "focused broadly on developing risk, economic, and decision analysis methods that are well-grounded in the natural and social sciences."[3] ExxonMobil's lobbyists derived from Harvard's work insights for their own Washington arguments.

The corporation's chemical division lobbyists urged federal regulators to adopt a "Hazard Index" approach to regulatory evaluation because it was a "defined, transparent methodology" that could draw upon mathematical analysis, as one of the corporation's PowerPoint packages put it.

Paul Thacker, a congressional investigator for a Republican senator who looked into the corporate uses of science to shape law and regulation, concluded that "the whole field of risk analysis has been compromised by the companies. . . . Risk analysis is like an op-ed," that is, just another form of argument, not a reliable or objective science in and of itself, as its proponents often suggested.[4] Yet ExxonMobil's lobbyists often enjoyed a much better command of facts about proposed bills or regulations than the generalist, harried congressional aides who worked on legislation.

By early 2008, the regulatory battle in which ExxonMobil Chemical's lobbyists were most heavily engaged was an unusual one. It concerned the corporation's defense of rubber ducks.

Phthalates are a man-made group of chemicals that were first introduced during the 1920s. Their use spread after they were added to polyvinyl chloride, a popular plastic, to make the vinyl softer and more flexible.

Exxon Chemical's phthalates business grew during the 1970s. A market opportunity arose when another class of chemicals sometimes used to soften plastics, polychlorinated biphenyls, or PCBs, were banned by Congress because of evidence that they were toxic to humans.

As American and European households used more and more plastics, phthalates became commonplace. They could be found in flooring, electrical wire casing, garden hoses, car seats, medical tubes, tape, pool liners, shoes, and cosmetics. Of particular concern was the presence of some phthalates in children's toys. The plasticizer used most commonly in vinyl toys was called diisononyl phthalate, referred to as DINP. ExxonMobil Chemical described itself as one of the world's leading makers of plasticizer chemicals, and in that role it had become one of the world's leading manufacturers of DINP. ExxonMobil did not make toys, but it sold its softening chemicals to worldwide companies that made vinyl balls, ducks, dolls, and bendable superhero action figures. (The bathtub ducks at issue were referred to colloquially as "rubber ducks," but they were not actually made from rubber; they were made from vinyl softened with DINP.)[5]

Scientific knowledge about a particular chemical's potential to harm people can be derived from a number of sources. Scientists may be able to infer the likely toxic properties of a chemical on the basis of detailed studies of other, similar compounds. In addition or separately, laboratory tests on rats or other animals may provide insight into whether a chemical may be carcinogenic or otherwise harmful. Another form of study is to track the effects of actual human exposure to a chemical over a long period of time. Such studies offer the promise of high accuracy, but are by their nature slow and expensive.

In the case of DINP, by the early 2000s, a number of scientific and animal studies had been conducted about its possible impact on human health, but there were no long-term human studies. The Centers for Disease Control and Prevention had discovered one striking fact by randomly examining humans—about three out of four people tested had some phthalates in their systems. The chemicals had become so ubiquitous in household and consumer products that they had become, in effect, a part of the human ecosystem; if it turned out that they were unsafe, it would be a real concern.[6]

On DINP in particular, the findings of animal studies, including one rat study carried out by ExxonMobil scientists at their own laboratory in New Jersey, were not particularly alarming, but the results were in some respects ambiguous, and their implications were disputed. Male rats exposed to phthalates in utero later exhibited abnormalities in their reproductive organs; this led some researchers to conclude that phthalates could interfere with testosterone. Some researchers and nongovernmental health lobbyists, such as those at the Breast Cancer Fund, feared that DINP might interfere with the development of reproductive organs if very young children were exposed as their bodies developed.[7]

In 1998, public interest health groups filed a petition at the Consumer Product Safety Commission demanding that it ban DINP in children's toys and issue a national advisory about the threat to child health. The commission opened a review. Its staff scientists concluded initially that regulation of phthalates was "worthy of additional future consideration" because of the way DINP seemed to act on human development, but that "more studies are needed."[8] The commission next convened a study known in Washington-speak as a C.H.A.P. (Chronic Hazard Advisory Panel). The panel acknowledged concerns about the genetic or reproductive effects of massive DINP exposure, but concluded that the actual exposure children might experience was so low as to be of negligible risk. The full commission voted in 2002 to deny the petition to ban DINP from toys.

Throughout the commission's review, ExxonMobil scientists and lobbyists argued that DINP was safe enough to be used because the dangerous dosages seen in the rat studies were much, much higher than those that would realistically be encountered by children. The only way young children would be likely to absorb DINP through toys would be by mouthing, teething on, or swallowing the toys. (Baby bottle nipples and teething rings typically were made from rubber or latex, because those materials provided a more natural feel; they generally contained no phthalates.) To follow up on this issue, after the pro-DINP vote, Consumer Product Safety Commission scientists sponsored an observational study in which they watched babies and toddlers handle rubber ducks and other toys to see just how often they stuck the toys in their mouths; they reaffirmed the commission's earlier finding that even the most oral children would not be at risk.

"A person would have to eat more than 3,400 rubber ducks made with DINP over their lifetime to exceed safe DINP exposure limits," PowerPoint slides left behind on Capitol Hill by ExxonMobil lobbyists declared. "If you took water and saturated it with DINP, an infant would have to drink more than 41,500 gallons to exceed safe DINP exposure limits."[9]

Given the lack of definitive human studies, phthalate regulation presented a test case pitting those—such as the authors of ExxonMobil's PowerPoint slides—who favored traditional risk analysis philosophies against those who favored the precautionary principle. Of course, even advocates of the latter had to make judgments about how much risk from a particular chemical was severe and irreversible enough to demand costly government action. In the case of phthalates, the issue was complicated by the fact that some versions of the compounds—those with low molecular weight—clearly were dangerous to human health, whereas the evidence about DINP, which had a high molecular weight, was more favorable. It required a knowledgeable and careful regulator or congressperson to hold in mind the distinctions among different phthalates.

In Europe, the precautionary principle won out. Although a European Chemicals Bureau study found that "no risk reduction" was required for DINP, in 2005, the European Union nonetheless banned the compound from children's toys that could be placed in the mouth. "Politics, not science, is the reason," the ExxonMobil PowerPoint slides circulated in Washington complained. "Politics," however, was in fact a synonym for the rise of the precautionary principle as a popularly supported basis of chemical regulation in Europe—and there was little reason to believe that philosophy would remain sequestered there.[10]

American environmental and public interest health groups had discovered that it was easiest to import European regulations inspired by the precautionary principle into the United States by starting first in the legislatures of more liberal states—California and Vermont, for example. The City of San Francisco formally adopted the precautionary principle as a framework for local regulations in 2003. The city's environmental regulators "discovered Europe has banned chemicals that the U.S. had not," a city environmental regulator recalled. "We said, 'If other governments have taken precautionary actions, then we can take that action as well.'" San Francisco adopted a phthalate ban that "precisely mirrored" the European ban.[11]

Legislators in the state capital of Sacramento promptly introduced proposals for a statewide phthalate ban. The Breast Cancer Fund, the Public Interest Research Group, the Natural Resources Defense Council, and other public interest health lobbies jumped in. Scientists and advocates at these groups had followed the phthalate issue since their failed effort to win a DINP ban at the Consumer Product Safety Commission during the late Clinton administration. The phthalate issue offered "a way to highlight the broken chemicals policy system that we have—to get people to pay attention to it because it's toys, it's things babies are putting into their mouths," recalled Gretchen Lee Salter, a policy manager at the Breast Cancer Fund. California legislators, urged on by public interest advocates like Salter, enacted the European Union standard in 2007. Minnesota, Connecticut, and Vermont legislators moved to do the same.[12]

ExxonMobil became alarmed enough to start lobbying directly in state capitals. "The one lobbyist we did hear from" was from Exxon-Mobil, recalled Virginia Lyons, a biologist and legislator who sponsored Vermont's ban. ExxonMobil's local advocate "stirred up the fishing and hunting community by saying if phthalates were banned in the state of Vermont, then fishing would be affected" because lures were made of soft plastics. In fact, the Vermont bill did not address fishing lures at all, only toys. Lyons finally outflanked the ExxonMobil lobby by adding to the final version of the bill a line declaring that nothing in the new law "should be construed to regulate firearms . . . hunting or fishing equipment." In Hartford, Connecticut, chemical industry lobbyists walked through the statehouse with Gumby toys in their pockets. They would grab a lawmaker, pull out the toy, and declare ominously, "They're going to ban Gumby!"[13]

The wave of phthalate regulation at last rolled into Washington in

2008. Breast Cancer Fund lobbyists briefed U.S. senator Dianne Feinstein about the issue, noting the San Francisco ban already enacted and the similar bill passed by the California legislature, which awaited action by Governor Arnold Schwarzenegger. Senator Feinstein saw an opening to advance the cause in Congress.

She owed that opportunity to unscrupulous toy makers in China. In 2007, Mattel, Inc., had recalled 967,000 toys—an Elmo Tub Sub, a Dora the Explorer backpack, and Giggle Gabbers shaped like Sesame Street's Cookie Monster and Elmo, among them—because the toys contained lead paint. China increasingly was the source of the toys America's children played with. The Mattel recall was just one in a series of revelations about the shoddy quality of some Chinese toy manufacturing. China bashing and child safety being two of the less controversial subjects in American politics, federal lawmakers scrambled to introduce bills that would tighten standards for the toys that American children enjoyed. As a Senate version of this toy safety bill glided toward passage early in 2008, California's two senators, Feinstein and Barbara Boxer, slipped in a floor amendment that incorporated a version of the European Union's phthalate regulation—one that would prevent American children from putting DINP-laden vinyl ducks into their mouths. The bill passed easily. It is not clear whether ExxonMobil's Washington office even understood what had happened until it was too late.[14]

The U.S. House of Representatives, now controlled by Democrats, also passed a companion toy safety bill. (It became known as the Consumer Product Safety Improvement Act of 2008.) The House version, however, did not address the phthalates question at all. The matter would be decided in a conference committee organized to reconcile the two bills for final passage. President Bush was unlikely to veto legislation designed to keep American children safe from faulty Chinese toys, so by the spring of 2008, it seemed clear that a final law would indeed be written, passed by both houses of Congress, and signed by the president. The question that mattered to ExxonMobil Chemical was whether that bill would ban DINP from toys.

ExxonMobil and public health lobbyists had been battling one another inconclusively over phthalates for a decade. The corporation's scientists had prevailed in the technocratic, relatively closed forum of the Consumer Product Safety Commission. Congress was a different venue, one that favored the public interest groups. Anticipating an intense summer of lobbying and a public relations struggle, the advocates at the Washington, D.C., offices of the Public Interest Research Group reached deep into their bag of advocacy tricks. They manufactured a twenty-five-foot-high inflatable rubber duck; they then carted the giant duck to Capitol Hill, to call attention to their sign-waving street demonstrations in favor of a DINP ban. In addition to the big duck, ExxonMobil's interest in the legislation was identified early on as an element of the public interest group's lobbying strategy. Noted Liz Hitchcock, one of the campaigners, "They're the perfect villain."[15]

Joe Barton, the Republican congressman who represented Texas's Sixth District, which lay just to the south of Dallas, bore a resemblance to Rex Tillerson, both in appearance and by his biography. Barton had been born in Waco, Texas, in 1949, three years before Tillerson's birth in Wichita Falls, two hundred miles away. Now in his late fifties, Barton combed his full head of silver hair in a style similar to that favored by ExxonMobil's leader; his face was similarly fleshy but fit looking. Barton's political creed had been derived from the same rural ethos that had shaped Tillerson's rise—a synthesis of Christianity, small-town values, and faith in free markets. A sign in Joe Barton's Capitol Hill office reads "Trust God; Tell the truth; Make a profit."

Barton had studied industrial engineering at Texas A&M University on a scholarship, then earned a master's degree in industrial administration from Purdue University. He worked in business for a decade and served as a consultant on gas deregulation issues for Atlantic Richfield Company, the large oil and gas company eventually absorbed by BP. By 1984, Barton was "an earnest man of thirty-five who had a strong desire to go to Congress and negligible prospects of getting there," as Mark Halperin and John F. Harris later wrote in their book, *The Way to Win*. But Joe Barton then made an excellent decision: He hired an Austin political consultant named Karl Rove to handle his direct mail as he challenged a better-known and better-funded opponent, Max Hoyt, in the Texas Sixth's Republican primary. Rove crafted mailers touting Barton as a "Proven Cost Cutter" who was "Educated to Lead" and had "Roots in Texas." He won— and after Barton reached Washington, he stayed.[16]

Barton rose to chair the powerful House Energy and Commerce Committee, which had long been the domain of Democratic representative John Dingell, of Michigan. Barton and Dingell swapped gavels as Republicans and Democrats traded control of the House. Barton fashioned a reputation as the leading expert on oil and gas issues in the House Republican caucus—he was a passionate believer in markets, but occasionally, too, a deal maker who could find legislative compromises with Dingell, who fiercely protected in Congress the interests of automakers and other industrial corporations in Michigan.

Barton's relationship with ExxonMobil was strong. In the years after the Mobil merger, Barton received more money from the ExxonMobil Political Action Committee than any other member of Congress—a total of $46,399. In general, ExxonMobil's K Street crew appreciated the Texas congressman's pro-market philosophy. In 2005, however, Barton had been infuriated when Lee Raymond and Dan Nelson, the ExxonMobil Washington office chief, declined to support a deal Barton had proposed to Dingell and other lawmakers to pass comprehensive energy legislation of the sort originally contemplated by Vice President Cheney's energy task force, but which had proved politically elusive during the first Bush term. One element of the 2005 deal was a prospective agreement by Congress to absolve oil companies from certain legal liabilities arising from spills of gasoline containing MTBE. (The deal was proposed just a year before ExxonMobil's disastrous spill of MTBE-laden gasoline at its gas station in Jacksonville, Maryland.) In exchange for legal protection, ExxonMobil and other oil companies would contribute to a $4 billion fund to pay nationwide MTBE cleanup costs. Barton thought he was close to an agreement to ensure payments to the fund, but Chevron, ExxonMobil, and the American Petroleum Institute balked. On the decisive weekend of negotiations, ExxonMobil's representatives, led by Dan Nelson, declared that the bill was just too much of a Washington mess for the corporation to support. Lobbyists for the major oil companies later said they had warned Barton that they would not pay more than $1.5 billion into the proposed cleanup fund. In any event, Barton was unhappy. He pulled all provisions for MTBE protections from the final bill and later issued a public letter protesting Lee Raymond's retirement package: "While we respect the right of corporations in America to set compensation packages as they see fit, it is hard to understand how, in light of most Americans paying nearly $3.00 per gallon at the pump, your board of directors can justify such an exorbitant payout."[17]

In 2008, as the lobbying fight over DINP and rubber ducks loomed, ExxonMobil needed Joe Barton again. Whether the final toy safety bill contained a Europe-inspired phthalate ban was a question that would now fall to Dingell and Barton. In 2007, after Democrats retook majority control of the House of Representatives, Dingell had ascended again to chair the Energy and Commerce Committee; Barton had returned to the role of ranking member, the committee's senior Republican. Because of the toy legislation's subject matter, members of the House Energy and Commerce Committee would staff the conference to determine the bill's final compromises. In theory, Dingell could go his own way, without including Barton, and then try to rely on the Democratic majority in the House to approve the final bill. But once the conference opened in June 2008, Dingell insisted that Barton sign off on the phthalate issue, whatever the final compromise turned out to be. Dingell was an old-school legislator who believed in forging agreements that made nobody happy and preferred when possible to work across the aisle. Also, ExxonMobil had supported Dingell over the years with steady campaign contributions. The corporation lobbied him and Barton simultaneously on the phthalate question. In any event, it would be better for Dingell if he had Barton's political cover on any final decision that favored ExxonMobil.

The Senate bill's phthalate provisions had effectively been written by two California liberals, importing European regulations. Dingell told his staff as the talks began that he would go along with whatever compromise Barton endorsed—but they had to keep the Texas Republican on board or there would be no deal. Barton, for his part, announced through his staff that he was simply not prepared to accept a ban on DINP in children's toys in any form.

"You're going to have to roll me," Barton told Dingell at one bicameral meeting.

"My friend," Dingell replied, "I don't want to have to do that."[18]

The stalemate took hold as summer descended—it was the capital's humid swamp season. ExxonMobil's Washington lobbying crew moved through Capitol Hill, seeking out the conferees and their staff. One Democratic lawyer on the staff of an Energy and Commerce member recalled "six old white guys in gray suits" who came to her office to hand out PowerPoints about phthalates. The essence of their presentation was, "DINP is not dangerous," she recalled, and that "they didn't want DINP singled out." The ExxonMobil lobbyists carried props designed to win the attention of young Capitol Hill staffers: iPod earbuds, which were made of materials softened by DINP. The message was that "phthalates are not harmful," recalled Valerie Baron, an Energy and Commerce Committee staffer who heard the briefings, "*and* they're in so many things, what would we even do without them? What would you do without your iPod headphones?"

The Senate bill banned six phthalates outright, but "DINP was the one that was most in play" during the conference negotiations, recalled a staffer involved in trying to write a final agreement. ExxonMobil's strategy focused heavily on presenting the history of scientific studies and risk analysis of phthalates, to emphasize that the evidence about DINP was far from definitive.

The conference unfolded as a series of hours-long meetings among staff in various congressional committee rooms; the group alternated among committee rooms in the House office buildings, along Independence Avenue; the Senate office buildings to the north of the Capitol; and inside the Capitol itself. "The way we got started was literally putting the two bills—the House bill and the Senate bill—side by side on the table and trying to marry the two together, paragraph by paragraph, section number by section number," a participant recalled. On the phthalates paragraphs, however, "Barton's orders to his staff were clearly not to budge."[19]

The consumer groups hauled out their big duck. "We actually stood outside on the steps, lobbying the old-fashioned way," flanked by the twenty-five-foot inflatable, Liz Hitchcock recalled. "Our strategy was to keep saying, 'Will Congress listen to Exxon or America's kids?'. . . The more we could say that, the more we could keep talking about Exxon, the better."[20]

Some of the consumer advocates criticized Joe Barton publicly for having accepted ExxonMobil campaign contributions in the past. That angered Dingell. "He didn't like the conduct from some of the consumer groups," a staffer involved said, particularly the "effort to demonize the other side—that was unhelpful." Dingell had been on the receiving end of similar attacks in the past and hadn't appreciated them. Barton and his staff dug in deeper.

The conference convened a "stakeholders meeting" on phthalates in a congressional hearing room—a semiformal session where consumer advocates and representatives from ExxonMobil, the American Chemistry Council, and the Consumer Product Safety Commission could all make their arguments in front of the key staffers negotiating the final bill.

The ExxonMobil scientists who specialized in phthalate lobbying turned up to represent the corporation. The congressional staff sat in committee member chairs, like judges. The industry scientists and the consumer group scientists and advocates took their places in the audience— on opposite sides. The setting felt "like one of those *Saturday Night Live* point-counterpoint debates," Janet Nudelman of the Breast Cancer Fund recalled.

"Pretty much every developed nation in the world has banned phthalates from kids' toys," Nudelman told the meeting. "But Congress is still debating." It was time to act, she said.[21]

Somebody presented a Gumby doll as an exhibit. Shannon Weinberg, Joe Barton's lead staffer on the issue, remarked, according to a participant, "If I were a mother, I'd never let my kids play with a Gumby." It was not obvious what the legislative implications of her comment were. Vinyl ducks made with DINP and ducks made without DINP were placed side by side and fondled; some staffers felt they could not tell much difference between the two kinds, despite being told by ExxonMobil lobbyists that DINP-less toys might be so hard and inflexible that they could pose a choking hazard.

ExxonMobil's scientists argued again that the amount of DINP in children's toys was so negligible as to pose no realistic hazard. "Any time there was a discussion," the conference participant remembered, "it always went back to, 'You'd have to eat five hundred thousand rubber ducks to have an impact'" (an exaggeration of the 3,400 ducks in the written materials). "It just reverted back to the . . . ducks. That's when the chaos started."[22]

In the end, the stakeholders meeting allowed all sides to be heard, but it did not precipitate a compromise. The stalemate dragged into July. The overall toy bill was popular among members and senators in both parties—it would set new standards for testing toys, and it would improve the quality of imported toys, defending America's toddlers from unscrupulous Chinese factory managers. It would be frustrating, conference staff felt, if the broader law failed to pass only because of the phthalates lobbying stalemate.

Later, several competing versions arose about how the final compromise originated. In any event, it was Solomonic: Three of the more obviously dangerous phthalates, out of the six outlawed in the Senate version, would be banned outright. DINP would be banned from children's toys and teethers in the United States—at least for now. To salve ExxonMobil's wounds, however, the bill would order the Consumer Product Safety Commission to convene another Chronic Hazard Advisory Panel to examine health effects from the full range of products and consider the "cumulative effect of total exposure to all phthalates in children's products." If the C.H.A.P. reached conclusions about DINP similar to those of a decade earlier, ExxonMobil's position might ultimately prevail and the temporary ban on DINP use in toys would be lifted. For the time being, however, newly manufactured vinyl ducks and other toys that might be mouthed by American children would be free of DINP. What mattered most about this compromise proposal was that Dingell was enthusiastic. He and his staff pressed it upon Barton; after some hesitation, Barton accepted.

President Bush signed the Consumer Product Safety Improvement Act into law on August 14, 2008. Its final provisions on phthalates could not be described as a triumph for ExxonMobil—the consumer lobbyists had gotten more of what they wanted than the corporation. Yet Joe Barton had hung in there and had won significant concessions. ExxonMobil had a long record of persuading the Consumer Product Safety Commission to see phthalate regulation its way, and now the future of DINP manufacturing would be back before the commission, with ExxonMobil's lobbyists once again involved in a detailed review of phthalate science and risk management.

The corporation apparently decided that Joe Barton deserved to be rewarded for his summer of stubbornness. In the world of political campaign contributions, there is a technique referred to as "bundling," by which employees of the same company, law firm, or other organized group simultaneously make contributions to the same political candidate, to create a booster effect with their money injection. The tactic is legal if the employees act voluntarily, without coercion. Within two weeks after Bush signed the final toy bill, nineteen high-ranking ExxonMobil executives began to make campaign contributions to the Congressman Joe Barton Committee, according to the dates recorded for public filing by Barton's committee.

On August 25, 2008, Mark Albers, of the upstream division, donated $350; Walter Buchholtz, the longtime lobbyist for ExxonMobil Chemical, donated $350; William Colton, the leader of Irving's Strategic Planning exercises, gave $350; Michael Dolan, another vice president, gave $350; Donald Humphreys gave $750; Richard Kruger gave $350; Stephen Simon, the leader of ExxonMobil's downstream businesses, gave $350; Sherri Stuewer, ExxonMobil's leading executive on climate change and environmental issues, gave $350; Andrew Swiger gave $500; and Theodore Wojnar, a general manager at ExxonMobil Chemical, gave $350.

About a week later, on September 2, 2008, ExxonMobil chairman and chief executive Rex Tillerson led a second wave of giving to the Barton committee, with a donation of $1,500; Sara Tays, a public affairs executive, gave $350; Stephen Pryor, the president of ExxonMobil Chemical, gave $350; Henry Hubble, in charge of the corporation's Wall Street relations, gave $350; and Ken Cohen gave $500. Jeanne Mitchell, ExxonMobil's lobbyist in the House of Representatives, soon gave another $1,000. When the recorded contributions stopped on October 2, about six weeks after the toy safety bill's passage, Barton had received $10,150.[23]

The sum might not be grandiose, but it did suggest a message: *We take care of our friends.*

"ExxonMobil does not collect, report, monitor, or track individual employees' personal political contributions," Alan Jeffers, a spokesman for the corporation, said in a statement prepared in response to inquiries about these donations. "ExxonMobil does not bundle contributions or in any way illegally facilitate the making of federal campaign contributions. . . . At the time you reference, Rep. Barton had been chairman of the U.S. House Energy and Commerce Committee and later minority ranking member of the committee, which placed him at the center of many policy issues affecting U.S. business and industry. It is incorrect and misleading to allege that legal donations by individual ExxonMobil employees were in any way tied to a single vote on a single issue among the many that Rep. Barton would have been involved with at the time."

Public records collated by the Center for Responsive Politics suggest that ExxonMobil took an especially strong interest in Barton during 2008, however. Barton attracted donations during each two-year campaign cycle from many corporations, but in 2008 ExxonMobil was one of the congressman's top three campaign donors for the only time in a decade. Combining corporate, ExxonMobil Political Action Committee, and individual employee contributions, ExxonMobil gave Barton $38,298 that year, more than 40 percent more money than in any other cycle since 2000.

In a telephone interview, Jeffers said he believed a fund-raising event had been held for Barton around the time the 2008 donations by Exxon-Mobil employees were recorded in public filings, but the spokesman later declined to respond to questions about the event's date, organizers, or the timing of invitations, in relation to the consumer bill's passage. Barton's office did not respond to requests for comment.

Twenty-three

"We Must End the Age of Oil"

The ExxonMobil Corporation Political Action Committee invested $722,000 in candidates for federal political office during the 2008 election cycle. Despite obvious signs of a strong Democratic electoral wave building across the United States at the end of the Bush administration, the corporation's political spending remained staunchly Republican. Only 28 of the 207 recipients of ExxonMobil P.A.C. contributions during the 2008 cycle were Democrats; in dollar terms, ExxonMobil gave just 11 percent of its money to Democrats. The corporate P.A.C. gave more heavily to Republicans than did the company's employees, when they made donations as individuals. Political contributions between 2000 and 2008 by individuals who declared an affiliation with ExxonMobil on disclosure forms—including Tillerson, Cohen, and other senior executives—totaled $1.22 million. Most employee contributions went to Republicans, but as a whole, employees gave more than twice as much to Democrats, as a percentage of their total, than the corporate P.A.C. did.[1]

"We are a business-oriented P.A.C.," an executive involved in the political spending decisions said. "So we are looking for candidates . . . who are pro-business. . . . Now when you apply that litmus, our P.A.C. is rightly criticized that we tend to give more money to Republicans than to Democrats, but it is a result of the [key vote system] approach we take and not a desired result."[2]

Joe Barton received more contributions from ExxonMobil than any member of Congress after 2000. Anne Northup, a Republican member of the House of Representatives from Kentucky, received the second most. Each of ExxonMobil's top ten recipients was a House Republican. "Whoever's in power in the House has almost dictatorial power," a Washington consultant who worked on oil industry issues said. "If you control what's going on in the House, you have huge influence over the final product; power is more diffuse in the Senate."[3] Moreover, low-tax, free-market ideology was ExxonMobil's "North Star," as a former executive involved put it, and the House Republican caucus increasingly offered the staunchest philosophical allies.[4] In the Senate, Ken Cohen told his colleagues in the K Street office, "it's all about the sixtieth vote," that is, making sure that ExxonMobil could block unfavorable proposed legislation by encouraging loyalists to invoke the peculiar Senate institution of the filibuster.

From an office in Arlington, Virginia, the ExxonMobil Citizen Action Team mobilized employees and retirees to reinforce ExxonMobil's lobbying positions and favored candidates. The team maintained a toll-free information line, playing on the corporation's stock exchange ticker symbol: 1-866-VOTE-XOM. It provided updates on the corporation's legislative priorities on a Web site and sent out glossy newsletters about pending legislation and upcoming campaigns to employees, retirees, and sympathizers. "Electing people who will pursue policies that are good for our industry and make sense for our families is an important responsibility," one of the mailings enjoined.[5]

As the 2008 presidential campaign began, however, the greatest risk that American politics posed to ExxonMobil arose from something the Citizen Action Team had trouble addressing: the corporation's general unpopularity and its attractiveness as a piñata for populists, left and right. ExxonMobil had never really broken free from the reputation for ruthlessness and self-interest Standard Oil had forged during the Gilded Age. As the Bush presidency entered its final year, retail gasoline prices rose toward four dollars a gallon. Quarter after quarter, higher global prices caused by Nigerian unrest and runaway Asian economic growth delivered jaw-dropping record profits to ExxonMobil. Its executives could hardly turn this cash flow off, but they had difficulty justifying it publicly in commonsense language. The problem was structural: Popular anger as pump prices spiked reflected a form of economic powerlessness among commuters and small business owners. Gasoline had become a necessity in the United States, like electricity, but its fluctuating price remained unregulated, in comparison with electric rates, which could be managed up and down more gradually, when underlying commodity prices changed. Moreover, when gasoline prices shot up very suddenly and without forewarning, the largest private corporation in the country made even more money, to add to its historically high profits. An American commuter or truck owner did not require socialist leanings to find this aggravating. "And how about this?" Jay Leno asked on *The Tonight Show*. "ExxonMobil today reported a profit last quarter—not last year, last quarter, one quarter—of $12 billion! . . . But in their defense, that money didn't come from the gas. That's just from the mini marts."[6]

O f all the candidates for president during the 2008 campaign, Barack Obama spoke most often and most pointedly about ExxonMobil. In debates and stump speeches, he offered none of the nuanced support for the sanctity of international oil contracts that he had voiced in private to Chad's dictator Idriss Déby while visiting N'djamena two years before. Obama and his speechwriters exploited ExxonMobil's unpopularity. They called out its profits and contrasted its wealth with the struggles of American working and middle-class families coping with long commutes and soaring gas prices. Obama's spin specialists sought to link their Republican opponent, Senator John McCain, to the pro-oil policies of Dick Cheney. Politically, this was an attractive target of opportunity: In the summer of 2008, 72 percent of Americans surveyed had a negative view of Cheney, and only 18 percent saw the vice president positively. "President Bush, he has an energy policy," Obama declared. "He turned to Dick Cheney and he said, 'Cheney, go take care of this.'. . . McCain has taken a page out of the Cheney playbook."[7]

That was campaign rhetoric; off the trail, Obama's evolving views about energy and climate policy were subtler. He had only begun to immerse himself as a senator. It took time to grasp the nuances of automobile mileage standards, international oil supply, clean coal technologies, balance-of-payments issues, and the myriad ways in which different forms of energy contributed to global warming and how the existing patterns might be altered by policy intervention or technological innovation. Obama's primary campaign involved a make-or-break commitment to Iowa, where ethanol subsidies were politically untouchable, so he embraced those with a particular emphasis on local ownership and populism. He embraced "energy independence" even though he knew that it was not achievable in a literal sense and perhaps not desirable, either. But the idea sold as nationalism, especially with independent voters in swing states. "There hasn't been a campaign in thirty years where the policy wonks don't cringe when 'energy independence' comes up, but the speechwriters get the last word," said Jason Grumet, one of Obama's energy policy advisers. Obama embraced "the smart version," Grumet said, "in a more metaphorical than a physical sense. . . . He knows the last barrel we use will be from Qatar or wherever it's cheapest . . . [but] he painted a bigger arc around the issue, that America can be great again, that it can renew itself."[8]

In June, as McCain emerged as the presumptive Republican nominee and Obama's grinding campaign against Senator Hillary Clinton neared a finish, McCain proposed a gas tax holiday to provide temporary relief to commuters burdened by soaring pump prices. The holiday would, in fact, help middle-class households hurting from unexpected gasoline expenses over which they had no control, but it had a chicken-in-every-pot opportunism about it and did nothing to address the glaring gaps in American energy and climate policies. Clinton embraced McCain's idea, but Obama announced, in effect, "This is stupid; the American people are smarter than this." Against the advice of his campaign pollsters, he used the issue to differentiate himself from McCain and Clinton, painting them as pandering politicians, and he came out fine.

The easiest case to make was to pound on Big Oil. To many independent voters and disillusioned Republicans, Obama's strategists knew, the symbolic ExxonMobil-Cheney complex pointed toward all that had gone wrong in the Bush years—the Iraq War, the rise in American economic inequality and economic insecurity at home, an economy that seemed grotesquely based on greed and was beginning to teeter, and a corrupt culture in Washington that reinforced these failings by favoring special interests. Obama therefore salted his campaign speeches with Exxon-Mobil references even when they were gratuitous—the very word "ExxonMobil" resonated in his favor. "It's not going to be easy to have a sensible energy policy in this country. ExxonMobil made $11 billion last quarter. They're not going to give up those profits easily," Obama said at an early primary debate. As the Democratic race narrowed to a campaign of attrition between Obama and Hillary Clinton, the two candidates competed in public about who opposed ExxonMobil more ardently. Then, after Obama's nomination became secure, he turned the same line of attack on McCain. When McCain announced a plan to reform corporate taxes, Obama's researchers figured out how much of the benefit would flow to ExxonMobil and immediately made that the focus of their criticism: "Think about that," Obama said at a rally in North Carolina. "At a time when we're fighting two wars, when millions of Americans can't afford their medical bills or their tuition bills, when we're paying more than four dollars a gallon for gas, the man who rails against government spending wants to spend $1.2 billion on a tax break for ExxonMobil. That isn't just irresponsible. It's outrageous!"[9]

The Bush administration and the McCain campaign studied polls that showed many Americans favored new offshore oil drilling as a strategy to reduce high gasoline prices. They announced coordinated plans to promote more domestic exploration in ocean waters: "Drill, Baby, Drill!" became a tongue-in-cheek-sounding chant at Republican rallies that summer. Obama promptly linked McCain's offshore plan to ExxonMobil's $11.68 billion in quarterly profits and denounced it as merely an "oil-company wish list." The Democratic nominee also embraced a plan to impose windfall taxes on oil corporations when global oil prices were about $80 per barrel, as they had been throughout 2008. Obama also declared again and again that if he were elected, he would push with new vigor, working with Democratic majorities in both houses of Congress to cap greenhouse gas emissions and to invest heavily in solar, wind, and biofuels. Privately, he recognized that the American oil industry "is of economic significance, and that greater domestic drilling reduces imports and the balance of payments problem," an adviser on energy issues said. But Obama wanted a comprehensive package that combined increased oil and gas production with progress on carbon pricing: "The idea was an 'all of the above' strategy that advanced both energy security and environmental goals." More than his profit bashing, Obama's climate and alternative energy policies— the ones he might actually be able to push through Congress as president— galvanized the attention of ExxonMobil's public affairs executives and K Street lobbyists. "We must end the age of oil in our time," Obama declared.[10]

"We felt like a candidate," one of the corporation's executives recalled. "Both parties were mentioning us by name. . . . So we were a candidate and we clearly knew that we were not electable." A Democratic congressman, Maurice Hinchey, even ventured that ExxonMobil's profits had "crossed the moral threshold of what is acceptable."[11] That sort of language—and the fervor and excitement that seemed to be coalescing around Obama's candidacy—suggested something more threatening to ExxonMobil than the usual cycle of congressional hearings and antitrust inquiries whenever gasoline prices spiked and American drivers howled. Trustbusters had broken up Standard Oil on moral as much as economic grounds.

ExxonMobil had no cause to fear anything so drastic. Yet the costs and taxes Democrats in Washington might impose on a corporation as unpopular as ExxonMobil in an emotional period of high gasoline prices and huge corporate profits could not be easily forecasted. In American history, a number of industries had been regulated and taxed during periods of popular revolt against corporate power on the grounds that the targeted businesses were "public callings" and served a public interest function: banking, telecommunications, transportation, electric energy, and natural gas transport, among them.

In mid-July, global oil prices crossed above $140 a barrel. If such prices persisted, Democrats were almost certain to seek relief for disadvantaged households trapped by long commutes to work and stagnant incomes. Tillerson and his corporate planning group could not figure out what was causing the 2008 price spike. The detailed internal analysis of global supply and demand produced by ExxonMobil economists during early 2008 suggested that global prices were moving much higher than market fundamentals would predict. Was hot speculative money pouring into commodity funds a factor? Were traders assessing political risk in West Africa or the Middle East as a more profound forward-looking bottleneck on supply than ExxonMobil did? Tillerson felt genuinely befuddled. On the one hand, the corporation's typical "one right answer" analysis suggested that some sort of bubble had inflated and that oil prices would eventually fall, relieving some of the reputational pressure faced by Big Oil that summer. On the other hand, ExxonMobil had forsworn price forecasts precisely because the corporation had learned over many years that oil prices could not be accurately predicted on the basis of the empirical methods employed by ExxonMobil's economists and planners. The truth about future prices was anybody's guess—in the meantime, the corporation's reputation was being pounded.

This was the kind of challenge for which Rex Tillerson had been chosen to lead ExxonMobil—a crisis that required more successful public communication than Lee Raymond had been able to deliver. Cohen approved the recommendation of his media specialists that Tillerson get out and speak more, to place ExxonMobil's profits "into context." Tillerson agreed. He invited a few members of the national press to visit him in Texas.

The 2008 presidential campaign, Jad Mouawad of the *New York Times* pointed out when he arrived to meet the chief executive, "has centered around expanding domestic drilling. But many say this will do nothing to reduce prices now because it takes ten years before any new production comes online."

"If you use that logic," Tillerson replied, "then we should not have any of the barrels that are available today. All of today's supplies were developed years ago. It is nonsensical for people to make that argument. It reflects the ongoing difficulty we have with people who don't understand the nature of the energy system."

"Sure, but the argument is that we should focus on the demand side of the equation and that we cannot drill out of the problem."

"Well, you can't conserve yourself out of this problem, either. You can't replace your fuels with alternatives out of this problem, either. The reason the United States has never had an energy policy is because an energy policy needs to be left alone for fifteen to twenty years to take effect. But our policymakers want a two-year energy policy to fit with the election cycle because that is what people want. The answer is you can't fix it right now."[12]

After that condescending and vaguely antidemocratic debut (energy policy had failed because it was "what people want"), Tillerson next sat for an on-camera interview with ABC's Charles Gibson. "Do you understand, and can you appreciate from your position, with the escalation of the price of a gallon of gas, why people are fed up, angry, indeed, disgusted with oil companies?"

"Well, I can understand why people are very upset and why they're very worried and concerned about their ability to deal with these high prices. In terms of where they should direct their anger, I don't think it's useful for me to comment on that; although it does bother me that much of that is directed at us. . . ."

"We in the media have made a lot of the profits that ExxonMobil has made, particularly in the last couple of quarters—more than $10 billion in profits first quarter this year; $11.68 billion in the second quarter of the year. When people, I don't know, complain about that to you, or say, 'How dare you? Those profits are obscene.' What's your best—in brief form—what's your best justification?"

"Well, I think it has to do with the ability to understand just the size of our business. Everything we do, the numbers are very large. I saw someone characterize our profits the other day in terms of $1,400 in profit per second. Well, they also need to understand we paid $4,000 a second in taxes, and we spent $15,000 a second in cost. We spend $1 billion a day just running our business. So this is a business where large numbers are just characteristic of it. . . ."

"John McCain, for his part—it's become a mantra for him: Drill now, drill here, drill immediately. Is that any kind of a solution?"

"Well, it's part of a solution. Again, I think this whole debate around someone looking for the solution is not a sensible approach. . . . We really should be developing all the supplies that are available to us, regardless of whether they come on now or whether they come on ten years from now. . . ."

"Senator Obama, he's calling for a windfall profits tax. . . . And when the public sees the kind of profits that the oil companies are making, and ExxonMobil in particular . . . isn't it fair that they wonder, 'Why not?'"

"Well, I guess the question is what's that going to solve? Are the American people going to be better off from an energy situation because we implement a windfall profits tax? Nowhere in a windfall profits tax do I see anything that addresses the problem. I understand that may be popular with some people because of how they view our current-day profitability. Certainly, again, if you put our profits in perspective, because of our scale and size, Charlie, on a unit-of-sales basis, our profits are way down the list. And so if we're going to institute, from a philosophic stand-point, a windfall profit [tax] on highly successful companies, who generate high profits, you're going to have to go after a lot of other industries and parts of our economy. . . ."[13]

That summer, Tillerson seemed to be channeling Lee Raymond's tonal scales. He struggled to find clarity. Televised interrogation about sky-high gas prices favored the inquisitor, and neither Tillerson nor his advisers had figured out quite how to translate their strategic intention—to create an ExxonMobil that was more accessible and more flexible about controversial public policy—into language that would stick or be trusted. The corporation's gigantic profits spoke amply for themselves. ExxonMobil's net profits, when expressed as a percentage of corporate revenue, were, in fact, not particularly great; it was the arithmetic of the corporation's global size that produced gargantuan raw numbers. ExxonMobil's effective rates of corporate tax paid also fell on the high side, in comparison with other large American-headquartered multinationals.[14] The problem was that Tillerson's reliance on these talking points missed the country's mood in 2008 by a wide margin: The American people were rapidly losing faith in the integrity of many large corporations, Wall Street banks, and their allies in the Republican Party. They were in debt, falling behind, and fed up.

Obama and his advisers had the surer sense of the corporation's political meaning in 2008. "We're talking about Joe the Plumber," McCain said during his third and final debate with Obama, on the eve of the presidential vote. McCain's remark came during a discussion of the tax issue.

Obama flipped over one of his mental file cards and said, before a television audience made of about a third of all American households: "In order to give additional tax cuts to Joe the Plumber . . . ExxonMobil, which made $12 billion, record profits, over the last several quarters, they can afford to pay a little more, so that ordinary families who are hurting out there . . . They need a break."[15]

In October, Cohen summoned the Democratic corporate responsibility specialist, Bennett Freeman, to another off-site summit meeting with ExxonMobil public affairs executives. Before a room of about a hundred of the corporation's managers, media, and political specialists, Freeman praised ExxonMobil for implementing the Voluntary Principles in its corporate security operations, and also for joining another voluntary regime, the Extractive Industries Transparency Initiative, which was designed to increase the visibility of oil corporations' payments to poor, dysfunctional governments in Africa and elsewhere. On the other hand, Freeman continued, the corporation needed to become more visible on human rights issues in Nigeria and Equatorial Guinea. "It's not your job to be Amnesty International," he said, but they had to advocate more clearly at the State Department in favor of human rights policies amid dictatorships where ExxonMobil pumped oil.

Then Freeman turned to climate change. As he spoke, Ken Cohen and other executives took notes.

"Look, with Lee Raymond, you were in the late nineteenth century,"

Freeman said. "With Tillerson, you've come a long way, but you're still just in the late twentieth century. This is the twenty-first century, and on climate change you need to change your tone and change your substance. You need to get behind cap and trade—or somehow advocate for setting and meeting carbon reduction goals. On alternative energy—whatever technological capability you have, whatever is most viable, you have to get on it and do it. This is a carbon-constrained future you're looking at. Whoever wins the election—probably Obama, but even if it's McCain—there will be a new landscape on climate change. You need to get into the mainstream on this. You need to send Tillerson to Washington, have him give a speech at the National Press Club. Do it in January. Don't do it now—it will get buried in election news. But you need to have Tillerson say, 'Exxon recognizes the reality of climate change.' And you need him to take a clear policy position, in recognition of that reality."[16]

The 2008 election results vindicated ExxonMobil's Washington strategy in at least one respect. A great Democratic wave had swept Obama to the White House. Democrats now controlled both houses of Congress. Yet in the Senate, members historically aligned with the oil industry's positions on taxation and climate, many of them longtime recipients of Exxon-Mobil's financial support, remained numerous enough to block unfavorable legislation. Obama's windfall tax proposal, in particular, looked dead on arrival—in reality, it had been mainly a tool of campaign rhetoric. On climate, however, there would now be new momentum for cap-and-trade laws that might raise the price of carbon use in the American energy economy and reduce greenhouse gas emissions. Bennett Freeman's analysis in October rang true, even if Tillerson and Cohen remained uncertain about whether to follow his advice.

They had been reviewing ExxonMobil's policy options. Obama clearly would now move a major climate bill, and it would have a fighting chance of passage. Did ExxonMobil want to be on the outside looking in during 2009, undertaking its usual campaign of opposition, in collaboration with the American Petroleum Institute and others? Or would the corporation be better off endorsing a carbon price at last, as Freeman argued, in part to have more credible access to whatever legislative negotiations ensued at Obama's instigation?

Dan Nelson and other public affairs executives who had fought along-side Lee Raymond against climate bills for so many years told their colleagues that fall of 2008 that they feared Rex Tillerson might be going weak in the knees. Nelson had worked with John Dingell to encourage the lawmaker to come out in favor of a revenue-neutral carbon tax, with few loopholes. But Dingell and Nelson both knew as the financial crisis descended late that year—the economy contracted more than 8 percent in a single quarter—that such a bill could never pass. Politicians in the Rust Belt would never vote for it while autoworkers faced massive layoffs and their employers stared down bankruptcy. If ExxonMobil changed its position on climate now, it would not actually win any new political friends in Washington, Nelson argued, but the corporation *would* anger and betray its allies on Capitol Hill and in the oil and coal industries. Such a reversal would be the lobbying equivalent of Rex Tillerson's summer media tour—not effective enough to change public or Democratic legislative opinion, but a source of unwelcome attention for an unpopular corporation.

A policy shift on carbon would also repudiate Raymond's legacy on a visible issue where he had stood firm against conventional wisdom under great pressure, as Raymond's friends and allies saw it. Why would Rex Tillerson consider such a public betrayal of Lee Raymond, particularly if it was not likely to bring ExxonMobil any real political or bottom-line benefits? Was Tillerson so anxious to be respected and accepted by ascendant Democrats (perhaps temporarily ascendant) that he would undertake such a significant change in corporate policy, one that directly repudiated his former mentor? *ExxonMobil succeeds because it does not compromise its core principles and hangs tough even when it is unfashionable:* This was the tone of the Raymond camp's argument inside the corporation that late autumn.

On the other side stood those—some independent members of the board of directors, younger scientists, and Obama voters at all levels of the corporation, who shared the country's sense of excitement and anticipation about the new president's election—who felt just as strongly that

Bennett Freeman was right, that the 2008 election signaled that it was time for ExxonMobil to modernize its political reputation and to break with its past intransigence on climate science. By moving early, the corporation could seek to prove, by accepting cap and trade or a carbon price at least in principle, that its new leadership recognized the seriousness of the risks of climate change, that ExxonMobil's communication strategy under Tillerson was not mere lip service.

Tillerson said later that he struggled over the question and went back and forth in his thinking as late as the Christmas break. He made one firm decision during the postelection period: He decided to clear out Dan Nelson from the Washington office and to replace him with Theresa Fariello, the registered Democrat who had served in the Department of Energy during the second Clinton term. From Irving after 2001, Fariello had served as Cohen's principal liaison to Democratic lobbyists such as Senator John Kerry's former chief of staff, David Leiter, and as a general source of Democratic-leaning political intelligence. By choosing Fariello, Tillerson made it clear that ExxonMobil would adapt to the Obama era, not fight it from the trenches. How far Tillerson was prepared to move was not clear. A change in communication and tone was one thing; a change in policy advocacy about carbon pricing would be something else altogether. Tillerson at least wanted a Washington office that would not actively resist him if he decided to move decisively on carbon. Tillerson offered Nelson the lead role in ExxonMobil's effort to win access to Iraqi oil fields, but Nelson decided to retire. Tillerson approved a lucrative retirement package; the former marine saluted and departed quietly. He remained loyal to Lee Raymond and told friends and colleagues privately that he feared Raymond's achievements and principles at the corporation might be at risk.[17]

Twenty-four

"Are We Out? Or In?"

A nton Smith, who was the chargé d'affaires at the United States embassy in Malabo, Equatorial Guinea, when Barack Obama was elected
president, possessed a streak of independence that sometimes made
it difficult for him to accept the conformity required by government service. He was a tall, lean, athletic man in his early forties, with green-gray
eyes and a flattened nose that looked as if it might have been broken in
a fight or a scuba-diving accident, possibilities that would not have surprised his friends. Smith had grown up in Arkansas, where he earned a
bachelor's degree in English at Henderson State University. Later, he
earned a master's degree at Georgetown University in Washington, D.C.,
as well as a second graduate degree at the U.S. Army War College, and he
joined the foreign service. His political views were difficult to categorize,
but he tended toward libertarianism and spoke favorably with friends
about Representative Ron Paul, a member of Congress who did not often
attract even glancing admiration from American diplomats. Smith moved
through diplomatic postings in war zones such as Iraq and the Balkans.
He also served for a year as a fellow on the staff of Senator Richard Shelby,
a conservative Republican from Alabama who held an influential position

on the U.S. Senate Committee on Banking. Although Smith had never served in Africa, his exposure to economic matters helped to qualify him for a posting in Equatorial Guinea. Malabo job openings, in any event, did not attract the swarm of internal State Department applications typical of, say, postings at the Barcelona consulate.

In 2007, Smith arrived as deputy chief of mission at the two-story rented concrete house that served as the U.S. embassy in Malabo. The ambassador, Donald C. Johnson, departed his posting before its scheduled end, leaving Anton Smith as chargé d'affaires, a designation that gave him the role of ambassador without its salary or full rank.

Equatorial Guinea remained a troubled country, but it was no longer the isolated, poor, malarial place that had seemed to induce occasional bouts of madness in previous generations of international diplomats. Modern hotels, housing complexes, hospitals, and freshly painted government compounds had sprung up in the tropical forests; the island capital and the mainland coast resonated with the mechanical roar of construction equipment. Internet connections were slow and balky, but they nonetheless brought Equato-Guineans into contact with worlds of information previously beyond reach. Oil-funded scholarships allowed more and more young people to study abroad. Many of the recipients were handpicked by the regime, but Western education endowed them with new ideas. When they returned to Malabo in polo shirts and baseball caps, they formed businesses or took up positions of responsibility in government ministries. The country remained anomalous in a number of respects—its government's failure under Teodoro Obiang, even after the resolution of the Riggs Bank fiasco of 2004, to adhere to international banking rules meant, for example, that Equatorial Guinea had no access to credit card facilities and therefore almost all local commerce arising from the oil boom had to be conducted in cash. Organized political opposition to Obiang remained virtually nonexistent, and the president periodically tried, jailed, and executed real and imagined conspirators against him.

Anton Smith took to Equatorial Guinea with a passion. Its remoteness and its lush landscapes spoke to his sense of adventure. On weekends he dove into the Atlantic to scuba dive or snorkel or swim through the warm saltwater lagoons. He made friends with an international cocoa farmer who spent weekends on a colonial-era seaside estate. Smith would turn up on some Sunday afternoons to help cook up paellas for eclectic bands of relaxing expatriates and Equato-Guineans. He spoke enthusiastically about introducing Equatorial Guinea to the international sport of bungee jumping. He met regularly with the country representatives of the American oil companies whose operations enriched Obiang's regime—ExxonMobil, Marathon, and Hess—and he tried to support their investments and policy priorities within the country and back in Washington. He developed friendships across the local community as well. Smith's long marriage was ending when he arrived in Malabo. He fell in love with an Equato-Guinean woman; she became pregnant, and they moved in together. (They later married.) Smith could occasionally speak about Africa in ways that struck some of his friends as misguided, but his intimate connection to a local family deepened his knowledge of the country he was professionally assigned to understand. Smith was self-conscious about the possibility that he was "going native," as diplomats refer to the tendency of envoys sent abroad to identify with their host countries, potentially at the expense of clarity about American interests. Yet if Anton Smith did seem, as the months passed, increasingly to resemble a character in a Graham Greene novel, this did not come at the expense of his professional devotion. He brought the same restless energy and willingness to defy convention to his role as America's principal liaison to the oil-endowed government of Equatorial Guinea as he did to the rest of his life. And the more time Smith spent in Malabo, the more he believed that American policy—its one-dimensional focus on past and present human rights abuses, its unwillingness to respond seriously to Equatorial Guinea's security needs, its reluctance to engage fully with Obiang's regime—was misguided, hypocritical, and self-defeating.[1]

Smith pointed out in sometimes-heated exchanges with colleagues at State Department headquarters that since the Second World War, during its search for oil security, the United States had entered into deep alliances with Saudi Arabia, Kuwait, and the United Arab Emirates, among other Middle Eastern oil producers. All were authoritarian states with dismal human rights records, particularly in the realms of free speech and assembly. Yet a diverse number of American presidents continually sold these regimes jets, tanks, and missiles so that they could protect their oil inheritance in an unruly neighborhood, and by doing so, assure supplies would be available to the United States. American military forces intervened directly to liberate Kuwait after Iraq's 1990 invasion, and the U.S. military provided an ongoing de facto defense of Saudi Arabia's oil fields. These geopolitical bargains had endured despite evidence that the Saudi government tolerated financial flows to violent anti-American Islamist radicals.

How did the case of Equatorial Guinea measure up by comparison? Rather well, Smith argued. Its government pledged full cooperation with American foreign policy. Its oil flowed a shorter distance across safer seas to American refineries. Its political economy was unattractive and its human rights record was poor, true, but at least Equatorial Guinea had the excuse that it had enjoyed a basis for economic modernization for only a decade—Saudi Arabia had been rolling in oil revenue since the 1960s, and it still had not liberalized its politics. Moreover, and perhaps most important, in Smith's judgment, the situation in Equatorial Guinea was improving. Obiang had made heavy investments in the social sector, particularly in housing. He had accepted international police training on human rights.

Yet the State Department seemed congenitally unwilling to acknowledge these incremental changes and the positive direction that they might suggest. In the Middle East, generations of diplomats had operated on the assumption that it was necessary to accept the limitations of Persian Gulf political economies in order to fuel America's economy with reliable, relatively cheap oil supplies. In Africa, by contrast, generations of diplomats had operated on the assumption that all that mattered were the continent's large, often intractable problems, such as disease and low-grade civil wars. When it came to setting policy priorities for Equatorial Guinea during interagency meetings at the National Security Council in Washington, human rights issues figured more heavily than they did in the cases of America's longtime Gulf allies. Saudi Arabia's oil production dwarfed Equatorial Guinea's, and the dangers of losing even unreliable allies such as Riyadh's royal family in a region where Iran's radical revolutionary government continued to foment upheaval might justify the different approach Washington took in the Middle East. But the policy differences were not subjected to careful scrutiny—the assumptions undergirding America's regional foreign and defense policies had become deeply embedded.

The United States had not formally imposed economic or military sanctions on Equatorial Guinea. But America's policies of restricted military aid and sales, and its constant harping on human rights and elections, amounted to "crypto-sanctions," as Smith called them in his private arguments with State colleagues. These constraints limited American influence in Malabo and threatened to drive Obiang into alliance with China, France, India, and other less squeamish oil-importing nations. This was the case even though the country produced 450,000 barrels of oil per day, and American oil companies, including ExxonMobil, had invested about $13 billion that would be at risk if Obiang fell from power or stopped trying to curry favor in Washington.[2]

During 2008, Smith struggled to persuade his superiors in Washington of his viewpoints. In May, Equatorial Guinea staged parliamentary elections in which Obiang's ruling party won ninety-nine out of one hundred seats. International observers raised doubts about the credibility of the polls, particularly given that the country's opposition leaders had been harassed, jailed, and forced into exile for years. Smith believed, citing his firsthand observations, that the vote itself should be certified as "free and fair." He failed to persuade Washington. Then, in November, Manfred Nowak, the United Nations special rapporteur on torture, visited Equatorial Guinea on an invitation from the government. Obiang's willingness to issue such an invitation struck Smith as a sign of progress. Nowak, however, inspected police stations and prisons, found evidence of recent abuses, and declared publicly that the country continued to use torture systematically against prisoners who refused to make coerced confessions. "Torture Is Rife in Equatorial Guinea's Prisons" was the headline on a U.N. news release about Nowak's inspection.[3] Smith criticized Nowak as a grandstander with preconceived ideas and argued that Obiang should be given some credit for openness. Not long afterward, Obiang was reelected to yet another term as president with 95 percent of the votes cast. Smith's superiors at Foggy Bottom reprimanded him during his annual review for failing to take policy direction from Washington.[4]

Smith decided that Barack Obama's inauguration and the changes of political appointees that accompany any new administration offered an opportunity to lay down fresh, reasoned arguments to reexamine American policy toward Equatorial Guinea. On February 27, 2009, he filed the first in a series of six analytical cables to Washington. Smith's series offered a comprehensive review of Equatorial Guinea's relations with the former African colonial powers of France and Spain; its recent business deals with oil-thirsty China; its internal clan politics; its struggles with corruption and government capacity; its security challenges; and the policy choices facing the Obama administration. Smith marked the cables Sensitive but Unclassified—as a practical matter there was little choice but to send diplomatic transmissions from Malabo through unclassified channels, as the embassy lacked cryptographic equipment. "Are We Out? Or In?" the chargé d'affaires asked in his first filing.

"Since at least Forsyth's 'The Dogs of War,' E.G. has been a favorite takeover target for both outside and inside plotters," he wrote. "President Obiang came to power himself in a coup likely assisted from the outside. He and his team know how it works. Though without official declaration, the country persistently operates under martial law-like conditions. This posture generates human rights concerns as documents are checked, guns are displayed, and foreigners get the fish eye." Still, Smith concluded, American policymakers would be better off if they recognized that Equatorial Guinea "is less a rogue state than it is a rudimentary one."

He argued, too, that American economic and energy interests required a new approach. "Under pressure from U.S. oil companies," he wrote, "Embassy Malabo was reborn in late 2003. . . . Nonetheless, our current state might still be better described as half-born than fully hatched. . . . The internal ambivalence of [the State] Department, the specter of a reluctant Capitol Hill and associated oppugnant human rights N.G.O.s, and the argument of scarce resources has kept Embassy Malabo on a drip-feed in these early years." He continued:

Despite open doors E.G.'s nasty reputation is sustained and our ability to address problems constrained. . . . We are behind the curve.
Unfortunately, while American oil companies are paying the bills
(NOTE: U.S. operators Marathon, Hess, and ExxonMobil are responsible for almost all current production and most of E.G. government
revenues), it is still more often the Chinese, the French, or the Egyp-
tians who get credit for assisting the country's development by undertaking high-profile projects. Our official allergy to E.G. apparently
acts as an appetite suppressant to most private U.S. companies that
might otherwise be interested.

"There are good guys and bad guys here," Smith wrote. "We need to
strengthen the good guys—for all his faults, President Obiang among
them. . . . We need to get serious about engagement," Smith wrote. "It's
time to commit. . . . It is time to abandon a moral narrative that has left
us with a retrospective bias and ambivalent approach to one of the most-
promising success stories in the region. . . . What do we want for Equatorial Guinea? Do we want to see the country continue to evolve in positive
ways from the very primitive state in which it found itself after independence? Or would we prefer a revolution that brings sudden, uncertain
change and unpredictability? . . .

"The latter has potentially dire consequences for our interests, most
notably our energy security."[5]

Since the failed mercenary coup attempt of 2004, which had been followed by Secretary of State Colin Powell's reassuring meeting with
Obiang at Foggy Bottom, the Bush administration had quietly trained
Equatorial Guinea's intelligence service, occasionally shared intelligence
about threats to the regime and to offshore oil operations, and permitted
training of Equatorial Guinea's onshore police and security forces by the
defense contractor M.P.R.I. (with human rights education as a part of the
curriculum). The Bush administration had also inaugurated naval training exercises among visiting U.S. Navy warships, the nascent Equatorial
Guinea navy and coast guard, and private security patrol boats operated

by ExxonMobil, Marathon, and Hess—exercises that were designed, in part, to speed the time required by armed Equatorial Guinean vessels to reach ExxonMobil's offshore oil platforms, if they came under assault. The administration dispatched General Charles Wald, deputy commander of European Command at the time, to meet with Obiang and assure him of American interest in the "shared responsibility" of protecting U.S. investments in the country. Not all of Bush's advisers fully accepted this proposition: Cindy Courville of the National Security Council told Obiang's senior aides that "because of market forces, the U.S. would benefit from Gulf of Guinea oil whether the U.S. had a good relationship with E.G. or not." The administration would obviously not lightly abandon Exxon-Mobil, Marathon, and Hess, however. As the Bush administration's quiet defense and intelligence cooperation with Equatorial Guinea deepened, Obiang pronounced that he considered the United States to be "our best ally," and he was "lengthy and effusive in his praise for the current state of U.S.-E.G. relations."[6]

The crypto-sanctions Smith complained about, as well as the limitations of Equatorial Guinea's military forces, meant that it was easier for American-headquartered oil corporations to handle aspects of Malabo's defenses than for Obiang to establish the capacity within his own Ministry of Defense. "The companies have better capacity to surveil than we do," Brigadier Francisco Nugua, a special adviser to Obiang for national security, said. ExxonMobil and its peers installed and operated sophisticated "domain awareness" equipment in the waters around Equatorial Guinea, radars and integrated communications that allowed the companies' security departments to track, identify, and monitor potentially threatening naval traffic. "If they see something, they communicate," Nugua said. Cooperation between the oil companies and the Equato-Guinean navy and coast guard evolved to the point where, by 2009, the oil firms monitored "not only a hostile invasion, but illegal immigration" into Equatorial Guinea by boat from neighboring, poorer countries such as Cameroon.[7]

As long as these security engagements remained secret, and ran on bureaucratic autopilot in Washington, the United States seemed prepared to deepen its partnership with Malabo. The difficulty was, every so often, news reporting or investigations by human rights groups would turn up fragmentary information about the growing security ties, and the disclosures would provoke angry denunciations by members of Congress and human rights advocates. The U.S. Navy, for example, conducted four or five training visits to Equatorial Guinea after 2008, under its Africa Partnership Station program, until publicity about the engagement led the State Department to demand that the training contacts be suspended until Obiang's human rights performance measured up.

Fortunately for Obiang, coup-prone African governments rolling in oil but lacking in arms and intelligence to defend their bounty had a discreet alternative to the Pentagon and the C.I.A. for defense support: Israel. Quietly, the Bush administration encouraged Obiang to enter into security and commercial ties with Tel Aviv.

During the cold war, the United States and Israel had occasionally collaborated covertly to shore up friendly governments in Africa. More recently, Israel had extended its global influence by using security partnerships to export its high-technology equipment and its hard-earned lessons in counterterrorism and defense against strategic surprise. Retired Mossad and Israel Defense Forces officers formed consultancies to sell electronic surveillance equipment, drones, gunboats, helicopters, and training packages to wealthy, insecure African regimes facing insurgencies or the threat of coup makers. These deals often had at least as much of a commercial as a security motivation, but even when profit figured to a great extent, the exported sales and training packages strengthened Israel by providing additional revenue for its defense and intelligence industries, and by building new global networks and political alliances. Israeli trainers and consultants peddling intelligence and defense systems quietly equipped Angola's oil-rich, formerly Marxist government and Nigeria's Joint Task Force, which battled unauthorized oil-thieving militants in the Niger Delta (as opposed to the authorized ones).[8]

In 2005, around the time the C.I.A. opened an intelligence liaison with Obiang's security service, a Mossad officer approached Ruben Maya, Obiang's national security adviser, according to a consultant to Equatorial Guinea's government. A meeting in Paris followed. The Mossad "suggested that Obiang make a quiet trip to Israel and they would scan his plane and review his security," the adviser recalled. This was a typical method by which Israel opened new intelligence supply relationships—it was the counterintelligence equivalent of a bank offering a television to customers willing to open a new account.[9]

After some false starts, the relationship between Malabo and Tel Aviv ripened. The more Obiang's ministers and advisers learned about Israel, the more they identified with the country. Equatorial Guinea, too, was a small country surrounded by enemies. A global power like the United States could provide useful advice about defense strategy, but Israel's advisers understood intuitively what it was like to be tiny, threatened, and unpopular. Among other problems, Equatorial Guinea struggled with coup-making conspiracies in West Africa that Obiang and his advisers believed were financed by networks of Lebanese traders. By retaining Israel for defense and intelligence advice, some of Obiang's advisers felt, they would be sending a message to all potentially hostile Lebanese: We know who you are.[10]

By 2009, the Israelis had sold Equatorial Guinea electronic surveillance equipment and taken orders for armed speedboats, to be delivered in 2011. The defense ministry also acquired two small Ukrainian frigates that could carry a helicopter to sea. Regime protection was a visible priority; bodyguards at Obiang's mainland palace practiced for would-be assassins at an outdoor shooting range mounted with human-size silhouetted targets. Some international diplomats in Malabo felt the Israeli trainers were not particularly helpful to their efforts to coax Obiang and his advisers toward political reform and the elimination of torture as a policing method; ex-Mossad officers doing business in Equatorial Guinea tended to exude a crush-your-enemies ethos that did not include much discourse about civil rights. The Israeli cooperation was not limited to defense and intelligence, however: Obiang also paid Israel to build and fully staff a modern hospital on the Equatorial Guinean mainland. Israeli doctors and nurses earning very healthy salaries provided international standards of care.

As his oil cash flow swelled, Obiang also bought attack Hind helicopters and rented out pilots and maintenance crews from Ukraine. The contracts provided that Ukrainian pilots would train local counterparts so that each helicopter crew would eventually have one Ukrainian and one Equato-Guinean at the controls in the event of hostile action against an invading force. Unfortunately, during one actual emergency, when gun-wielding Nigerian criminals arrived in Equatorial Guinea in fishing boats to rob local banks, the Equato-Guinean copilot did not turn up fast enough to give chase. A fully armed Hind followed the escaping bank robbers' boat into the Atlantic with only a Ukrainian in the cockpit. According to reports that circulated afterward in Malabo, when Equato-Guinean officers speaking by radio ordered the pilot to fire, the Ukrainian declined, declaring that the direct, trigger-pulling use of lethal force was not provided for in his training contract.[11]

I n Washington, Obiang's lobbyists at Cassidy & Associates tried to keep the momentum of cooperation with the Bush admnistration moving. At State, their strategy was to argue, referring to Equatorial Guinea's president, who had now been in power for more than twenty-five consecutive years, "I'm not telling you this is a good guy; I'm telling you that he's willing to change." To prove the point, during one meeting with high-ranking State officials, in late 2005, Obiang awkwardly handed over a check for $4 million to pay for development work in his country that would be carried out by the U.S. Agency for International Development. Normally, U.S.A.I.D. used American taxpayer funds to assist poor countries; Obiang said he would be willing to cover America's expenses if the agency would help to improve his country.[12]

The gesture—and diligent pushing by Cassidy's advocacy team—helped win Obiang a second visit to State Department headquarters, this time to meet with Condoleezza Rice, Bush's secretary of state during the president's second term. Obiang still maintained a residence in suburban Maryland and frequently visited the United States. A date was set for an April morning when Obiang was in the capital.

To prevent unwelcome publicity, both sides agreed that while Rice would receive Obiang formally in her office on State's seventh floor, and two official photographers would record the meeting so that Obiang could display authenticating pictures back home, the meeting would otherwise be private—it would not be listed on Rice's public schedule, and there would be no press notification.

As Obiang climbed into his car to ride to Foggy Bottom, one of his lobbyists' cell phones rang. It was a reporter: "I hear your guy is going to be at a press conference." The Cassidy team was stunned: Privacy had been negotiated and agreed upon, they thought.

There were three places at State headquarters where the secretary, by protocol, could greet an official visitor: outside, at the dignitary's car door; inside the main lobby, at the elevator; or up on the seventh floor, at the secretary's office. Cassidy and Bush administration officials had agreed that to maintain secrecy, Obiang would travel up to Rice's office to be met there, out of sight. Why, then, would Rice's office have notified reporters of the meeting?

The reason, Cassidy's lobbyists were told later, was that the Bush White House had decided that morning that Rice needed urgently to make a public statement about Iran, and that she needed to do so in the morning, Washington, D.C., time, so that it would be carried on evening news broadcasts in the Middle East, which was roughly eight hours ahead. The Obiang visit, scheduled for 10 a.m., presented the best opportunity for Rice to get her statement out. So the word had gone out to attract the press to an event that was supposed to be off the schedule.

Obiang and Rice did talk privately, at first. Seated in the secretary's office, with other State officials and note takers present, Rice asked for Obiang's help with diplomatic efforts to calm violence in Sudan's Darfur region. She asked, too, if Equatorial Guinea would vote to support Guatemala's candidacy for membership on the United Nations Security Council.

Obiang told her that he had sought closer ties with the United States government "for a long time."

Rice asked Obiang about his "top priorities," and the president mentioned health, sanitation, and education. The secretary "commented favorably on Equatorial Guinea's improvements on the education front," as a summary of the conversation put it. (In fact, according to United Nations statistics cited by the World Bank, Equatorial Guinea's rates of enrollment and completion in primary school had declined between late 1994, before oil's discovery, and 2009, when the country's national per capita income was about $19,000.)

Obiang invited Rice to visit Equatorial Guinea. She said that if Obiang continued "with reforms in human rights and democratization and invested in the social sector . . . Equatorial Guinea could indeed become a model country in Africa."

It was the sort of pablum that had been exchanged between Obiang and senior Bush administration officials ever since the 2004 coup attempt; like Colin Powell, Rice had gently mentioned human rights as an American priority, but she had not pressured or scolded Obiang.

They rose to meet the reporters waiting outside. Photographers snapped them standing together, smiling. Rice called Obiang a "good friend" of the United States. She went on to make her comments about Iran.[13]

The next day, the *Washington Post* pointedly reported Rice's remark about America's friendship with Obiang. Other journalists, human rights groups, and congressional aides soon denounced Rice for naively coddling a dictator. As the unfavorable commentary accumulated, Rice was "embarrassed, pissed off, angry," an adviser involved in the episode recalled. It was yet another setback for Obiang and his regime in their effort to become West Africa's Kuwait in Washington's perception. Cassidy & Associates advised Obiang to wait for the next American president before pushing for more engagement.

ExxonMobil, Marathon, and Hess lobbied the Bush administration hard, arguing that Equatorial Guinea was an important country too often neglected, according to an adviser involved in the effort. But the companies did not want to be tainted by accusations that they ignored human rights violations any more than Condoleezza Rice did. As a fallback, it was easier for both ExxonMobil and the Bush administration to quietly encourage Obiang to build up his defenses with Israel.[14]

Wayne Clark, an ebullient Texan, served as ExxonMobil's country representative in Malabo as the Obama administration took office. ExxonMobil country managers typically rotated every few years; Clark's predecessor in Equatorial Guinea, Jim Spears, had recently moved on to Angola. The corporation's production from its offshore Zafiro field had leveled out to about 200,000 barrels per day, but at $80 per barrel, that still amounted to a business with about $5.8 billion in annual revenue. ExxonMobil maintained its low profile in Malabo; its elevated compound along the airport road had undergone none of the expensive renovations visible elsewhere in the capital after 2007. Oil workers shuttled to and from the corporation's offshore platforms by helicopter and they moved in and out of Equatorial Guinea while interacting only minimally with local citizens.

ExxonMobil's policy about working in Equatorial Guinea, approved by headquarters in Irving and distributed as talking points to public affairs officers worldwide, echoed the arguments that Anton Smith made in his cables to the incoming Obama administration. The policy was reflected in 2009 talking points on human rights that were issued for use when uncomfortable questions arose. The talking points reflected the evolution of ExxonMobil's attitudes toward human rights matters since the days of the Aceh civil war, but also the continuity of its commitment to work with any government that would agree to acceptable legal contract terms, and that was not subject to prohibitive sanctions under American or international law. "We publicly condemn the violation of human rights in any form and actively express our views to governments around the world," the prepared talking points said. "We have been dealing with these issues for many years and believe that our efforts improve the quality of life in communities where we operate. ExxonMobil is very concerned about human rights. However, we also believe that engagement enhances the cause of human rights far more than political isolation. Our practices are designed to ensure respect for human rights in our sphere of influence, which may by example have its effect on others."[15]

It was in ExxonMobil's interest for Obiang's government to be able to defend itself from invaders. It was also in ExxonMobil's interest for Equatorial Guinea to improve its reputation.

Simeon Moats, the former State Department official who worked on Africa policy in ExxonMobil's Washington office, developed a PowerPoint presentation entitled "Business Practices & Transparency," which described, in simple and graphic form, the corporation's philosophy about the alloca-tion of responsibilities in poor and troubled countries. ExxonMobil would contribute "taxes and royalties" as well as "transparency" to the host Afri-can government. The government would in turn take responsibility for the "rule of law, health, education" and the "business environment." Ex-xonMobil's disciplined, modern "business practices" could also strengthen the national economy. "What's missing?" Moats's slide presentation asked. One slide offered a troubling list in answer: rule of law, predictable regu-lations, the sanctity of contracts, transparency of transactions, an educated and healthy workforce, and basic infrastructure. This was the gap African governments would have to fill while ExxonMobil carried out its profit-making role and provided taxes and royalties in support.[16]

ExxonMobil had enough trouble trying to educate the African despots it worked with about the basic functioning of the global oil industry—persuading these leaders to emulate Singapore seemed unrealistic. When oil prices fell sharply after the global financial crisis of 2008, for exam-ple, ldris Déby, in Chad, accused ExxonMobil of deliberately holding down oil production to ride out the low prices. The charge lacked logic—ExxonMobil was suffering financially, too—but Déby was deeply sus-picious of Stephane de Mahieu, Ron Royal's successor as ExxonMobil lead country manager in N'djamena. Déby implied that ExxonMobil might be coming under pressure from "third parties" to hold down production so as to weaken Déby's regime and aid the president's enemies. ExxonMobil arranged a meeting between Déby and some of the corporation's Africa hands in Washington. The ExxonMobil team "told Déby that it was firmly committed to maximizing its own profits, which implied maximization of production, and that ExxonMobil would not permit any pressure by any third party to change that policy," according to an account provided later by De Mahieu. The team noted that ExxonMobil "had never abandoned operations despite Chad's past political instability," including multiple rebel invasions. Déby gradually backed off.

Nonetheless, as Moats's PowerPoint suggested, there was one element of anti-oil campaigning after 2000 that aligned somewhat with Exxon-Mobil's philosophy. This was the campaign for transparency about the management of oil revenues by governments. The underlying idea was that if citizens in poor countries had more information about how oil money flowed to their governments, they could improve governance and check corruption. The movement gained visibility as an initiative of George Soros's Open Society programs under the rubric of "Publish What You Pay." Later, with support from British Prime Minister Tony Blair and the government of Norway, it had evolved into a formal voluntary compact among governments, corporations, and nonprofit campaigners: the Extractive Industries Transparency Initiative. Under the initiative's complex and voluntary rules, countries and companies each pledged to disclose publicly data about oil revenues and royalties in particular countries, and to engage with civil society groups about how the revenue would be used to benefit the public. In theory, at least, such transparency would reduce the ability of corrupt politicians to steal oil proceeds.

Soros had written to Lee Raymond about his ideas in 2002. Raymond took the approach seriously. Raymond disagreed with Soros about some of the campaign's principles—he thought the Open Society approach was discriminatory and too far-reaching, particularly because it sought compliance from corporations that traded on stock markets but had no means to enforce the same rules against competitors that were privately owned. But under Tillerson, ExxonMobil developed formal positions about the transparency campaign through its issues management process, and it assigned a public policy manager, John Kelly, a thirty-year veteran of the corporation, to lead the effort. Kelly joined the Extractive Industries Transparency Initiative's board of directors in 2007.

"We think it [membership in E.I.T.I.] helps demonstrate that we are opposed to corruption in any form and committed to honest and ethical behavior wherever we do business," an ExxonMobil executive involved in the effort said. (The executive acknowledged, "I guess that message hasn't sunk in with everyone.") Kelly and other executives devoted extensive time and travel to organize the Equatorial Guinea regime's candidacy for membership, which, if achieved, would provide the country with a rare seal of international legitimacy. In general, the ExxonMobil executive said, "We try not to get out in front of countries" on public policy issues, "so we are not going to go and bang on their door and say, 'Okay, it's time.'" However, because ExxonMobil itself had made a formal decision to participate, "we are willing to . . . talk to them about why we support E.I.T.I. and why we think this would be a good idea—and we certainly have done that in lots of countries, including Equatorial Guinea."[17]

There was self-interest in this altruism: "If you are helping to hold a government accountable for the use of revenues," the ExxonMobil executive continued, "that's just helping to enhance the business investment climate in there, because you don't have a lot of dissatisfied citizens that can find other ways of expressing their opinion about what's going on in the government," such as by joining insurgencies or coup plots. "And if in fact the government is using the money to benefit the citizens, that's got to be for poverty reduction, for infrastructure, for lots of other things that help improve the investment climate."[18]

Early in 2009, Equatorial Guinea's prime minister, Ignacio Milam Tang, flew to Doha, Qatar, to attend the Extractive Industries Transparency Initiative's annual conference and deliver a speech at the Ritz-Carlton hotel. He spoke in a plush, wood-paneled conference room, at a lectern embossed with the luxury hotel's emblem. Transparency activists; government officials from Europe, Africa, and Central Asia; and oil and mining company executives made up the audience. Tang wore a dark suit and put on reading glasses to see his text. "I have come here to assert the commitment of our president and our people to abide by the commitments" made to the Transparency Initiative regarding Equatorial Guinea's disclosure of oil revenues. He described his country's history and the impact of the oil boom on national life and wealth. "A lot of work needs to be done," he said. "But we are confident we will meet a fruitful success."

The political scientist Benedict Anderson compared nation-states to "imagined communities." In the case of Equatorial Guinea, by 2009, there were several. There was the country that seemed normal and functional enough to dispatch a prime minister to a conference in Qatar to read a speech that pledged adherence to international treaty norms. This was the Equatorial Guinea preferred by—needed by—ExxonMobil and other hopeful supporters of the Obiang regime. The country they imagined had a difficult past but was gradually mustering the will to improve and normalize. There was also the Equatorial Guinea imagined by the British mercenary coup leader Simon Mann, his coconspirators, and the security consultants from Israel and M.P.R.I.—a place that resembled less a nation-state than a museum housing crown jewels, but one lacking robust fencing, alarms, and armed guards.

There was also the Equatorial Guinea known by most of its citizens and internal political contestants: a family-run enterprise dominated by a particular mainland clan of the Fang ethnic group and presided over by an aging godfather who ruthlessly punished usurpers. The norms of this Equatorial Guinea—the family business—did not conform to the norms of the Transparency Initiative. As Anton Smith wrote in one of his cables to Washington, "Among the Fang, family comes first."

P resident Obiang acknowledged fathering forty-two children. Of these, two sons, Teodorin ("Little Theodore") and Gabriel, wielded the most influence by 2009. They had different mothers. Teodorin's, who was referred to as Obiang's "church" or formal wife, belonged to an Equato-Guinean family that had risen alongside the president during the early oil era, grabbing up land and businesses. Gabriel's mother came from the tiny neighboring island nation of São Tomé; she suffered from the lack of prestige that arose from being a foreigner. By kinship and political tradition, Teodorin, therefore, was the son who had the most natural claim to succeed his father as president, should his father's health ever falter. The heir showed little interest in education or government work, however. Gabriel, on the other hand, had returned from college in the United States as a well-spoken, serious young man adept at PowerPoint presentations and the vernacular of international oil and gas deals. He told oil industry representatives that he had been influenced, in thinking about Equatorial Guinea's future, by the American film *Field of Dreams* and its spiritual catchphrase "If you build it, he will come." Gabriel hoped to construct an oil-led development strategy that would help the country develop a technically competent middle class and a more diverse economy akin to Singapore's. ExxonMobil's, Marathon's, and Hess's representatives fantasized that Gabriel might emerge as Obiang's successor as president and lead a final drive toward modernization, protecting their investments along the way. But Gabriel lacked the Fang clan base necessary to secure his rise in politics. As one student of the country and the region put it coldly: After President Obiang's death, Gabriel was more likely to be assassinated than promoted.[19]

That left Teodorin, the first son, who had been born on June 26, 1969. As he entered middle age, he was a broad-shouldered man with dark chocolate skin and puffy cheeks, who had developed a taste for luxury goods and properties. Before oil, Equatorial Guinea's most lucrative resource was the lumber in its jungle forests. Teodorin's mother secured an appointment for her son as minister of agriculture and forestry; in that capacity, he told American officials, he was "granted" a concession to sell timber. During the 1990s, a Malaysian contractor retained by Teodorin brought in forty teams of lumberjacks who clear-cut and shipped out whole logs to Asian markets, leaving the minister with a multimillion-dollar grubstake while he was still in his twenties. Later, his family's access to Equatorial Guinea's oil wealth evidently enriched him further, although despite the Transparency Initiative, Obiang never made clear exactly how he allocated "national" revenues to Teodorin or other family members. What seemed plain was that as he neared middle age, Teodorin had easy access to many tens of millions of dollars. He began to travel and to spend.

By 2007, according to investigations by the U.S. Immigration and Customs Enforcement division and French police, Teodorin's international property holdings included two luxury speedboats; a Gulfstream V private jet valued at $38.5 million; and a collection of almost three dozen luxury cars, including two $1.5 million Bugatti Veyrons, a $990,000 Maserati MC12, a $530,000 Rolls-Royce Phantom, a $256,000 Ferrari 512M, a $213,000 Ferrari 550 Maranello, and a $115,000 Maserati Coupe F1.

When American investigators came into possession of a check register from one of Teodorin's California companies, Beautiful Vision Inc., they encountered the following list of expenditures for a single month: $82,900 to Naurelle for furniture; $137,313 to Ferrari of Beverly Hills; $63,326 to the Soofer Gallery for a carpet; another $332,243 to Ferrari of Beverly Hills; $51,288 to Dolce & Gabbana; $121,977 to Fields Piano; another $50,000 to Ferrari of Beverly Hills; another $59,850 to the Soofer Gallery;

$280,409 to Autostar Signature for another Ferrari; $338,523 to Lamborghini Beverly Hills; and $181,265 to GlobalJet Corp.[20]

Teodorin moved in and out of the United States on an A-1 diplomatic visa, often carrying more than $1 million in cash, which he routinely failed to declare to U.S. customs officers as required by law, according to the I.C.E. investigators. Los Angeles and New York were among his favored destinations; at one stage, Teodorin set up a record company in Los Angeles specializing in rap and hip-hop music, and for a while he traveled in the company of the glamorous rapper Eve. His spending and his migration into the hip-hop business suggested an air of danger and urban sophistication, but in person he could just as often come across as unworldly.

"He showed up in my office with his entourage—four or five people," all from Equatorial Guinea, recalled an attorney in Los Angeles who worked with him. "His English wasn't very good; he had other people who would do the language for him. He was willing to pay whatever was required. He would ask, 'How much do you want?' And the check was right there."[21]

Teodorin could be enthusiastic about laying out money for a new purchase or project, but his managerial follow-up was often lacking. Unpaid bills and civil lawsuits accumulated in Los Angeles County civil courts. Teodorin relied on American lawyers, real estate agents, personal assistants, bankers, bodyguards, and freelance fixers and hangers-on to manage his bills and his chaotic consumer habits. His American advisers took note that the luxury automobiles abandoned in Teodorin's California garage would by themselves finance a respectable hospital back home. It was not unusual for Teodorin to lose track of his cars; in one case, he asked an adviser to fly across the country to Los Angeles to move one of them from a parking garage where he had absentmindedly abandoned it. Former assistants filed lawsuits in Santa Monica civil court alleging that Teodorin failed to pay overtime as required by California law; the defendant often missed court appearances and depositions.[22]

After the failed coup attempt led by Simon Mann, Teodorin turned up at the Beverly Wilshire Hotel to do some house hunting. Through one of his Los Angeles attorneys, he summoned a Hollywood Hills real estate agent, Neal Baddin, to his suite. Baddin thought Teodorin seemed "bigger than life." Baddin agreed to serve as his real estate agent, and over the next seventeen months, he helped negotiate the purchase of a $30 million Mediterranean estate on a fifty-foot bluff overlooking the Pacific Ocean, in Malibu. Teodorin's neighbors in the gated community included James Cameron, the director of *Titanic* and *Avatar*, and the comedian Dick Van Dyke.

As the purchase closed, one of Teodorin's Los Angeles attorneys contacted Paul Finestone of the Finestone Insurance Agency, seeking to buy a policy for the Malibu estate and thirty-two cars that would be housed there. The attorney explained that a number of American insurance companies had so far refused to sell Teodorin insurance; he asked Finestone to find a company willing to do business with the Obiang family.

When one insurer asked why Teodorin intended to employ armed security guards at his Malibu home, Finestone explained that his client was an "investor and collector" who was "independently wealthy" and needed guards to protect against kidnapping.

American International Group, Inc., which would gain notoriety for its role in the 2008 global financial crisis, refused to sell to Teodorin after learning about his background. Finestone wrote to A.I.G. to challenge the insurer's decision. Equatorial Guinea, he wrote, "is a major supplier of oil to America and a critical interest of American energy needs." President Obiang, Teodorin's father, "is no better and no worse than the Saudi Royal family. . . . We insure billions and billions of dollars of Saudi property bought with our oil money here in America and A.I.G. has no problem handling a great deal of that business."[23]

America's oil dependency required even Los Angeles insurance brokers to consider the relative virtues of corrupted oil alliances.

Several weeks after President Obama's inauguration, the United States received intelligence reporting from Nigeria that some sort of an attack was being planned on high-level targets in Equatorial Guinea. Such reports were increasingly common. Piracy, oil smuggling, and speedboat militancy carried out mainly by armed Nigerians under the brand name of the Movement for the Emancipation of the Niger Delta (M.E.N.D.) continued to spread throughout the Gulf of Guinea.

That winter, ExxonMobil's West African operations were on particularly high alert. In December, in Nigeria's Akwa Ibom State, armed gangs had shot up an ExxonMobil caravan, apparently seeking to kidnap expatriate workers; Nigerian security guards returned fire and repelled the attackers. A month later, the guards were in action again, around the same housing compound in Eket where the traumatic kidnapping of 2006 had occurred; again, the ExxonMobil security force managed to ward off the assailants before they could reach any oil workers. Militants in speedboats also attacked an ExxonMobil oil platform in the ocean waters off Akwa Ibom. Malabo and its harbor lay only eighty-five miles by boat from Eket, straight across the Bight of Biafra—for speedboat-equipped militants and robbers, it was an easy commute.

In Equatorial Guinea, ExxonMobil, Marathon, and Hess had developed an e-mail system to distribute warnings about impending coups, invasions, or waterborne bank robberies in which armed men in speedboats arrived at Equatorial Guinea's coastal cities to hold up banks and escape by sea. The oil companies' security departments tended toward caution and often ordered lockdowns at their Malabo and mainland compounds on receipt of even fragmentary intelligence reports.

The United States continually earned credit with President Obiang— and partially compensated for the harping it made Obiang endure about human rights—by sharing warnings about invasions or coups. Typically, Obiang reacted to the warnings by erecting checkpoints around the capital, detaining foreigners, and otherwise tightening his already-heavy police deployments. These visible precautions taken after American intelligence warnings likely prevented some of the threatened attacks from going forward as planned. There was a boy-who-cried-wolf problem inherent in the repeated warnings and preemptions, however, particularly because there had been no serious coup-making attack on Obiang, beyond the plotting stage, in several years.

In the darkness of February 17, 2009, speedboats bearing armed men arrived in Malabo. Three boats entered the harbor; three others arrived on the eastern side of an adjoining peninsula. The attackers disembarked, unopposed, and headed toward the presidential palace, where, as it happened, the president was not home. (When Washington passed on its latest round of attack or coup warnings, President Obiang had quietly slipped out of Malabo to his better-fortified palace on the African mainland.) The Israeli trainers had planned for this moment—their Equato-Guinean charges were supposed to swarm in and counterattack to protect the palace and repel the invaders. In the event, the response was more ad hoc than the Israelis would have hoped. Senior ministers and generals who were supposed to lead the counterattack failed to turn up when the shooting started. Several younger officers did respond, however, and they fired vigorously, killing at least one raider, arresting others, and, after a two-hour gun battle, chasing the remainder of the group back to sea in their speedboats.

One of the younger Equato-Guineans who defended the presidential palace that night took a bullet in the hand. A few days later, as calm returned, President Obiang celebrated the soldier in public as a national hero. Obiang appointed a businessman to accompany the wounded hero to New York, to seek out the finest American surgeons available to repair the soldier's hand. Lacking American health insurance cards, Obiang provided the soldier's businessman escort with $125,000 in cash to pay for medical expenses. At John F. Kennedy International Airport in New York, however, U.S. Customs officers discovered the cash in the businessman's luggage. The money had not been properly declared, and the businessman was arrested on money-laundering charges. After some confusion and delay, the case was eventually cleared up.

The raiders, it turned out, had been Nigerian militants who had ties to a section of Obiang's exiled political opposition in Spain, and who had been trying to sell protection services to sections of Equatorial Guinea's government. The militants did not feel that their offer of protection was being taken seriously enough, so they had decided to mount a demonstration project in Malabo, to show that their services were indeed required if ExxonMobil and the other oil companies wanted to operate in security. What the attackers might have done if they had penetrated the presidential palace and found Obiang at home was not clear.[24]

B arack Obama's pronouncements about foreign policy during the 2008 election campaign suggested that he was prepared to rethink the Bush administration's approach to governments that were hostile to the United States or that did not conform to American ideals about democracy and human rights. The Obama administration seemed to be signaling that it sought "dialogue and engagement," as an ExxonMobil executive put it after the president's inauguration. "They are saying that about Russia, they are saying that about China, they are saying that about Iran. . . . That is the cornerstone of their foreign policy."[25] Why not Equatorial Guinea, too? That was the basic question that Anton Smith had presented in the six analytical cables he filed from Malabo during the late winter and spring of 2009, hoping to redirect Obama administration policy toward deeper engagement.

Secretary of State Hillary Clinton's advisers included some energy "realists" such as those who had shaped her husband's second-term policies aimed at securing oil supplies from Central Asia. Clinton named as a special energy policy envoy David Goldwyn, who, before joining State, had organized a business group designed to support Libyan leader Muammar Gaddafi's plans to reopen the Libyan oil business to international corporations—among them, ExxonMobil. To run Africa policy, Clinton named Johnnie Carson, a longtime foreign service officer who had served as U.S. ambassador to Kenya, Uganda, and Zimbabwe before his appointment by President Bush as national intelligence officer for Africa. Between Goldwyn's background as an oil industry consultant and Carson's deep experience of engagement with flawed African governments, Anton Smith's arguments about Equatorial Guinea found at least some influential readers inclined to his views.

The human rights community saw an opportunity to mark a new course, too, but in a very different direction: "The new Obama administration has an opportunity to show that energy security does not have to come at the expense of human rights and good governance," Human Rights Watch argued in a major report about Equatorial Guinea released that July. It recommended investigations to seize and repatriate to Equato-

Guinean citizens' assets in the United States "obtained through corruption," and it recommended that Obama "ensure through new or existing laws and regulations that U.S. companies do not become complicit in the corruption and abuses that mar resource-rich countries like Equatorial Guinea."[26] The formulation suggested that ExxonMobil was not already complicit. Anton Smith attended a launch event around the Human Rights Watch report, where he said he "did not recognize" the Equatorial Guinea described by the report's investigators, who had not visited the country in recent years, in part because it was difficult to do so without official sponsorship. Smith's adversaries at Human Rights Watch and the advocacy group EG Justice were appalled by his remarks and his defense of the Obiang regime, and they argued privately to State officials that Smith was unfit to represent the United States in Malabo because he had evolved into an apologist for the regime.

As in other areas of foreign policy, the Obama administration proved conflicted about whether to pursue "realist" engagement with Equatorial Guinea or pursue a more liberal, activist agenda of the sort recommended by Human Rights Watch. The department did agree during 2009, as Smith had recommended, to upgrade its representation in Malabo by appointing a full complement of liaisons to Obiang: an experienced ambassador, a deputy chief of mission, and a defense attaché from the Pentagon. To some degree, Smith's arguments prevailed: The Obama administration continued the policies of security, intelligence, and limited military engagement with Equatorial Guinea that the Bush administration had forged after the 2004 coup attempt. Yet American policy changed only in increments. There was no fundamental reexamination.

On May 4, 2009, Ken Cohen wrote to Human Rights Watch to describe and defend the corporation's policies in Equatorial Guinea. "ExxonMobil is committed to being a good corporate citizen wherever we operate worldwide," Cohen wrote. "We maintain the highest ethical standards, comply with all applicable laws and regulations, and respect local and national cultures." At the same time, "the practical realities of doing business in developing countries are challenging. . . . E.G., like many developing nations, has a limited number of local businesses and a small population of educated citizens. . . . Many businesses have some family relations with a government official, and virtually all government officials have some business interests of their own, or through a close relative. . . .

"While it may be virtually impossible to do business in such countries without doing business with a government official or a close relative of a government official, it is still possible—indeed, it is expected—that we do business ethically and comply with all U.S. and local laws."[27]

Twenty-five

"It's Not My Money to Tithe"

O n the morning of January 8, 2009, twelve days before Barack Obama's inauguration as president, Rex Tillerson arrived at the Ronald Reagan Building on Pennsylvania Avenue, two blocks from the White House, to announce ExxonMobil's new lobbying position on climate change. He made his way to the rear of the cavernous Reagan building, which housed several government agencies. Upstairs, at the Woodrow Wilson International Center for Scholars, a government-supported think tank, Tillerson strode into an amphitheater where about one hundred scholars, researchers, and journalists had gathered. He took the podium and unfolded a printed speech.

He ticked through the corporation's positions on American energy policy: Washington needed "long-range thinking"; rising global demand for oil and gas, through 2030, was inevitable; America, therefore, needed to develop "all our energy resources"; and it was particularly urgent to open up offshore and other domestic territory to drilling. Normally, the Exxon-Mobil chairman resisted arguments that pandered to the American yearning for "energy independence," since he regarded the very idea as misguided.

Yet if measured appeals to American nationalism were necessary to win approval for domestic oil drilling, he was willing to make them: "There is enough oil and natural gas offshore and in non-wilderness and non-park lands to fuel fifty million cars and heat nearly one hundred million homes for the next twenty-five years," he declared.

He referred to climate change impassively as an "important global issue." The incoming Obama administration proposed to reduce greenhouse gas emissions by enacting a cap-and-trade system in which polluters could buy and sell pollution credits under an overall "cap." Tillerson argued that Europe's similar system, inaugurated several years earlier, did not work well and had introduced "unnecessary cost and complexity," while creating "problems with verification and accountability."

In Beijing in 1997, Lee Raymond had delivered a landmark speech in which he argued that the evidence suggested global warming was not taking place at all. Ever since, ExxonMobil's leaders had criticized public policies to reduce greenhouse gas emissions, such as the Kyoto Protocol, cap-and-trade proposals, and alternative energy subsidies. That morning in Washington, however, Tillerson's speech took an unexpected turn: For the first time in ExxonMobil's century-long history, its chairman went on to advocate that the government impose higher taxes on oil and gas use, to reduce the risks posed by climate change:

> There is another policy option that should be considered, and that is a carbon tax. As a businessman, it is hard to speak favorably about any new tax. But a carbon tax strikes me as a more direct, a more transparent, and a more effective approach. . . . Such a tax should be made revenue neutral. In other words, the size of government need not increase. . . .[1]

The idea of a carbon "sin" tax, comparable to the excise taxes imposed on tobacco products, had a distinctive history. Then-senator Al Gore proposed a version of the tax in his 1992 bestselling book, *Earth in the Balance.* Gore suggested that revenue from a carbon-based-fuels tax be used to reduce payroll taxes on salaried Americans—tax pollution, not work, was his rhetorical flourish. Advocates at some ardent environmental lobbies, such as Greenpeace and the Rainforest Action Network, advanced Gore's proposal in the years afterward. Some of their leaders and thinkers preferred a broad carbon tax to regulator-heavy cap-and-trade systems; the latter allowed some polluters to buy their way out of accountability for their emissions. On the ideological right, some free-market tax economists, such as Kevin Hassett at the American Enterprise Institute, also endorsed the carbon tax after Gore proposed it, on the grounds of its relative economic efficiency. Tillerson and right-leaning economists argued that such a tax should be neutral; that is, revenues raised should be returned to taxpayers, perhaps by a reduction in the payroll tax. At a time of fiscal strain, however, a carbon tax also had the potential to shore up federal finances: At $20 to $25 per ton, the range around which there was the greatest political support, a tax could raise at least $100 billion annually. By the time of the 2008 presidential campaign, however, the carbon tax had become a politically marginal and quixotic proposal.

Cap and trade's intellectual history, too, reflected compromises between conservatives and environmentalists. The first Bush administration embraced the system as a market-based way to control acid rain—and succeeded. Many of the large corporations that would be directly affected if carbon taxation of any kind was imposed—electric utilities that burned coal, for example—had concluded that they could manage their interests most successfully by lobbying for a tailored cap-and-trade program that eased their transition to higher carbon costs. The very efficiency of a carbon tax caused some coal-dependent utility executives to shudder because such a pure tax would hit every corporation in proportion to its polluting activity. By comparison, a global corporation of ExxonMobil's profitability could absorb the financial hit, and in any event, it did not produce coal, the greatest greenhouse gas offender. A modest-size, locally regulated American coal utility might see its profits and market value shrink traumatically under a carbon tax; this explained the breadth of business support for cap and trade.

ExxonMobil had already conducted detailed reviews of cap and trade versus a direct carbon tax in 2006, as part of the climate strategy review

Tillerson had ordered after becoming chief executive. Ken Cohen and other executives recoiled from cap-and-trade systems because of the systems' susceptibility to manipulation by speculators and other distorting complexities. They also loathed the idea of a new federal regulatory system with which they would have to comply.

They had leaned toward the conclusion that if they had to endure a higher carbon price, they would continue to oppose cap and trade, but might support a straight carbon tax. Between 2006 and 2008, following the internal review, ExxonMobil quietly began to test out this change of lobbying position. In private discussions at the American Petroleum Institute, and at think tanks such as the Brookings Institution and the American Enterprise Institute, ExxonMobil executives rehearsed arguments in favor of a carbon tax, without openly endorsing the proposal. A few reports in the business press hinted that ExxonMobil might be leaning toward a straight carbon tax. The corporation also explored what a Washington lobbying strategy in support of such a tax might look like.

Justin Peterson, who had worked on Senator Elizabeth Dole's staff and on the 2000 Bush-Cheney presidential campaign, served as managing partner at the D.C.I. Group, one of the outside lobbying firms in Washington that worked for ExxonMobil. Peterson supported a lobby coalition, the U.S. Climate Task Force, founded in 2008 and staffed by a former Gore aide, Elaine Kamarck. The task force sought to advance a carbon tax as an economically efficient alternative to cap and trade. The group received funding from a business coalition, called The Future 500, which tried to induce major American businesses not directly involved in carbon-intensive industries—Nike, Coca-Cola, Intel, Kraft Foods, and Hewlett-Packard, for example—to come out for a carbon tax. The task force tried to develop a campaign that could also attract major oil companies like ExxonMobil that opposed cap and trade.[2]

ExxonMobil participated in the task force and similar efforts, indirectly and quietly. "We had determined that a carbon tax was a better approach, in our mind, but our engagement on that issue was below the radar," an ExxonMobil executive involved said. "We knew that if we came out and we said, 'ExxonMobil says that a carbon tax is the way to go,'" it

would backfire and the corporation would be accused of trickery. Exxon-
Mobil would be accused of bad faith "because they know that no one's
going to vote for it, or they are just trying to slow down action, blah, blah,
blah. . . . We didn't want to come out publicly for that very reason. We
just thought there would be a lot of baggage."[3]

Obama's election persuaded Tillerson to change tack. The ExxonMobil chairman first had to overcome internal objections, however. At the
time of Obama's election, Dan Nelson, the Lee Raymond protégé, was
still running the corporation's Washington office. According to an executive who heard Nelson's arguments, he dissented from the plan to openly
back carbon taxation; he argued to colleagues that Tillerson would only
annoy ExxonMobil's political friends, incite its opponents, and confuse
everybody else, without actually changing public policy. At the October
2008 off-site meeting organized by Ken Cohen, Bennett Freeman urged
ExxonMobil's public affairs executives to endorse a higher carbon price,
but between Election Day and the eve of Obama's inauguration, Tillerson
hesitated.

He resisted less for the reasons Nelson cited than because, reflexively,
as a onetime Texan neighbor of Richard Armey's who donated regularly
to Republican political candidates, it pained Tillerson to endorse any tax
increase. As he spoke at the Wilson Center in Washington that January
morning, Tillerson looked out at the cluster of dark-suited ExxonMobil
executives in the audience just as he was about to read out the change in
the corporation's lobbying position. "I still wasn't sure, at that moment,"
he told them later.

"Why are you making this carbon tax proposal now?" a reporter asked
him afterward.

"If we are going to take a view, take a position, we need to do it now,"
he said. "Because the debate is going to get under way again. . . . I've been
chewing on this one for about three years—cap and trade versus carbon
tax. What I've really been saying is, 'There has to be a third option.' And
I haven't been able to identify one. . . .

"We have tried to get down into the details of, if you are going to
design a cap-and-trade system in the United States, what is it going to look

like? It's pretty scary. When you think about the enormous new bureau-cracy that would have to be created—it would be bigger than the I.R.S. . . . The default fact is that we've got [to] have something that is simpler."

Another reporter pressed him: "Exxon and other producers will be facing a Democratic administration, a Democratic Congress. What kind of reception do you expect for the next four years? Chilly?"

Tillerson laughed. "We still have friends on both sides of the aisle. As I said, we work with the government that is here, just like we work with the government in whatever country we are dealing with around the world. . . . We are going to engage, and we hope that they value our input."[4]

Barack Obama's most influential advisers on climate politics and policy—including Carol Browner, the former Environmental Protection Agency administrator, and John Podesta, the president's transition chief—gave virtually no consideration to a carbon tax that autumn and early winter. A Democratic Party–led coalition focused instead on the development of a big cap-and-trade bill that would be introduced in Congress early in the Obama presidency. The corporate center of this lobbying push was the United States Climate Action Partnership, an advocacy group in Washing-ton that had attracted Shell, Dow Chemical, Ford Motor Company, and major coal-dependent utility companies, as well as powerful environmen-tal groups such as the Natural Resources Defense Council.

The negotiations within the Climate Action Partnership about what sort of cap-and-trade bill might be acceptable to the group's diverse cor-porate and environmental members had become a kind of private dress rehearsal for the lobbying scrum expected on Capitol Hill once Obama and the new Democratic congressional leadership settled in. As prepack-aged coalition politics and congressional lobbying, the Climate Action Partnership "was a very developed piece of work," Browner recalled.[5] It was, however, a fragile coalition—the oil company lobbyists involved felt that their industry's interests were often neglected in comparison with coal utilities from political swing states such as Virginia, West Virginia, and

Ohio. The Climate Action Partnership was a rare example, nonetheless, of a powerful business-environmentalist alliance focused on a major environmental policy reform that would impose costs on business—it was an association that had slowly taken form after a very long lobbying struggle over climate policy in Washington dating back to the second Clinton term.

The coalition had gathered momentum after 2006, amid economic growth and low unemployment. Those conditions no longer prevailed. In September 2008, the Wall Street bank Lehman Brothers collapsed, triggering a banking panic that froze up credit lines and paralyzed the global economy; the United States plummeted week by week into its deepest recession since the 1930s. Collapsing production and rising joblessness challenged every assumption about policy and politics that Obama had relied upon to win office—including climate policy.

On December 16, in Chicago, Obama met with Browner and his top economists. The depths of the economic crisis made clear to them that they would now have to push for a large stimulus bill, to use rapid federal government spending to prevent a full-blown depression. Obama and his advisers decided that day to design the stimulus to make a down payment on their major domestic priorities—particularly clean energy. Franklin Roosevelt's stimulus during the Depression years had built national park facilities; Obama's bill, they concluded, should launch a new era of investment in solar energy, wind power, other clean energy technology, "smart" meters to regulate home electricity use more efficiently, upgrades to the national electric grid transmission system, home weatherization, and energy efficiency programs. These expenditures ultimately would total $80 billion. The renewable energy advocates around Obama recognized, however, that the long-term economic viability of solar and wind power would depend on whether dirtier, cheaper sources of energy such as oil and coal would be taxed—directly or through cap and trade. If carbon-heavy fuels like gasoline and coal did not become more expensive, the rate of adoption of solar and wind would slow, and the dangers of climate change would remain unacceptably large, they believed.

The greatest obstacle facing Obama on climate regulation as he prepared for inauguration, then, was hardly ExxonMobil. With Chevron and

Shell in the cap-and-trade lobbying coalition, the oil industry had been split and weakened as a lobbying force on climate policy. The challenge was whether the cap-and-trade lobbying coalition would hold together at all under the mounting pressure of the 2008–2009 recession. "Fundamentally, if you're going to have an economy-wide cap-and-trade system, you need to trust government and Wall Street," said one of the president's outside energy advisers. That trust was collapsing even faster than the Obama team understood.

ExxonMobil stood apart. The corporation, said a second Obama adviser involved, "seemed to me to follow a track that was quite different from the other [oil] majors—being firmly fixed in the 'Fuck you, no apologies, oil-is-here-to-stay mode.'" The corporation saw itself as merely carrying out its own global environmental and economic policy advocacy. It dispatched public affairs officers to explain its position to foreign governments with which it partnered to produce oil, lest those governments be confused about ExxonMobil's thinking. Its briefings early in 2009 emphasized that "cap-and-trade is complex, unpredictable, cumbersome and expensive, making it difficult for firms to plan long-term investments. In comparison, ExxonMobil believes the predictability of a progressive carbon tax would encourage new investment in carbon reduction technologies."[6]

Abraham Lincoln was a hero of Rex Tillerson's boyhood. As a child in small-town Texas, Tillerson read books on great leaders of the type favored by the Boy Scouts of America. Lincoln was a member of this canon, notwithstanding the complications his presidency created for the Republic of Texas and its successor Confederate state. Tillerson remained "personally fascinated and inspired by Lincoln" as he came of age at the University of Texas and as a young ExxonMobil executive. He particularly admired Lincoln's "ability to confront adversity with courage, find inspiration in challenges both personal and political, and shape leadership through the strength of diversity, with extraordinary grace."[7]

During the second week of February 2009, Tillerson flew by corporate jet into Washington to attend a ribbon-cutting ceremony at Ford's

Theatre, where John Wilkes Booth had shot Lincoln dead. A few years before, Tillerson had agreed to chair a $50 million campaign to renovate Ford's. ExxonMobil had contributed, as had one of its major business partners, the State of Qatar. (That an undemocratic kingdom with limited personal freedoms had funded the restoration of a theater devoted to memorializing the American president who emancipated slaves was an observation politely avoided by the speakers.) "Working on this campaign has been a labor of love for me," Tillerson said.[8]

The next day he arrived with a few colleagues at the Eisenhower Executive Office Building, next door to the White House's West Wing. Tillerson and his team made their way to Room 157, where Carol Browner, now Obama's chief White House climate and energy policy adviser, joined them, along with some of Browner's staff. Tillerson had asked for the meeting.

Tillerson and Ken Cohen did not know whether Obama's anti-oil populism during the campaign would carry on once the president had to govern. They decided to approach their lobbying during the early Obama administration on an issue-by-issue basis. Perhaps the most realistic opportunity involved offshore drilling. Tillerson wanted to push Obama for decisions that might open up the Gulf of Mexico for further exploration and drilling. Polling during the 2008 campaign had shown that voters supported domestic drilling—perhaps Obama would respond in office, even as he pushed simultaneously for cap and trade. The Irving team assumed that Obama's advisers would welcome their perspective, notwithstanding ExxonMobil's heavy spending in the past on the president's political opposition. As an ExxonMobil executive put it, "Why wouldn't the administration want the views of the country's biggest energy company?"

There was rarely anything personal or intimate about an ExxonMobil lobbying meeting. As Tillerson had put it in January, the corporation managed Washington with the same PowerPoint-enabled educator's mind-set that it brought to bear in Abuja, N'djamena, and Malabo. In Room 157, Tillerson laid out to Browner ExxonMobil's principal policy priorities in the United States in 2009: He urged the administration to loosen the

congressional moratoria on drilling in American ocean waters and the Gulf
of Mexico. On climate, he ticked through ExxonMobil's reasons for endorsing a carbon tax over a cap-and-trade regime. "He was just shopping
the idea that there was a better way" to raise carbon prices in America,
recalled a participant.

Browner had immersed herself deeply in the coalition-building politics of cap and trade, however; the idea of starting over with a carbon tax
proposal was, at best, politically impractical. Obama's chief energy policy
adviser concluded after the meeting that Tillerson "was happy to have a
position that nobody was going to embrace," as the participant put it.[9]

George W. Bush had narrowed the list of points he wanted to make to
his successor during his private handoff conversations with Obama that
winter. One topic he emphasized privately to Obama was the importance
of America's alliance with Saudi Arabia, and particularly, the quality of
the personal relationship between the American president and the Saudi
king. After the shock of September 11, Bush had invested great effort to
rebuild trust with King Abdullah; Bush talked with Obama about how to
manage that bond.[10]

Obama's White House team was turning away from traditional, geopolitical thinking about oil and power, however. As his national security
team assembled, for example, the president's closest advisers turned aside
suggestions that he establish a special energy geopolitics section at the
National Security Council, similar to the one that had managed Eurasian
pipeline politics during the late Clinton administration. At the Department of Energy, Obama appointed a cautious scientist with no background in oil and gas, Steven Chu, as secretary. At almost every decision
point, Obama emphasized renewable energy investments and greenhouse
gas limitations as the pillars of his energy policy. In bilateral meetings
with the Saudis, the Obama energy policy envoys stressed solar power
cooperation that could feed the sun-saturated desert kingdom with sustainable electric power. If Obama had thought much about oil pipeline
routes in the Caucusus, freedom of maneuver for oil tankers in the Gulf

of Guinea, or European natural gas supply security, his instinct seemed to be to set aside that sort of strategizing.

One month after his inauguration, Obama flew to Ottawa, Canada's canal-laced capital city. By tradition, new American presidents made their first foreign trip to Canada. As it happened, that winter, one of the biggest issues in U.S.-Canadian relations involved the geopolitics of oil and the security of American oil supply. The matter was also of deep importance to ExxonMobil.

Canada was by a wide margin America's largest single supplier of imported oil, at 1.9 million barrels per day in 2008. It was doubtful that many Americans could recite this fact; their ignorance reflected the fact that Canada posed almost no political risk to the United States, and so its role as an oil spigot for American consumers was inconsequential. That was certainly true in comparison with, say, the role of Saudi Arabia, America's second-largest supplier, located in a rough neighborhood far away, with a record of funding Islamist radicals and imposing oil embargoes over foreign policy disputes. Canada's underpublicized oil bounty included conventional reserves, but also a vast treasure of "crude bitumen," as ExxonMobil referred to it. Environmental activists often referred to these bitumen reserves as "tar sands oil," evoking images of a sticky mess of a sort that might have trapped unsuspecting dinosaurs eons ago. The dueling language reflected a profound disagreement about the oil's value.

The reserves in question lay 50 to 150 feet underneath the sands of the McMurray Formation, near the Athabasca River in northern Alberta Province, in western Canada. Lakes, streams, and boreal forests of stubby trees had covered the sands for centuries. By 2007, as new technology made it easier to separate the oil from its earthen mix at a reasonable cost, *Oil & Gas Journal* estimated that Alberta held 175 billion barrels in total oil reserves, which amounted to the third-largest national oil reserve in the world, after Saudi Arabia and Venezuela.[11]

ExxonMobil's Canadian affiliate, Imperial Oil, had been producing oil from the Alberta sands since 1978, through a joint venture called Syncrude. The operations required open-pit mining to dig out the oily sand with mechanical shovels fifty feet high. Hot water or caustic soda then washed the sand to separate out the bitumen. "Upgraders" similar to those ExxonMobil had installed in the Orinoco basin of Venezuela refined the remainder into a synthetic blend that imitated the refinery-friendly characteristics of light, sweet crude.[12]

The final product was highly desirable in world oil markets, but the manufacturing process was environmentally destructive, expensive, and energy intensive. It also required immense water use. Syncrude stripped forests, dug out peat and dirt, and then attacked the sands below with its giant shovels. Environmental investigators documented toxic pollution runoff from the mining operations. Moreover, the industrial processes required to extract and manufacture oil from bitumen required burning more carbon-based fuels than would be burned to drill a normal oil well. As climate change gradually emerged as a global threat, environmental groups also campaigned against the Alberta operations because of the extra polluting energy that was needed to dig out and refine the bitumen. Oil from Alberta, barrel for barrel, contributed among the highest greenhouse gas emissions of any source of oil in the world. Carbon sequestration technology might eventually allow Alberta's producers to capture greenhouse gases around the giant shovels and the upgraders and inject those gases into underground storage caverns, but that technology remained immature, unproven, and expensive.

"Climate Leaders Don't Buy Tar Sands" read a banner draped across an Ottawa bridge when President Obama's motorcade rolled in to the Canadian capital.

Over lunch with Stephen Harper, the pro-business prime minister, the Canadian cabinet team that worked on climate and energy was wary. The Bush administration had expressed no concerns at all about the pollution caused by companies operating in the oil sands; to the contrary, senior officials such as Energy secretary Samuel Bodman traveled regularly to Ottawa to convey the message, in effect, as a Canadian official involved put it, "Produce as much of this oil as you can—we'll buy all of it." Obama's position seemed unclear. Prime Minister Harper's advisers prepared for Obama a giant map of North America depicting—with drawings of bubbles of various sizes—the greatest industrial sources of greenhouse gas emissions on the continent. The map's biggest bubbles showed that American coal-fired electric power plants were the greatest climate change offenders. By comparison, the oil sands were relatively minor contributors to global warming, the map showed. Of course, the map avoided emphasizing that the sands were, in fact, Canada's greatest source of greenhouse gas emissions, by far, and would be for the foreseeable future.[13]

"What do you think? Is it dirty oil?" a Canadian Broadcasting Corporation interviewer asked Obama.

"What we know is that oil sands create a big carbon footprint," Obama answered. "So the dilemma that Canada faces, the United States faces, and China and the entire world faces, is how do we obtain the energy that we need to grow our economies in a way that is not rapidly accelerating climate change?"[14]

Obama had been among those American senators who had previously endorsed laws to limit oil imports from Canada derived from the Alberta sands, on environmental grounds. The 2007 Energy Independence and Security Act contained a provision, known as Section 526, that restricted U.S. federal agencies from procuring transportation fuel derived from any oil source with an unusually heavy carbon footprint. Section 526 was almost comically complicated because it defined the banned fuel sources through reference to statistical greenhouse gas emission averages that had never before been calculated for such a reason. Around the same time that Section 526 was enacted, California lawmakers also adopted a "low carbon fuel standard" for gasoline or other fuels used in that state. Lawsuits ensued. It remained unclear, for example, whether, under the North American Free Trade Agreement, California or the federal government had the legal right to limit Canadian oil exports in this way.

ExxonMobil had a big stake in the Alberta sands, but so did BP, among other oil multinationals. In a world of rising resource nationalism, Canada's oil, however dirty, offered the largest single reservoir in a free-market economy that could be easily acquired and owned by American or British corporations and their shareholders. They would not yield its potential casually.

The Washington lobbyists of the major companies banded together after Obama's return from Ottawa. Through the American Petroleum Institute, they launched a public relations and lobbying campaign to support expanded production from the Alberta sands. In Houston, the oil companies formed a front organization, the Consumer Energy Alliance, to fight against any proposed restrictions on Canadian oil. The campaigners appealed above all to American nationalism; by keeping out Canadian oil, lawmakers would only ensure that Alberta's oil was sold to China or Japan, leaving the United States even more dependent on unreliable suppliers from the Middle East and Venezuela. An ad taken out by the Consumer Energy Alliance in the *Weekly Standard* showed Caucasian schoolgirls at play, presumably Canadians: "Energy Security? The Answer Just Might Be Closer Than You Think."

ExxonMobil put its corporate shoulder into the lobbying campaign. David P. Bailey, an in-house specialist on the subject, joined a task force at the Council on Foreign Relations that was organized early in 2009 to produce a definitive study on the issue. The study group's advisory committee included, besides Bailey, a Chevron representative, business consultants, academics, and environmentalists. The final report, "The Canadian Oil Sands: Energy Security vs. Climate Change," was rigorous and thorough. It recommended addressing the greenhouse gas emissions problem posed by the Alberta sands through linked cap-and-trade systems in the United States and Canada. The report also expressed skepticism about Section 526 and similar fuel restrictions aimed at Canadian oil. "Tread carefully with any low-carbon fuel standard," it recommended. "Resist the misuse of other U.S. environmental regulations to constrain oil sands." Apart from its endorsement of cap and trade, the council report generally sided with ExxonMobil's positions.[15]

Throughout the oil sands lobbying struggles of 2009, ExxonMobil's executives were indignant about the issue, an oil industry lobbyist involved recalled. The corporation's internal analysts and lobbyists saw Section 526 and Obama's consideration of similar limits on Canadian imports as "more of an affront to the industry" than a legitimate public policy dilemma.

Canada's politics concerning the oil sands were complicated, but as a practical matter, there was virtually no chance that Alberta's provincial politicians or the country's national leaders in Ottawa would seriously limit Canada's production in the years ahead. Too much national wealth was at stake. For its part, ExxonMobil's Syncrude subsidiary possessed licenses to operate in Alberta through 2035; its holdings totaled at least 734 million barrels as of the end of 2008. A separate oil sands project called Kearl in which ExxonMobil was invested would soon produce another 110,000 barrels per day in bitumen.

All this meant, from ExxonMobil's perspective as a global corporation, that it did not necessarily have a Washington lobbying issue concerning the Alberta sands at all: If the United States were dumb enough, in the corporation's estimation, to restrict Canadian imports, then ExxonMobil would just sell the same oil to Asia. It did not even seem obvious to ExxonMobil why the oil industry should spend so much time and money lobbying on the issue in Washington—building consumer front groups, buying ads in newspapers, enduring accusations that it was out, once again, to destroy the environment—so as to convince the Obama administration and Congress to see Canada as a strategic friend and a source of energy security. To ExxonMobil's representatives, Canada's relative desirability as an oil supplier seemed one of the more obvious propositions in foreign and energy policy. If the Obama team did not agree, let it explain its thinking to the American public. "The whole industry was pretty pissed off about it," the lobbyist involved recalled.[16]

In March, Theresa Fariello, Dan Nelson's successor as head of Exxon-Mobil's K Street office, settled on a $1.4 million, four-bedroom townhome in a tree-shaded, secluded section of Georgetown. Her arrival as a Democrat representing Irving's interests in Obama's Washington marked the most significant investment Tillerson had yet made in a political strategy of his own. Fariello had grown up in modest circumstances, and some of her acquaintances thought of her ascension as ExxonMobil's chief Washington lobbyist as an American success story. During the late Reagan administration, the first Bush presidency, and much of Clinton's presidency, Fariello had lobbied for Occidental Petroleum in Washington. She had been active as well in the pro-business wing of the Democratic Party. One of the party's leading deal makers, perennial presidential candidate and New Mexico governor Bill Richardson, had hired Fariello at the Department of Energy during Clinton's second term. As Democrats left federal office after George W. Bush's disputed election, ExxonMobil recruited Fariello to its public affairs operation. She moved to Irving and worked closely with Ken Cohen during the Bush presidency. More than any ExxonMobil executive in Cohen's departments, she managed channels to Democratic lobbyists and allies during the Bush years.

While considering the dilemmas that Obama, Harry Reid, and Nancy Pelosi posed to the corporation, Fariello and Cohen had developed a multifaceted plan to bring ExxonMobil in from the political cold and to make the corporation more relevant in an age of Democratic ascendancy. Tillerson's carbon tax endorsement was just one prong. Cohen authorized contributions to the Clinton Global Initiative, to advance and publicize ExxonMobil's global programs on women's rights. He designated Lorie Jackson, an African American lobbyist in the Washington office with graduate degrees from Stanford University and Harvard University, to represent this charitable and policy push at public events.

ExxonMobil also announced that summer a potential $600 million investment in algae-derived biofuels that might replace gasoline one day, in partnership with the genetic researcher J. Craig Venter. The investment plan involved deep scientific uncertainties, but it did conform to ExxonMobil's criteria in considering alternative energy initiatives, namely, that it would invest only if the payoff would be relevant to its core businesses and would be potentially scalable and transformative. The algae announcement won widespread and uncritical news coverage; it appeared that ExxonMobil had finally internalized BP's example that, in calculating the full costs of alternative energy investments, unpopular oil companies should factor in the marketing benefits of free favorable publicity.

In Washington, Fariello invited State Department officials, ambassadors, congressional staff, and other influential arrivals to Obama's administration to hear ExxonMobil's energy futures briefing, its gospel of

2030. The corporation purchased billboard space in the Washington Nationals' new baseball stadium, on the Anacostia River, and erected image ads depicting ethnically diverse ExxonMobil scientists and engineers surrounded by photographs of molecules and other scientific signifiers. The strategy was working, Cohen told the Management Committee in Irving: ExxonMobil's favorability ratings in its internal American and Canadian public opinion polling soared during 2009, from about 30 percent favorability the year before to about 50 percent. The internal polls showed that all of the big oil companies—ExxonMobil, BP, Chevron, and Shell—had recovered some of their public reputation in the United States, more or less in tandem. Perhaps the ratings were best understood as simply a reflection of public feeling about retail gasoline prices, which plunged during 2009 because of the recession. The strategy overall, said an executive involved, was premised on the belief that Democrats in Obama's Washington "don't want B.S. They don't want greenwashing. . . . Let's establish a tone. . . . We are prepared to disagree and hopefully we can think about it in a way that's mutually respectful."[17]

ExxonMobil was partially constrained, however, by the fact that its Democratic Party connections were heavily located in the failed presidential candidacy of Hillary Clinton. Theresa Fariello had worked actively to support Hillary Clinton's candidacies for senate and president. Moreover, the principal Democratic lobbyist Fariello worked with in Washington, David Leiter, a former chief of staff to Senator John F. Kerry (D-Massachusetts), happened to be married to Tamera Luzzatto, Hillary Clinton's chief of staff in her Senate office during the 2008 campaign.

The Clinton universe was in general more connected to Fortune 500 executive suites than the Obama campaign had been. While raising campaign funds throughout the 1990s, Bill Clinton had developed friends in virtually every American industry. When he first ran for the U.S. Senate in 2004, by contrast, Obama had not needed decisive access to national corporate funding because his main Republican opponent dropped out early in the race, following a sex scandal; Obama won effortlessly. In his run for the White House, Obama had tapped Hollywood and Wall Street for support, and he attracted allies in the executive suites of large technol-

ogy companies such as Google, but he did not have the breadth of connections to the Fortune 500 during the primary campaign that Hillary
Clinton enjoyed.

In Congress, early in 2009, allies of Obama's overthrew one of ExxonMobil's most stalwart Democratic allies, Representative John Dingell
(D-Michigan), chairman of the House Energy and Commerce Committee, where all climate and energy legislation originated. Representative
Henry Waxman (D-California), a pit bull of a liberal whose West Los
Angeles constituency adamantly supported aggressive action on climate
change, succeeded Dingell. Obama's White House team cheered the coup
on Energy and Commerce because they believed it would make possible
a big cap-and-trade bill, which Dingell might have resisted.

David Leiter and other ExxonMobil lobbyists, such as Kelly Bingel, a
former chief of staff for Senator Blanche Lincoln (D-Arkansas), called
Democratic members and staff in the House frequently to arrange meetings with ExxonMobil's staff lobbyists and executives. A staffer for one
centrist Democrat in the House estimated that, whereas ExxonMobil
might come by at most once a year in previous eras, its lobbyists were
seeking meetings in 2009 almost at a rate of once per month. "They don't
come in to be combative," the staffer said. "Three quarters of the time they
are defensive on something—cap and trade or tax issues. . . . I think people sense that Exxon's scared," he continued. "They need to find new
friends."[18]

Sherri Stuewer, one of the relatively few women at ExxonMobil to rise
to senior management, now served as the corporation's lead adviser on
climate issues. Her title was vice president of environmental policy and
planning; she was a confident-looking woman in her fifties with shoulder-
length auburn hair. During the first week of June 2009, she arrived at a
farm along the Florida-Georgia border, White Oak, an unusual conference
center built by the heir to a paper fortune, Howard Gilman. He had populated his land with endangered African and Asian species, including rhinos, giraffes, okapi, tigers, cassowary, bongos, and guars, cowlike beasts

from the Indian subcontinent. Safari vans allowed guests to view the animals. A stuffed polar bear loomed over the conference center's game room. Adding to the Jurassic Park-inspired atmosphere of eccentricity, Gilman had also established, on the same estate, a dance facility for the ballet maestro Mikhail Baryshnikov.

"Power Politics: Moving Americans to a Clean Energy Future" was the private three-day meeting, sponsored by the Washington think tanks Third Way and the Center for Policy Innovation, that drew Stuewer. Other corporate representatives from companies that supported cap and trade arrived, as did Governor Joe Manchin of West Virginia, a staunch advocate of coal interests, and environmentalists from influential groups such as the Natural Resources Defense Council and the Environmental Defense Fund.

A cap-and-trade bill known as Waxman-Markey, after its sponsors, was on the verge of passing the House. The bill was lengthy and complicated, but it brought some Democrats from coal states such as Virginia into political alliance with West Coast environmentalists such as Waxman himself. However, the legislation's chances in the Senate, where the filibuster created a de facto requirement for a sixty-vote supermajority, looked doubtful. ExxonMobil had changed its public posture by endorsing a carbon tax, but it did not want cap and trade to succeed. The corporation and its allies had also kept all restrictions on Canadian oil imports out of the Waxman-Markey bill in the House; they wanted to ensure nothing of the kind resurfaced from the Senate. Part of the purpose of the White Oak meeting was to outline strategies that might push some sort of carbon cap or price through the Senate, despite the obstacles.

Stuewer seemed overwhelmed by work and did not hang out much in the game room or the bar, but she participated in the break-out sessions. Tony Kreindler of the Environmental Defense Fund "got into a spirited discussion with her," as he recalled it. Kreindler knew all about ExxonMobil's support for a carbon tax, and in his circles, "there are two explanations—one is paranoid—that they were reading the tea leaves and proposed a poison pill they knew would never pass. The other explanation, which I'm inclined to believe, is that Exxon believed some [carbon pricing] mechanism was inevitable, and they took a very hard look at their business model and decided they could simply out-compete everyone else if the policy were a carbon tax."

Kreindler and Stuewer talked in detail about changes to the Waxman-Markey formulas that might bring it closer to something ExxonMobil could accept, by structuring the cap-and-trade regime so that it looked and worked more like a straight-up carbon tax. "We need certainty," Stuewer said, according to Kreindler. That is, ExxonMobil could accept a carbon price if it knew that the price would not gyrate wildly.

Kreindler pressed her. Why were they here? "Because we need to reduce emissions or because you need price certainty?"

Stuewer pushed back: The solution ExxonMobil recommended would, in fact, reduce emissions; cap and trade was a less certain mechanism.

"Why don't you support something that would actually make a difference?" Kreindler asked. Stuewer insisted that she thought she was.[19]

As the scrum over climate policy on Capitol Hill took form that summer, ExxonMobil lobbyists made two arguments to fence-sitting congresspeople and senators. Even if cap and trade were phased in, so that its costs did not hit the United States until the economy recovered fully from recession, they said, the system would nonetheless destroy jobs and growth. A Brookings Institution study showed, in fact, that the Obama and Waxman proposal might take about 2.5 percent out of American gross domestic product during the next forty years—the equivalent of one year's economic growth—as the cost for removing between 110 and 140 billion metric tons of CO_2 from the atmosphere. That was not a daunting price if climate change was accepted as a grave national danger, but among congresspeople whose constituents in 2009 suffered from personal bankruptcies, mortgage defaults, and even homelessness, it was not an easy trade-off to accept. The same Brookings study showed that cap and trade would destroy about 15 percent of jobs in the oil and coal industries by 2025, although it would have virtually no effect on employment outside of the energy sector. ExxonMobil emphasized those forecasted job losses to members of the Senate from the oil and coal states that would be hardest hit.[20]

The corporation's representatives also pointed out that the public did not understand cap and trade very well. Among those who had heard of the policy proposal at all, more than twice as many of those surveyed had negative as positive assessments. During 2009, ExxonMobil lobbyists left behind in congressional offices PowerPoint slides documenting a private Hart Research Associates poll, "Energy and Climate Change Policy," that had been commissioned by the U.S. Climate Task Force, the pro–carbon tax coalition. The poll showed that Americans preferred a carbon tax to cap and trade, and they strongly favored a tax when educated about the arguments in favor of each approach.[21]

O n the morning of September 23, 2009, Rex Tillerson arrived at the Sheraton on Seventh Avenue to participate in the "Investing in Girls and Women" Plenary Session at the Clinton Global Initiative. Delegates in the ballroom sat with name tags dangling from their necks. The decor was tinted blue, purple, and red; two large screens on the sides of a stage flashed photographs of dignified Third World poor.

Bill Clinton took the stage wearing a brown suit; it was an early hour, and his baggy eyes and raspy voice hinted at a late night. "It was very important to me that this issue of women and girls be highlighted" at this year's initiative, he told the audience. He invited a number of "commitment makers" to the stage to congratulate them on their philanthropy before the audience. Muhammad Yunus, the Nobel laureate and founder of Grameen Bank, was among the honored; so was Lorie Jackson of ExxonMobil.

Clinton introduced Diane Sawyer. She summoned to the stage members of a panel to talk about women's issues; the panel included Tillerson and Lloyd Blankfein, the chief executive of Goldman Sachs, the investment bank with a public reputation as much in need of repair as ExxonMobil's. The ballroom atmosphere suggested the laying on of liberal, globalized hands to cleanse sinful multinational corporations. "These are some of the power hitters," Sawyer said of Tillerson and Blankfein.

Tillerson talked about ExxonMobil's charitable initiatives to support girls and women in some of the poor countries where the corporation extracted oil. "Technology comes very natural to ExxonMobil," he said. "What are the technologies that will provide them [girls and women] capabilities to undertake their daily activities in a more effective and efficient way?"

Sawyer later asked him: What is the responsibility of a multinational corporation to make the world better through charitable activity? Is it a tithe of 10 percent? How much?

"Ultimately," Tillerson said, "this is our shareholders' money we're spending. It's not my money to tithe. It's not the corporation's. It's our shareholders'."[22]

By late 2009, whatever anxieties Tillerson and Cohen might have possessed as the Obama administration took office, it had become apparent that ExxonMobil would prevail, again, on the public policy issues that mattered most to the corporation. Cap-and-trade legislation died a slow death in the U.S. Senate; its proponents could not construct a filibuster-proof majority. In Copenhagen, in December, representatives of the world's major economies failed to agree on post-Kyoto rules that would deliver serious reductions in greenhouse gas emissions. (ExxonMobil sent its astrophysicist and climate policy advocate, Brian Flannery, to attend the Copenhagen negotiations; Flannery sought to educate media and delegates about the issues, as the corporation saw them.)

In Irving, ExxonMobil's corporate forecasters monitored the Copenhagen talks, but they had by now concluded that even if some sort of international protocol on climate were reached, it would not actually affect emissions very much. "The implementation becomes extremely complicated, extremely political, and it's hard to see that expanding on a really, really wide scale," one of the executives involved said. Transformations as China industrialized further would do much more to determine the world's climate future than negotiations such as those at Copenhagen, ExxonMobil's analysts concluded.

Tillerson and his colleagues shifted their image advertising and lobby-

ing messages to emphasize jobs, which their internal polling showed resonated strongly. Tillerson returned to Washington in October to speak at
the Economic Club, at the luxury Ritz-Carlton hotel.

On ExxonMobil's carbon tax proposal, which Tillerson had unveiled
in Washington almost a year earlier, the chairman said, "I hope you see it
shows how serious we are about this issue. . . . We're engaged heavily. . . .
We need to get this as right as we can."[23]

Twenty-six

"We're Confident You Can Book the Reserves"

Before 2003, the marble floor in the entrance lobby of Baghdad's flag-ship Al-Rashid Hotel contained an inlaid tile mosaic of America's forty-first president—not a convincing likeness, but a recognizable one. The mosaic rendered George Herbert Walker Bush in a suit, tie, and pocket chief, accompanied by a caption: "Bush Is Criminal." After the invasion of Iraq led by America's forty-third president, George W. Bush, the Al-Rashid's ownership redecorated.

The hotel was a tall, rectangular building with concrete balconies, surrounded by green lawns and palm trees; it presented an inviting target to the mortar squads of the anti-American insurgents who embroiled Baghdad in violence after the invasion. The Al-Rashid lay within the walled Green Zone, where American and Iraqi authorities sought to gov-ern and stabilize the country. Rocket attacks, parades of demanding inter-national visitors on tight schedules, and repetitive conferences aimed at arresting Iraq's downward spiral took a toll on the Al-Rashid's staff and ambience.

On June 29, 2009, Richard Vierbuchen, an ExxonMobil vice presi-dent in charge of upstream operations in the Middle East, made his way to the Al-Rashid's ballroom, which had been decorated emphatically in green, the color of Islam. Rows of green cloth chairs faced a stage backed by a green curtain; on the stage, officials from Iraq's oil ministry sat at a long table draped with shiny green bunting. To one side, propped on a stand, stood an empty glass box about the size of an aquarium. Television lights saturated the room in brightness.

Vierbuchen was a tall, slim, athletic-looking man in his fifties. He held a doctoral degree in applied geophysics from Princeton University and had worked in the Arab world and Central Asia for the corporation for many years. After 2003, ExxonMobil's Global Security and political risk analysts had advised Tillerson that the corporation's executives should stay away from Iraq, as the threats of kidnapping and violence were too great. Vierbuchen and his colleagues had talked with Iraqi government officials about the possibility of ExxonMobil's bidding for work in Iraqi oil and gas fields, but these meetings took place outside the country—in Jordan, Turkey, the United Arab Emirates, Europe, or the United States.

Now, at last, a somewhat calmer Iraq was on the verge of putting some of its immense reserves—the second- or third-largest combined oil and gas bounty on the planet, after Saudi Arabia and alongside Russia—up for bids by international oil companies. Iraqi politics and pride demanded that companies wishing to participate in the auction send executives in person to Baghdad. Vierbuchen could be assured, at least, that thousands of American soldiers still remained in the Iraqi capital, providing security to the Green Zone and other critical installations.

After the Second World War and before the rise of resource nationalism in the Middle East, ExxonMobil had shared in a lucrative oil concession in Iraq. The corporation had owned and produced some of the country's richest, sweetest reserves, located near the head of the Persian Gulf. Saddam Hussein had thrown Exxon out in 1972.

The Al-Rashid auction had the trappings of a game show. Prime Minister Nouri Al-Maliki delivered a speech. The American ambassador, Christopher Hill, came to listen and watch. Live on national television, one by one, oil executives and Iraqi officials dropped envelopes of varying colors into the square glass box onstage. Screens in the ballroom tabulated cer-

tain scores for each bid, using guidelines issued by the Iraqi Ministry of
Oil. The scores sought to measure which oil company offers would most
benefit Iraq. The large, invited ballroom audience applauded when Iraqi
officials opened some of the envelopes.

The day's purpose was to provide Iraq a much greater reward from
its oil wealth. Six years after the American invasion, despite the Bush
administration's investment of $4 billion of U.S. taxpayer funds to reha-
bilitate the country's oil infrastructure, and the loss of countless lives, Iraqi
crude production remained stuck at about 2.5 million barrels per day,
below its peak during the Saddam years.[1] Moreover, Iraq's anti-American
insurgency had evolved by 2006 into a vicious civil war along Iraq's Sunni-
Shia sectarian fault line. The war was one reason why the oil industry had
underperformed. Decrepit pipelines dating to the Saddam era, inadequate
electricity, and field production technologies that had barely been updated
since the 1970s meant that even when violence gradually subsided, after
the "surge" of U.S. forces President Bush ordered into Baghdad in 2007,
Iraq still lacked the means, on its own, to raise oil production quickly.
Outside estimates of the investments needed to bring Iraqi production up
to 6 million barrels per day or more—a target easily within reach, if geology were the only factor—ranged from $25 billion to $75 billion. Iraq had
organized the Al-Rashid auction to accept bids from potential foreign
investment and technology partners—not only oil corporations from the
free-market West, but state-owned companies from Russia, China, and
India as well. The glass box and the television cameras were intended to
assure the public that the outcome would be transparent and not subject
to the sort of corruption and official theft that was otherwise prevalent in
the country.

Richard Vierbuchen had been involved in the negotiations and paperwork to "prequalify" ExxonMobil for the auction. He regarded the reopening of Iraqi oil to outside investment as one of the world's great oil
opportunities—a unique prospect, given the undeveloped size and proximity to sea-lanes that Iraq's big southern fields offered. The oil was of
exceptionally high quality, "almost like engine oil," as an American government official who worked with Iraq on its oil industry put it.[2]

Vierbuchen rose from his chair when ExxonMobil's turn arrived. He ascended the stage and walked toward the glass box. He carried a tan envelope. The bid inside contained, in accordance with rules set by the Iraqi oil ministry, just two numbers: a target for increased production in the particular oil field for which ExxonMobil was bidding, and the price per barrel, in U.S. dollars, that the corporation wanted to be paid. Vierbuchen smiled awkwardly for the cameras and dropped his envelope in the box.

ExxonMobil's bid on that summer's day in 2009 was the product of six years of patience, lobbying, and a willingness within the corporation's leadership to work at times outside of the normal ExxonMobil playbook. The uniqueness of Iraq's position as a war-scarred oil giant that had vast reserves and easy export routes demanded creativity. It also required a subtle understanding of the ways that Iraq's oil future would run through Washington.

On July 26, 2006, a typically hot and muggy summer's day in the American capital, Oliver Zandona, who followed Iraq oil issues in ExxonMobil's K Street office, arrived at 1000 Independence Avenue, the headquarters of the United States Department of Energy. George W. Bush's friend and secretary of energy, Sam Bodman, had arranged a meeting between Iraq's visiting minister of oil, Hussain Al-Shahristani, a gray-bearded Shia politician, and representatives of major oil companies, both American and European.

Zandona and other oil company representatives peppered Shahristani with questions about when Iraq might finally pass a law specifying the legal rights of international companies and when it might be in a position to provide adequate security. They advocated production-sharing agreements and "stressed that certainty and consistency in laws and stable taxing regimes are important," according to minutes of the session.

Shahristani said he was hopeful that partnerships with international oil giants might soon be possible. He had a message, however. The oil companies should not talk about contracts to anyone in the Iraqi government or in Iraq's diverse political parties but him, the ministry official in charge. Companies should not attempt to cut side deals that were not directly negotiated by the oil ministry.

Shahristani raised his index finger in the air. "Clean games, gentlemen," he told the company lobbyists and executives. "Clean games."[3]

It was an aspiration, at least.

Zandona helped to shape ExxonMobil's basic strategy for Iraq after the 2003 invasion. Essentially, the strategy was to wait out the war and maintain a healthy public distance from the tainted Bush administration, while also remaining in close private contact with American officials to stay up to date and to push for policy that would lead to foreign investment in Iraq's oil fields. "From Exxon's perspective, Iraq is the last of the easy oil," said a State Department official who interacted with the corporation frequently.[4]

Immediately after the U.S.-led invasion, ExxonMobil and other international companies bought or "lifted" Iraqi crude at southern terminals, at the head of the Persian Gulf. These direct oil sales to well-known international oil companies were designed and authorized by the Bush administration's occupation authority in Baghdad, the Coalition Provisional Authority. Saddam Hussein's regime had sold oil legally under the United Nations Oil-for-Food Programme (although many of these deals were compromised by payoffs and corruption) and also off the books, by smuggling through middlemen. When the United States established power in Baghdad, its occupation leaders wanted to assure Iraqis that the days of Saddam Hussein's shady oil smuggling and corruption were over, and that profits from Iraqi crude would not be skimmed off by fly-by-night traders and middlemen. Iraq did not have the capacity to ship its own oil to spot markets in Europe, where cargoes of oil were bought and sold for cash, or to refinery customers in Asia; its oil had to be sold to middlemen, for onward sale. The postwar oil sales to well-known Western oil companies benefited ExxonMobil's downstream and oil-trading operations, but the margins on such trades were thin, and the deals were a far cry from the big and profitable prize of owning a piece of Iraq's upstream oil production over the long run.[5]

As it sold ExxonMobil Iraqi crude in 2003, the Bush administration also urged the corporation to open a Baghdad office to lend support and credibility to what its occupation leaders hoped, naively, would be a rapidly normalizing country. Lee Raymond declined. By October of that year, violence was spreading across Iraq. The Coalition Provisional Authority increasingly was not in a position to make international oil deals; it was struggling to provide Iraq with adequate supplies of gasoline and electricity. An Iraqi government that would be credible enough with the Iraqi people to bargain on a subject of such symbolic and material importance as oil production partnerships with foreign companies looked a long way off.

"If you don't feel like you can protect your people, [if] you put them in a position where they are really vulnerable, then you don't have any business being there," recalled Lee Raymond. "It just seemed for a time, with everything going on in Iraq, [it would be] a long time before anybody could meet the threshold of putting people in there."

The corporation ran a regional office in Amman, Jordan, where Iraqi businessmen and refugees gathered as the war darkened, and another in Dubai. ExxonMobil used this forward position to monitor events and meet with Iraqi officials and executives from its historically state-owned oil companies. The corporation hired a retired U.S. Army colonel who had run security for the oil policy section of the Coalition Provisional Authority to help advise on security decisions. In Houston, Scott Ingersol, who ran ExxonMobil's Iraq program, assembled a team of experts, including former Bush administration officials who had learned about Iraq's oil during the war—lawyers, business development specialists, and political risk analysts. If and when there was a way in, they would be ready to move quickly.

ExxonMobil had scant knowledge about the leading executives and managers in Iraq's decaying oil industry. Aging technocrats—many trained in the United States, the United Kingdom, and Europe before Saddam Hussein's coup—ran oil production, but they had been isolated and monitored by Saddam, who distrusted all but his most loyal clansmen. The corporation gradually rebuilt personal contacts within the Iraqi oil sector, particularly with senior technicians trained in the West and with promising younger engineers who wanted up-to-date training on technologies. In 2005, spurred by the Bush administration, Iraq signed memoranda of un-

derstanding with about thirty international oil companies, including ExxonMobil. The companies agreed to provide training to Iraqi oil technicians
and to carry out studies.

"Basically," said an American official involved in constructing these
initial deals, ExxonMobil and its American and European peers told the
Iraqis, "we're going to spend $20 million on you guys over the next years.
All you need to give us is people. What's in it for us? Well, hopefully, we
can develop a relationship with you, impress you with what we've got,
and do business in the future."

In Washington, Oliver Zandona, Robert Haines, and others in the K
Street office met continuously with American officials working on Iraq
economic and oil policy.

"Your job is to promote U.S. companies" was the essence of Exxon-
Mobil's message, recalled a State Department official who heard it regularly. "Do your best to create a level playing field. If you put us up against
any state-run company" from Russia or China, "on a level playing field,
we'll win." Beyond that, the corporation wanted the Bush administration
to "stay as far away from the oil sector as possible," the official recalled,
because otherwise, the United States would be "perceived as meddling,
and [would] be a P.R. problem."[6]

Then, too, fierce Iraqi nationalism after the American invasion ensured
that no elected government in Baghdad could blithely trade away ownership of Iraqi oil to foreign companies, even if doing so might enrich the
government's coffers. Iraqis ratified a new constitution that declared that
the country's oil belonged solely to the Iraqi people.

Shahristani's "clean games" visit to Washington in 2006 signaled, however, that the Iraqi oil ministry had gradually come to recognize it could
not restore Iraqi oil production to acceptable levels on its own. After
the talks with Energy secretary Bodman and the oil company lobbyists,
Shahristrani soon negotiated what he called "technical service agreements"
with ExxonMobil and other companies. Under its deal, ExxonMobil
would evaluate how to raise the rate of oil production in particular Iraqi
fields.

The corporation used its access to develop and present to Iraqi officials an ambitious, unpublicized plan for ExxonMobil's reentry into the

country: a $100 billion "transformative" program under which Exxon-
Mobil would lead huge, staged investments in Iraq's oil and gas fields.
These upstream investments would be integrated with similarly ambitious
downstream projects—refineries and petrochemical complexes. In essence, ExxonMobil proposed a strategic position in the Iraqi oil industry
comparable to its dominant position in Qatar. The scale of the plan was
bold—and also politically unrealistic. International oil investments in Iraq
brought to the surface all of the ills and paralysis of Iraq's postinvasion
politics. ExxonMobil's PowerPoint slides with photographs of shiny infrastructure, arrows showing the benefits of integration, and lots of big dollar
numbers could not change that.[7]

Whatever deals emerged, Shahristani's oil ministry wanted to pay the
companies in oil, not cash, so as to avoid the complications of negotiating
in Baghdad with a separate political fiefdom at the Ministry of Finance.
It wasn't clear, however, that companies that received oil from the deals
could sell it freely, as they had under authority of the now-defunct Coalition Provisional Authority. Kuwait's government had won a war reparations award against Iraq's government, to pay for losses caused by Saddam
Hussein's 1990 invasion. The enforcement of this award meant that cash
from the sale of Iraqi oil normally ran through United Nations–authorized
accounts in New York, where a portion was garnered to pay Kuwait. It
was not clear whether ExxonMobil's sales of Iraqi oil might be subject to
the same garnishing. Zandona and other ExxonMobil lobbyists repeatedly
pressed the Bush administration to sort out the issue at the United Nations and with Kuwait and Iraq, so that the corporation could lift Iraqi
crude without any legal ambiguity.

If Iraq's greatest sensitivity was sovereignty, ExxonMobil's imperative
was reserve replacement—the need to be able to continually book oil
reserves in a manner approved by the Securities and Exchange Commission. If Iraq's oil opening were to make a difference to the corporation,
deals had to be structured so that ExxonMobil could book reserves.

Initially, after the war began, ExxonMobil and other major American
oil companies pushed the Bush administration to persuade Iraq's government to adopt production-sharing agreements or other contract terms
that would allow private oil companies to book Iraqi reserves for Wall

Street. An early study of Iraq's oil industry carried out for the Coalition Provisional Authority by the consulting firm BearingPoint, Inc., suggested to Iraqi oil officials the benefits of production-sharing deals; the study cited nations such as Azerbaijan as examples.

In Baghdad and Washington, Bush administration officials told their Iraqi counterparts that contract matters were up to them, said a State Department official involved, but they also made clear to the Iraqis, as the official put it, "You lack capital, you lack technology, and your workforce needs to be reeducated. You have to offer some incentives to allow the companies to come in. Of course, the whole issue is booked reserves."[8]

What hope ExxonMobil had to own a share of Iraq's oil rested on the long, costly efforts of the Bush administration to remake Iraq's oil policies after the invasion. From the beginning, executives from the American oil industry, almost all of them from Texas, had led the Bush administration's efforts; they volunteered for the mission, uncompensated. Philip Carroll, the former Shell U.S.A. executive and friend of George H. W. Bush, who had been recruited before the 2003 invasion to advise Iraqi and American officials about Iraq's oil sector, had resisted the Bush administration's most ardent free-market ideologists, particularly those who advocated outright privatization of Iraq's state-owned oil companies. When analysts at the conservative Heritage Foundation called the privatization of Iraq's oil "a no-brainer," Carroll quipped, "It's a no-brainer: Only someone with no brains would think of that."[9]

After Carroll left the Coalition Provisional Authority, a succession of other American oil industry veterans arrived in Baghdad one after another to serve as senior advisers on oil restoration and policy: Rob McKee, who had run upstream operations at Conoco, and Mike Stinson, another former Conoco executive. They slept in trailers in the Green Zone, worked in dusty sections of the Republican Palace, endured nightly mortar rounds, and braved ambushes when they rolled in armored sport-utility vehicles across town to the oil ministry for meetings. Robert Morgan, a British oil industry veteran, died in one such insurgent ambush. Diaries the American oilmen kept during their deployments as advisers depict a dizzying atmosphere in which a few patriotic, talented Iraqi oil technocrats struggled in an environment of political dysfunction, rumors of corruption, and constant violence. On the American side, McKee and Stinson, who considered themselves conservative patriots called to public duty, and who carried firearms outside the Green Zone to protect themselves, found their idealism sapped by petty bureaucratic infighting in Washington and reams of federal government paperwork.

The American oil advisers in Baghdad after 2003 knew well what their colleagues in the international oil business wanted from Iraq: legal and financial clarity, and contract terms that allowed bookable reserves. If they had any doubts about that, oil executives arrived regularly in Baghdad to lobby them. Sam Laidlaw, a Chevron executive, turned up in Baghdad early on "to see if he could represent a consortium of American companies to put possible [production-sharing contract] terms in front of the Iraqis" and to "find the right kind of financial vehicle that would get around the U.N." controls on Kuwaiti claims.

When Iraqi political leaders heard that sort of talk about foreign ownership of the country's oil, it only inflamed the "politics or kind of the grassroots emotional issues of ownership of a national resource," recalled Norm Szydlowski, a Chevron veteran who worked with McKee and Stinson as an adviser in Baghdad. "Many people [were] convinced that the U.S. presence and the coalition [was] here to 'take their oil.' "[10]

The American oil advisers were wealthy men at the ends of their careers; they were not in Baghdad to make money, but rather to have an adventure and serve the Bush administration. Their outlook about the bookable reserve question was that, first, it was up to the Iraqis to decide, but second, in the long run, what was good for Iraq probably would also be good for Western oil companies. It seemed to the American advisers that Iraq's democratic government should first establish a strong national oil company, comparable to the ones in Saudi Arabia, Kuwait, and many other Middle Eastern nations. A unified, integrated Iraqi national oil company and its political masters could provide Baghdad's politicians with a sounder footing to decide whether, as in Africa and some parts of Southeast Asia and Central Asia, the government wanted to invite foreign companies in and allow them to book some of Iraq's oil, in exchange for capital and up-to-date technology, or try to go it alone, perhaps with service companies such as Halliburton and Schlumberger. "Although I'm a free-marketer and a capitalist," Rob McKee observed, "it was very apparent to me and everybody else that Iraq needed a central energy policy."[11] That had to precede decisions about sharing reserves with international companies.

The invasion had shattered the old Iraqi state, effectively breaking its politics into sectarian and regional pieces tied together by a weak, continually renegotiated, inexperienced regime in Baghdad, one lobbied heavily not only by the United States, but also by Iran. Year after year, the country's new leaders talked about oil policy reform, but they could not agree on a national oil law.

During this interregnum, while ExxonMobil and the other international majors hunkered down in Amman and Dubai, small-time operators and freelance wildcatters—Israelis working through Turkish front companies, Russians, and American and British adventurers—turned up in the Green Zone looking for fast deals. Some individual wildcatters drove themselves by car from Jordan to Baghdad, barreling down insurgent-controlled highways, "dumb and courageous at the same time," as McKee put it.

The adventurers were aided by splits in Iraqi decision making about how to sell oil. Coalition Provisional Authority Order Number 39 banned new contracts that would exploit Iraq's natural resources until an elected government in Baghdad could be seated under a new democratic constitution. Iraq's northern Kurdistan minority, however, which had organized itself under an entity it called the Kurdish Regional Government, ignored this edict.

The best Iraqi oil fields lay in the Shiite-dominated south of the country and employed export pipelines leading to the Persian Gulf. The northern Kurdish regions contained large fields of heavier oil and natural gas, but options for overland export were limited. After 2005, Kurdish leaders interpreted Iraq's new constitution to mean that oil production would be "locally managed by the people living in the regions," as the Kurdish Regional Government's minister of natural resources, Ashti Hawrami, put it.[12] The Shiite-dominated government in Baghdad declared that Kurdish decision to be illegal and unconstitutional, but it found itself powerless to stop the K.R.G. from letting contracts. To attract risk takers to its legally disputed fields, the Kurds let production-sharing contracts with fully bookable reserves on terms designed to excite Western capitalists. Early takers included a small Norwegian producer, D.N.O., and a firm called Hunt Oil, headquartered in Dallas, which had multiple personal ties to the Bush White House.

Shahristani and other oil ministry officials in Baghdad warned Exxon-Mobil, Chevron, Shell, BP, and other major international firms that if they tasted Kurdish crude, they would never, ever see a drop of Iraq's bigger oil fields in the south. They also declared the production-sharing contracts let to D.N.O. and Hunt to be illegal under Iraqi law.

Oliver Zandona and other industry lobbyists pushed the Bush administration in Washington after 2005 to sort out the Kurdish mess. "The policy question before us was, 'What do you do with these contracts?'" recalled a Bush administration official involved in the discussions. "I have to say, I think we really screwed that one up."[13] By failing to make absolutely clear that Western oil companies should not cut independent deals with the Kurds, and by winking at the Kurdish contracts that had already been negotiated, the administration sent mixed signals to companies and Iraqi political leaders alike, and may have prolonged the country's debilitating stalemate over oil sharing.

The Kurds were the most pro-American faction in Iraq, however. There was a natural bias inside the Bush administration to cut the Kurdish Regional Government slack because of all the ways Kurds cooperated with Bush's state-building project.

During 2005 and 2006, Meghan O'Sullivan, the influential National Security Council official overseeing Iraq policy for the Bush administration, shepherded through a formal, classified, interagency policy review on Iraq strategy, including a subsection on Iraq's oil. The aim was to promote oil as a means of finance and political unity for the struggling Iraqi state. That policy was formalized in 2006, briefed to President Bush, and dispatched to embassies worldwide by State Department cable. The Bush administration urged Iraq to negotiate a single national oil law that would account for Kurdish autonomy, clarify the legality of the Kurdish wildcatter contracts, and provide a basis for large-scale investment in the future by the likes of ExxonMobil.

As a practical matter, however, because there was no sanction placed on companies such as D.N.O. and Hunt Oil that were cutting side deals to pump Kurdish crude, the Bush administration effectively signaled neutrality on the matter—as if what was at issue in the Kurdish sidebar contracts was just a business dispute, not a potential fault line through Iraq's ethnic and sectarian fabric. A firmer policy would have sanctioned D.N.O. and Hunt for jeopardizing Iraqi unity or banned their dealings outright under American law.

The Hunts were colorful adventurers whose private, family-run firm had chased oil dreams in hard places since 1934. The Hunt brothers famously cornered the world's silver market at one stage, precipitating a crash. Ray L. Hunt, a conservative maverick, became chief executive in 1974; later, he turned operations over to his son, Hunter L. Hunt, though the father remained an active player. By the time of George W. Bush's Iraq War, Hunt Oil described itself as a firm that specialized in "unconventional" resource deals; it owned and produced oil in the tribal hinterlands of Yemen, among other places. Risky Kurdish production-sharing contracts appealed to the firm's traditions and instincts.

The Hunts also enjoyed deep ties to the Bush White House. Ray L. Hunt served on the president's Foreign Intelligence Advisory Board, which reviewed intelligence issues and operations. Through the board, Hunt had access to classified assessments about Iraq's politics and economy. In addition, Jeanne Phillips, a Bush fund-raiser whom the president rewarded in 2001 with an appointment as U.S. ambassador to the Organization for Economic Cooperation and Development in Paris, France, had moved from the administration to become a senior vice president at Hunt Oil. Eric Otto, a young Republican campaign activist who worked on oil industry issues in Baghdad at the Coalition Provisional Authority, made his way from the Green Zone to Hunt Oil's Dubai office, where he helped Hunt put together the Kurdish deals. Given the political pedigree and backgrounds of these Hunt executives, it would be entirely reasonable for the Kurdish Regional Government to assume that Hunt had some sort of sanction from Washington, even as State Department spokesmen clucked

that Hunt's search for Kurdish oil deals wasn't "particularly helpful" because it undermined efforts to negotiate a national oil law, as urged by
ExxonMobil and other larger corporations.

Ray Hunt flew into Erbil, the seat of the Kurdish Regional Government, in September 2007 to meet with K.R.G. president Masoud Barzani;
Hunt thereafter emerged as the first U.S.-headquartered oil company to
win a share of Iraqi crude, through a production-sharing contract that
would allow booked reserves and provided Hunt with lucrative profit
margins.

David McDonald, a Hunt executive, said the company hadn't asked
permission from the Bush administration to make this deal: "We do not
seek advice as to whether or not Hunt should proceed with an exploration
contract and we were never advised not to do so." President Bush said
publicly, "I know nothing about the deal." A subsequent State Department
investigation did not contradict Bush, but it found that administration
policy about the Kurdish oil contracts was "ambiguously articulated" and
lacked the force of law. Hunt had exploited the Bush administration's
foggy policies and benefited from its divided attitudes.[14]

The lure of quick oil profits in Kurdistan attracted liberal opportunists,
too. Peter Galbraith, a son of the economist John Kenneth Galbraith, was
a longtime Democratic foreign policy aide who served as ambassador to
Croatia under President Bill Clinton. Galbraith had long campaigned for
Kurdish rights and autonomy. After the 2003 invasion, he quietly became
a shareholder and partner in D.N.O.'s Kurdish oil deals; his stake grew to
tens of millions of dollars. As he profited from Kurdish oil, Galbraith advised Kurdish officials during constitutional negotiations with Baghdad
that concerned the disposition of Kurdish oil fields, wrote essays in the
New York Review of Books, and delivered speeches promoting Kurdish
autonomy without disclosing his own financial interests. Galbraith subsequently apologized to editors and readers of the New York Review; he
wrote that while his business arrangements were subject to confidentiality
agreements, "I should have stated that I had business interests in Kurdistan. I regret not having done so."[15]

Meghan O'Sullivan left the Bush administration before the Hunt deals, in September 2007. She became a professor at Harvard University. Late in 2008, she accepted a consulting agreement with Hess Corporation to provide political risk assessments about regions worldwide. These included briefings on the Middle East and Iraq, where she also provided introductions for Hess executives. O'Sullivan taught at Harvard about the geopolitics of oil, and she was interested in high-level strategic consulting, but she did not want to profit from particular oil deals in the country where she had overseen war policy, although Hess offered her that opportunity. She recused herself from Kurdish contract negotiations and declined compensation linked to specific transactions in Iraq. Hess eventually announced an agreement to produce oil in Kurdistan.[16]

The tan envelope Richard Vierbuchen dropped into the glass box at the Al-Rashid did not win a prize. For a moment, it seemed as if it might. There was "a buzz the room," as a U.S. embassy cable put it, because ExxonMobil proposed a large increase in output at Iraq's prized Rumaila field—the third largest oil field in the world—and said it would accept only $4.80 per barrel in what Iraq referred to as the "remuneration fee" that would be paid to international companies for their work, once certain criteria were met. Some oil pros in the audience said they were shocked that ExxonMobil would accept such a low per-barrel fee; according to one calculation, the bid would allow Exxon only a 9 percent rate of return, far below its normal global targets, and a profit margin normally accepted only in highly stable, low-risk political environments.

A staff member from the Ministry of Oil soon appeared with a red envelope, "which turned out to be the key envelope in this venture," as an ExxonMobil executive put it later. The red one contained a maximum per-barrel price—the maximum remuneration fee, or MRF—to be paid by Iraq's government to any international bidder. This turned out to be only $2 per barrel. "There was stunned silence," the American cable reported. "Giggles were heard. . . . Minister Shahristani grimaced on several occasions as he opened the bid envelopes and saw how far apart most bids were from his MRFs."[17] It became clear that ExxonMobil and virtually all of the international companies that had turned up at the Al-Rashid auction had asked for more money per barrel than Iraq's government was willing to pay, even though ExxonMobil and many of the others had reduced their target profit rates in order to get an initial share of Iraq's oil bounty. The auction, therefore, failed—except, perhaps, as reality television for entertainment-starved Iraqi audiences.

The event soon triggered a new round of direct negotiations between international companies and the Iraqi government designed to bridge the price gaps, however. The auction had been so transparent and regimented, in the opinion of ExxonMobil executives, that it left out subtler issues such as the potential benefits to Iraq of technology transfers, job training, and the like. The inability to negotiate and talk about these other issues at the Al-Rashid auction was the "dark side of transparency," as Exxon-Mobil's executives told their Iraqi counterparts, jokingly.[18] An Exxon-Mobil executive complained to the U.S. embassy in Baghdad that many aspects of Iraq's contract plans had not been finalized or made clear, and that in any event, "transparency and discretion should not be mutually exclusive." The executive declared that ExxonMobil would seek to "close" contracts outside of the public auction system. The corporation revived the ideas first presented in its 2006, $100 billion "transformative" plan for Iraq. The Bush administration's diplomat feared that ExxonMobil's private negotiations would "harm the perception that oil and gas contracts will be transparently awarded," a linchpin of Bush's oil policy in Iraq.

ExxonMobil complained to the Baghdad embassy that the Iraqi government had rejected a bid for the West Qurnah oil field "even though it would have generated up to $50 billion in investment, up to $600 billion in revenue to the [government of Iraq], and up to 200,000 direct and indirect jobs." The corporation's terms would have given Iraq "a 98 percent share of gross revenue," compared to a global average in the 70 percent range for such contracts. "The bid failed," an ExxonMobil executive said, because the Ministry of Oil "demanded a 99 percent share." He added that the corporation would seek to "educate" Iraq's government about global standards.[19]

Iraq's government had taken advice from sophisticated oil consultants and bankers to protect its interests and achieve its production objectives.

The provisional, proposed contracts it offered in private talks during the months after the Al-Rashid auction would pay out only after ExxonMobil reached a production target—the corporation would be paid only on barrels it produced above that threshold.

Vierbuchen and his supervisors in Houston and Irving accepted that principle. They had two issues to press in the weeks after the Al-Rashid auction, however. One was the price per barrel that ExxonMobil would be paid on oil produced after the threshold was crossed—in effect, the corporation's profit margin. The second was whether the form of the agreement would allow ExxonMobil to record the barrels with which it expected to be paid in the future as proven reserves. Otherwise, Exxon-Mobil's role in Iraq would look a lot like a service contractor such as Halliburton or Schlumberger, and Rex Tillerson made clear that that was not an acceptable contract model.

On price, because of the attractions of Iraqi oil and the enormous long-term potential of its reserves, Tillerson approved terms that would involve less gross profit than ExxonMobil normally sought—as little as $2 per barrel. He seemed to accept the judgment of analysts such as Nordine Ait-Laoussine, a former Algerian oil minister who had reopened his country to foreign investment before Iraq. In the Middle East, at least, Ait-Laoussine said, "The days of 12 percent to 18 percent returns are gone. Perhaps they should lower their expectations on earnings." In ExxonMobil's case, that would be something to consider only if it could add to the corporation's resource base. On reserve booking, the corporation turned to its lawyers.

The question was whether ExxonMobil could develop contract terms that would allow it to book Iraqi oil as proved reserves without forcing Iraq's government to accept production sharing or other contract forms typically rejected in the Arab world on nationalistic grounds. As far back as the 1980s, after Saudi Arabia nationalized its oil industry, Exxon lawyers had worked on contract formulations that might allow the kingdom to claim full ownership over its oil before its people, while structuring Exxon's position so that, even short of outright ownership, the S.E.C. rules would nonetheless recognize the corporation's right to book reserves for Wall Street. Now ExxonMobil and other international oil companies in final negotiations with Iraq's oil ministry returned to some of these earlier contract ideas. They agreed, for example, to be paid for their technical work in the southern oil fields in the form of oil, rather than in cash. (By now they were confident that they could sort out the Kuwait reparations matter by having the United Nations issue an international legal opinion affirming that their oil should not be garnished.) ExxonMobil wrote up terms describing their rights to that oil in ways that they believed conformed to S.E.C. requirements and talked these through with Baghdad's negotiators.

Late in 2009, Vierbuchen and the president of ExxonMobil Upstream Ventures, Rob Franklin, at last closed a deal—after months of private negotiations with Iraqi officials—for exclusive access to Iraq's West Qurna Phase One oil project. The field contained at least 8.7 billion barrels—a behemoth by industry standards. ExxonMobil agreed to apply modern technology and techniques to raise the field's production from its current 300,000 barrels per day to more than 2.3 million barrels per day within six years, taking a gross profit of only $1.90 per barrel—below the $2.00 per-barrel remuneration fee Iraq had specified at the first unsuccessful auction. The deal was structured so that ExxonMobil first had to increase the field's production to 10,000 barrels per day higher than the peak production achieved under Saddam Hussein. After that, ExxonMobil was permitted to take its $1.90 premium as oil in kind. ExxonMobil kept secret the full terms of the deal, so it was difficult to judge how profitable it might ultimately become. Deutsche Bank predicted that ExxonMobil might earn a 19 percent rate of return from West Qurna when all factors were considered—comfortably within the corparation's targets for profitability worldwide.

An initial agreement between ExxonMobil and Iraq stalled; the corporation "asked for advocacy" from the Obama administration's Baghdad embassy to resolve the impasse. "ExxonMobil has again approached Post [the U.S. embassy], seeking renewed advocacy to Prime Minister Maliki," the embassy reported to Washington. "Post will continue to press the issue at senior levels."

The Obama administration's lobbying helped; the deal went through. Eighteen months later, ExxonMobil had crossed the contract threshold and was loading Iraqi crude into supertankers in the Persian Gulf that could hold 2 million barrels at a time. The U.S. embassy in Baghdad estimated that the work of ExxonMobil's partnership in West Qurna alone would raise Iraq's oil production by 2 million barrels per day within six to eight years.[20]

Royal Dutch Shell was ExxonMobil's junior partner; BP worked nearby, under very similar terms, in the Zubair field. Iraq's government hoped that the country's long-promised potential as a 6-million-barrel-per-day power in global oil markets would soon be realized.

Just more than seven years after American soldiers and marines poured over the Kuwaiti border into Iraq, ExxonMobil shareholders owned, on paper at least, a small slice of the country's oil reserves. Seven years was almost precisely the length of time Lee Raymond had predicted, when the war began, that it would take for Iraq to be calm enough for big oil companies to enter. "I wouldn't say the profit margins have unlimited potential," said the corporation's Rob Franklin. On the other hand, he noted, "we're confident you can book the reserves."[21]

Twenty-seven

"One Plus One Has Got to Equal Three"

B ob Simpson was a tax accountant who wore his slacks stuffed inside his cowboy boots. When he was a young boy, an aunt brought him periodically to downtown Fort Worth, Texas, to shop at Leonard's department store, a wonderland of toys, sporting goods, and furniture. Crumbling brick buildings dating to the Texas oil boom of the early twentieth century surrounded Leonard's. Simpson grew up in modest circumstances in a small town nearby, graduated from Baylor University, took an accounting job in Fort Worth, and never left. When he began to earn big money, he bought up and restored many of the decrepit buildings he had seen as a child. On one occasion he paid $160,000 for the grand-champion steer at the Southwestern Exposition and Livestock Show and donated the animal to the Fort Worth Zoo. Increasingly, he was one of the city's most active patrons.[1]

Simpson was a numbers man. He kept books and organized tax returns for others until 1986, when he founded Cross Timbers Oil. Over the next two decades he built the company into a Wall Street darling. He acquired onshore American natural gas fields abandoned by the large international oil companies as they moved overseas and into deep-water offshore oil drilling in search of large new reserves. He also managed operations and financial strategy very tightly; Simpson became a master at growing through acquisitions.

He renamed Cross Timbers as the more ticker-friendly XTO; its profits grew very rapidly, from $186 million in 2002 to $1.9 billion in 2008, which vaulted XTO to number 330 on the Fortune 500 list of the largest stock market–traded corporations headquartered in the United States. *Barron's* named Simpson one of the thirty most-respected business leaders in the world for four consecutive years, alongside Warren Buffett and Steve Jobs.

His thinning hair had turned gray, and as he reached his sixties, he grew a not-so–Wall Street white beard. He gave up day-to-day management responsibilities at XTO, while remaining chairman, and the beard hinted that he might be ready for a further change of lifestyle. XTO now employed three thousand people, all of them in the United States, a third of them in Fort Worth. Simpson's stock option–incented executives and his Wall Street shareholders had become used to rates of profit growth that could not go on forever, certainly not in an industry whose performance was tied to volatile commodity prices.

In the summer of 2009, Simpson and XTO's senior executives and directors attended the corporation's annual management retreat at the Fairmont Chateau Whistler, tucked beneath the mountains of British Columbia, Canada. Simpson repaired to the hotel bar with Jack Randall, an XTO director who was a partner in an investment bank that specialized in oil and gas mergers. As they munched bar food, they talked about the industry and options for future strategy, including the possibility of a merger or an acquisition of XTO by one of the oil majors.[2]

The American natural gas business was in the midst of a historic boom as new drilling techniques unlocked huge reserves of domestic "shale" gas—natural gas trapped in shale rock formations—and other unconventional sources. XTO was a leading producer of shale and unconventional gas. It owned positions in most of the major shale gas plays in the United States, including the Marcellus Shale on the East Coast, which was exciting interest. The corporation's headquarters in Fort Worth stood near the Barnett Shale, one of the country's best-known shale gas reserves, where

XTO owned a large and lucrative position. A natural gas rush gripped Fort Worth as drillers, land men (who specialize in leasing land for drilling), and financiers scoured the region to grab positions. The nationwide boom atmosphere meant that natural gas production would likely rise and gas prices would fall. The financial crisis and recession of 2008 to 2009 had also dampened total energy demand, at least temporarily. Also, some of XTO's past success had been due to Simpson's financial wizardry in the futures and derivatives markets—his ability to enhance profit by locking in hedges on high gas prices, to guarantee strong cash flow and protect against market price declines. If prices fell for a prolonged period, hedging wouldn't produce the same degree of benefit. Big international oil majors continued to look at unconventional gas companies like XTO with avarice, despite the falling prices, because the majors had largely missed out on the domestic gas boom that XTO had ridden. For a wily numbers man like Simpson, these factors—prices past a peak, a boom mentality in the industry, and hungry, cash-rich corporate buyers—all flashed "sell."

Who would be an ideal purchaser, Simpson and Randall wondered? Chevron was in the midst of a leadership transition, and the corporation was being sued in Ecuador over an oil spill that might produce a major financial liability—at a minimum, the lawsuit was a wild card. They considered Shell, too, which was active in onshore gas plays, and a few less likely contenders. Before the check for their snacks came, they had settled on BP and ExxonMobil, both cash-rich and highly interested in the unconventional gas market. Simpson told Randall to approach both corporations to see if they might be interested in a merger or other combination with XTO.[3]

Randall owned a significant amount of stock in XTO—nothing as large as Simpson's holding, but enough to motivate him. Simpson also agreed to pay Randall's firm, Jefferies Group Inc., a transaction fee of $24 million if a merger were completed.[4] Randall had previously worked at Amoco for fourteen years, landing in the company's mergers and acquisitions group. He left to form an oil and gas advisory firm that later became part of Jefferies, an investment bank. He and his fellow directors at XTO had been thinking for years about how the corporation might eventually find an acquirer; almost all successful independents in the oil and gas business ultimately merged or were acquired. That was also the common exit strategy for a founder like Simpson, and a deal now would allow all of XTO's shareholders to benefit from his foresight. Like a marriage broker of old, Randall had already been cultivating a courtship between Bob Simpson and Rex Tillerson at ExxonMobil.

Randall had a personal tie to Tillerson: They had belonged to the same marching band fraternity at the University of Texas. Randall played trumpet; Tillerson played drums. They had both been engineering students in the marching band—that is, double nerds. Randall was a couple of years ahead of Tillerson at U.T., and the men had not known each other well at the time, but the shared history reinforced their professional relationship when Randall became a Houston-based broker of oil and gas properties. At industry and university luncheons, Tillerson and Randall would occasionally run into each other and catch up on oil and gas matters or reminisce about university days.

Around 2007, Randall had suggested that Tillerson invite Bob Simpson on a hunting trip, so the two men could get to know each other better. Tillerson agreed, and he and Simpson spent a few days shooting together on ExxonMobil's vast ranch near Alice, Texas. They got along. Each had been reared in unglamorous circumstances in rural Texas and had now achieved transforming wealth and success. Each had put down roots in the Dallas–Fort Worth area and reveled in the region's history and ranch culture. Each regarded himself as a disciplined leader devoted to operational perfection. Their corporations occasionally partnered on deals and worked compatibly.

After his Fairmont Chateau Whistler bar summit with Simpson, in late July, Jack Randall telephoned Tillerson.

"Rex, I need to come to see you," he said. "It's very, very important. It's very confidential."

Tillerson invited him to Irving, to meet in his office. When Randall arrived on August 6, he explained that Bob Simpson was thinking about a "strategic combination" between XTO and ExxonMobil. Might Exxon-Mobil be interested?

"Yes, I think we'll be interested," Tillerson answered. "Let me take some time to soak on it."[5]

In 1976, as a young Exxon engineer on his second assignment, in East Texas, Rex Tillerson was asked to work on a drilling technique known as hydraulic fracturing, which employs pressurized fluids to shatter rocks and unlock natural gas buried in complex geological formations. The drilling and engineering problems he wrestled with anticipated the shale gas boom that undergirded XTO's success. In part because of his early, direct experience, Tillerson felt he understood the unconventional gas business. Not everyone in or around ExxonMobil thought Tillerson had the analysis right, however.[6]

For most of Tillerson's career, the exploitation of American onshore natural gas beds had not been a major priority for Exxon and other international oil companies. Alaska's large gas fields attracted their attention, but the regulatory and political approvals necessary to pipe the gas to the Lower 48 continually eluded them. In Lee Raymond's era, the big opportunities in oil and gas seemed to lie overseas, in new territories opened up by the cold war's end, in Saudi Arabia, and in ocean waters, where a corporation like ExxonMobil could bring technological advantage to bear. For ExxonMobil, apart from its large projects in Qatar and Aceh, managing natural gas was often a by-product of exploiting oil. Natural gas associated with oil reserves in deep water and elsewhere could be a challenge because the gas was often "stranded" at the oil wellhead—there was no economical way to pipe it to a customer. One way to dispose of such stranded gas was to burn or "flare" it. ExxonMobil flared gas routinely at its offshore African wells. That exacerbated greenhouse gas emissions from oil operations, however, and it wasted a natural resource that might otherwise fuel, say, Nigeria's moribund electricity sector. In some places, international oil companies, including ExxonMobil, built plants to extract from stranded gas commercial products known as gas liquids. In other cases they built plants to create liquefied natural gas that could be shipped globally. The focus had been on creating additional value (and as climate change worries rose, reducing pollution) from associated gas worldwide, not searching for new freestanding supply at home.

Shell, Chevron, and BP largely followed similar strategies—they ignored onshore, complex gas reserves in the United States, Europe, and elsewhere. They invested instead in liquefied natural gas. L.N.G. could soak up both associated offshore gas and bring some large, stranded "non-associated" gas fields to market, such as the North Field in Qatar, while gradually creating a global gas market that looked reassuringly similar to the free-flowing, globally integrated oil market.

For years, to the international majors, the kinds of Texas and Oklahoma shale gas fields that Bob Simpson had scooped up while building XTO after 1986—and the kind of field in East Texas that Tillerson had been assigned to early in his own career—looked picayune, expensive to produce, and of doubtful long-term profitability. Still, ExxonMobil and other majors fiddled around some in these American gas fields over the years—they took leases and they drilled wells, but they did not invest at anything like the scale of their overseas L.N.G. and gas liquids projects.

Lee Raymond had declared in 2003 that American natural gas production had probably peaked. The Energy Department predicted that the United States might run out of domestic gas supplies, which were used mainly for heating and electric power generation, in just two decades. Alan Greenspan, educated by private conversations with Lee Raymond, urged Congress to consider fast-tracking the construction of liquefied natural gas import terminals around the United States to address this coming, widely predicted gas shortfall.

As it turned out, Lee Raymond had been wrong. Within a few years of his declaration, because of the emergence of unconventional gas drilling techniques that proved cost effective, the Energy Department revised its forecasts and now predicted that the United States had about a century's worth of natural gas reserves. ExxonMobil and its international competitors had missed this mother lode lying beneath American soil.

Around the time of his visit to Tillerson in Irving, Jack Randall also met with an executive of BP's division in the United States, headquartered in Houston. He asked his BP contact to explore whether the corporation might be interested in acquiring or merging with XTO to leap forward in the onshore American gas business. The executive told him, "Let us think about it."

When the BP executive called back, however, he reported, "We actually like your gas assets better than we like our gas assets. But the timing is just bad for us."

Tillerson called in mid-August. "I think we are seriously interested. What do you think the next step is?"[7]

Simpson and Tillerson booked a private during room at the Fort Worth Club, in an early-twentieth-century building on Seventh Street. On a drizzly evening, they staggered their arrivals so club members might not notice them. If word of their discussions leaked, XTO's share price would soar, making a merger price negotiation all but impossible.

One question about power and prerogatives in a merged company proved easy to set aside. Randall assured Tillerson, "Bob isn't looking for a job at ExxonMobil."

The founder's departure would clear the way for ExxonMobil to take full control of XTO, as was its traditional method. Yet unlike in the case of the Mobil merger a decade earlier, Tillerson made clear that for the deal to work, he needed XTO's top management and technical talent, other than Simpson, to stay on.

"I've got to do something about natural gas," Tillerson explained. All of ExxonMobil's corporate forecasting pointed toward rising gas demand during the next two decades and beyond, in the United States and globally. Shale or unconventional gas discoveries had upended American markets, flooding the country with apparently durable sources of supply. New discoveries were being announced around the world. ExxonMobil had no global organization dedicated to the full gamut of the emerging unconventional gas challenge—exploration, technology, engineering, drilling, finance, and marketing. Unconventional gas required new thinking in many of these disciplines. "I've either got to build my own or I've got to buy somebody with expertise," Tillerson said. They discussed how ExxonMobil's vast financial resources could bankroll a worldwide expansion of the business and drilling strategies Simpson had developed in the United States.

With XTO, ExxonMobil would buy some attractive American gas properties, yes, but the larger purpose would be to convert the acquired corporation into a new gas division inside ExxonMobil. The deal would not be driven by prospective cost savings. It would be a way to buy depth in the natural gas sector faster than ExxonMobil might create such capability on its own.

"One plus one has got to equal three or more," as Simpson and Randall put it during the early talks, summing up the shared Exxon and XTO view of the merger's goal.[8]

An agreement to merge with XTO would be the most important decision so far of Rex Tillerson's tenure as ExxonMobil chief executive. The corporation had not made an acquisition worth more than $2 billion since the $81 billion merger with Mobil a decade earlier.[9] An XTO deal would likely be worth only about half of the nominal value of the Mobil transaction, before accounting for inflation since then, but even so, it would constitute a major bet placed on behalf of ExxonMobil shareholders.

Until now, Rex Tillerson had presided competently over strategies, projects, and plans bequeathed to him by Lee Raymond. Arguably, the only major strategic shift Tillerson had steered since taking over was in politics and public policy, by repositioning ExxonMobil on climate change and carbon pricing, and by seeking, however quixotically, to improve ties to Washington's ascendant Democrats. Tillerson could be sure of one thing: Once news of his talks with XTO became public, his strategic business judgment would be scrutinized as never before.

By the time Tillerson and Simpson moved into full-blown merger talks during the fall of 2009, it had become common for industry analysts to attribute the unexpected American natural gas boom to technological innovation—that is, the discovery, refinement, and implementation of new techniques to extract gas previously thought unrecoverable. There was truth in this, but the "eureka" explanations masked a long history. Engineers at Exxon and many other companies had known for decades that the United States had large amounts of gas trapped in sand, shale rocks, and coal beds. They had also long known that certain unconventional

drilling techniques—horizontal drilling and techniques to inject pressurized fluids to fracture rocks to release and join isolated pockets of gas—
might allow these reserves to be exploited. The obstacles to refining these
techniques mainly had to do with their costs. During the 1980s and
1990s, the wellhead price of natural gas in the United States hovered at
or below two dollars per thousand cubic feet. The drilling techniques
required to unlock unconventional gas were often too expensive to justify
at that price.

If the United States had possessed a national energy policy that emphasized domestic supply even when such supply might cost extra, the
government might have stepped in to conduct advanced research. There
was no such policy. The government-funded institute that studied unconventional onshore gas drilling technologies and techniques—the Gas
Research Institute—had withered by 2000, for lack of industry, congressional, and White House interest. After 2001, American natural gas prices
moved up, toward four dollars and then five dollars per thousand cubic
feet, and later toward seven dollars. The price rises, not any fresh thinking
in Washington, changed incentives.

One of the Gas Research Institute's directors was a Texas natural
gas wildcatter named George P. Mitchell, the founder of Mitchell Energy.
His firm produced gas from a conventional field in the Barnett Shale, in
North Texas, but his field was aging and its rates of production were in
decline. Mitchell knew there was more gas beneath his leased ground, but
the gas was trapped in shale rocks. As American gas prices finally rose, he
galvanized his engineering staff, with aid from the research institute, to
revive and improve drilling techniques to fracture rocks and pull gas from
difficult beds. As he succeeded and proved the viability of this approach,
others joined in—among them, XTO. Record-high gas prices and tax incentives that allowed for recovery of research costs forgave expensive
learning and mistakes.[10]

Unconventional gas drilling damaged the environment. The techniques could contaminate groundwater, if carried out improperly, by causing chemical-laced drilling fluids and natural gas to leak into aquifers.
Drilling companies did not typically disclose the chemical makeup of
fluids used to fracture rocks, for competitive reasons, so the public could

not easily judge whether the fluids were dangerous to human health. The onshore gas rush also had sizable impacts on land use and development in rural areas—it turned pristine spaces into industrial zones. In the early days of the onshore gas boom, however, the drilling took place mainly in oil-patch states like Texas, Oklahoma, and Louisiana, whose populations and political classes had long ago decided that the economic benefits of oil and gas exploitation, properly managed, outweighed the environmental risks.

Geologists wielding modern computer software and ground penetration radar had not previously devoted themselves to looking for "tight" or trapped unconventional gas beds in the United States. When they did in earnest, after 2003, they reported large finds. As early as 2003, the Gas Technology Institute, successor to the Gas Research Institute, revised past estimates upward to report that America's total natural gas resource base was about 2,000 trillion cubic feet.[11] Americans consumed a little more than 20 trillion cubic feet of natural gas in 2003, roughly the equivalent of 8 million barrels of oil per day, or nearly the amount of the country's actual liquid oil imports. (These numbers explained the very rough, back-of-the-envelope forecast that the United States had a century's gas supply under the ground: One hundred years of consumption at 20 trillion cubic feet per year equals 2,000 trillion cubic feet.) As the years passed and other government and industry panels considered the matter, they published similar top-line figures. But the estimates proved shaky; there was no doubt that there was a lot of unconventional gas in the United States, but exactly how much could be recovered as commercial fuel involved engineering questions that had barely been studied. The most bullish forecasts sounded like hype because they lacked a solid scientific basis.

How long unconventional gas resources might truly last would depend, for example, on the pace of geological depletion in gas beds. This was a matter with which drillers had relatively little experience because the techniques were so new. Other factors would include the pace of demand for natural gas in electricity generation, particularly as a substitute for coal; the future of carbon pricing and greenhouse gas regulation; the trajectory of natural gas prices; and the pace of technical innovation. The idea that the United States truly had enough of its own gas to last a

century seemed optimistic, but equally, the forecast in 2003 by Alan
Greenspan that America might have only two decades of domestic supply
remaining, and that "we are not apt to return to earlier periods of relative
abundance and low prices" had clearly been proved incorrect.[12]

ExxonMobil reentered American unconventional gas exploration and
production on a modest scale after prices rose enough to meet the Management Committee's rigorous return-on-capital guidelines. After Tillerson took charge, he pushed into onshore unconventional gas leasing more
aggressively. The environmental issues did not seem to concern him
greatly. He conceded that there had been cases where the handling of
fluids used to fracture rocks had "not been done as well as it could be,"
but the "incidents" constituted a "very, very, very small percentage" in the
context of total production. He also declared that the threat to underground drinking water from such drilling was "very low." Tillerson's emphatic tone echoed Lee Raymond's early confidence about the evidence
on climate change, but he was unabashed. Within ExxonMobil, there was
controversy about shale gas, but it did not involve environmental issues.
It concerned the company's strategies for replacing the amount of oil and
gas it pumped and sold annually.[13]

ExxonMobil's huge investments in liquefied natural gas showed the
corporation's bias toward "manufacturing drilling," a phrase that referred
to producing oil and gas through industrial prowess rather than wildcatter
guile. "We had become a very big company that did very big projects," said
a former executive. To win in unconventional gas, could ExxonMobil now
adapt to the more classical land scouting, exploration, and entrepreneurial
tactics required to outfox sellers and competitors? Some of the prospective challenges in unconventional gas played to ExxonMobil's strengths
in manufacturing—such as the need to develop engineering innovation
that would improve the rates of early depletion in unconventional gas
fields. But to apply its skills ExxonMobil needed big properties at a reasonable price.

ExxonMobil's profitability reflected in part the deliberate, return-on-
investment-driven decision making of its Management Committee. The
upside was rigor and high rates of return on capital invested; the downside
was caution and missed opportunity. How much of a flyer was Exxon-

Mobil willing to take to get in on the gas rush, and how fast could the corporation move? Was it really possible for the corporation to replace reserves, capture the sudden emergence of the domestic unconventional gas play, and raise worldwide oil and gas production each year, all while demanding exceptionally high rates of return for every new project investment? In a perfect world, an oil corporation with cash flow like ExxonMobil's would pour its cash into new oil and gas reserves when commodity prices were low and milk them when prices were high, as Raymond had done with the Mobil merger. But the opportunity emerging in American unconventional gas seemed to be now—when prices were high. Should ExxonMobil compromise its profit standards at least a little to make a strategic shift?

Tim Cejka, a round-cheeked Pittsburgh native who had studied geology and risen through Exxon's exploration division, ran the company's upstream operations at the time of the XTO merger talks. Cejka was an oil and gas hunter who had served as an exploration adviser to various ExxonMobil divisions before reaching the Management Committee. He oversaw ExxonMobil's leasing in search of unconventional gas loads— 250,000 acres in the Horn River Basin in British Columbia, 400,000 acres in Hungary, and 750,000 acres in the Lower Saxony Basin in Germany.

Cejka knew that the risks in such exploration ran high and that ExxonMobil's record was unproven. About the Hungary leases, he told Russell Gold of the *Wall Street Journal* in July 2009, just as secret talks between ExxonMobil and XTO were about to start, "Depending on how that goes, we'll either be patting ourselves on the back or walking away."[14] Tillerson kept the XTO negotiations so secret that even Cejka did not know about them. Cejka kept working to compete with XTO on North American gas leases even as the merger talks ripened.

Tillerson faced a clear choice: Would it be smarter to keep trying to find North American unconventional gas, or would it be better to use ExxonMobil's massive cash and treasury share positions to buy in?

ExxonMobil brimmed with cash. The corporation carried more than $30 billion in cash on its balance sheet. Plus, by 2009, it held in its "treasury" more than 3.2 billion shares of its own stock, with a market value of more than $220 billion, which it had repurchased over the years from the open market and set aside for possible use in acquisitions.[15] During the great recession and financial panic that followed the collapse of Lehman Brothers in 2008, many American banks, corporations, and their employees worried week by week about whether their businesses might go under. Rex Tillerson's greatest worry during the dark September of Lehman's collapse, he later confessed, was whether ExxonMobil's massive cash deposits were parked in banks that would survive the crisis. As the global financial system teetered, ExxonMobil shuffled its billions to safe havens and waited for the economic storm to pass.

Good morning, and I want to thank all of you for joining us today," Rex Tillerson said into a speaker set before him. "ExxonMobil and XTO Energy Inc. have announced an all-stock transaction valued at $41 billion. . . .

"This is not a near-term decision, obviously. This is about the next ten to twenty to thirty years of what we believe has now emerged as a very important part of the global resource portfolio. . . . It's going to be important to meeting energy supply, and that's the real value creation that we see."[16]

It was December 14, 2009. The secret talks with Simpson and the senior team at XTO had not leaked. ExxonMobil retained the investment bank J.P. Morgan and the Wall Street law firm of Davis Polk & Wardwell to lead its side of the negotiations; XTO retained Barclays Capital and the longtime mergers law firm Skadden, Arps, Slate, Meagher & Flom. Tillerson initially proposed to pay a modest 15 percent premium over the market price for XTO shares; Simpson said that "would not be acceptable." Each side prepared valuation ranges based on forecasts of varying natural gas prices in the future, and their merger bankers prepared charts showing prices paid in comparable mergers in the energy industry and in other sectors. Once the deal became public, another acquirer might swoop in to try to overbid ExxonMobil, so in one of their periodic meetings, Tillerson extracted an agreement from Simpson that XTO would pay a breakup fee of $900 million to ExxonMobil if the merger were not completed. In the end, they circled in on a price agreement by which Exxon-

Mobil would pay about 25 percent above XTO's average stock market price during the month before the announcement. That seemed an uncontroversial compromise—it was the median premium above-market price paid in U.S. corporate transactions greater than $10 billion since January 1, 1998, according to a Barclays analysis. In a tax-free exchange of shares, ExxonMobil effectively paid $51.69 per share for XTO. During the final weeks, they had also wrestled over the employment terms required to retain top XTO executives and engineers, who had grown accustomed to the get-rich stock options doled out by Bob Simpson; they would now have to adjust to the more conservative compensation rules at ExxonMobil. To retain XTO's top five executives long enough to manage a smooth transition, Tillerson restructured their compensation contracts and wrote rich new consulting agreements that linked performance to millions of dollars in stock and cash over the next several years, including a total of $84 million for Simpson; $48 million for XTO chief executive Keith Hutton; and $37 million for senior executive Vaughn Vennerberg.[17]

Tillerson said he decided to buy XTO in part because ExxonMobil's corporate planning department forecasted rising natural gas demand. Climate change legislation in Congress was collapsing, and it was not easy to see when it might be revived, but in the medium run, higher carbon prices imposed by regulators—as already had been laid down in Europe and announced in Australia—still seemed very likely. If enacted, they would hurt coal and help natural gas. Mandates in the United States for more renewable energy such as wind and solar power also complemented natural gas investments because gas-fired electric plants could address, with relatively low emissions, the "intermittency" problem posed by renewables. (Intermittency referred to the fact that the wind did not always blow and the sun did not always shine, and so electricity generated from those sources could be erratic. Complementary gas-fired electricity could keep currents flowing on calm, rainy days.) Also, the megawatt-per-hour cost of gas-generated electricity looked favorable when compared with nuclear and unsubsidized renewable sources.

Tillerson insisted that ExxonMobil's shift toward natural gas through the XTO purchase was not a "deliberate strategy" to favor natural gas over oil. In fact, however, ExxonMobil was nearing the point where it would own, on its books, more natural gas than oil. During the decade leading to 2010, ExxonMobil had replaced, on average, only 95 percent of the oil it pumped out and sold each year, but it had replaced, on average, 158 percent of the gas it extracted and sold. After incorporating XTO's reserves, 45 percent of ExxonMobil's reported reserves would be gas.[18] Tillerson claimed that ExxonMobil's disciplined systems could extract high profits from either oil or gas, but in the industry, gas was often less profitable to produce than oil, for a host of reasons—not least, the low prices plaguing American gas producers after 2008. Conoco forecasted that American gas prices would remain mired at relatively low levels and would not return to the boom prices of 2007 and 2008 anytime soon. Shell's forecasters were a little more optimistic, but cautious.

A PowerPoint produced by analysts at the Society of Petroleum Evaluation Engineers in Houston noted that ExxonMobil's purchase of XTO was "based on the assumption that much higher natural gas prices" were coming in the future, and yet, there was "considerable risk in shale plays" because of uncertain geological and commercial factors. "Reserves are overstated," the presentation continued. "Costs are understated. . . . The gold rush mentality destroys capital and ensures the rule of expediency over science and risk management."

Uncertainty and skepticism of this kind leached out from geological engineers in the form of unfavorable press reporting, some of which went so far as to ask whether the American shale gas boom was some sort of Ponzi scheme in which early investors bid up faulty assets and lured in big-money suckers like ExxonMobil. Unconventional gas wells behaved unlike other wells, and their decline and production rates could be hard to calculate—much about the drilling patterns in these fields still remained to be discovered. An individual gas well might lose its productivity much more rapidly in the first year of drilling than an oil well would, "but the decline rate on the [total] field is nil, because you continue to drill" in other sections of the field, as Shell's Simon Henry put it. Yet there was evidence to support the doubters, too. At a minimum, shale gas producers were going to have to communicate with investors more forthrightly than they had done early on about their costs, risks, and profit potential.[19]

Wall Street swiftly made clear that it did not approve of Rex Tillerson's decision to buy XTO. It looked to analysts and investors that Tillerson had overpaid for Simpson's company and that ExxonMobil had made risky assumptions about future natural gas prices. Investors hammered ExxonMobil's share price, relative to its peer group, in a way the corporation had not experienced for many years. Instead of the premium price that ExxonMobil shares had long enjoyed, ExxonMobil stock soon sold at a discount. As analysts at Reuters *Breakingviews* pointed out, during the seven months after the merger announcement, adjusting for the average 4 percent decline in the share prices of its peers Royal Dutch Shell and Chevron, ExxonMobil shareholders saw $41 billion disappear from the corporation's total market price—an amount that eerily matched the price Tillerson had paid for XTO.[20]

Had ExxonMobil unwisely bought XTO at or near the top of the boom? It was certainly becoming clear that the peak years of 2007 and 2008 had led to reckless overinvestment in American gas leases by large, debt-burdened companies such as Chesapeake Energy. As that excess investment unwound, there would likely be opportunities for bottom-feeders to sweep up unconventional gas leases at lower prices than were reflected in the price ExxonMobil paid for XTO. That didn't necessarily mean the merger was a mistake. That would depend on how ExxonMobil exploited XTO's properties and expertise over time. Yet it was another basis for doubt. John Watson, the chief executive of rival Chevron, slipped the knife in: "We saw valuations for unconventionals that were a bit out of line with our view of value," he told Wall Street analysts. "So our view wasn't so much that shale gas wasn't a good place to be. It was just the valuations at the time [of ExxonMobil's purchase of XTO] were strong, so we waited."[21]

Mark Gilman, the oil industry analyst at Benchmark Capital, regarded the XTO purchase as a sign that ExxonMobil's long-term failure to build upstream reserves—which Gilman laid mainly at Lee Raymond's door—was at last coming home to roost. Tillerson had little choice but to buy new reserves in a high-price environment because otherwise, he would be presiding over a shrinking corporation, which could reduce ExxonMobil's share price, which could further limit its ability to buy its way out of its dilemma. The price paid for XTO might mean a reduction in ExxonMo-bil's historical rates of return, but that, too, was inevitable and even wel-come, in Gilman's view, if it led to a more successful long-term performance in reserve replacement. On XTO's purchase price, "I don't fault Rex," Gilman said. "It's what you have to do when you have a weak hand." He objected, however, to the specific choice of Simpson's company, which he believed Tillerson had selected too much for "cultural and geographic" reasons, meaning the similarities in Tillerson's and Simpson's personal backgrounds, and the Fort Worth location of XTO headquarters. There were other unconventional gas owners—Devon Energy, for example—that might have paid off better.[22]

Dissent bubbled about the XTO deal within important sections of ExxonMobil's executive ranks and alumni networks. Like many ac-quisitions in commodity industries, the deal's payout would depend substantially on future prices, which nobody could forecast with certainty. According to a valuation prepared by Barclays Capital, without account-ing for ExxonMobil's potential to extract extra value from XTO's re-serves through engineering prowess, if natural gas prices remained as low as $5 per thousand cubic feet through 2014 and beyond, XTO might be worth only between $21 and $30 per share, a fraction of what Exxon-Mobil had paid. At least a few current and former senior executives wor-ried about whether ExxonMobil could produce XTO's gas profitably, even if gas prices did break out of their doldrums.[23]

Privately, according to some accounts, Tim Cejka argued that if he had been allowed to pay for exploration leases at the high per-unit prices that ExxonMobil had accepted in the price it paid for XTO, he would have more "organic" or ExxonMobil-discovered gas to show for his efforts. Cejka denied in a brief telephone interview that any serious dispute de-veloped over this rate-of-return issue. In any event, ExxonMobil's record during his time as head of exploration, at least toward the end of his tenure, was poor, whether it was his fault or not. By late 2009, it became apparent that Tim Cejka's big forays into exploration and land leasing in Europe, at least, would not produce any early bonanzas. ExxonMobil's early drilling yielded many dry holes. As Tillerson admitted, "Quite frankly, no one has enough information at this point to know" whether European unconventional gas would ever pan out.[24] Overall, the corporation's struggle in exploration and development showed no signs of turning around—its well-drilling failure rates rose by more than a third during 2007 and 2008. Cejka retired, leaving the company soon after the XTO deal closed.

"The mainstream belief that shale plays have ensured North America an abundant supply of inexpensive natural gas is not supported by facts or results to date," wrote an analyst at *The Oil Drum*, an independent online energy journal. "The supply is real but it will come at higher cost and greater risk than is commonly assumed. The arrival of ExxonMobil and other major oil companies on the shale gas scene is positive because they will not follow the manufacturing approach, and will do the necessary science that should make shale plays more commercial. This does not, however, ensure success. ExxonMobil has come late to the domestic shale party. . . . It is also possible that XTO has already drilled the best areas in more mature shale plays, while the potential of newer plays has not yet been established."[25]

An unsigned memo carrying similar doubts circulated among retired ExxonMobil executives. "It is a really tough job to figure out if Exxon-Mobil management is doing a good job of enhancing shareholder value, given the inherent limitations of its already huge size and inevitable momentum," the memo noted. "Sure, you can make comparisons with competitors (which ExxonMobil has tended to lag in recent years) but given that ExxonMobil is fully a third larger than its nearest competitor, one is dealing with apples and oranges to some extent."

The memo continued, "One has to respect and acknowledge the positive things that ExxonMobil does on a daily basis, such as:

- After the lessons of the *Valdez* . . . the enormous efforts and expense the company puts into avoiding even the smallest oil spills.
- The terrific and expensive commitment to employee and contractor safety. . . .
- The vigor with which the company polices employee expense reports to insure that employees are not stealing from the company.
- The integrity of its bidding processes in avoiding fraud in its purchasing function.

- The commitment to reduce costs in its general business operations. . . .
- The engineering quality in its refineries and its production facilities.
- The exhaustive capital investment process.
- Its industry-leading return on assets."

On the other hand, "one has to ask, do the shareholders pay Rex Tillerson $29 million a year to be a caretaker? . . . Lee Raymond, former ExxonMobil C.E.O., notwithstanding his dour personality and penchant for trying to control every detail of a huge company's operations . . . at least knew that when oil prices were at nine dollars a barrel, it was time to buy a company with good upstream assets, which he did when he bought Mobil corporation. Rex Tillerson, on the other hand, with less exquisite timing, agreed to pay . . . an expensive 25 percent over market premium [for XTO]. Had the deal been struck earlier, at the end of March 2009, the purchase price, with the same market premium percentage, would have been a very palatable $38.23 a share."

The memo concluded: "The stock's performance in recent years accurately reflects their less than mediocre business capabilities. To call them incompetent may be to go too far, but it is close . . . mighty close. . . . Given the peaceful slumber this Board of Directors has enjoyed for the last twenty years, one has to ask a closing question: Why would anyone want to be an ExxonMobil shareholder?"[26]

W as this criticism of Rex Tillerson's leadership fair? During 2010, Tillerson completed his fifth year as chief executive. That was long enough to begin to judge his record. The numbers showed a mixed but far from disastrous performance. Many of the critical questions about his decision making post-Raymond would require a decade or more to measure. The fairest grade was probably "incomplete." Whether the price Tillerson paid for XTO was too high or not, his essential theory of the purchase was the same as Lee Raymond's theory about the enormously successful Mobil merger: Exxon would exceed Wall Street expectations over time by extracting value from the acquired assets that no other company knew how to extract. Raymond had paid a 15 percent premium for Mobil's shares at a time when oil prices were so low that oil doomsayers ruled, just as doomsayers about shale gas were prominent in late 2010. Perhaps the XTO properties would yet perform under Exxon's management as the Mobil properties had.

ExxonMobil earned $30.5 billion in profits during 2010, short of the Tillerson-overseen record of 2008, but stunning nonetheless. The corporation had earned more profit than any publicly traded corporation in America in each year of Tillerson's reign so far. In a sign of the times, ExxonMobil jockeyed occasionally with PetroChina, the state-owned oil company, for the status of the world's largest corporation by stock market value, but ExxonMobil was valued highest more often than not. Much of the corporation's top-line profit reflected soaring commodity prices over which it had little control. Yet ExxonMobil also remained at the top of its industry class, judging by return on capital employed, or R.O.C.E., the metric by which the corporation preferred to compare itself with its closest American peers, Chevron and Conoco, and the most closely comparable overseas competitors, Royal Dutch Shell and BP. The corporation's R.O.C.E. was 22 percent during 2010, about where it was after the Mobil merger, and higher than the next-best performer, Chevron, by 5 percent. In all, the numbers showed Tillerson had not allowed financial, investment, or operating discipline to slip during his five years in charge.

ExxonMobil's lead over one competitor, Chevron, had narrowed, however, to the point where, by market and financial performance measurements, the two companies were about tied. By 2010, ExxonMobil's R.O.C.E. topped Chevron's largely because Exxon's huge chemical and downstream operations performed twice as well as Chevron's did. Certainly Tillerson and his team deserved credit for maintaining the high margins Raymond had delivered in these notoriously difficult businesses. Yet the downstream business looked increasingly uneconomic in the long run because governments in emerging economies were installing new refineries and petrochemical complexes, backed by state subsidies, for reasons other than profit making—to ensure energy security, for example, or in the case of Saudi Arabia, to create better jobs and promote scientific education. This glut of subsidized capacity would challenge ExxonMobil in the long run. Yet in the oil and gas upstream, where the great majority of profits earned by fully integrated oil companies resided, and where the greatest future profit opportunities lay, Chevron had now about caught up with ExxonMobil; Chevron's upstream R.O.C.E. in 2010 was a robust 23 percent. The average barrel of oil or equivalent amount of gas produced by Chevron was more profitable than a barrel produced by Exxon, according to Chevron's calculations. Moreover, using other metrics often highlighted by Wall Street analysts—total stockholder return and cash flow per share, for example—Chevron now substantially outperformed ExxonMobil. Chevron's shareholders did better than ExxonMobil's during 2010. (The rest of the peer group lagged.) Tillerson and his colleagues might rationalize their slippage by blaming a short-term herd mentality on Wall Street that turned hostile to ExxonMobil's shares because of the XTO deal, and indeed the corporation's shares did bounce back after the initial XTO hangover, but the numbers spoke clearly enough of a tightening competition.

Tillerson deserved credit for accomplishments not visible on Exxon-Mobil's balance sheet. His Hamlet-like performance on carbon taxation and climate change did him little credit, but he had led a determined drive to reduce the greenhouse gases emitted by the corporation's own operations and had delivered real improvements. On his watch, ExxonMobil had reduced gas flaring—the wasteful burning of natural gas produced during oil extraction, which contributed to global warming—by more than half. In Nigeria and other countries with weak governments, the corporation had missed announced targets for the elimination of flaring; it blamed the failure of its partner regimes. Still, between the progress it did make and greater energy efficiency, ExxonMobil had reduced its total direct greenhouse gas emissions by eleven million metric tons, a significant achievement.

Tillerson had also taken steps to address ExxonMobil's fudging about whether the corporation was finding enough oil and gas each year to replace the amount it pumped and sold. Under Raymond and again during Tillerson's first years, ExxonMobil had declared publicly through press releases and at Wall Street analyst presentations that it had found enough new "proved reserves" of oil and gas to replace each year's production and sales. But in making this claim, the corporation ignored the accounting methods required by the Securities and Exchange Commission. In some years, ignoring the S.E.C. reporting rules allowed the corporation to side-step embarrassment. In 2008, using S.E.C. rules, and based on the corporation's limited public disclosures, ExxonMobil's reserve replacement would have been below 75 percent, an alarming rate. But instead of accounting forthrightly for this failure, the corporation issued a press release that quoted Tillerson boasting, "ExxonMobil . . . has replaced an average of 110 percent of production over the last ten years."[27] That was a defensible claim only if one preferred ExxonMobil's self-regulation to federal rules.

On December 31, 2008, the outgoing Republican-led Securities and Exchange Commission revised its reserve reporting rules to allow the counting of oil sands, shale gas, and other previously banned categories of reserves. The commission also changed other reporting rules that Raymond and Tillerson had found objectionable. The changes, achieved by oil industry lobbying, liberated ExxonMobil from spinning. The corporation ceased double counting: From now on, it would report only numbers authorized by the S.E.C. It did not retract its previous claims to Wall Street and the public, however, noting only that its long reserve replacement "streak" was based on some years when the S.E.C. rules were not used.

The cleaner 2010 reserve replacement numbers looked good on the surface, but were concerning underneath. When the XTO gas properties were incorporated into ExxonMobil's resource base, the corporation reported that it had replaced an extraordinarily strong 209 percent of the oil and gas it produced that year. Yet XTO's purchased properties accounted for four fifths of the corporation's new reserves. Without XTO, according to Barclays, ExxonMobil would have replaced only *45 percent* of its 2010 oil and gas production—a performance so abysmal that if it continued for a prolonged period, ExxonMobil would be on a path to liquidation. By comparison, Conoco's "organic" or internally generated reserve replacement rate in 2010 was 138 percent. Shell's was 133 percent.[28] Of course, ExxonMobil had always been better at buying other people's oil than at finding it. Arguably, from a shareholder's perspective, it made no difference whether the oil and gas ExxonMobil pumped and sold so profitably each year had been discovered because of geological genius or bought with piles of cash generated by financial and operating acumen. If Tillerson could maintain the financial performance that made the XTO acquisition possible, he might continue to buy what he could not find. But at a minimum, the numbers made clear how important the XTO purchase would be to Tillerson's legacy on Wall Street and in the oil industry: If the deal underperformed, the corporation would be hard-pressed to maintain its superiority.

Tillerson promised when he took charge to increase ExxonMobil's annual production of oil and gas to 5 million barrels per day by 2009. The actual number was 3.9 million—more than 20 percent short. Tillerson promised again that ExxonMobil's production would grow steadily until 2014, but the trailing numbers showed the corporation in a long, flat pattern—its annual production in 2001, after the Mobil merger closed, was 4.3 million barrels per day.[29] Tillerson had not cracked the challenge of reserve replacement that had also daunted Raymond.

Before Tillerson, dissent and hard feeling inside ExxonMobil often traced to Lee Raymond's blunt manner. Under Tillerson, ExxonMobil might be a kinder, gentler place to work, yet some of the old guard feared a loss of the toughness and discipline they had valued in Raymond. Retired executives of the Raymond era took one another out to dinner in Houston, Dallas, and elsewhere and talked about whether Tillerson had enough of the guts and firmness that Raymond had mustered to drive ExxonMobil's financial performance.

Tillerson's remarks to Wall Street analysts increasingly made it clear that he was aware of these dissenters. It required the equivalent of Kremlinology to perceive Tillerson's public replies to these dissenting factions, but his rejoinders were detectable. At analyst meetings, Tillerson started to use 2006, the year he took the top job, as the basis for reporting about— and boasting about—ExxonMobil's financial performance. He ignored the Raymond years, and he went so far as to explain how his leadership had extracted profitability from one tough project, the Kearl oil sands play in Canada, because he had made flexible analytical judgments about projected rates of return that would not have been taken "five, six, eight years ago," when Raymond was in charge.

On April 19, 2010, Tillerson arrived at the Hilton Americas-Houston hotel and convention center to receive the Jesse H. and Mary Gibbs Jones Award for contribution to the international life of Houston. It was a typical appointment in an oil industry chief executive's diary—a short hop on a corporate jet, a prepared speech before a sympathetic audience, a lunch of Cornish game hen and vegetables, and a roundtable talk with students from the University of Houston, Rice University, and the University of St. Thomas.

As the students snapped pictures of him on their cell phones, Tillerson was relaxed, giddy, and self-deprecating about his looks. One student asked if it was true that he and his wife rode motorcycles for fun.

"Pass," Tillerson said, smiling.

Another asked about solar and wind power.

"ExxonMobil is not really against renewables," Tillerson replied mirthfully. "We sell a lot of lubricant oil to the windmill operators. . . . The more windmills are built, the more oil we sell."[30]

To the larger audience of about 750 Houstonians seated at banquet tables, Tillerson read out a philosophical defense of capitalist private enterprise and an explanation of ExxonMobil's mission in the world. Job losses, bank layoffs, and housing foreclosures had swept the American heartland since 2008. Tillerson's words reflected his Boy Scout optimism and Christian faith.

"The 'service we render' and the ongoing investments we make from our earnings are critical," Tillerson said. "Simply put, delivering energy in a safe, secure and responsible manner improves the lives and opportunities of billions of people the world over. . . .

"When government and industry respect the rightful role of the other—and trust each other to faithfully fulfill their respective roles—progress is possible. . . . Deepening understanding and building trust between the public and private sectors is more important than ever. . . .

"Service . . . responsible . . . respect . . . trust . . . faithfully . . ."

He seemed to wish for ExxonMobil to be considered as a kind of public trust. He used the words "trust" or "mistrust" five times.

"Often the policy changes that are most damaging to entrepreneurs and innovation flow from a fundamental mistrust in the private sector," Tillerson declared. He concluded, "Leaders in the private and public sector both have a responsibility to challenge the basis and perceptions for the mistrust."[31]

He took in applause, shook more hands, and departed the Hilton in the midafternoon.

The next morning, April 20, 2010, at 8:52 a.m., on an offshore oil rig called the *Deepwater Horizon* in the Gulf of Mexico, a drilling engineer working for BP, Brian Morel, e-mailed his office in Houston, not far from where Tillerson had delivered his speech about trust between government and business.

"Just wanted to let everyone know the cement job went well," Morel wrote.

David Sims, a BP drilling operations manager, e-mailed Morel and his colleagues at 10:14 a.m.: "Great job guys!"[32]

Twenty-eight

"It Just Happened"

R andy Ezell reached over from his bunk and touched a button to illuminate his electronic alarm clock. It read 9:50 p.m. He picked up his ringing telephone.

"We have a situation," Steve Curtis told him. "The well is blown out. We have mud going to the crown."

"Do y'all have it shut in?" Ezell asked. Ezell carried the title "senior tool pusher"; he was one of the more experienced hands aboard the rig that night. Curtis was an assistant driller.

"Jason is shutting it in now," Curtis said. "Randy, we need your help."

"Steve, I'll be—I'll be right there."

In the darkness, the *Deepwater Horizon* floated on 4,992 feet of seawater in the Gulf of Mexico. It had been commissioned nine years earlier, one of a new generation of seagoing industrial robots designed to drill for oil in unprecedented saltwater depths. It was a towering, brightly lit metallic behemoth—almost 400 feet tall and 250 feet across. The rig had hovered for weeks over a BP-managed prospect called Macondo No. 252. The name referred to a fictional town in the Gabriel García Márquez novel *One Hundred Years of Solitude*. One hundred and twenty-six men and women were aboard the rig that night. Only six worked directly for BP; the rest, like Randy Ezell, worked for BP's contractors and subcontractors, including two of the largest corporations in the global oil service industry, Switzerland's Transocean and America's Halliburton.

Ezell got up and lurched into the hallway. The explosions began: They were "take-your-breath-away explosions, shake-your-body-to-the-core explosions, take-your-vision-away explosions," one of his coworkers said later. Ezell knew roughly what had happened. Drilling any oil well required managing the risk that trapped oil and gas under extreme pressure in the ground, when punctured by a drill bit, might escape uncontrollably and ignite. Since 2001, the workforce drilling for oil in the waters of the Gulf of Mexico—about 35,000 people altogether—had endured 60 deaths, 1,550 injuries, and 948 fires and explosions.

The blasts bounced Ezell off the bulkhead and left him trapped on the floor beneath debris. He twice tried to raise himself, but could not. The third time adrenaline jolted him. "I told myself, 'Either you get up or you're going to lay here and die.'" He raised himself.

Methane shooting through well pipes whooshed eerily in the darkness outside. Fire rolled in waves across the platform. Ezell heard calls for help and stayed behind as more explosions rumbled. He tended an injured coworker and then carried the wounded man to safety. He found coworkers lowering lifeboats and rafts into the water. He joined the exodus with Curtis.

On another part of the platform, Mike Williams, the *Deepwater Horizon*'s chief electronics technician, stood on the rig's edge with Andrea Fleytas, twenty-three, one of three women working aboard. They watched as the life rafts and rescue boats pushed away onto water now illuminated by fire and searchlights. The two of them seemed to be marooned.

"It's okay to be scared," Williams told her. "I'm scared, too."

"What are we going to do?" she asked.

Williams said they could either stay on the platform and burn to death or jump more than one hundred feet into the sea and hope for the best.

Williams jumped. He fell "what seemed like forever," plunged into the water, resurfaced in a pool of greasy fuel, swam free, and found a rescue boat. The boat carried him to a larger vessel, the PSV *Damon B. Bankston*, which had responded to distress calls and pulled up nearby to collect survivors.

On the *Bankston*, Williams discovered Andrea was alive and well. Back on the rig, she had seen a last life raft lowering from the platform and had leaped in.

A roll call confirmed that eleven men were missing and presumed dead. As dawn approached, the *Deepwater Horizon* workers watched from the *Bankston* as the drilling ship burned and listed. It would soon sink to the bottom of the Gulf's seabed.

BP later investigated the *Deepwater Horizon* accident and issued a report concluding, "A complex and interlinked series of mechanical failures, human judgments, engineering design, operational implementation and team interfaces came together to allow the initiation and escalation of the accident. Multiple companies, work teams and circumstances were involved."

Mike Williams put it this way: "All the things they told us could never happen, happened."[1]

BP's catastrophe soon surpassed the *Exxon Valdez* wreck as the worst oil spill in American history. The *Valdez* had released 257,000 barrels of oil into Prince William Sound. The amount of oil released by the *Deepwater Horizon*'s blown well proved harder to measure, but eventually, the best scientific estimates held that almost 5 million barrels spilled before the well could be plugged. The *Exxon Valdez* had jolted America's largest oil corporation to remake its safety, operations, and management systems. Over the ensuing two decades, within ExxonMobil, the wreck on Bligh Reef provided a kind of origins myth for internal reform and redemption, one repeated at employee meetings and safety minute rituals, as well as to journalists and shareholder audiences. If ExxonMobil regarded itself now as straighter than straight, the corporation's narrative went, it was only because it had known firsthand the terrible consequences of failed risk management.

ExxonMobil's K Street staff often extrapolated the corporation's relatively strong safety record into an argument to members of Congress and oversight agencies that industry self-regulation can work well, and that ExxonMobil's self-regulation, in particular, was highly credible and should be relied upon by government and the public.

Twenty-one years and twenty-seven days after the *Exxon Valdez* struck Bligh Reef, the *Deepwater Horizon* blowout exposed what the bipartisan national commission that investigated the disaster would call "such systematic failures in risk management that they place in doubt the safety culture of the entire industry." Deep-water oil exploration and drilling, in particular, involved "risks for which neither industry nor government has been adequately prepared."[2]

The chain of errors that destroyed the *Deepwater Horizon* also exposed deep failures within BP and its contractors that were obviously not ExxonMobil's doing or responsibility. Yet as had been the case in Prince William Sound two decades earlier, ExxonMobil and BP were linked in one critical aspect of the risk management system designed to protect America's ocean ecosystems from oil disasters: response and cleanup.

The Mississippi River dumped sand, dead plants, and other precursors of fossil fuels into the Gulf of Mexico over millions of years. Freshwater delta flows sometimes produced pressurized traps of oil and gas offshore, as was the case off the Niger Delta in West Africa. The Gulf of Mexico's salt domes held fossil fuels, as it turned out, as a result of many of the same ancient geological processes that had endowed onshore Louisiana and East Texas with oil riches. Early American oil geologists did not take long to follow the oil trail from Texas into the Gulf. Prospectors drilled the Gulf's first offshore well in 1938.[3] The returns proved compelling. Oil exploration in shallow Gulf waters delivered high rates of success—new wells struck oil twice as often as on Texas or Louisiana lands, and the volumes uncovered were often much greater. For a while, the only wells that made technical and economic sense were those that could be drilled in waters shallow enough to support a drilling platform anchored by pilings in the seabed. Profits incentivized innovation. In 1962, Royal Dutch Shell announced that it had invented a floating drilling contraption that would allow oil exploration in waters too deep to support a traditional platform. The deep-water oil era began—and it, too, boomed. Floating platforms spread from the Gulf to the Pacific Ocean and Alaska. Production from the Gulf of Mexico alone soared from 348,000 barrels per day in the year of Shell's announcement to 915,000 barrels by 1968, almost 10 percent of America's domestic total.[4]

On January 28, 1969, Union Oil Platform A-21 blew out in the Santa Barbara Channel. The spill soaked thirty miles of California beaches in oil. It was the early age of color television and dramatic visual news—moon landings, Vietnam jungle firefights, and televised presidential debates. Images of dead seagulls coated in oil and California beach enthusiasts mucking in tar beamed across national newscasts night after night, adding momentum to America's burgeoning environmental movement. President Richard Nixon's secretary of the interior, Walter Hickel, imposed a moratorium on all drilling and production in California waters.[5]

That decision initiated what became an undeclared, ad hoc system for controlling offshore drilling in American waters. If the waters to be leased for risky oil drilling adjoined states with tourism-dependent economies or voters who supported tight environmental regulation, drilling would be banned. But if the waters adjoined states with pro-oil politics, such as Texas and Louisiana, offshore drilling would be permitted and even encouraged. (Alaska, with its gung-ho pro-oil politics and its vast stretches of protected public lands and waterways, was a special political case; some offshore leasing proceeded there, but it was often contested.) Under a federal law enacted in 1953, individual states own and manage resources beneath ocean waters for three nautical miles from the shore, although Florida and Texas own nine nautical miles' worth, because of old treaty claims.[6] The federal government owns and manages the rest of America's territorial waters, through the Department of the Interior. Even President Ronald Reagan, who was elected with a sweeping mandate to deregulate industry and spur economic growth, could not overcome America's strangely Balkanized politics of offshore oil drilling. Reagan's secretary of the interior, James Watt, initiated plans to lease oil in all of America's oceans, but in the end, aggressive drilling went forward only in the waters off Texas, Louisiana, Mississippi, and Alabama. The eco-minded West

Coast states of California, Oregon, and Washington wanted no part. Florida's tourism and coastal real estate industries could not abide the risks of a Santa Barbara–scale spill, even though the state's voters sometimes leaned Republican. Thus even Governor Jeb Bush, a scion of oil, would oppose offshore drilling for a time. On the Atlantic coast, Virginia, North Carolina, and New Jersey occasionally flirted with leasing Atlantic Ocean tracts for oil drilling in exchange for royalty revenue, but none of these states ever produced governors and political constituencies strong enough to go ahead.

After the *Deepwater Horizon* blowout, it became commonplace to observe that Big Oil had captured and weakened Washington's regulation of offshore drilling, by influencing and outfoxing the weak and underfunded unit of the Department of the Interior, the Minerals Management Service, or M.M.S., which oversaw leasing and drilling in federal waters. It was certainly true that the oil industry outmatched M.M.S. regulators and that the industry muscled through an oversight system that relied heavily on self-regulation. But the weak regulatory system was also a consequence of the segregated American politics of offshore drilling.

The most powerful national environmental lobbies—the Natural Resources Defense Council, the Environmental Defense Fund, the Nature Conservancy, and the Sierra Club—did not focus heavily on the technical, regulatory, and risk management issues surrounding the Gulf's Red State deep-water drilling operations. To the extent that the environmental lobbies worked on offshore oil issues, they focused more on preventing new leasing in Alaska or in the eastern Gulf, off Florida, where the oil industry sought to expand.

Also, as drilling boomed, Interior became the conduit for annual royalties that reached $23 billion in 2008, $17.3 billion of which was funneled to the deficit-burdened United States Treasury.[7] This cash flow reinforced congressional complacency. Besides, in most years, shipping accidents accounted for much more oil pollution leaching into ocean waters than offshore drilling or associated pipeline leaks. All this created a pressure-free atmosphere around Interior's Minerals Management Service. Compared with food safety, toy safety, mountaintop coal mining, climate change, air quality, or water quality, the rigors of deep-water drilling operations and worker safety in the western and central Gulf of Mexico did not attract great scrutiny—as evidenced by the high numbers of deaths and injuries that took place on offshore platforms.

Deep-water drillers "succumbed to a false sense of security," as the national commission put it. One warning sign was the well-documented unreliability of blowout preventers. These were contraptions meant to function as last-ditch fail-safe devices to smother uncontrolled wells before they could blow. A Norwegian firm, Det Norske Veritas, published a paper that examined fifteen thousand offshore wells operating between 1980 and 2006. It found eleven cases where teams drilling deep-water wells, fearing a blowout, had switched on their preventer devices. In only six cases did the wells come under control—an apparent failure rate of *almost 50 percent*.[8] The Department of the Interior commissioned studies by WEST Engineering Services in 2002 and 2004 that looked in detail at the workings of certain types of blowout preventers, including that deployed on the *Deepwater Horizon*, and found that in many cases, the preventers did not work as advertised. The findings illustrated, the authors of one of the Interior-commissioned studies wrote, "the lack of preparedness in the industry" to manage "the last line of defense against a blowout."[9]

Complacency ran to the top of the American political system. While seeking the White House, Barack Obama excoriated John McCain for proposing to expand offshore drilling, because this would have "long-term consequences for our coastlines but no short-term benefits, since it would take at least ten years to get any oil. . . . When I'm president, I intend to keep in place the moratorium." In office, he did not. An Interior Department review overlooked the evidence of weak fail-safe systems and implausible cleanup preparations, and, noting the lack of serious accidents to date, Interior secretary Ken Salazar recommended new leasing to the White House in early 2010. In a political trade-off made to advance climate change legislation pending in Congress, legislation that would never pass, Obama announced plans to consider drilling in previously closed areas of the south and mid-Atlantic Ocean and the eastern side of the Gulf of Mexico—if it were approved, this would amount to the largest geographical expansion of domestic offshore leasing in a generation. "I don't agree with the notion that we shouldn't do anything," the president explained. "It turns out, by the way, that oil rigs today generally don't cause spills. They are technologically very advanced."[10]

In the sections of the Gulf open to drilling, Exxon was a laggard. Its annual reports and public affairs campaigns boasted about the cutting-edge technologies that made deep-water drilling so promising, but the truth was that ExxonMobil had for long missed many of the big opportunities in the Gulf. Humble Oil's early exploratory offshore wells, drilled off Texas during the 1970s, were expensive busts. By 2010, Chevron had drilled many more deep-water wells worldwide, proportionate to its size, than ExxonMobil. At the time of the *Deepwater Horizon* accident, ExxonMobil had only a single drilling project under way in the region. The corporation's major deep-water projects lay offshore in Equatorial Guinea, Angola, and Nigeria. This imbalance was not by design; ExxonMobil had simply missed out on the Gulf's early deep-water boom.

As a condition for leasing tracks in the Gulf of Mexico, the Minerals Management Service required companies to prepare and file spill response plans that addressed a long list of questions laid out by the regulators. These plans included how boats would be deployed, how chemical dispersants might be managed, how injured wildlife would be cared for, and above all, how oil would be cleaned up. In its filings to the Department of the Interior, ExxonMobil reported that it had developed experience and systems that would allow it to respond forcefully to even a major blowout in the Gulf of Mexico, one that might threaten the economies of built-up coastal areas from Tampa to Galveston.

"ExxonMobil's primary focus remains the prevention of incidents which might cause pollution" the corporation's 2009 filing to the regulators declared, "but in recognition that complete elimination of risk is impossible, the Oil Spill Response Plan describes the resources and procedures that would be used to mitigate potential impact."[11]

To write and file spill response plans, ExxonMobil turned to The Response Group. So did BP, Shell, Chevron, Conoco, and other Gulf drillers. The Response Group, a business dedicated to providing "effective emergency preparedness and response solutions," was a small planning and regulatory paperwork consultancy located in a tree-shaded two-building office park in Cypress, Texas, to the northwest of Houston. The firm specialized in helping oil, chemical, and other industrial firms develop, write, and file the disaster response plans required by state and federal regulators. The result was that all of the major American companies operating in the Gulf of Mexico filed essentially the same five-hundred-page boilerplate plan describing their emergency preparedness, even though each of the companies drilled as lead operator in different conditions and had distinct corporate capabilities. Each plan filed with the M.M.S. promised blithely that the driller could handle a spill even larger than the one that began on April 20, 2010, with the blowout of the *Deepwater Horizon*. Each plan declared that the driller would rely on response equipment that, as listed, was transparently inadequate to fulfill the plan's claims.[12]

After the *Exxon Valdez* wreck, Lee Raymond reflected, "The lesson learned here was to try and make sure that there were procedures both in the company and in the respective governments that they knew and we knew that if an incident were to happen, exactly what to do and how to do it." The lesson had not been learned; living up to Raymond's exhortation would have involved more extensive investments in boats, planes, and predeployed equipment than either the oil companies or the government was prepared to make. The plans on file with Interior did contain credible organizational charts, analysis of oil spill response procedures, and emergency planning manuals. Few of these were linked to real-world capabilities, however. The plans also contained howlers—ExxonMobil's plan, and several others evidently prepared from the same Response Group text, referred to preparations that had been taken to protect walruses, although walruses have not swum in the Gulf of Mexico for about three million years. The same plans listed as a marine wildlife expert a Florida Atlantic University professor, Peter Lutz, who had been dead for several years. (Tillerson later acknowledged that this was an "embarrassment," but he added, by way of justification, "The fact that Dr. Lutz died in 2005 does not mean his work and the importance of his work died with him.") The Department of the Interior accepted the filings as adequate.[13]

The spill response plans by BP and ExxonMobil on file at the time of the *Deepwater Horizon* blowout were almost identical, except for one feature. ExxonMobil's plan contained a forty-page appendix K, entitled "Media." The media management appendix was more than four times longer than the plan for oil removal, and eight times longer than the plan for "resource protection."[14]

The appendix provided a snapshot of ExxonMobil's uniform systems of public information management. It instructed ExxonMobil public affairs officers that information requested by a reporter during an oil spill emergency should be sorted into four categories. The document provided examples of types of information in each category. With Category A information, for example, it was permissible to say, at any time, on any occasion, "ExxonMobil said today no details were yet available." In Category D, the most sensitive, the example listed in the Interior response plan was "Global Warming." The document instructed, "All response statements and media releases from Category D are to be issued from" headquarters. If a reporter asked a question on this subject, the correct response was, "We will have someone from our Corporate Headquarters contact you to discuss any impact on global warming."

The appendix also provided employees and contractors with thirteen draft press releases that might be used. These included a "holding statement," a statement for "facility fire/explosion," a "product spill [reported]," a "product spill [actual]," an "employee fatality," and a "public fatality/serious injury." In the latter case, the canned statement read, "We are greatly saddened by this tragic event and express our deepest sympathy to the families of those affected. We are working with [APPROPRIATE AUTHORITIES] at the site to investigate the cause of the incident."

If criminal charges were even a remote possibility, the correct statement would be, preemptively, "We believe that there are no grounds for such charges. This was clearly an accident and we are working to respond to the immediate needs of the incident."[15]

Flying over Prince William Sound twenty-three years earlier, as the *Valdez*'s oil spread into dark shapes, BP's Lord Browne had reflected about how the oil industry would now be "measured by its weakest member, the one with the worst reputation," that is, Exxon. After long years of resentment and competition between the two companies, the tables had turned.

Browne was no longer around to face criticism. He had resigned as BP's chief executive on May 1, 2007, after admitting that he had made false statements in a British court document about the origins of his relationship with a Canadian man, Jeff Chevalier, with whom he had been romantically involved. The pair had been dining and social companions, court records showed, of Prime Minister Tony Blair; Peter Mandelson, Blair's controversial adviser; and other former luminaries of New Labor in Britain. Lord Browne's resignation from BP had seemed, in 2007, only a distasteful coda to the end of credulity about the Blair era. By 2011, it was plain that BP's corporate culture, more focused on hubristic global strategy than on day-to-day execution, had helped to set conditions for the *Deepwater Horizon* accident.

When Browne took charge of BP in 1995, he inherited a bloated, government-influenced corporation in deep trouble. Collapsing oil prices threatened BP's viability. Browne had responded much as Lee Raymond and Lawrence Rawl had done when confronting oil price declines in the 1980s, during the years leading up to the *Valdez:* He slashed costs and reorganized departments aggressively. After timely acquisitions of Amoco and Atlantic Richfield, he cut the corporation's combined costs by one fifth. "For us, it was clear that it [scale through merger] could permanently change the cost structure of the company. . . . It opens up opportunities to do new things because they're cheaper."[16] His financial engineering turned the corporation's profitability around and vaulted BP to the top of the global oil tables, positioning the company to compete anywhere. Browne also oversaw a successful push into the Gulf of Mexico. In 1999, with Exxon as a minority partner, BP discovered a billion-barrel offshore Gulf field called Thunder Horse, which would pump, after costly delays, 250,000 barrels of oil a day by 2009, or almost 5 percent of all American production. BP became the largest holder of deep-water leases in the Gulf and had exploration wins in American waters where ExxonMobil had struggled. At a prospect called Tiber in the Gulf, BP found a field it estimated to contain about 5 billion barrels of oil.[17]

Cost cutting and management redesign undermined BP's safety culture, however. "We have never seen a site where the notion 'I could die today' was so real," Telos, a consulting firm that inspected BP's Texas City, Texas, refinery, reported in 2005.[18] Two months later, fifteen workers perished in an explosion there. Browne vowed reform, but neither he nor his successor, Tony Hayward, who styled himself as a work boots–and–coveralls antidote to Browne's slick globalism, actually delivered. The year after the Texas City disaster, a BP pipeline broke and dumped 200,000 gallons of oil in Alaska. Inspectors fined a BP refinery in Ohio $3 million because it had persisted with dangerous practices that had contributed to the Texas City explosion.[19] The Center for Public Integrity found that between 2007 and 2010, BP refineries in Texas and Ohio were responsible for 97 percent of the "willful, egregious" safety violations documented by the federal Occupational Safety and Health Administration, the American regulator in charge of workplace safety. Many citations involved BP's failure to live up to previous settlements and commitments. The corporation racked up 760 violations in the willful category; during the same period, ExxonMobil had 1.[20]

On the morning after the *Deepwater Horizon* caught fire, if a pollster had telephoned executives or engineers in the American oil industry, described the circumstances of the blowout, and asked which major oil corporation was most likely to have been the platform's operator, an overwhelming majority probably would have blurted out, "BP." The corporation had partnered over the years with all of its major competitors and in these projects had earned a reputation for poor operations management in project after project.

BP paid $34 million in March 2008 to lease Mississippi Canyon Block 252, about nine square miles, which it renamed Macondo. Houston project managers dispatched the *Deepwater Horizon* to drill a single well, confirm the presence of oil, and if the prospect looked promising, cap the well and return later to begin production. In such deep water, high pressure at the seabed and temperatures at the bottom of the well that can reach 240 degrees Fahrenheit require extraordinary vigilance. Macondo was troublesome from the start. A drill bit stuck in rock. By mid-April, the project was running six weeks behind deadline and more than $58 million over budget. The pressure to finish the well, cap it, and move on intensified.[21]

The medley of errors that led to the April 20 blowout "can be traced back to a single overarching failure—a failure of management," the national commission concluded. That management failure encompassed BP and its two largest contractors on the project, Halliburton and Transocean. Supervisors at each company made errors that if avoided might have prevented the blowout. During the final critical hours, Transocean's team failed to monitor the well properly. Halliburton provided cement to seal the well that tests later showed was probably unstable—a problem that Halliburton knew about from its own internal testing, but failed to report. BP's project managers, who had ultimate responsibility, made a series of decisions apparently aimed at reducing costs that made failure more likely. Government regulators also "lacked the authority, the necessary resources, and the technical expertise" to prevent these transgressions, the commission found.[22]

As the oil poured into the Gulf, Tillerson pushed ExxonMobil's talking points into Washington's political ecosystem. He characterized the accident as a "dramatic departure from the industry norm in deep-water drilling." ExxonMobil would never have made the mistakes BP made, he said. Moreover, the recklessness of BP's operation should not catalyze intensive new regulation because BP's failure was so unusual. Tillerson spoke so forcefully about BP's apparent errors that he sounded as if he might be auditioning to appear as an expert witness at BP's liability trials. "It appears clear to me that a number of design standards were—that I would consider to be the industry norm—were not followed," Tillerson declared. "We would not have drilled the well the way they did."[23]

Tillerson arrived one evening that summer at the Metropolitan Club, near the White House, for a private dinner with influential editors and writers sponsored by the Center for Strategic and International Studies. Over roast beef and Yorkshire pudding, Tillerson went after BP's management. There were many warning signs during the Macondo operation, but BP suffered from a "culture" of looseness and rule bending, Tillerson said. BP had fine engineers and was technologically impressive, Tillerson added, but the corporation did not emphasize safety or individual accountability. "They've always been an outlier," he said. "We work with them all over the world and we've seen this."

Tillerson spoke blithely about the potential ecological damage that might result from the accident. Because Americans reacted so skittishly to the possibility that seafood might be poisoned, commercial fishing in Gulf waters slowed sharply during 2010, even outside of areas restricted by the government, so fish populations would flourish. "You like to fish? The best fishing in the world's going to be in the Gulf next year," he predicted.

If Washington now overreacted to the *Deepwater Horizon* and limited offshore drilling for many years, this would be bad policy, he said, but it would pose no strategic threat to ExxonMobil's business. "We've got opportunities all over the world."[24]

The swagger was vintage Exxon, but the public policy at issue was the corporation's philosophy of risk management. Just ten days after the *Deepwater Horizon* exploded, a ruptured ExxonMobil pipeline dumped about a million gallons of oil in coastal areas of eastern Nigeria, soiling shorelines dotted by impoverished seaside villages. The affected area lay far from American television news bureaus, and its kidnapping gangs made it a risky place to travel in any event. The spill barely registered. Not all accidents can be prevented, Tillerson and ExxonMobil's lobbyists acknowledged. Even if one accepted that ExxonMobil's own safety and self-regulatory record was exemplary, relative to peers, and even if one assumed that the corporation's relatively vigilant internal practices would endure indefinitely, without ever deteriorating again, how did Exxon propose to ensure that every other corporation in the oil industry adopted its standards, if not by government regulation? Tillerson volunteered that ExxonMobil's safety systems were "not proprietary" and he would share them with other companies, but it was neither practical nor appropriate for the corporation to police its competitors.

In comparison with other regulatory schemes, supervision of oil drilling and transport involved an unusual challenge: The incentive to find new oil in a constrained world drove all of the major companies to risky frontiers. Resource nationalism, the rise of global state-owned companies with favored positions in their home countries, and the struggle for annual reserve replacement at gigantic corporations like ExxonMobil had led them to deep water, to weak and conflict-ridden states with vulnerable populations, and increasingly to the Arctic ice, where cold temperatures might render conventional spill cleanup techniques inoperable. The national commission concluded that BP's blowout drilling in pioneering conditions in the Gulf of Mexico was not a "statistical inevitability" because sound management and regulation could have prevented the accident. Yet the record of oil accidents worldwide over thirty years was one of repetitive spills and failures, even at the best practitioners, such as ExxonMobil after the *Valdez*. In commercial aviation, idiot-proof safety systems and close regulatory inspection had reduced accidents to an overall nuisance level, although they were obviously devastating when they occurred. Marketplace incentives played a crucial role in commercial aviation. The public demanded protection from reckless airplane operators and pushed airline companies into compliance—crashes repelled customers. By comparison, in oil's case, the environmental consequences of a single accident could be very severe, but they did not threaten the lives of oil customers or change their purchasing behavior. The damage was typically remote, and for consumers gasoline remained a necessity. Marketplace incentives did work constructively in one respect—the high financial and reputational costs of the *Exxon Valdez* and the *Deepwater Horizon* served as a powerful deterrent to corporate recklessness at drilling sites—but an occasional catastrophic error could be managed and survived, as ExxonMobil had demonstrated and BP probably would. And because the need to find oil in hard places pushed corporations into greater risk taking, the overall effect was very different from aviation: It was as if United Airlines, to remain profitable and viable in the long run, had to fly faster and higher each year, while managing all the risks that came along with that stretching of its capabilities.

Tillerson rejected the national commission's finding that the BP blow-out placed "in doubt the safety culture of the entire industry." He argued that the commission "did not investigate the entire industry," and so their finding "seems to ignore years of record of good performance."[25] And yet the commission reached its conclusion in part because "the record shows" that in the absence of effective federal regulation, "the offshore oil and gas industry will not adequately reduce the risk of accidents, nor prepare effectively to respond in emergencies."

Tillerson could hardly reject the criticism about preparedness. Under his leadership, ExxonMobil had not invested in accident response capabilities in proportion to the new risks created by deep-water drilling. The corporation was not especially active in the Gulf, but it was moving to increase its presence, and it pledged preparedness in regulatory filings. The costs of preparing aggressively for a rare blowout such as the *Deepwater Horizon*'s—acquiring and positioning adequate equipment, rehearsing and planning to the same level of precision that the corporation brought to drilling operations—might be high, but they were far from prohibitive. The record showed that ExxonMobil had not made these investments nor urged that others in the industry do so.

"Your [accident response] plan is written by the same contractor that BP's is," Bart Stupak, a Michigan congressman, reminded Tillerson at a hearing that summer, as oil continued to pour into the Gulf. "So if you can't handle 40,000 [barrels of spilled oil a day], how will you handle 166,000 per day," as ExxonMobil's plan claimed could be managed?

"The answer is that when these things happen, we are not well equipped to deal with them," Tillerson admitted.

"So when these things happen, the worst-case scenarios, we can't handle them, correct?"

"We are not well equipped to handle them. There will be impacts, as we are seeing. And we've never represented anything different than that. . . . That's just a fact of the enormity of what we're dealing with."

"But they do happen."

"It just happened."[26]

A containment dome fitted onto the Gulf's floor above the spewing Macondo wellhead capped the blowout on July 15, 2010, just under three months after it began. The crisis faded rapidly. Americans had other problems on their minds. Voters angry about public debt, busted mortgages, Wall Street greed, and high unemployment went to the polls less than four months later and replaced the Democratic majority in the House of Representatives with Tea Party–influenced conservative Republicans devoted to smaller, less intrusive government. The *Deepwater Horizon* blowout reinforced popular anger toward Big Oil, but would produce no new politics to threaten the status quo in American energy policy. The accident's economic victims—commercial fishermen in Louisiana, coastal hotel owners, offshore oil workers—lived mainly in the states whose citizens voted repeatedly to accept the ecological risks of deep-water drilling.

Eighty-five percent of the world's energy—to fuel cars and trucks, to run air conditioners, to keep iPhone-tapping legions fully charged—still came from taking fossil fuels out of the ground and burning them. The likelihood that this would change anytime soon appeared slight.

ExxonMobil faced serious trials as a business in the years ahead—annual reserve replacement, maintaining its share price by extracting full value from XTO's unconventional gas holdings, and global competition—but its place at the heart of America's energy economy, as a bastion of fossil fuel optimism, remained unchallenged.

Forecasting "peak oil," the moment when world supply will reach its height and begin to decline, is a fool's errand, the long record of inaccurate past forecasts would suggest. At a minimum, there appears to be enough oil left in the world to meet projected rates of demand for several decades, and likely longer. Gas and coal supplies are even more abundant. Mongolia alone reports probable coal reserves of 152 billion tons, enough to fire every smoke-spewing power plant in China for half a century. Russian, Qatari, and Iranian natural gas deposits should last many decades, and the United States may be able to meet its own gas demand from domestic supply, if unconventional reserves fulfill their promise. Fossil fuels that emit carbon dioxide when burned are therefore likely to remain embedded in the world economy for at least half a century longer, barring a radical scientific breakthrough that allows a renewable energy source to compete economically at gargantuan scale.

It seems just as likely that the costs imposed on American society by fossil fuel dependency will remain high for an indefinite time. Between 2004 and 2009, the United States ran a deeper trade deficit—between $186 billion and $414 billion each year—to import oil and gas than it did to import goods from China.[27] The regimes in receipt of these outbound dollars—Saudi Arabia, Russia, Iran, and Venezuela, to name four—were chronically unfriendly. Rising global oil prices, usually caused by wars, strikes, or other upheavals overseas, have preceded ten of the last eleven American economic recessions, including the Great Recession that began in 2008. (That downturn was caused by a financial and housing bubble that would have burst disastrously even if every American drove a magical self-powering wind mobile, but the spike in fuel prices on the eve of the bubble's reckoning weakened confidence and household balance sheets at a turning point.) Rising oil demand from two-car families in the world's rising economies such as China and India, combined with the chronic instability of oil exporters from the Middle East to West Africa to Venezuela, means that oil supply and price shocks are likely to recur more frequently, adding to the multiple sources of American economic insecurity.

All of these economic costs of oil dependency have been evident since the 1970s, yet American democracy has produced no politics to reduce them. The lobbying power of oil corporations is hardly the only factor. Oil prices gyrated during the 1980s and 1990s; at the bottom of these cycles, gasoline was often a trivial segment of many household budgets. During the late 1990s, gasoline expenses averaged as little as 2 percent of American pretax household income. That made it relatively painless for American voters to ignore oil dependency's indirect costs and to reject the higher gasoline or carbon taxes that would be required to incent change. By the summer of 2011, gasoline expenses approached 10 percent of household income at a time of widespread economic pain. The opposite kind of policy paralysis now took hold: To change the gasoline pricing system would impose heavy new costs on working- and middle-class families suffering the most in the aftermath of the Wall Street–primed housing bust.

The threat of climate change presents the most serious danger yet to arise in the long age of fossil fuels. But global warming's victims—future generations—do not vote. Durable political majorities in advanced democracies have often been willing to impose economic costs on themselves to address current pollution that endangers living generations—smog, acid rain, poisoned water, and toxic runoff from manufacturing. Persuading those same voters to impose costs now to protect their grandchildren from climate risks that can be described only in outline has proved much more difficult.

The British economist Nicholas Stern credibly forecasted that reducing carbon dioxide emissions enough to avert potentially catastrophic global warming would cost 1 to 2 percent of global gross domestic product now, while failing to act may eventually cost five to twenty times that amount. That seemed a more politically plausible trade-off in the economic boom year of 2006, when Stern announced his findings, than it did in 2011, in the maw of stock market panics, European sovereign debt crises, flat growth across many industrialized democracies, and rising income inequality.[28]

Britain and continental European democracies have already taxed themselves to ease the climate risk faced by future generations. Coal-dependent Australia, after long resistance, has adopted a carbon price. In the United States, most of the major oil corporations that had earlier undermined the findings of climate science, including ExxonMobil, now accept, if reluctantly, that a price on carbon is coming, and that it might be justified. The near-bipartisan deal on climate policy in Congress during 2009 suggests that America will likely enact some carbon price, but only a relatively modest one, and only after the American economy recovers from recession and stagnation, which may take five or more years.

In Washington, higher taxes on carbon-based fuels will inevitably come later than they might have due to the resistance campaigns funded by oil and coal corporations—particularly ExxonMobil's uniquely aggressive influence campaign to undermine legitimate climate science during the late Clinton administration and the early Bush administration. With its ideological allies, ExxonMobil funded the promotion of public confusion about climate science by means that future employees and executives of the corporation are likely to look back on with regret.

The climate risks future generations will inherit will pass to them from many authors, of course, and hardly just from ExxonMobil. Even if ExxonMobil began immediately to invest all of its lobbying and public policy expenditures to help enact an aggressive carbon price in the United States, the West's ability to persuade China, India, and other poor, industrializing countries to adopt and enforce adequate emission reductions would still appear doubtful.

Early in 2011, the research group Climate Central, working from BP forecasts about future energy demand, calculated that stabilizing carbon dioxide concentrations by 2050 at five hundred parts per million (about a quarter higher than current levels) would require reducing average emissions per unit of energy used in the world by 4.2 percent per year. The analysts noted, "The highest previously recorded rate of decarbonization in a country probably took place in France between 1975 and 1990, when that country's nuclear power system expanded very rapidly," and yet even in that extreme instance, France's emissions fell by only 2.6 percent annually.[29]

The numbers argue that global warming on a scale scientists describe today as dangerous will occur.

On July 1, 2011, ExxonMobil's Silvertip pipeline, running from Wyoming to the corporation's refinery in Billings, Montana, sprang a leak and poured about 1,000 barrels of oil into the majestic Yellowstone River. The corporation estimated that cleanup and payments for damaged property would cost $42.6 million. "We deeply regret this incident has happened," corporate spokesman Kevin Allexon said.[30]

On the same day, Baltimore County jurors deliberating in the second of two civil lawsuits filed over the massive leak of gasoline from the former Jacksonville, Maryland, Exxon station—the case filed by Peter Angelos, Stephen Snyder's archrival in the Baltimore plaintiff's bar—returned a verdict of actual and punitive damages of $1.5 billion, ten times greater than the award Snyder had won for his clients. ExxonMobil vowed to appeal; the corporation continued to appeal Snyder's award of damages, too.[31]

A week later, on July 8, the United States Court of Appeals for the District of Columbia reinstated the lawsuit filed by villagers in Aceh, Indonesia, who alleged that ExxonMobil bore responsibility for torture and killings they had suffered at the hands of Indonesian soldiers guarding the corporation's gas fields. ExxonMobil's lawyers had challenged the case at every turn; the lawsuit had now been pending without trial or settlement for more than a decade. The corporation again filed an appeal. A month later, ExxonMobil announced that it was placing its interests in Aceh's gas fields and liquefied natural gas operations up for sale. A spokesman said the sale had "nothing to do with the Aceh lawsuit," but was the result of routine reviews of worldwide holdings.[32]

In Russia, later that summer, Rex Tillerson flew to the Black Sea resort of Sochi to meet with Vladimir Putin. Before television cameras, the two men sat on opposite sides of a horseshoe-shaped table and announced a new partnership between ExxonMobil and Rosneft, the Russian oil company. The oil firms agreed to invest at least $3.2 billion to develop oil beneath the Arctic Kara Sea; if the deal survived the backtracking and disputes that disrupted so many other Russian oil deals, the total investment in the project could reach $500 billion. The United States Geological Survey estimated that the Arctic held about 90 billion barrels of recoverable but undiscovered oil and about as much natural gas as Russia's onshore supplies, which were the world's largest.[33] Most of the Arctic's oil and gas is believed to lie in areas controlled by Russia. In the Kara Sea, where ExxonMobil agreed to drill, oil development has become easier because of the rapid retreat of Arctic sea ice, most likely due to global warming. In 2011, on a typical August day, the amount of Arctic sea ice was about 40 percent less than had been present on an average August day between 1979 and 2000.[34] At the announcement in Sochi, Vladimir Putin spoke approvingly of ExxonMobil's "unique technology" and the corporation's ability to operate in the Arctic's "difficult conditions." As to the scale of the investment planned, Putin added, "It's scary to utter such huge figures."[35]

Mikhail Khodorkovsky, the oil and banking tycoon who had drawn Lee Raymond into negotiations designed to transform the U.S.-Russian energy partnership during the first term of the Bush administration, languished in a remote Russian prison, under sentence until at least 2017.

In West Africa, ExxonMobil's managers continued to bring oil to market despite the coup plots, kidnapping raids, corruption, and factionalism menacing the corporation's host regimes in Nigeria, Chad, and Equatorial Guinea. After millions of dollars in expenditures on Washington lobbyists, Equatorial Guinea's president, Teodoro Obiang, managed for the first time to have his photograph taken at the side of an American counterpart: Barack Obama, who agreed to pose with Obiang at a museum reception in New York. Obama's administration continued to license military and police trainers to support Obiang's regime, although the White House, cautioned by human rights activists and congressional critics of Equatorial Guinea, held back from partnership. Obama's Justice Department filed a civil lawsuit against Obiang's free-spending son, Teodoro, seeking forfeiture of his Malibu mansion and other assets on the grounds that Obiang's money came from "foreign official corruption." ExxonMobil produced more than 600,000 barrels of oil and gas liquids per day from West Africa; the corporation produced roughly the same amount of oil from the Gulf of Guinea as it produced from the United States and Canada. In 2011, Walter Kansteiner, the assistant secretary of state for African affairs during the first term of the George W. Bush administration, joined ExxonMobil as a senior adviser on the corporation's Africa strategies.

In Iraq, ExxonMobil followed adventurous Hunt Oil into Kurdistan, in defiance of Baghdad's government and despite discouragement from the Obama administration, which feared, as the Bush administration had, that oil deals struck independently with the Kurds would worsen Iraq's ethnic conflicts. ExxonMobil's decision risked stirring the ire of Iraq's Shia-led national government, which had awarded the corporation a contract to raise production in its massive West Qurna field in the south of the country. Tillerson undertook his gambit without informing the Obama administration in advance. After ExxonMobil signed agreements concerning six Kurdish oil fields, Tillerson arranged a conference call with senior State Department officials, and told them, "I had to do what was best for my shareholders."

The Obama administration announced plans in the summer of 2011 for the first sale of oil leases in the deep waters of the Gulf of Mexico since the *Deepwater Horizon* blowout. Interior secretary Ken Salazar said twenty million acres would be put up for lease—all in the Gulf's western waters, nearest to Texas, Louisiana, Alabama, and Mississippi. "We have strengthened oversight at every stage of the oil and gas development process," Salazar said. The sales were an "important step toward a secure energy future."[36] The administration also considered proposals for a pipeline that would transport oil from Canadian sands to refineries in the United States. Obama initially rejected the arguments of environmentalists and climate scientists who fear the pipeline will lock in energy-intensive oil production, and by doing so exacerbate global warming. The president later put the decision on hold.

It remained arguable how "American" ExxonMobil's private empire was, given its global reach. Yet in its strategies and systems the corporation remained recognizably a descendant of the American icon John D. Rockefeller and his Standard Oil. And of all the banking, industrial, and transportation giants birthed by America's Gilded Age, none could look back on a winning streak and a record of durability comparable to those of ExxonMobil's.

The more recent heights of ExxonMobil's profitability and political influence during the Raymond and Tillerson eras reflected in part the growing relative power of corporations in the American political and economic system. Corporate profits in 2011 made up a larger share of American national income, when compared to workers' wages and small business income, than at any time since 1929, when such statistics were first recorded. The United States Supreme Court, in its landmark decision in *Citizens United vs. Federal Election Commission*, reaffirmed in 2010 the freedom of corporations to fund political advocacy. In the years after the Mobil merger, Raymond and Tillerson oversaw more spending on direct lobbying in Washington than all but two other American companies, General Electric and Pacific Gas & Electric. ExxonMobil had evolved into the most profitable corporation headquartered in the United States—and one of the most politically active—in an era of corporate ascendancy.

On July 28, 2011, ExxonMobil announced its profits for the first half of the year. The total came in at $21.3 billion, a whisker under the amount the corporation reported during the same period in 2008, when it set a record for the most nominal profit earned by any corporation in American history.

Eight days later, on August 5, 2011, Standard & Poor's announced the first-ever ratings downgrade of the bonds issued by the United States Treasury, marking them down from a AAA rating to AA-plus. The Standard & Poor's downgrade meant that ExxonMobil, one of only four American corporations to maintain the AAA mark, now possessed a credit rating superior to that of the United States.

Standard & Poor's received intense criticism for its judgment that the American government's ability to repay its lenders might be in any doubt. Yet the fiscal trajectories of the United States Treasury and ExxonMobil had certainly diverged. In 1999, the year that Exxon's acquisition of Mobil closed, the federal government and the corporation each took in more money annually than was required to meet expenses. Their paths then divided. In an era of terrorism, expeditionary wars, and upheaval abroad, coupled with tax cutting and reckless financial speculation at home, one navigated confidently, while the other foundered. From the day of the Mobil merger closing until the day of the S&P downgrade, the net cash flow of the United States—receipts minus expenditures—was approximately negative $5.7 trillion. ExxonMobil's net cash flow from operations and asset sales during the same period was a positive $493 billion.[37]

Acknowledgments

This book would not have been possible without the generosity of the scores of people who agreed to provide interviews, sometimes about sensitive or controversial subjects. Some of those to whom I owe the most cannot be named, but I am grateful for their trust and assistance. None of my sources, researchers, or collaborators should be judged accountable for errors or misjudgments in the text; the responsibility is mine.

Steve LeVine, Susan Murcko, and Ann O'Hanlon provided careful, multiple readings of early drafts, made important suggestions, rescued me from errors, and otherwise made the book better.

S. C. Gwynne contributed biographical insights about Lee Raymond and Rex Tillerson early on. Ian Gary generously opened his boxes and files about African oil. Ben Lando conducted interviews and provided insights about Iraq's oil. Rob McKee and Mike Stinson shared their fascinating diaries and archives about their service as American oil advisers in Baghdad. Chris Goldthwait was equally helpful with his collection of letters from Chad. Dan Freifeld offered valuable insights. Vince Crawley arranged interviews at Africa Command in Stuttgart, Germany. Sally Donnelly was a great help at the Pentagon. Miriam Elder contributed energetically from Moscow. Robert Becker provided excellent legal advice about the Freedom of Information Act. I owe gratitude to Angue Ondo for her aid during my travel to Equatorial Guinea. Sam Olukoya and the resourceful Chris Ewokor assisted my travel in Nigeria. Esseme Eyiboh provided hospitality and high spirits during a memorable journey to Akwa Ibom. In Chad, I owe thanks to Besba Tong-pa Raoutouin, Michelle Bonnardeaux, and Celeste Hicks. In Jakarta and Aceh, Miki Salman and Sidney Jones helped greatly. Alan Jeffers at ExxonMobil endured my inquiries cheerfully.

I again enjoyed the fortune of partnership with Ann Godoff, Susan Petersen Kennedy, and the team at Penguin Press. I have now worked with Ann continually for two decades. As ever, she encouraged the most serious, ambitious work possible, and her editing notes were perceptive and important. Thanks as well to Tracy Locke, Lindsay Whalen, Ben Platt, and Deborah Weiss Geline.

Jon Wallace contributed research, organizational skill, and high morale. Thanks also to Christina Satkowski for her skill and reliability, and to Victoria Collins for her excellent work.

Alexandra Coll, Emma Coll, Maxwell Coll, and Rory Steele helped with research, chronologies, and other aspects of the family business, and were inspiring for many other reasons, too.

During the life of this project, I served as president of the New America Foundation, a nonprofit, nonpartisan public policy research institution headquartered in Washington, D.C., with an annual budget of about $16 million. About three fifths of New America's revenue comes from philanthropic foundations such as the Ford Foundation, the Rockefeller Foundation, the Bill & Melinda Gates Foundation, the Macarthur Foundation, the Smith-Richardson Foundation, the Peter G. Peterson Institute for International Economics, and the Open Society Institute. Most of the rest of the revenue comes from contributions by individual philanthropists, such as Eric and Wendy Schmidt, Bernard and Irene Schwartz, Jeffrey and Cal Leonard, Gus and Rita Hauser, David and Katherine Bradley, Chip Kaye, William Gerrity, and Boykin Curry. The foundation also receives government grants and corporate donations. For the latest full year available, each of these sources of revenue provided less than 5 percent of the total. During 2009, my colleague Steve Clemons, while overseeing programs on American foreign policy, solicited and received from ExxonMobil two contributions, of $25,000 and $80,000, respectively, to support conferences he ran. I recused myself from those discussions and activities. A full listing of the foundation's financial supporters is available at www .newamerica.net.

Colleagues at the New America Foundation contributed to the research and writing of this book in many ways, especially by challenging and refining my understanding of geopolitics, and by generously allowing me the time to travel and interview abroad. Thanks in particular to Simone Frank, Eric Schmidt, and Rachel White; the book would not have been possible without them. Thanks also to Liaquat and Meena Ahmed, Amjad Atallah, Peter Beinart, Peter Bergen, David Bradley, Steve Clemons, Jeannette Clonan, Reid Cramer, Michael Crow, Boykin Curry, Patrick Doherty, James Fallows, Sheri Fink, Brian Fishman, Frank Fukuyama, Joel Garreau, Atul Gawande, Bill Gerrity, Tom Glaisyer, Tim Golden, Eliza Griswold, Lisa Guernsey, Ted Halstead, Rita Hauser, Laurene Powell Jobs, Fred Kaplan, Zachary Karabell, Chip Kaye, Andrew Lebovich, Jeffrey Leonard, Flynt Leverett, Daniel Levy, Michael Lind, Maya MacGuineas, Lisa Margonelli, Andres Martinez, Kati Marton, Danielle Maxwell, Mary-Ellen McGuire, Walter Russell Meade, Sascha Meinrath, Lenny Mendonca, Evgeny Morozov, Bob Niehaus, Amanda Ripley, Nicholas Schmidle, Troy Schneider, Bernard Schwartz, Sherle Schwenninger, Anne-Marie Slaughter, Katherine Tiedemann, Laura Tyson, Robert Wright, Tim Wu, Dan Yergin, Fareed Zakaria, and Jamie Zimmerman, as well as the many other staff and fellows who saw to my continuing education.

Thanks, at the *New Yorker*, to David Remnick, Dorothy Wickenden, Nicholas Thompson, Virginia Cannon, Amy Davidson, Pam McCarthy, Nandi Rodrigio, Seema Gauhar, Tim Farrington, Jane Mayer, and Larry Wright, as well as many other terrific colleagues. Thanks as well to Susan Glasser at *Foreign Policy*, David Plotz and Jacob Weisberg at *Slate*, Carlos Lozada at the *Washington Post*, and Robert Silvers at the *New York Review of Books*.

Melanie Jackson, my literary agent for the past twenty-seven years, again made it all work.

Michael Abramowitz, Luke Albee, Rick Atkinson, Jane Atkinson, Dean Baquet, Phil Bennett, Eric Cohen, Dan Coll, Geoffrey Coll, Steve Fierson,

David Finkel, Bart Gellman, Jane Getter, Bradley Graham, the Greenhouses, John Harris, Adam Holzman, Monica Klien-Samanez, Dylan Landis, Dafna Linzer, David Maraniss, Linda Maraniss, the Morrises, Lissa Muscatine, Janice Nittoli, Ama Nkrumah, the Reisses, Joanne Reynolds, Anthony Spaeth, Valerie Strauss, the Tulchins, and Alexandra Viets provided friendship, hospitality, entertainment, and other support to the Colls during the long life of this project. Thanks as well to Robert and Shirley Coll, and to John, Joan, Marian, and Mel. I owe it all again to Susan; her enthusiasm for this work did much to bring the book to the finish.

Notes

Private Empire is based primarily on interviews with more than 450 people in the United States and abroad; some agreed to be interviewed multiple times. The interview subjects included current and former ExxonMobil executives, executives at competing corporations, lobbyists, scientists, lawyers, diplomats, military officers, intelligence officers and analysts, government policymakers, former guerrilla leaders, congressional staff, Wall Street analysts, energy industry consultants, environmentalists, and social activists. The narrative also benefited from the release of about eight hundred pages of documents—mainly State Department cables—that were provided to me in response to Freedom of Information Act requests concerning ExxonMobil's recent activities in Indonesia, Russia, Equatorial Guinea, Chad, and elsewhere in West Africa. The full release of Wikileaks' collection of State Department cables from 2003 to early 2010 provided additional valuable insights and details, particularly about ExxonMobil's activity in Chad, Nigeria, and Venezuela. Wikileaks cables—as opposed to those released in response to my F.O.I.A. requests—are indicated below by (W). Court records and trial and deposition transcripts from *Exxon Valdez* litigation in Alaska; the Jacksonville, Maryland, gasoline spill case *Jeff Alban et al. v. Exxon Mobil Corp.*; the litigation concerning the corporation's involvement in the Aceh conflict, *John Doe I et al. v. ExxonMobil et al.*; and ExxonMobil's court hearings concerning its operations in Venezuela provided much additional, valuable testimony by ExxonMobil executives, as well as excerpts from corporate documents and e-mails. Documents and diaries shared by former American oil advisers to the Coalition Provisional Authority in Iraq were very helpful. Researchers at the Center for Responsive Politics provided guidance and support for analysis of the center's important data sets on campaign contributions and lobbying. Investigative files on Equatorial Guinea produced by the Permanent Subcommittee on Investigations of the Committee on Governmental Affairs of the United States Senate provided unique banking and financial records. The extensive investigations into climate science policy carried out by the House Oversight and Government Reform Committee brought forward numerous internal government records. The Union of Concerned Scientists, Greenpeace, and other environmentalist investigators have also obtained and published important government and industry documents on climate policy, from which I was able to draw. Oxfam, Global Witness, Catholic Relief Services, Human Rights Watch, Amnesty International, Coventry Cathedral, and the International Crisis Group have published valuable investigations of conflicts and corporate responsibility issues in Africa and Asia that I sought to explore. Securities and Exchange Commission filings provided extensive data about ExxonMobil's oil and gas production, reserves, and financial reporting. I am in great debt to the published work of many other reporters, scholars, and international affairs analysts, as the notes that follow reflect.

Many of the interviews for this book were conducted on the record. Where an interview subject spoke on condition that he or she would not be named, the notes provide as much information as possible, consistent with these agreements. On-the-record interviews conducted by researchers who worked with me on the book are indicated by the presence of the researcher's initials in parentheses following the source information. To conduct interviews and field research, I traveled to Alaska and throughout the United States, as well as to Indonesia, Equatorial Guinea, Chad, Nigeria, Europe, the Middle East, and elsewhere abroad. Other international interviews were carried out by telephone and by local researchers contracted for the purpose.

ExxonMobil authorized eight current executives and managers to provide background interviews and briefings, which were helpful but limited in scope. None of those interview subjects agreed to be quoted by name. The corporation declined other requests for interviews in the United States and abroad. The corporation's chief executive, Rex Tillerson, declined several requests for interviews. The corporation did provide credentials so that I could attend a number of events where Tillerson spoke and took questions. His predecessor, Lee Raymond, agreed to be interviewed. More than four dozen other current and former ExxonMobil executives, directors, managers, employees, consultants, and contractors also provided interviews. Some of these people spoke without authorization or addressed sensitive subjects and therefore requested anonymity, as the notes reflect.

After the manuscript was substantially drafted, with researcher Haley Cohen, I attempted to fact-check material about current ExxonMobil executives and the corporation by submitting memoranda totaling more than one hundred pages to ExxonMobil's public affairs department. Two current executives responded initially to fact-checking questions, but ultimately, spokesman Alan Jeffers said ExxonMobil would offer no additional response to the fact-checking questions. The corporation was the only party of the dozens reached during the fact-checking process that declined to participate. I also submitted to ExxonMobil for formal comment sixteen questions concerning controversies, lawsuits, and other matters. The corporation declined to reply to all of these questions except one, concerning 2008 contributions by ExxonMobil executives to the campaign committee of Representative Joe Barton (R-Texas), as is reflected in chapter 22.

The chapter-by-chapter notes below provide the sources for quotations, numbers, and narrative incidents recounted in this book.

PROLOGUE: "I'M GOING TO THE WHITE HOUSE ON THIS"

1. Joseph Hazelwood descended: Keeble, *Out of the Channel*, p. 41. Yearbook motto, I.Q. score, Stonewall Jackson, and Oscar Wilde: Coyle, *Outside*, October 1997. Coyle's extraordinary profile of Hazelwood is the best single published source on the former captain's life and on the impact of the grounding on him; Hazelwood's trial testimony is also bracingly direct.
2. "midlife crisis": Joseph Hazelwood's trial testimony, *Baker v. Exxon*, No. 04.35182, United States Court of Appeals, May 10, 1994. "detected . . . take care of it": Ibid. Ordering beer:

John Donvan et al., *Turning Point*, ABC News, June 15, 1994. The ABC News documentary is an exceptional work of television journalism. Also, Deposition of Lee R. Raymond, United States District Court for the District of Alaska, A-89-095, November 19, 1992. Two or three vodkas: Coyle, op. cit. When Coyle interviewed Joseph Hazelwood extensively in 1997, Hazelwood was employed at a maritime marine insurer in New York and said he had given up alcohol. In 1994, he testified at trial that "the last drink I recall having is March 23, 1989."

3. Salary: Joseph Hazelwood trial testimony, *Baker v. Exxon*, May 11, 1994. B.P. field party: Roderick, *Crude Dreams*, p. 124. July 28, 1977: Ibid., p. 417.

4. 1,264,155 barrels: There are several published estimates of the ship's load, all within a fairly narrow range. This is the number published by the Alaska Resources Library and Information Services. More than one hundred times: ABC News, op. cit.

5. Author's visit to Prince William Sound, June 2010. Every four seconds: Coyle, op. cit. "Judging . . . over": ABC News, op. cit.

6. Coffee break: Ibid. Radar, blood tests: Keeble, op. cit., p. 43.

7. Quotations from Gregory Cousins's trial testimony: Keeble, ibid., p. 44.

8. "Wasn't a compelling reason": Coyle, op. cit.

9. For a thorough reconstruction of what occurred on the bridge, drawn from trial testimony and testimony before the National Transportation Safety Board, see Keeble, op. cit., pp. 45–47. "Serious trouble": From Cousins's testimony, Coyle, op. cit.

10. "Vessel . . . fucked": Keeble, op. cit., p. 48. Vomited, "breadbasket . . . an end": Joseph Hazelwood's trial testimony, *Baker v. Exxon*, op. cit. "We fetched up . . . a while": ABC News, op. cit.

11. "I've got . . . heart attack": Steve McCall oral history, in Bushell and Jones, *The Spill*, p. 47. "You could . . . the crew": Mark Delozier oral history, ibid., p. 29.

12. "may have . . . with it": Steve Cowper oral history, ibid., p. 41. "The game rules . . . previously": Keeble, op. cit., p. 51. Exxon employment cuts: *New York Times*, April 2, 1989. Profits per employee in 1987: *BusinessWeek*, July 18, 1988. For N.T.S.B.'s assessment of Exxon's culpability, see the letter of its chairman, James Kolstad, to Exxon's chairman, Lawrence Rawl, September 18, 1990. The N.T.S.B. found that Cousins was working on too little sleep because of crew scheduling and that "evidence indicated that watch-keeping safeguards . . . had been compromised because of the manning level" aboard the tanker. See also, "Grounding of U.S. Tankship *Exxon Valdez* on Bligh Reef, Prince William Sound Near Valdez, AK, March 24, 1989," N.T.S.B. Report no. MAR-90-04.

13. "It was hard . . . unmanageable": Don Cornett oral history, in Bushnell and Jones, op. cit., p. 98. Transcripts of telephone recordings from the Alyeska Emergency Center, from 4:57 a.m. on March 24, 1989, and on March 26, 1989: Transcribed by the Alaska Resources Library and Information Services. Cornett worked on Exxon's public relations challenges stemming from the spill for the next seven years. Despite his initial enthusiasm for the *Valdez* media battle, he later reflected that "this was not a job that any sane person would ever seek."

14. "Chagrined . . . for Exxon": Trial testimony of Lee Raymond, *In Re* Exxon Valdez *Oil Spill*, August 25, 1994.

15. Dennis Kelso oral history, in Bushnell and Jones, op. cit., p. 62.

16. Senior Coast Guard officer: Interview with Admiral Paul Yost. "a lot of cleanup equipment . . . oil spill specialist": *New York Times*, April 2, 1989. "There is . . . opposed": Raymond deposition testimony, op. cit., November 19, 1992.

17. Ibid.

18. All quotations from Admiral Paul Yost oral history, Bushnell and Jones in op. cit., pp. 124–127, and from an interview with Yost.

19. Transcript, *The MacNeil/Lehrer NewsHour*, July 27, 1989.

20. "didn't get along . . . Go ahead": Admiral Paul Yost oral history, in Bushnell and Jones, op. cit.

21. Ibid.

22. Keeble, op. cit., p. 186.

23. "no matter . . . envisioned": Lee Raymond deposition testimony, op. cit.
24. All John Browne quotations and biography citations: Browne, *Beyond Business*. Valdez episode, Browne's flight and reflections: Ibid., p. 39. "That oil company was now Exxon": Ibid., p. 40.
25. "A Conversation with Lee Raymond," *Charlie Rose* PBS, May 6, 2004.

CHAPTER ONE: "ONE RIGHT ANSWER"

1. Associated Press, April 30, 1992; *New York Times*, May 1, 1992.
2. *United States of America v. Arthur D. Seale* and *United States of America v. Irene J. Seale*, findings of the United States Court of Appeals, Third Circuit, 20 F.3d 1279 decided April 7, 1994. "This tragic allegation": *New York Times*, May 8, 1992.
3. "Wherever he is": *New York Times*, May 11, 1992. "If you interfere . . . soldiers in war": *New York Times*, July 24, 1992.
4. *New York Times*, June 20, 1992. *United States v. Arthur D. Seale*, op. cit.
5. Interview with Lee Raymond.
6. Arthur and Irene Seale biography: *New York Times*, June 21, 1992; June 28, 1992; and July 1, 1992.
7. Surveillance, kidnapping, shooting: *United States v. Arthur D. Seale*, op. cit. "More like a closet": Arthur Seale's interview with ABC News, released November 12, 1992.
8. Deposition of Lee Raymond, United States District Court for the District of Alaska, A-89-095, November 19, 1992.
9. Interview with Lee Raymond.
10. New safety regime, Raymond Rule: Ibid. Joseph R. Carlon: Interview with a former Exxon manager involved with corporate security.
11. Prizes: Interview with a twenty-eight-year retired Exxon manager. "If we have a whole lot of paper cuts": Glenn Murray, Exxon corporate safety program manager, in the *Dallas Morning News*, June 20, 2010.
12. Interview with former managers. Africa: "Electronic Monitoring of Driving Safety Performance," Esso Exploration & Production Chad/Cameroon Development, Project Update no. 23, p. 23.
13. "The only way": Fair Disclosure, transcript, ExxonMobil Corporation Analyst Meeting, March 4, 2003.
14. All quotations, "Operations Integrity Management System," www.exxonmobil.com, examined and typed, June 24, 2010.
15. Interview with an Exxon executive.
16. Interview with Kathleen Cooper.
17. Author's visits to Irving, Texas, 2008–2009. "God Pod" and "Death Star": Interviews with former Exxon employees.
18. Tarbell, *The History of the Standard Oil Company*, p. 274.
19. Exxon's relative size: Trial testimony of Lee Raymond, *In re* Exxon Valdez *Oil Spill*, August 25, 1994.
20. "Lots of wrong ways": Interview with Ed Chow. "prickly as partners": Interview with a competing executive. "Fundamentals": Interview with Lee Raymond.
21. Interviews with two former ExxonMobil managers.
22. "You could have . . . before it stops": Interview with an executive familiar with the New York office. $477 million: *New York Times*, April 2, 1989.
23. Interview with an executive who served as a director of the corporation during the Lee Raymond era.
24. Ibid.
25. Executives noticed a process analogous to natural selection: Interview with an Exxon executive. "dog eat dog": *New York Times*, May 9, 1982.
26. "You don't like them . . . in the room": Interview with a Washington-based executive at a competing firm.

27. Fortune 500 rankings from 1996: http://money.cnn.com/magazines/fortune/fortune500_archive/full/1996/.

28. "What you're hearing . . . focus": "Presentations and Q&A Session," ExxonMobil Corporation Analyst Meeting, March 9, 2005.

29. "Exxon's attitude . . . hope not": Interview with a U.S.-based competing executive. "strong corporate culture": Rockefeller family members' press conference, April 30, 2008. "Self-referential . . . not good citizens": Interview with Robert Monks. (BVH). "It doesn't take you": Interview with a former White House official.

30. "We don't run this company . . . principles": Interview with Fadel Gheit, a longtime energy industry analyst at the investment firm Oppenheimer & Co. "relentless pursuit of efficiency": *Forbes*, April 21, 1997.

CHAPTER TWO: "IRON ASS"

1. Aircraft, flight operations: Affidavits of Patricia W. Andrews, Charles W. Cone, and James W. Johnson, *Equal Employment Opportunity Commission v. ExxonMobil Corporation*, 30-6-cv-1732-K, United States District Court, Northern District of Texas. Aviation Services operations, what Raymond said about Charlene: Interviews with former ExxonMobil employees.

2. Interviews with former ExxonMobil employees, California and Arizona property records.

3. Interviews with former ExxonMobil employees.

4. Interview with a party guest and Exxon employee.

5. "I can envision": Interview with a former ExxonMobil employee.

6. "out-Rawling Rawl": *BusinessWeek*, April 2, 1990.

7. "Crude oil . . . around here": Interview with Lee Raymond.

8. Solar panels: *New York Times*, August 9, 1981. Clifton Garvin regretted his experiment with alternative energy: "I've got a backyard full of these damn collectors, and they look like hell. . . . The thing works great when the sun shines, but you'd be amazed at how much the sun's not out around here."

9. "Patience pills": Interview with Peter Townsend.

10. "I don't think Mickey": Fair Disclosure transcript, ExxonMobil Corporation Analyst Meeting, March 4, 2003.

11. Interview with Lee Raymond.

12. Exxon Corporation, Securities and Exchange Commission, Form 10-K405, for the fiscal year ended December 31, 1997, and Form 10-K for the fiscal year ended December 31, 1973. The numbers used reflect worldwide oil liquids and liquid natural gas production. If other natural gas production were included on an oil-equivalent basis, the numbers in each year would be higher, but the basic portrait of shrinkage would remain.

13. Interview with Lee Raymond.

14. Exxon Form 10-K405, March 18, 1998; "Exxon Replaces 121% of Production in 1997," *M2 PressWire*, February 5, 1998. These observations draw on Steve LeVine's groundbreaking research into the differences between reserves reported by Exxon in 10-K filings and those reported in its annual press releases about reserve replacement.

15. As quoted in Securities and Exchange Commission RIN 3235-AK00, "Concept Release on Possible Revisions to the Disclosure Requirements to Oil and Gas Reserves," December 12, 2007.

16. Interview with Fadel Gheit.

17. Browne, *Beyond Business*, p. 69.

18. Geographical distributions of proved reserves are as of December 31, 1998, as described in Securities and Exchange Commission 10-K forms filed by Exxon, Mobil, Chevron, BP, and Shell.

19. Browne, op. cit.

20. "We need to face . . . they're done": Transcript of Exxon-Mobil merger press conference, December 1, 1998. "Worried . . . geologically": *New York Times*, November 16, 2008.

21. Browne, op. cit., p. 70.
22. "Oil Markets: The Dynamics of Structural and Financial Change," Edward L. Morse, Power-Point presentation, Columbia University, September 2008.
23. "Maybe we should . . . other things": *New York Times*, December 2, 1998. "Guess who . . . massive anxiety": Interview with a senior executive involved. "Could smoke . . . broken a dam": Browne, op. cit., pp. 70–72.
24. Interview with a senior executive involved in the merger discussions.
25. Ibid.
26. *International Oil Daily*, May 25, 2005.
27. *Financial Times*, August 21, 2011.
28. Interview with a former Exxon employee involved.
29. Browne, op. cit., p. 72.
30. "the proverbial": *Washington Post*, December 2, 1998. "Competition . . . changes": Merger press conference transcript, December 1, 1998.
31. Ranked forty-fifth as an economic entity: "World Investment Report 2002: Transnational Corporations and Export Competitiveness," United Nations Conference on Trade and Development, 2002, p. 104. G.D.P. figures from International Monetary Fund, World Economic Outlook database, September 2002. "economy of scale": Interview with Lee Raymond. "It's a great time": *BusinessWeek*, April 9, 2001.

CHAPTER THREE: "IS THE EARTH REALLY WARMING?"

1. "Conversation with Lee Raymond," *Charlie Rose*, PBS, May 6, 2004.
2. Center for Responsive Politics, OpenSecrets.org database.
3. Interview with an energy industry executive.
4. Interview with Joseph A. Gillan.
5. Federal disclosure filings collated by OpenSecrets.org.
6. "when they come": Interview with an Exxon executive.
7. Interviews with individuals familiar with ExxonMobil's lobbying.
8. From "This is what the corporation believes" to "don't do it very well": quotations from interviews with Republicans and oil industry officials in Washington.
9. Interview with Lee Raymond.
10. Dressler and Parson, *The Science and Politics of Global Climate Change*, p. 51.
11. "No reason . . . too late": "Carbon Dioxide and Climate: A Scientific Assessment," p. viii.
12. "Embrace the uncertainty": *Wall Street Journal*, June 15, 2005. After the I.P.C.C. shared the Nobel Peace Prize with Vice President Al Gore in 2007, Exxon executives remarked sardonically that because their scientists had participated in the assessments, the corporation could now claim that it shared a Nobel for its global warming endeavors.
13. *Wall Street Journal*, June 15, 2005.
14. Hansen et al., "Global Climate Changes as Forecast by Goddard Institute for Space Studies Three-Dimensional Model," August 20, 1988.
15. "IPCC Second Assessment: Climate Change 1995," p. 5.
16. Lee Raymond, Plenary Address, Fifteenth World Petroleum Congress.
17. "eight hundred . . . anything else": Ibid. Cooney: Deposition of Philip Cooney, U.S. House of Representatives, Committee on Oversight and Government Reform, March 12, 2007.
18. "the most effective": Interview with Kert Davies (BVH). "the promotion": Cooney deposition, op. cit.
19. "Global Climate Science Communications Action Plan," April 3, 1998. Cited in "Political Interference with Climate Change Science," House Committee on Oversight, op. cit.
20. Greenpeace database, citing Exxon Education Foundation Dimensions 1998 report and the ExxonMobil Foundation's Internal Revenue Service filing for 2000.
21. All shareholder meeting quotations: "Excerpts from the May 31, 2000, ExxonMobil Annual General Meeting," transcript produced by Campaign ExxonMobil. "If the data

were compelling . . . No?": "A Conversation with Lee Raymond," op. cit., November 8, 2005.

22. All George W. Bush quotations from transcripts of the presidential debate at Wake Forest University, Winston-Salem, North Carolina, October 11, 2000.

23. Interview with Artistides Patrinos.

24. Ibid.

25. Haley Barbour memo first reported in the *Los Angeles Times*, August 26, 2001. Christine Todd Whitman and Paul O'Neill: Suskind, *The Price of Loyalty*, p. 123.

26. Randy Randol memo released in response to a Freedom of Information Act request made by nongovernmental organizations.

CHAPTER FOUR: "DO YOU REALLY WANT US AS AN ENEMY?"

1. "So must ExxonMobil" and "very close to closing down": Jakarta to Washington, March 7, 2001. The author obtained this and other State Department cables and documents cited in this chapter, except where otherwise noted, through a Freedom of Information Act request.

2. About a fifth of worldwide upstream revenue: Estimate derived from interviews with analysts and ExxonMobil's investor information department. Profits in 1998, 1999, and 2000: ExxonMobil's estimates provided in discovery, cited in Plaintiff's Memorandum, July 18, 2008, *John Doe I et al. v. ExxonMobil Corporation et al.*, 01-1357 (LFO) United States District Court for the District of Columbia. Government of Indonesia's take was $1.2 billion: Jakarta to Washington, March 12, 2001. Indonesia's oil and gas revenue was about $5 billion; its government revenue before International Monetary Fund loans and other international aid was about $19 billion.

3. "A quixotic . . . exaggeration" and "circulating . . . diplomacy": William Nessen, "Sentiments Made Visible: The Rise and Reason of Aceh's National Liberation Movement," in Reid, *Verandah of Violence*.

4. Lost contract to Bechtel: Kristen E. Schulze, "Insurgency and the Counter-Insurgency: Strategy in the Aceh Conflict, October 1976–May 2004," in Reid, ibid. Doral Inc.: Interview with Nezar Patria. "lived long enough" and "mid-life crisis": Nessen, in Reid, op. cit.

5. Declaration of independence: Nezar Patria, "Islam and Nationalism in the Free Aceh Movement, 1976–2005," International History Department, London School of Economics. "foreign oil companies . . . immediately": Schulze, in Reid, op. cit. "They have Mobil Oil . . . support us": Interview with Nezar Patria.

6. Jakarta to Washington, March 12, 2001.

7. Ransom of $100,000, "heightened concern": Jakarta to Washington, March 7, 2001. Newspaper advertisement proposal, "not advisable": Jakarta to Washington, February 23, 2001.

8. Jakarta to Washington, March 9, 2001.

9. Interview with Robert Gelbard (HC).

10. "We'll know it": Jakarta to Washington, March 14, 2001.

11. Jakarta to Washington, March 8, 2001.

12. Interview with Ifdhal Kasim.

13. *BusinessWeek*, December 28, 1998.

14. Ibid.

15. "confusion and ambivalence": Interview with Agus Widjojo.

16. State Department findings: 2000 Country Reports on Human Rights, www.state.gov/g/drl/rls/hrrpt/2000/eap/707.htm. "The companies . . . dilemma in Aceh": Jakarta to Washington, July 7, 1999, obtained by the author from a source.

17. Voluntary Principles on Security and Human Rights: www.state.gov.www/global/human_rights/001220_fsdrl_principles.html.

18. "They just didn't see the relevance": Interview with Arvind Ganesan. "We don't sign on": Interview with an ExxonMobil executive.

19. All quotations from ExxonMobil internal documents from excerpts cited in Memorandum & Opinion, August 27, 2008, and Plaintiff's Memorandum, July 18, 2008, *John Doe I v. ExxonMobil*, op. cit.

20. Plaintiff's Memorandum, ibid.
21. Interviews with individuals familiar with ExxonMobil's Global Security department. Also, declarations of Ron Wilson, John Alan Connor, Michael Farmer, Robert Haines, and Lance Johnson, *John Doe I v. ExxonMobil*, op. cit.
22. "nothing as dramatic," "physical safety": Interview with an individual involved with the issues in Aceh. Foreign Corrupt Practices Act dilemma and reviews: Interview with another individual involved.
23. Memorandum & Opinion, op. cit.
24. "Ludicrous": Interview with an individual involved. "deployment . . . security affairs": Memorandum & Opinion, op. cit.
25. Mike Farmer: Excerpts from his deposition, Memorandum & Opinion, ibid. Document quotations: Ibid., and the Plaintiff's Memorandum, op. cit.
26. "Such a formal move": Abuja to Washington, August 12, 2003 (W). Other quotations from interviews with two individuals involved.
27. All quotations from Colin Powell-Alwi Shihab meeting: Secretary of state to Jakarta, March 16, 2001.
28. Robert Haines, Karen Brooks, and Skip Boyce: Interviews with individuals familiar with the discussions. All quotations from Ron Wilson-Luhut Pandjaitan meeting: Jakarta to Washington, March 19, 2001.
29. Interview with Robert Gelbard (HC).
30. All quotations from the Banda Aceh meeting: Jakarta to Washington, April 12, 2001.
31. "seemed to support . . . burials": Interview with Nordin Abdul Rahman. "ExxonMobil land . . . unacceptable": Interview with Munawar Zainal.
32. Interview with Robert Gelbard (HC).
33. All quotations from Skip Boyce's meetings in Banda Aceh: Jakarta to Washington, April 26, 2001.
34. Stockholm to Washington, June 11, 2001.
35. Jakarta to Washington, June 19, 2001.
36. Jakarta to Washington, August 23, 2001.
37. Jakarta to Washington, July 11, 2001.
38. Interview with Terry Collingsworth (BVH).
39. Complaint, *John Doe I v. ExxonMobil*, op. cit.

CHAPTER FIVE: "UNKNOWN INJURY"

1. Interview with Mandy Lindeberg. Design of Lindeberg's study: "Estimate of Oil Persisting on the Beaches of Prince William Sound 12 Years After the *Exxon Valdez* Oil Spill," *Environmental Science and Technology*, 2004.
2. Interviews with Auke Bay scientists and ExxonMobil consultant David Page (MR).
3. Agreement and Consent Decree, October 9, 1991. *State of Alaska v. Exxon Corporation*, and *United States of America v. Exxon Corporation*, A91-082 CIV and A91-083 CIV.
4. Interview with Jeffrey Short.
5. Ibid. See also, "Petroleum Hydrocarbons in Caged Mussels Deployed in Prince William Sound After the *Exxon Valdez* Oil Spill," American Fisheries Society Symposium, 1996.
6. Salmon study: "Sensitivity of Fish Embryos to Weathered Crude Oil: Part II. Increased Mortality of Pink Salmon (*Oncorhynchus gorbuscha*) Embryos Incubating Downstream from Weathered *Exxon Valdez* Oil," *Environmental Toxicology and Chemistry*, 1999. Damage to fish hearts, future research, single-generation effects: Interviews with Jeffrey Short and Stanley "Jeep" Rice. Financial damages: Interview with Craig Tillery, deputy attorney general, state of Alaska (MR).
7. All quotations, interview with Jeffrey Short.
8. Ibid.
9. "PCB Exposure in Sea Otters and Harlequin Ducks in Relation to History of Contamination by the *Exxon Valdez* Spill," *Marine Pollution Bulletin*, June 2010. Interview with Jeep Rice.

10. Interviews with Peter Hagen, Jeffrey Short, Jeep Rice, and Mandy Lindeberg.
11. *Anchorage Daily News*, January 31, 2002.
12. Interview with Jeffrey Short.
13. Interview with David Page (MR).
14. All quotations from interviews with Jeffrey Short and David Page (MR).
15. Ibid.

CHAPTER SIX: "E.G. MONTH!"

1. Interview with Frank Ruddy.
2. This summary is drawn from three overlapping accounts of Equatorial Guinea's short, brutal history: Roberts, *The End of Oil*; Ghazvinian, *Untapped*; and Maass, *Crude World*.
3. Quotations and details: Len Shurtleff, "A Foreign Service Murder," *Foreign Service Journal*, October 2007.
4. 10 percent: Teodoro Obiang Nguema quoted in SecState to Yaounde, September 14, 2001. Except where indicated, the author obtained the State cables quoted in this chapter through a Freedom of Information Act request.
5. Roberts, op. cit.; Ghazvinian, op. cit.; and Maass, op. cit.
6. Interview with Teodoro Obiang Nguema.
7. Interview with Frank Ruddy.
8. "Thanks": Interview with Teodoro Obiang Nguema. That Juan Olo negotiated without bankers or lawyers: Interview with an adviser to Obiang.
9. Documents, including extensive internal Riggs memos and e-mails, obtained and published by the Permanent Subcommittee on Investigations, Committee on Governmental Affairs, United States Senate, July 15, 2004. Hereafter "Riggs documents."
10. Yaounde to Washington, July 13, 1999, obtained by the author.
11. Abuja to Washington, September 9, 2005 (W).
12. "We should put": Interview with the adviser to Obiang. Five thousand dollars: Riggs documents, op. cit.
13. Riggs documents, ibid.
14. SecState to Monrovia, February 24, 2001; Yaounde to Washington, March 2, 2001.
15. SecState to Yaounde, March 2 and March 16, 2001.
16. SecState to Yaounde, March 16, 2001. Africa Global appears to have wound down its operations. Calls to a Louisiana public affairs firm that manages the name plate on its Web site were not returned.
17. All quotations, SecState to Yaounde, April 12, 2001.
18. Ibid. Also SecState to Madrid, July 27, 2001.
19. All quotations, SecState to Yaounde, September 14, 2001.
20. Yaounde to Washington, October 28, 2002.
21. Yaounde to Washington, January 30, 2003.
22. All quotations, Yaounde to Washington, January 29, 2002.
23. Ibid.
24. *Wall Street Journal*, January 10, 2006.
25. Up to $3 million in cash; "The president . . . P.R. firm": Riggs documents, op. cit.
26. Riggs documents. Plane features: Roberts, *The Wonga Coup*.

CHAPTER SEVEN: "THE CAMEL AND THE JACKAL"

1. "Poor to non-existent . . . corruption": N'djamena to Washington, February 8, 2006, from a State Department cable released to the author in response to a Freedom of Information Act request. All other quotations from Christopher Goldthwait's collection of letters, which were made available to the author by Ambassador Goldthwait. Hereafter "Goldthwait letters."

2. Fifteen hundred doctoral degree holders: *Fortune*, April 16, 2001. Chari floodplain: Exxon-Mobil PowerPoint slides, released by Esso Chad. Upstream skills groups and reorganization: Interview with a former ExxonMobil scientist.
3. "Visit of Chad President," *State Department Bulletin*, 1987.
4. English translation of the 1988 convention made available to the author by Ian Gary. "You don't have time . . . he wanted": Interview with Salibou Garba.
5. Rosemarie Forsythe as prodigy: LeVine, *The Oil and the Glory*, p. 211. Forsythe's role: Interviews with oil industry and former Clinton administration officials. April planning exercise: Interviews with ExxonMobil executives.
6. Interviews with ExxonMobil executives, ibid.
7. "The good Lord didn't see fit": Dick Cheney, Cato Institute, June 23, 1998. Simeon Moats: Interviews with former government and intelligence officials who consulted with Moats after he joined ExxonMobil. Hank Cohen and quotations: Interview with Hank Cohen.
8. All Tom Walters quotations: Testimony Before House Subcommittee on Africa, April 18, 2002. Rex Tillerson quotation: *New York Times*, May 16, 2001.
9. "The notion . . . go down well": Skjærseth, "ExxonMobil: Tiger or Turtle?" "The biggest thing . . . comfortable": *BusinessWeek*, April 9, 2001.
10. Remarks at "The Politics of Development and Security in Africa's Oil States," Johns Hopkins University School for Advanced International Studies, April 2, 2009.
11. For a thorough account of the World Bank plan, see Gary and Reisch, "Chad's Oil: Miracle or Mirage," Catholic Relief Services.
12. Philippe Le Billon: Remarks at "The Politics of Development and Security," op. cit. "The Chadians came in": Interview with an industry security executive. "There were regular . . . the abuses": N'djamena to Washington, July 6, 1999 (cable obtained by the author from a confidential source).
13. Remarks at "The Politics of Development and Security," op. cit.
14. "The oil project . . . pristine poverty": Testimony by Donald Norland before House Subcommittee on Africa, op. cit.
15. Author's correspondence with Hassaballah Soubiane, translated from his original French.
16. Ibid.
17. Lawrence Summers's letter of August 14, 2000, made available by Ian Gary.
18. N'djamena to Washington, July 6, 1999.
19. All quotations, Goldthwait letters, op. cit.
20. Ibid.
21. Harry Longwell: *Fortune*, April 16, 2001; Gary and Reisch, op. cit.
22. Interviews with two individuals familiar with the N'djamena station.
23. Ibid.
24. Interview with Karen Kwiatkowski (BVH).
25. National Intelligence Council: "Global Trends 2015," http://www.dni.gov/nic/NIC_global trend2015.html. Walter Kansteiner: From author's transcription of the conference.
26. Quotation from author's interviews with a U.S. official involved.
27. Interview with Salibou Garba.
28. All quotations from Goldthwait letters, op. cit.
29. "Snapshot Summary," Esso Exploration & Production Chad, Inc., Chad Export Project Update no. 13, 2003.

CHAPTER EIGHT: "WE TARGET OIL COMPANIES"

1. Frank Sprow; all biographical details and quotes: Interview with Frank Sprow.
2. ExxonMobil's $100 million contribution: ExxonMobil Web site, examined and typed.
3. All quotations, interview with Kert Davies.
4. Ibid.
5. Ibid.

6. Letter and memo to John H. Marburger, assistant to the president for science and technology, from Brian P. Flannery, March 18, 2002.
7. The ExxonMobil Foundation.
8. Oreskes, "Behind the Ivory Tower."
9. Dressler and Parson, *The Science and Politics of Global Climate Change*, p. 3.
10. Details of the raid: "Greenpeace Unlawful Entry to ExxonMobil Corporation Headquarters," letter from Kenneth P. Cohen to Lon Burnam, July 1, 2004; Associated Press, May 27, 2003; Fort Worth *Star-Telegram*, May 28, 2003. Tom Cirigliano: *Dallas Morning News*, May 28, 2003.
11. Public Interest Watch: Form 990, August 1, 2003.
12. *New York Times*, April 19, 2004.
13. Securities and Exchange Commission, "Concept Release on Possible Revisions to the Disclosure Requirements Relating to Oil and Gas Reserves," December 12, 2007.
14. "This marks the tenth year": "ExxonMobil Replaces Production for Tenth Year in a Row," February 18, 2004, press release. "Continued high-quality additions": "ExxonMobil Replaces Production for Ninth Year in a Row," January 28, 2003, press release.
15. "A well-established": ExxonMobil Form 10-K, February 28, 2007. Also, interviews with managers and executives who participated in the reserve counting system.
16. Interview with an executive involved with the process. Interview with Mark Gilman.
17. ExxonMobil 10-K filings and annual press releases, op. cit.
18. Steve LeVine, *ForeignPolicy.com*, February 16, 2011.

CHAPTER NINE: "REAL MEN—THEY DISCOVER OIL"

1. Interview with Lee Raymond.
2. Arbogast, *Resisting Corporate Corruption; Time*, July 28, 1975.
3. *Forbes*, August 5, 2009.
4. Interview with Lee Raymond.
5. "Liquefied Natural Gas: Understanding the Basic Facts," Department of Energy, August 2005.
6. "Balancing Natural Gas Policy: Fueling the Demands of a Growing Economy," National Petroleum Council, September 25, 2003. http://fossil.energy.gov/programs/oilgas/publica tions/npc/03gasstudy/npcgas03_preface.pdf.
7. Reuters, June 21, 2005.
8. Coll, *Ghost Wars*, p. 398.
9. *New York Times*, December 2, 1998. "sour lemon": Author's interview with a person who attended.
10. "I have to say . . . more talks": *BusinessWeek*, July 14, 1975.
11. Bloomberg, April 3, 2011.
12. Course of gas negotiations: Interviews with individuals familiar with the Saudi gas project.
13. Details about the Al-Hamra attack: National Association for Business Economics, May 15, 2003; *Los Angeles Times*, May 17, 2003.
14. Interview with Lee Raymond.
15. Ibid.
16. The causes of the failure of the Saudi negotiations have been reported in the *New York Times*, June 24, 2002, and the *Washington Post*, June 6, 2003. Raymond's outlook, the final meeting with Al-Faisal, and all quotations from interviews with individuals familiar with the negotiations.
17. All quotations from interviews with executives and directors involved.
18. Ibid.
19. Ibid.

CHAPTER TEN: "IT'S NOT QUITE AS BAD AS IT SOUNDS"

1. Interviews with current and former ExxonMobil executives.
2. Ibid.
3. Ibid.
4. Interviews with a former ExxonMobil manager.
5. Interviews with Lee Raymond.
6. Interviews with current and former ExxonMobil executives.
7. "the charm of a con man": Sidney Jones, writing an obituary in the *Independent*, May 15, 2003; Jendrzejczyk died at age fifty-three. "meeting over": Interview with an activist who met with Exxon executives during this period. "We're doing a great job": Interview with Arvind Ganesan.
8. Interview with Arvind Ganesan.
9. All quotations from "Notes for a Presentation to ExxonMobil," October 27, 2002, provided to the author.
10. Interview with Bennett Freeman.
11. Interview with a former ExxonMobil executive familiar with Lee Raymond's reaction.
12. Interview with a former senior BP executive.
13. *Chief Executive*, April 2001.
14. Interviews with ExxonMobil executives.
15. *Chief Executive*, op. cit., and Browne, *Beyond Business*, pp. 197 and 234.

CHAPTER ELEVEN: "THE HAIFA PIPELINE"

1. All quotations from "Post-War Planning," statement by Douglas J. Feith, Senate Foreign Relations Committee, February 11, 2003. Examined and typed at: http://policy.defense.gov/sections/public_statements/speeches/archive/former_usdp/feith/2003/february_11_03.html. "the fucking stupidest": Woodward, *Plan of Attack*, p. 281.
2. "literally nothing": Donald Rumsfeld's remarks during an Infinity Radio call-in program, November 15, 2002, as reported by CBS News. War aims, Douglas Feith statement: Senate Foreign Relations Committee, op. cit.
3. Daniel Yergin: *Washington Post*, December 8, 2002.
4. 115 billion barrels of proven reserves: The Energy Information Administration of the U.S. Department of Energy cited this figure in 2007; it attributed the estimate to the *Oil & Gas Journal*. The E.I.A. noted in its "Country Analysis Brief" for Iraq that the 115-billion-barrel estimate dates to 2001 and was based "largely on 2-D seismic data from nearly three decades ago." Updated technology may reveal "an estimated additional 45 to 100 billion barrels" of recoverable oil, the E.I.A. said. Iraq's pre-invasion production hovered around 3 million barrels per day; Iraqi planners believed that figure might be raised to 6 million barrels per day with substantial investment. Saudi Arabia announced plans in 2005 to raise its production capacity to about 12.5 million barrels per day; in 2008, it produced just under 11 million barrels per day.
5. Interview with Tariq Shafiq.
6. "We're not here for the oil": Interview with an administration official who received the talking points. Paul Bremer quotations: *Washington Post*, June 23, 2003.
7. Interview with Philip J. Carroll Jr.
8. Future of Iraq Project, Oil and Energy Working Group, Subcommittee on Oil Policy, Summary Paper, April 20, 2003, p. 5. A redacted version of the paper was declassified and published by the National Security Archive: http://www.gwu.edu/~nsarchiv/NSAEBB/NSAEBB198/FOI%20Oil.pdf.
9. Arile Cohen, and Gerald P. O'Driscoll Jr., "The Road to Economic Prosperity for a Post-Saddam Iraq," Heritage Foundation Backgrounder no. 1594, September 24, 2002.
10. "broad-based, mass privatization": "Moving the Iraqi Economy from Recovery to Sustainable

Growth," U.S.A.I.D., quoted in Chandrasekaran, *Imperial Life in the Emerald City*, p. 115. "This was part . . . resurfacing": Interview with an administration official involved in the discussions.

11. Interviews with Gary Vogler and Michael Makovsky.

12. Interviews with Gary Vogler, Michael Makovsky, and a third Bush administration official involved.

13. Philip Carroll would resign: Interview with a former National Security Council official. All quotations: Interview with Carroll.

14. A copy of the memo was provided to the author by a C.P.A. official other than Philip Carroll.

15. Feith's history and thinking about oil security are from interviews with Douglas Feith.

16. Both quotations from "Some Historical Lessons from the World Oil Market," Pentagon PowerPoint Briefing, April 28, 2005, provided to the author by Douglas Feith.

17. Condoleezza Rice, Colin Powell, and Paul Wolfowitz: Interviews with former Bush administration officials, and *Michigan Daily*, March 30, 2011; *Sunday Times* (London), March 4, 2011.

18. Doubled to $18.5 billion: "China, Africa, and Oil," Council on Foreign Relations Backgrounder, June 6, 2008. $44 billion: Presentation by Bo Kong, at "The Politics of Development and Security in Africa's Oil States," Johns Hopkins University School of Advanced International Studies, April 2, 2009.

19. Interview with a former intelligence official involved in the review.

20. Interview with David Gordon.

21. "Some Historical Lessons from the World Oil Market," op. cit.

22. Interview with Aaron Friedberg. Interview with David Gordon.

23. Interview with Aaron Friedberg.

24. Interview with a Bush administration official.

25. Interview with a former National Security Council official.

26. ExxonMobil's scenario planning, "an element of surge . . . really happen": Rex Tillerson remarks, "A Conversation on Energy Security," Council on Foreign Relations, March 9, 2007.

27. Ibid.

28. All quotations, "A Conversation with Lee Raymond," *Charlie Rose*, PBS, May 6, 2004.

29. "A Conversation on Energy Security," op. cit.

CHAPTER TWELVE: "HOW HIGH CAN WE FLY?"

1. "Was fortunate at this critical time" and "transform the relationship": BBC News, November 15, 2001. "looked the man . . . soul": BBC News, June 16, 2001.

2. Don Evans and Vladimir Putin, all quotations: Interviews with former Bush administration officials.

3. Moscow to Washington, June 3, 2002. The State Department cables relied on for this chapter were obtained by the author through a Freedom of Information Act request.

4. Early thinking about Russia strategy: Interviews with former Bush administration officials involved. "I think all of us at the senior level": Interview with Spencer Abraham.

5. U.S. Energy Information Administration: http://www.eia.doe.gov/pub/oil_gas/petroleum/ analysis_publications/oil_market_basics/sup_image_worldprod.htm#Former%20Soviet%20 Union. The E.I.A.'s historical estimates of Soviet production do not break out separate figures for Russia, but assuming that non-Russian production was similar in 1988 to what it was immediately after the Soviet breakup, Russia's peak production would have been about 10 million barrels a day.

6. Moscow to Washington, June 3, 2002.

7. Don Evans's conversations with Russian counterparts: Interviews with former Bush administration officials familiar with the energy dialogue.

8. All quotations from an interview with Leonard Coburn.

9. All quotations from an interview with a former ExxonMobil executive involved in the Sakhalin negotiations.

10. April flight, rehearsing for negotiations with Vladimir Putin: Interviews with ExxonMobil employees. Bush visit timed to coincide with deal announcements with ExxonMobil and Chevron: Moscow to Washington, June 3, 2002, op. cit.

11. ExxonMobil's $140 million contract: *International Oil Daily*, May 23, 2002. Joint Statement: Office of the Press Secretary, White House, May 24, 2002.

12. Moscow to Washington, June 3, 2002.

13. The summary of Mikhail Khodorkovsky's rise is drawn from Goldman, *Petrostate*, op. cit.; Hoffman, *The Oligarchs*; and Baker and Glasser, *Kremlin Rising*.

14. Interview with Vladimir Milov conducted by Miriam Elder.

15. Interview with Bruce Misamore.

16. Ibid.

17. Ibid.

18. All quotations, Browne, *Beyond Business*, p. 145.

19. Interview with Bruce Misamore.

20. *Washington Post*, November 3, 2003.

21. Library of Congress: Ambassador Alexander Vershbow to Washington, November 7, 2002. Table one at the prayer breakfast: Interviews with former Bush administration officials.

22. The PowerPoint and a video of Mikhail Khodorkovsky's presentation are available on the Web site of the Carnegie Endowment for International Peace: http://www.carnegieendow ment.org/files/2003-02-07-khodorkovsky-presentation.pdf.

23. Moscow to Washington, November 7, 2002.

24. Interview with Bruce Misamore.

25. Interview with former Bush administration officials involved in the dialogue.

26. Interview with Bruce Misamore.

27. Interviews with executives familiar with the negotiations.

28. All quotations from Mikhail Khodorkovsky's presentation at the Carnegie Moscow Center seminar on global energy, June 17, 2003, from author's files.

29. Interviews with executives familiar with the negotiations.

30. The author David Hoffman conducted an interview with a Yukos executive in Moscow on July 1, 2003, as the Kremlin began to ratchet up pressure on Mikhail Khodorkovsky; Hoffman generously shared his notes.

31. Moscow to Washington, July 14, 2003.

32. All quotations from interviews with an executive familiar with the detailed account of the conversation briefed to ExxonMobil afterward.

33. All quotations, interview with Bruce Misamore.

34. Goldman, *Petrostate*, pp. 111–12, and an interview with a former ExxonMobil executive involved in the negotiations.

35. Interview with Bruce Misamore.

36. Interview with Leonard Coburn.

37. "Absurd . . . enforcement system": U.S. embassy in Moscow to Washington, October 29, 2003.

38. Moscow to Washington, October 23, 2003.

39. Interview with Vladimir Milov conducted by Miriam Elder.

40. "Everyone ought . . . long-term industry": *Petroleum Intelligence Weekly*, November 10, 2003. "There are some things there": *Charlie Rose*, PBS, May 6, 2004.

41. Moscow to Washington, May 14, 2004.

42. Vladimir Putin's job offer to Don Evans, the Russian president's conversation with John Snow, and all quotations from interviews with former Bush administration officials.

CHAPTER THIRTEEN: "ASSISTED REGIME CHANGE"

1. Theresa Whelan's biography, all quotations: Interview with Theresa Whelan. Deborah Avant, a political science professor at George Washington University, reported a brief sum-

mary of some of Whelan's remarks at the November 19 dinner in written testimony before the House Armed Services Committee, April 25, 2007.

2. Roberts, *The Wonga Coup*, p. 79. Roberts's superb book provides the definitive account of Greg Wales, Simon Mann, and their activities in Equatorial Guinea, and is a foundation of the narrative in this chapter. Wales provided some of the contracts and other documents to Roberts, who was helpful to the author.

3. Interview with Theresa Whelan.

4. All quotations, ibid.

5. Roberts, *The Wonga Coup*, p. 83. The author's efforts to locate Greg Wales for comment in Britain and South Africa were unsuccessful.

6. *Reuters*, July 8, 2008, from coverage of Simon Mann's trial in Malabo. Restless in 2003: Roberts, *The Wonga Coup*, p. 15.

7. Text of joint appearance: The *Guardian*, March 17, 2003.

8. Roberts, *The Wonga Coup*, pp. 140–41.

9. "The advance group . . . government": From a State Department cable declassified in response to the author's request under the Freedom of Information Act, Yaounde to Washington, March 11, 2004. That cable attributes to Du Toit the assertion, during his appearance before diplomats, that he would "receive $5 million" for his assistance in this plot. Other sources such as Roberts's *The Wonga Coup* (p. 76) put the figure at $1 million. Du Toit had been imprisoned for several days in the notorious Black Beach prison at the time he made his statement, but the March 11 cable, approved by U.S. ambassador George Staples, reported that "Du Toit showed no signs of abuse or coercion when relating his story."

10. Meetings and quotations from memos and minutes in the Riggs documents (see chapter six, note 9).

11. Interview with J. R. Dodson.

12. Ibid.

13. Roberts, *The Wonga Coup*, p. 184.

14. "Problems had arisen": Yaounde to Washington, September 3, 2004. The cable is based on reporting of Nick du Toit's trial in Malabo.

15. Yaounde to Washington, March 11, 2004.

16. Roberts, *The Wonga Coup*, pp. 195–99.

17. Riggs documents, op. cit.

18. Interview with Teodoro Obiang Nguema, op. cit.

19. All quotations from interviews with advisers to Teodoro Obiang Nguema.

20. Interview with Theresa Whelan.

21. These and all other quotations from the Colin Powell meeting are from SecState to Yaounde, June 26, 2004.

22. The road map, meetings: Interviews with former Bush administration officials and advisers to Teodoro Obiang Nguema. Talking point quotations: SecState to Yaounde, October 5, 2004.

23. All quotations from interviews with former Bush administration officials and advisers to Teodoro Obiang Nguema.

24. Ibid.

25. Malabo to Washington, March 12, 2009.

26. "Money Laundering and Foreign Corruption: Enforcement and Effectiveness of the Patriot Act," hearing of the Permanent Subcommittee on Investigations, United States Senate, July 15, 2004.

CHAPTER FOURTEEN: "INFORMED INFLUENTIALS"

1. Interviews with four former Bush administration officials who worked on energy policy during this period.

2. Bloom to Hutto, e-mail released under F.O.I.A., April 14, 2005.

3. All quotations from "ExxonMobil 2005 Media Brief," Universal/McCann, as referenced in the Union of Concerned Scientists' 2007 report: "Smoke, Mirrors & Hot Air: How Exxon-Mobil Uses Big Tobacco's Tactics to Manufacture Uncertainty on Climate Change."
4. "A Conversation with Lee Raymond," *Charlie Rose*, PBS, November 8, 2005.
5. "We cannot forecast the price . . . fundamentals are right": Interview with Peter Townsend, former chief of investor relations at ExxonMobil. Also, interviews with other ExxonMobil executives. Rex Tillerson continued this forecasting practice; he told *Fortune* in 2007, "We tell the organization, 'Folks, we really don't have a clue what the price of oil is going to be, and so given that, how should we run this business?'"
6. Interviews with ExxonMobil executives. Also, "The Outlook for Energy: A View to 2030." ExxonMobil has published revised versions of this forecast each year since 2004; the details of the forecast as it was published and briefed early in 2005 are derived from transcripts of presentations by ExxonMobil executives at that time.
7. "The Outlook," ibid., and interviews with ExxonMobil executives.
8. *Independent*, February 21, 2005.
9. Pacala and Socolow, "Stabilization Wedges." Also, Robert H. Socolow and Stephen Pacala, "A Plan to Keep Carbon in Check," *Scientific American*, September 2006.
10. Oreskes, "Behind the Ivory Tower."
11. Freudenburg, "Seeding Science, Courting Conclusions."
12. All quotations, ibid.
13. Lunch, succession, Palm Springs: Interviews with current and former ExxonMobil executives.
14. Interview with a former ExxonMobil director.
15. *Fortune*, September 15, 2003.
16. Interview with Lee Raymond.
17. *Economist*, December 24, 2005.
18. "Energy Prices and Profits," U.S. Senate Committee on Commerce, Science and Transportation, and the Committee on Energy and Natural Resources, November 9, 2005.
19. Associated Press, March 26, 2002; interview with Spencer Abraham.
20. All quotations from the full "Energy Prices and Profits" hearing record, op. cit.
21. Abu Dhabi to Washington, September 28, 2005 (W). Abu Dhabi to Washington, February 17, 2003 (W).
22. "What in the hell": Interviews with ExxonMobil executives. "Major Outstanding Issues" and "ExxonMobil would prefer": Abu Dhabi to Washington, November 7, 2005 (W). "Mostly about money": Abu Dhabi to Washington, September 28, op. cit. "Exxon's technical": Abu Dhabi to Washington, November 21, 2005 (W).
23. "Lee Raymond Retires: The Lessons of History," brief by Adam Sieminski and Paul Sankey for Deutsche Bank Securities, Inc., December 29, 2005.
24. ExxonMobil Annual Report, 2005. Interview with Mark Gilman.
25. Interview with Mark Gilman. Amy Myers Jaffe: "Is Wall Street Quite Wrong When It Comes to Big Oil?" *New York Sun*, February 10, 2005.
26. ExxonMobil Annual Report, 2005.
27. Interviews with former ExxonMobil executives. Estimate of the total retirement package, including restricted stock and options, is from Jad Mouawad, *New York Times*, April 14, 2006.

CHAPTER FIFTEEN: "ON MY HONOR"

1. Interviews with ExxonMobil executives familiar with the transition planning. "Remains unchanged": ExxonMobil Corporation Analyst Meeting, New York, March 8, 2006. "It's true . . . bum rap": Remarks at the Northwest Texas Council, Boy Scouts of America, 100th Anniversary Dinner, November 11, 2010, in response to a question from Ann O'Hanlon. "Let me assure you": *Dallas Morning News*, May 31, 2007.
2. The author is indebted to the biographical research on Rex Tillerson and his family carried

out for this book by S. G. Gwynne. This section draws on and incorporates language drafted by Gwynne in a research memo.

3. Gwynne, ibid., and research in Wichita Falls by Ann O'Hanlon.
4. Gwynne, ibid. Fraternity hazing: Author's interview with a participant. Ayn Rand: *Scouting Magazine*, September 2008.
5. Gwynne, op. cit.
6. Interview with a former ExxonMobil manager.
7. *Fortune*, April 30, 2007.
8. The account of the committee's appointment and review work is from interviews with five current and former managers involved. ExxonMobil has referred only obliquely to the work in public.
9. *MIT News*, May 19, 2009.
10. Interview with an individual familiar with the K Street office.
11. Interviews with five current and former managers involved.
12. "In terms of showing . . . equaled again": Remarks at the Northwest Texas Council, November 11, 2010, op. cit.
13. "They have a self-righteousness" and "tobacco industry": *Fortune*, April 17, 2006. The author is indebted to Ann O'Hanlon's thorough review and analysis of oil corporation campaign filings between 2000 and 2008, from which these comparisons are drawn. Walter Buchholtz: *Washington Post*, September 1, 2004.
14. Interviews with individuals familiar with the ExxonMobil Washington office.
15. Theresa Fariello biography: Meridian International Center Board of Directors.
16. Louis Finkel: *New York Times*, December 2, 2006.
17. The account in this section of the Airlie Center Dialogue is drawn from interviews carried out by Ann O'Hanlon with seven participants, as well as e-mails and notes taken during the meetings.
18. All quotations, ibid.
19. Ibid.
20. "U.S. Climate Change Policy," e-mail from Kenneth P. Cohen, January 11, 2007. "We know our climate is changing": *Fortune*, April 30, 2007.
21. "ExxonMobil's Top Executives on Climate-Change Policy," euractiv.com, February 14, 2007. This extensive interview with Cohen and Stuewer provides a very thorough account of the policy and communication strategy that emerged from the 2006 climate policy work.

CHAPTER SIXTEEN: "CHAD CAN LIVE WITHOUT OIL"

1. 368 wells in 2006: "Chad/Cameroon Development Project, Project Update no. 21, 2006," Esso Exploration & Production Chad, Inc. Landscape: Author's travel to southern Chad. Rules, Guantánamo, entombed: Interviews with employees and other individuals familiar with ExxonMobil's operations in Chad. Twelve thousand square miles, nine camps, 450 oil production sites: N'djamena to Washington, June 5, 2007 (W).
2. N'djamena to Washington, ibid.
3. Author's travel and local interviews. About 2,500 security guards: Interviews with two executives at a company involved in the security operations.
4. Ibid.; N'djamena to Washington, op. cit.
5. Interviews with U.S. officials and other individuals familiar with ExxonMobil's operations in Chad. PowerPoint slides: "Benefits to USA from Chadian Crude Oil Operations," circa 2008.
6. Land payments, price rises, micro lending: Interviews with Catholic Church monitors in Doba who were involved with the World Bank's micro lending project. Wages skimmed, settlement: Interviews with Doukam Ngartandoh and Djim Ngaro Michel of the Association for the Defense of the Interests of Former Esso Workers, in Doba. "Regarded as a model": N'djamena to Washington, February, 8, 2010 (W). "The influx . . . before the oil": "Interagency Support on Conflict Assessment and Mission Performance Planning for Chad," March 20, 2006, courtesy of Ian Gary, Oxfam.

7. Author's travel to village and local interviews. N'djamena to Washington, 2007, op. cit.

8. ExxonMobil demurs when asked for infrastructure support: Interview with Adoum You-nousmi, Chad's minister for national infrastructure. Blaming World Bank: Interviews with ExxonMobil executives.

9. Interview with Boukinebe Garka.

10. Flint and De Waal, *Darfur: A New History of a Long War*, pp. 114–15.

11. The numbers here are U.S. government estimates of Chad's defense capabilities, obtained from interviews with U.S. officials who asked not to be further identified.

12. "Chad/Cameroon Development Project, Project Update no. 24," 2008, p. 77, Esso Exploration & Production Chad, Inc.

13. Interview with Mahamat Hissène.

14. N'djamena to Washington, February 6, 2006 (W).

15. Paul Wolfowitz's thinking about Africa, "wasn't helped . . . failing place it is": Remarks at American Enterprise Institute, "Does Africa's Future Depend on Global Financial Institutions?" April 24, 2009. "Chad has a sovereign right": Wolfowitz described what Déby said to him during their phone call during a conference call with reporters, *New York Times*, January 7, 2006.

16. N'djamena to Washington, January 31, 2006 (W).

17. N'djamena to Washington, February 6, 2006 (W).

18. N'djamena to Washington, January 9, 2006. This and other cables, as indicated, were released to the author under a Freedom of Information Act request. Some of these cables were written around the same time as cables released by Wikileaks, but are not in that online collection.

19. N'djamena to Washington, July 12, 2006.

20. Robert Zoellick telephoned Wolfowitz: Secretary of State to N'djamena, January 19, 2006. The document provides a redacted account of a meeting among Zoellick, other State Department officials, Chad's foreign minister Ahmad, Allam-mi, and other Chadian officials, January 10, 2006. "The government brought . . . revenues to date": World Bank, "Chad-Cameroon Petroleum Development and Pipeline Project: Overview," December 2006.

21. Pascal Yoadimnadji communiqué: N'djamena to Washington, April 15, 2006. Wall's meeting with Déby and Ito, all quotations: N'djamena to Washington, April 15, 2006, and N'djamena to Washington, April 16, 2006.

22. N'djamena to Washington, April 26, 2006.

23. N'djamena to Washington, August 7, 2005.

24. Interview with Mahamat Hissène.

25. Ibid.

26. N'djamena to Washington, August 30, 2006.

27. All quotations, N'djamena to Washington, September 6, 2006. The embassy cable describing Barack Obama's meeting with Idriss Déby was marked sensitive but unclassified when it was written; its contents were cleared at the time by Obama's staff; and it was released without redactions in response to the author's Freedom of Information Act request.

28. Interview with Mahamat Hissène. $281 million, lump sum within seven days: N'djamena to Washington, October 9, 2006, F.O.I.A. N'djamena to Washington, October 9, 2006 (W).

CHAPTER SEVENTEEN: "I PRAY FOR EXXON"

1. *Jeff Alban et al. v. ExxonMobil Corp.*, 03-C-06-010193 2OT, Baltimore County Circuit Court, Maryland, pleadings and trial testimony.

2. Quoted in plaintiffs' opening statement, *Alban v. ExxonMobil*, ibid. Also, trial testimony of Dean Jerome, written logs from Jacksonville Exxon of January 13, 2006.

3. "A very big majority" and "working properly": Trial testimony of David Schanberger, *Alban v. ExxonMobil*, op. cit.

4. All quotations, trial testimony of Russell Bowen, *Alban v. ExxonMobil*, ibid.

5. Ibid.

6. Trial testimony of Steven Polkey, ibid.

7. Talking points: Testimony of Russell Bowen, op. cit. Sign: Quoted in plaintiffs' opening statement, op. cit.
8. Remediation sites: Cited in plaintiffs' closing statement, *Alban v. ExxonMobil,* op. cit. That Exxon alone operated 62,000 sites three decades earlier: *Fortune,* April 23, 1990.
9. D. L. Clarke to Carmine DiBattista, bureau chief, Bureau of Air Management, state of Connecticut, January 27, 2003.
10. School grades, "I don't think you could hire me": "The Amazing Snyder-Man," in *Maryland Super Lawyers,* 2008.
11. "How did I do?": To author, Maryland Special Court of Appeals, January 3, 2011. "I just wish . . . never enough": *Baltimore Sun,* April 29, 1999.
12. *Baltimore Sun,* ibid. *Maryland Super Lawyers,* op. cit.
13. Steve Tizard: Maryland *Daily Record,* March 30, 2009.
14. Plaintiffs' opening statement, *Alban v. ExxonMobil,* op. cit.
15. All quotations, ibid.
16. All quotations, defendants' opening statement, ibid.
17. All quotations, transcript of trial testimony, ibid.
18. All quotations, plaintiffs' closing statement and defendants' closing statement, ibid.
19. *Baltimore Sun,* March 12, 2009.

CHAPTER EIGHTEEN: "WE WILL NEED WITNESSES"

1. Remains and identity card discovered near Cluster II in 2005: Interviews with Kaharuddin's widow and other relatives. Landscape and relatives searching: Author's travels in Aceh and interviews with human rights researchers there and in Jakarta.
2. Interview with Hendra Fadli, the Banda Aceh–based coordinator for Kontras Aceh, a human rights research group that photographed and investigated the bones at the time of their discovery. The Indonesian magazine *Tempo* on June 21, 2007, quoted an Acehnese man, Saiful Bahri, as saying that he and his brother had been detained at the Supply Chain Building and that his brother remained missing. "We will need witnesses" is from the *Tempo* account.
3. "If we keep . . . bury everything": *Frontline,* "Indonesia: After the Wave." Documentary aired June 26, 2007. See "Extended Interviews" at http://www.pbs.org/frontlineworld/stories/indonesia605/interview_darmono.html.
4. See Chapter Four.
5. Size of Indonesian military deployments: *Jakarta Post,* September 9, 2002. Human rights violations: See annual U.S. State Department reports, as well as "Aceh Under Martial Law: Inside the Secret War," Human Rights Watch, December 2003.
6. Author's interview with Baharuddin.
7. Interview with a former employee familiar with the campaign.
8. All quotations, ibid.
9. Interviews with lawyers involved with the case: "It's another one of those pesky . . ." E-mail from David P. Stewart to Frank M. Gafney, May 17, 2002, released under a Freedom of Information Act request.
10. Letter from William H. Taft IV, July 29, 2002, *John Doe I et al. v. ExxonMobil Corporation et al.,* 01-cv-01357 United States District Court for the District of Columbia.
11. All quotations, transcript of status hearing, May 24, 2005, ibid.
12. All quotations, transcript of status hearing, May 1, 2006, ibid.
13. Kevin Murphy, ExxonMobil's rollout of the O.I.M.S.-based new rules: Interviews with ExxonMobil executives involved. See "Framework on Security and Human Rights," published after 2006 by ExxonMobil on its Web site. "It's like implementing": Interview with Arvind Ganesan.
14. "They wanted to grind us . . . two people": Interview with Agniszka Fryszman.
15. Bill Scherer at the White House, what he was told: Interview with Terry Collingsworth (BVH).

16. All quotations from interviews with participants.
17. Memorandum & Opinion, July 18, 2008, *Doe I v. ExxonMobil*, op. cit.

CHAPTER NINETEEN: "THE CASH WATERFALL"

1. The account of the dueling art exhibitions is from interviews with a former ExxonMobil manager involved.
2. Caracas to Washington, March 4, 2004 (W).
3. All quotations, Caracas to Washington, August 11, 2004 (W).
4. Ibid.
5. BBC News, *Hard Talk*, June 14, 2010.
6. Associated Press, February 12, 2003. As of that time, Hugo Chavez had fired 11,917 out of 37,942 P.D.V.S.A. employees, according to a former executive, Juan Fernandez.
7. Losses and loan amounts: "Freezing Injunction," *Mobil Cerro Negro, Ltd. v. Petróleos de Venezuela, S.A.*, Claim no. 2008, Folio 61, January 24, 2008, High Court of Justice, Queen's Bench Division, Commercial Court, Great Britain.
8. Caracas to Washington, January 12, 2005 (W). Associated Press, January 8, 2007.
9. Lee Raymond: Fair Disclosure transcript, ExxonMobil Corporation Analyst Meeting, March 10, 2004, New York. Rex Tillerson: "A Conversation on Energy Security," Council on Foreign Relations, March 9, 2007. The quotation is in direct response to a question about Mexico, but follows similar comments about Venezuela and global energy security in general.
10. Ibid.
11. George W. Bush to Rex Tillerson: Interview with an individual familiar with reports of the conversation and ExxonMobil's response.
12. Rex Tillerson, "A Conversation on Energy Security," op. cit.
13. All oil and gasoline price statistics cited are from the U.S. Energy Information Administration.
14. Simmons, *Twilight in the Desert*.
15. Morse, "Low and Behold."
16. Slides: Presentation by Stuart McGill, Goldman Sachs Global Energy Conference, January 11, 2005. Rex Tillerson: ExxonMobil Annual Shareholder Meeting, May 27, 2009.
17. $3 trillion windfall: Christof Ruehl, chief economist, British Petroleum, address at the World Bank, March 3, 2009. Oil expenditures as a percentage of gross domestic product: Henry Lee, "Oil Security and the Transportation Sector," in Gallagher, *Acting in Time on Energy Policy*.
18. Associated Press, January 13, 2007.
19. Caracas to Washington, January 24, 2007 (W).
20. Interviews with former ExxonMobil executives and State Department officials. Also, Caracas to Washington, March 29, 2007 (W).
21. Caracas to Washington, May 25 and June 26, 2007 (W).
22. Interview with a former ExxonMobil manager.
23. Declaration of Hobart E. Plunkett, *Mobil Cerro Negro v. PDVSA Cerro Negro*, ibid.
24. Rex Tillerson: Fair Disclosure transcript, ExxonMobil Corporation Analyst Meeting, March 5, 2008, New York.
25. Interview with a former ExxonMobil manager.
26. "Complaint for Order of Attachment," *Mobil Cerro Negro. v. PDVSA Cerro Negro*, op. cit.
27. Joseph D. Pizzurro, "Memorandum of Law of Defendant PDVSA . . . " January 24, 2008, *Mobil Cerro Negro v. PDVSA Cerro Negro*, op. cit.
28. Ibid.
29. Declarations of Steven K. Davidson and J. R. Massey, December 27, 2007, *Mobil Cerro Negro v. PDVSA Cerro Negro*, op. cit.
30. Attorneys for ExxonMobil and P.D.V.S.A. provide slightly contradicting accounts of the minute-by-minute sequence of the bond repurchase closing and the service of freezing order papers; the account here is from a declaration of Mitchell Seider, of the ExxonMobil

law firm Latham & Watkins, who was present in the Curtis conference room during the afternoon of December 28.
31. Pizzurro, "Memorandum of Law of Defendant PDVSA," op. cit.
32. All quotations, "Oral Argument," February 13, 2008, *Mobil Cerro Negro v. PDVSA Cerro Negro*, op. cit.

CHAPTER TWENTY: "MOONSHINE"

1. Text as delivered, CQ Transcript Wire, *Washington Post*, February 1, 2006.
2. In 2003, according to the Energy Information Administration, the United States consumed just under three gallons of gasoline per person per day. The rest of the industrialized world—mainly Europe—consumed less than half as much per person. Poorer countries consumed even less per person. American relative overconsumption has persisted since then—only the oil-rich, lightly populated emirates of the Persian Gulf come close, and even their profligate consumers guzzle less gasoline per person than Americans.
3. Rex Tillerson: *Fortune*, April 17, 2006.
4. Interview with a former Bush administration official.
5. Remarks by Rex Tillerson, Woodrow Wilson International Center for Scholars, Washington, D.C., January 8, 2009.
6. Ibid.
7. Interviews with multiple ExxonMobil executives and former executives. It is not clear which year ExxonMobil first incorporated an assumption about carbon pricing in the United States into its forecasts, but it appears to have been around 2007, after the Democrats took control of the House of Representatives and began to gather momentum in anticipation of the 2008 presidential election.
8. "The Hydrogen Economy: Opportunities, Costs, Barriers and R&D Needs," National Academy of Sciences Press, 2004.
9. Interview with Sally Benson.
10. "Hydrogen and Fuel Cells: Opportunities and Challenges," slides presented in Brussels, Belgium, November 25, 2003, author's files and www.exxonmobil.com/files/PA/Europe/Blewisfinal_Amcham.pdf.
11. Lee Raymond: Fair Disclosure transcript, ExxonMobil Corporation Analyst Meeting Overview, March 4, 2003. Michael B. McElroy, "The Ethanol Illusion," *Harvard Magazine*, November–December 2006.
12. Appearing at a Cambridge Energy Research Associates conference in February 2007, Rex Tillerson said, speaking of ethanol, "I don't know how much technology I can add to moonshine."
13. Transcript of conference call with Emil Jacobs and genetic researcher J. Craig Venter, July 14, 2009.
14. Ken Cohen quotations: Interview with EurActiv, February 14, 2007. "We just ignored it": Interview with an ExxonMobil executive.
15. All quotations, interview with the ExxonMobil executive, ibid.
16. All quotations, ibid.
17. Interview with a second executive involved.

CHAPTER TWENTY-ONE: "CAN'T THE C.I.A. AND THE NAVY SOLVE THIS PROBLEM?"

1. Eket landscape: Author's research travel, 2009. ExxonMobil's Nigerian production in 2006: Securities and Exchange Commission filings. Kidnapping narrative: Aberdeen *Evening Express*, Aberdeen *Press and Journal*, *The Scotsman*, *Sunday Express*, *Daily Mail*, October 4, 2006 to October 23, 2006.
2. Energy Information Administration import statistics, 2006–2008.
3. Chaplin perspective and quotations: Lagos to Washington, December 7, 2004; September 9, 2005; February 26, 2006; and March 12, 2007 (W).

4. Lagos to Washington, March 12, 2007, ibid.

5. Interview with John Campbell. Also Abuja to Washington, June 17, 2008 (W).

6. *This Day* (Nigeria), March 10, 2006.

7. Lagos to Washington, March 17 and March 31, 2006 (W).

8. Interviews with Chief Samingo Etukakban and Chief Nduese Essien in Akwa Ibom. Lagos to Washington, May 3, 2006 (W).

9. Graeme Buchan spoke at a press conference in Scotland after his release, as reported in the *Daily Telegraph*, October 23, 2006. "Uneasy": Interview with a State Department official. "Fidgety": Lagos to Washington, May 3, 2006 (W).

10. Interview with Victor Attah.

11. *Sunday Express*, October 22, 2006.

12. Interview with Victor Attah.

13. Interview with a senior State Department official.

14. Paula Smith: *Sunday Express*, op. cit. Paul Smith: Press conference, as reported in the Aberdeen *Evening Express*, October 23, 2006.

15. http://hdr.undp.org/en/reports/global/hdr2006/.

16. *Washington Post*, February 12, 2009. Also, *United States of America v. Jason Edward Steph*, United States District Court, Southern District of Texas, H-07-307, indictment filed July 19, 2007.

17. All quotations, author's interview with a Western diplomat in Nigeria.

18. From the outstanding report on Delta militancy authored by Stephen Davis for the Coventry Cathedral: "The Potential for Peace and Reconciliation in the Niger Delta," February 2009.

19. Interview with an American official.

20. Interviews with two senior security consultants in Lagos who serve the international oil industry.

21. Estimates from participants in "Nigeria: Prospects for Peace in the Niger Delta," Center for Strategic and International Studies, June 15, 2009.

22. Abuja to Washington, September 9, 2005.

23. All quotations, remarks by Rex Tillerson, Woodrow Wilson International Center for Scholars, January 8, 2009.

24. Azaiki, *Oil, Gas and Life in Nigeria*, p. 51.

25. When I traveled to conduct research in Akwa Ibom during 2009, I was hosted by Nigerian House of Representatives member Esseme Eyiboh. We shared the protection of a single Mobil Police gunman for several days. The policeman had a flying Pegasus emblazoned on the breast pocket of his uniform.

26. Godswill Akpabio supporters, Qua Iboe landscape: Author's travels. Spy police lawsuit: *Globe and Mail*, November 22, 2004.

27. Interviews with Esseme Eyiboh and Nduese Essien.

28. Interviews with Victor Attah and Nduese Essien.

29. Remarks at the Woodrow Wilson Center, op. cit.

30. Godswill Akpabio: "Words That Build a Nation," op. cit. U.S. official and Esseme Eyiboh: Author's interviews.

31. Jendayi Frazer, Keynote address, Gulf of Guinea Maritime Safety and Security Ministerial Conference, November 15, 2006, Cotonou, Benin.

32. Interview with Connie Newman.

33. Lagos to Washington, March 31, 2006; May 12 and June 18, 2007.

34. Interviews with U.S. officials.

35. Lagos to Washington, February 23, 2010 (W).

36. Interview with a security consultant in Lagos.

37. All quotations, history of U.S. aid, piracy statistics: From author's interviews with multiple U.S. officials and corporate security specialists who worked on the subjects described after 2006.

38. Lagos to Washington, June 3, 2008, and January 15, 2009. Abuja to Washington, November 26, 2008 (W).

39. Interviews with U.S. officials.
40. Ojo Maduekwe: Center for Strategic and International Studies event, op. cit. Theresa Whelan: Testimony before the Subcommittee on National Security and Foreign Affairs of the House Committee on Oversight and Government Reform, July 15, 2008.
41. Interview with Major General Michael Snodgrass.
42. Lagos to Washington, September 19, 2008 (W).
43. Interview with an officer at Africa Command.
44. All quotations, author's interviews with U.S. officials.

CHAPTER TWENTY-TWO: "A PERSON WOULD HAVE TO EAT MORE THAN 3,400 RUBBER DUCKS"

1. ExxonMobil 10-K, 2007.
2. *New York Times Magazine*, December 9, 2001.
3. http://www.hcra.harvard.edu, examined and typed, January 30, 2011.
4. "Hazard Index": ExxonMobil PowerPoint presentation. Interview with Paul Thacker (MR).
5. "Phthalates 101": PowerPoint presentation left by ExxonMobil lobbyists with Capitol Hill offices, circa 2008. Also, ExxonMobil's "Response to CPSC's Request for Information," January 12, 2009. The author is grateful for the outstanding work of Megha Rajagopalan, who obtained these and other materials, and conducted many of the background interviews cited.
6. To see the Centers for Disease Control and Prevention's Fact Sheet on phthalates, which contains links to its exposure studies, see http://www.cdc.gov/exposurereport/Phthalates _FactSheet.htm.
7. For an accessible but detailed review of scientific research into the health effects of phthalates in plastics, see the Congressional Research Service Report for Congress, "Phthalates in Plastics and Possible Human Health Effects," updated July 29, 2008.
8. Memorandom to Michael A. Babich, Consumer Product Safety Commission, August 31, 1998.
9. "Phthalates 101," op. cit.
10. "No risk reduction": ExxonMobil, "Approach to Cumulative Risk," op. cit. "Politics": "Phthalates 101," ibid.
11. Interview with a San Francisco regulator.
12. Interview with Gretchen Lee Salter (MR).
13. Interview with Virginia Lyons (MR). 'They're going to ban Gumby!'": Interview with Sarah Uhl, Clean Water Action, Hartford, Connecticut (MR).
14. Mattel recall: *New York Times*, August 2, 2007.
15. Interview with Liz Hitchcock (MR).
16. Halperin and Harris, *The Way to Win*, pp. 206–8.
17. *Wall Street Journal*, May 27, 2006. Also, interviews with individuals familiar with the 2005 energy legislation episode.
18. Interview with a conference participant.
19. "Six old white guys . . . not to budge": All quotations from interviews with congressional staff involved.
20. Interview with Liz Hitchcock (MR).
21. Interview with Janet Nudelman (MR).
22. All quotations, interviews with participants.
23. All dates and contribution amounts from Federal Election Commission records.

CHAPTER TWENTY-THREE: "WE MUST END THE AGE OF OIL"

1. Ann O'Hanlon invested extraordinary time and effort to analyze and collate Federal Election Commission records to document the insights about ExxonMobil's political strategy relied upon in this chapter.

2. Interview with an ExxonMobil executive.

3. Interview with the consultant quoted, as well as interviews with other individuals familiar with ExxonMobil's Washington strategies.

4. Interviews with individuals familiar with ExxonMobil's Political Action Committee.

5. "Electing people": From Citizen Action Team mailings provided by recipients, author's files.

6. *The Tonight Show with Jay Leno*, August 1, 2008.

7. National Public Radio, August 5, 2008, www.npr.org.

8. Interview with Jason Grumet.

9. "It's not going to be easy": Transcript of MSNBC Democratic debate, February 26, 2008. "Think about that . . . outrageous": Associated Press, June 9, 2008.

10. "Oil company wish list": Politics USA, July 31, 2008. "Is of economic significance": Interview with Jason Grumet. "We must end the age of oil": Barack Obama used this formulation on the campaign trail a number of times between the spring and summer of 2008. Among them: ABC News, August 5, 2008.

11. "We felt like a candidate": Interview with an ExxonMobil executive. Maurice Hinchey: Newhouse Newspapers, February 26, 2008.

12. *New York Times*, July 19, 2008.

13. ABC News, August 13, 2008.

14. ExxonMobil's effective corporate tax rate of 47 percent in 2009 was among the highest among large corporations.

15. Third presidential debate, *New York Times*, October 15, 2008.

16. All quotations, interviews with Bennett Freeman.

17. Interviews with individuals familiar with ExxonMobil's deliberations. Tillerson said later: Rex Tillerson's remarks at the Woodrow Wilson Center for International Scholars, January 8, 2009.

CHAPTER TWENTY-FOUR: "ARE WE OUT? OR IN?"

1. The portrait of Anton Smith is drawn from interviews with diplomats, nonprofit activists, human rights activists, business representatives, government officials, and other individuals in Equatorial Guinea and the United States. His views are also contained in the series of analytical cables Smith composed early in 2009.

2. U.S. companies invested $13 billion: Malabo to Washington, February 27, 2009.

3. "Torture Is Rife . . .": http://www.un.org/apps/news/story.asp?NewsID=28998&Cr=tortur e&Cr1=rapporteur.

4. Interviews with people familiar with Anton Smith's tour in Malabo, op. cit.

5. All quotations are from the linked series of cables, Malabo to Washington, February 27, 2009; March 3, 2009; March 10, 2009; March 12, 2009; March 30, 2009; and May 21, 2009.

6. Wald meeting, "Shared responsibility": Washington to Yaounde, June 24, 2005. Cindy Courville, "because of market forces": Yaounde to Washington, March 29, 2006. "our best ally": Malabo to Washington, November 23, 2006. "lengthy and effusive": Youande to Washington, October 6, 2006. These cables were released to the author in response to a Freedom of Information Act request.

7. All quotations, interview with Francisco Nugua.

8. Interviews with Africans and other sources familiar with Israel's defense consulting in the region. Israeli aid to the Joint Task Force in the Niger Delta is described in Lagos to Washington, February 23, 2009 (W).

9. Interview with an adviser to Equatorial Guinea familiar with the meetings.

10. Interviews with several advisers to Equatorial Guinea and other individuals familiar with the Israeli arrangements.

11. Ibid.

12. Interviews with individuals familiar with Equatorial Guinea's Washington strategy. Check for $4 million: Yaounde to Washington, February 22, 2006.

13. Background to the meeting, negotiations about visibility: Interviews with people familiar

with Equatorial Guinea's Washington strategy. All quotations from the private meeting with Condoleezza Rice: Washington to Yaounde, April 18, 2006.

14. Spokesmen for Marathon and Hess declined comment.

15. ExxonMobil Corp. Regional Oil Spill Response Plan—Offshore Operations, Appendix K, "Media," p. K-34.

16. Simeon L. Moats, ExxonMobil Corporation, "Business Practices & Transparency," Power-Point presentation, September 28, 2006.

17. Interview with an ExxonMobil executive.

18. Ibid.

19. Teodoro Obiang Nguema acknowledged forty-two children: Malabo to Washington, March 10, 2009. *Field of Dreams*, Gabriel Obiang's development plans: Interviews with individuals familiar with Gabriel's statements and thinking.

20. Global Witness, "Undue Diligence: How Banks Do Business with Corrupt Regimes," March 2009. United States Senate, Permanent Subcommittee on Investigations, Committee on Homeland Security and Governmental Affairs: "Keeping Foreign Corruption Out of the United States, Four Case Histories," Majority and Minority Staff Report, February 4, 2010.

21. Interview with a lawyer who worked with Teodoro Obiang.

22. *Lily Panayotti v. Sweetwater Management, Inc.*, Los Angeles County Superior Court, West District, SC-099588. *Veronique Guillem v. Sweetwater Management, Inc.*, Los Angeles County Superior Court, West District, SC-104747. *Dragan Deletic v. Sweetwater Management, Inc.* Los Angeles County Superior Court, West District, SC-104745.

23. All quotations from, "Keeping Foreign Corruption Out," op. cit.

24. This account is drawn primarily from interviews with individuals familiar with the February 17 attack. Also, BBC reporting on the events, February 17 through February 24, 2009.

25. Interview with an ExxonMobil executive.

26. Human Rights Watch, "Well Oiled: Oil and Human Rights in Equatorial Guinea," July 9, 2009.

27. Letter from Ken Cohen to Human Rights Watch, May 4, 2009, attached as an appendix to "Well Oiled," ibid.

CHAPTER TWENTY-FIVE: "IT'S NOT MY MONEY TO TITHE"

1. All of Tillerson's quotations are from his remarks at the Woodrow Wilson International Center for Scholars, January 8, 2009, recorded and transcribed.

2. ExxonMobil's review after 2006: Interviews with executives involved. Task Force: Interview with Elaine Kamarck (AO).

3. "We had determined . . . baggage": Interview with an ExxonMobil executive.

4. All Tillerson quotations from the Woodrow Wilson Center appearance, op. cit.

5. Interview with Carol Browner.

6. Ibid., and interviews with other Obama advisers involved. Abu Dhabi to Washington, April 13, 2009 (W).

7. Remarks by Rex Tillerson at Ford's Theatre, February 11, 2009, recorded and transcribed.

8. Ibid.

9. "Why wouldn't the administration want": Interview with an ExxonMobil executive. "He was just shopping the idea": Interview with a participant in the meeting.

10. Interview with a former Bush administration official involved in the transition briefings.

11. Canadian exports to the United States, Alberta reserve estimates cited by *Oil & Gas Journal:* From the U.S. Energy Information Administration.

12. For an excellent account of the issues and manufacturing processes involving sands oil, see Robert Kunzig, "Scraping Bottom," *National Geographic*, March 2009.

13. "Produce as much of this oil as you can," the map of North America prepared for Barack Obama: Interview with a Canadian official involved.

14. Canadian Broadcasting Corporation, February 17, 2009.

15. "The Canadian Oil Sands: Energy Security vs. Climate Change," Council on Foreign Relations Special Report No. 47, May 2009.
16. Interview with an industry lobbyist.
17. "don't want B.S.": Interview with an ExxonMobil executive.
18. Interview with a House staff member involved with energy legislation.
19. Interviews with multiple participants at the White Oak conference. Tony Kreindler and Sherri Stuewer: Interview with Tony Kreindler (AO). ExxonMobil did not respond to requests for comment.
20. Presentations by Walter McKibbin, Adele Morris, and Peter Wilcoxen, "The Economic Impact of Climate Change Reduction Strategies," Brookings Institution, June 8, 2009.
21. Hart Research Associates, "Energy and Climate Change Policy: A Survey Among American Voters Conducted September 2009 for U.S. Climate Task Force," slide deck obtained by the author.
22. Scene and all quotations, Clinton Global Initiative, September 23, 2009.
23. Scene and quotation, Economic Club of Washington, October 1, 2009.

CHAPTER TWENTY-SIX: "WE'RE CONFIDENT YOU CAN BOOK THE RESERVES"

1. The 2004 Iraq Reconstruction and Rehabilitation Fund appropriated by Congress provided $4 billion over three years to repair Iraq's oil and electricity infrastructure following the looting that occurred during the U.S.-led invasion. The amount was not nearly enough to account for the decay in Iraq's oil industry during Saddam Hussein's rule, but it was intended, a U.S. General Accounting Office official said in an interview, to "jump-start" the industry's reconstruction.
2. Interview with an American official involved. The account of the Al-Rashid auction and Richard Vierbuchen's role is from interviews with ExxonMobil executives, State Department and White House officials involved with Iraq oil policy during the Bush administration and the Obama administration, and contemporary news coverage of the auction. For the auction scene in the Al-Rashid ballroom, see ReutersVideo, June 29, 2009, among other sources: http://www.youtube.com/watch?v=5pBjh1iOnss.
3. Interview with an American official who was present at the meeting and who recorded the minister's comments in handwritten notes. Muttitt, *Fuel on the Fire*, p. 214, quotes from minutes obtained under a Freedom of Information Act request.
4. Author's interview.
5. The account of American oil policy after the invasion is drawn in part from extensive interviews with five American officials who worked on Iraqi oil policy at the Coalition Provisional Authority and its successors in Baghdad, as well as diaries and contemporary records describing their work.
6. All quotations, author's interviews.
7. "Transformative" Program: Baghdad to Washington, August 26, 2009 (W).
8. Bearing Point report: "Options for Developing a Long Term Sustainable Iraqi Oil Industry." Sector study presented to U.S.A.I.D., December 19, 2003. Quotations from interviews with an American official involved.
9. Interview with Philip J. Carroll.
10. Chevron: From contemporary records provided to the author. Norm Szydlowski: Author's interview.
11. Interview with Rob McKee.
12. Interview with Ashti A. Hawrami, K.R.G. minister of natural resources. "Oil and Gas Rights of Regions and Governorates," June 12, 2006, Kurdish Regional Government Web site: http://www.krg.org/articles/detail.asp?smap=02010100&lngnr=12&asnr=&anr=18704&rnr=223.
13. Interviews with an American official involved.
14. "We do not seek advice": E-mail from Jeanne Phillips to Ben Lando, October 11, 2007. Also,

e-mail from Hunt executive David McDonald after a meeting with State Department officials, cited by Senator Carl Levin in a letter to Bush administration national security adviser Stephen Hadley, quotes McDonald as concluding that the Bush administration had "no policy, neither for nor against." George W. Bush: CBS News, June 26, 2009.

15. The *New York Times* and the Norwegian newspaper *Dagens Naeringsliv* first reported on Peter Galbraith's Kurdish oil holdings in 2009. "I should have stated": "A Statement by Peter W. Galbraith," *New York Review of Books*, January 14, 2010.

16. Interview with an American official who worked with Meghan O'Sullivan during her Hess consultancy. Also, O'Sullivan's published online biography at Harvard University.

17. Baghdad to Washington, June 30, 2009 (W).

18. All quotations, notes from background briefings provided by ExxonMobil executives in Washington, D.C., obtained by the author.

19. Baghdad to Washington, August 26, 2009 (W).

20. Deutsche Bank: Muttitt, op. cit., p. 326. Also, Baghdad to Washington, November 15, 2009 and January 5, 2010 (W). Agreement details, crossed threshold: Interviews with State Department officials who have worked with ExxonMobil on its Iraq investments, background briefings by ExxonMobil executives, and materials published for investors by ExxonMobil.

21. Rob Franklin: "Exxon Exec Optimistic in Iraq Entry," *Iraq Oil Report*, February 4, 2010.

CHAPTER TWENTY-SEVEN: "ONE PLUS ONE HAS GOT TO EQUAL THREE"

1. Fort Worth *Star-Telegram*, April 2, 2006, and December 15, 2009.

2. Interview with Jack Randall.

3. Ibid., as well as disclosures made by both corporations to the Securities and Exchange Commission about their negotiations, Form S-4, ExxonMobil Corporation, February 1, 2010.

4. Form S-4, ibid., p. 96.

5. Interview with Jack Randall, Form S-4, ibid., pp. 39–49. Also, *TendersInfo*, December 16, 2009.

6. Rex Tillerson's testimony to Subcommittee on Energy and Environment, House Energy and Commerce Committee, January 20, 2010.

7. Interview with Jack Randall.

8. Ibid.

9. *TendersInfo*, op. cit.

10. Interviews with industry executives and researchers.

11. "Meeting Gas Supply and Demand in a Deregulated Environment," PowerPoint presentation, Gas Technology Institute.

12. Alan Greenspan's testimony to the House Energy and Commerce Committee, June 10, 2003.

13. Transcript from ExxonMobil's Analyst Meeting, March 9, 2011.

14. Interviews with executives and analysts familiar with ExxonMobil's natural gas strategy. Gold: *Wall Street Journal*, July 10, 2009.

15. *New York Times* (*Deal Book Blog*), December 14, 2009.

16. Conference call transcript, December 14, 2009, produced by ExxonMobil and filed with the Securities and Exchange Commission.

17. Form S-4, op. cit.; for consulting agreements and compensation, see p. 94.

18. *Business Wire*, February 15, 2011.

19. "Doubts About Shale Plays: Implications of ExxonMobil Acquisition of XTO Energy," Society of Petroleum Evaluation Engineers, February 2010.

20. *New York Times*, July 30, 2010.

21. Transcript from Chevron Corporation's 2011 Security Analyst Meeting, March 14, 2011.

22. Interview with Mark Gilman.

23. Barclays valuation: Form S-4, op. cit., p. 58. Interviews with former ExxonMobil executives and consultants.

24. Transcript from ExxonMobil's Analyst Meeting, March 9, 2011, op. cit.
25. *The Oil Drum*, February 22, 2010.
26. Memo obtained by the author. Also, interviews with former ExxonMobil executives familiar with the discussions after the XTO deal.
27. *Business Wire*, February 16, 2009.
28. *Financial Times*, February 16, 2011.
29. Promise to increase production: *Fortune*, July 28, 2006. Actual production in 2009: Exxon-Mobil 10-K, filed February 26, 2010.
30. Interviews by Ann O'Hanlon with eight students and World Affairs Council staff who participated.
31. "The Future of Energy and the Role of Corporate Citizenship," remarks by Rex Tillerson, April 19, 2010.
32. "Deep Water: The Gulf Oil Disaster and the Future of Offshore Drilling." Report to the President, National Commission on the B.P. Deepwater Horizon Oil Spill and Offshore Drilling, January 2011, p. 4.

CHAPTER TWENTY-EIGHT: "IT JUST HAPPENED"

1. Safina, *A Sea in Flames*, pp. 37–43, and "Deep Water: The Gulf Oil Disaster and the Future of Offshore Drilling," Report to the President, National Commission on the B.P. Deepwater Horizon Oil Spill and Offshore Drilling, January 2011, pp. 8–17.
2. "Deep Water," ibid., p. vii.
3. Ibid., p. 23.
4. Ibid., pp. 25–26.
5. Ibid., pp. 28–29.
6. Outer Continental Shelf Lands Act: Ibid., pp. xl and 58.
7. *U.S. Department of the Interior News*, November 20, 2008: http://www.boemre.gov/ooc/press/2008/pressDOI1120.htm
8. *New York Times*, June 20, 2010.
9. WEST Engineering Services, "Shear Ram Capabilities Study for U.S. Minerals Management Service," September 2004.
10. "Long-term consequences . . . moratorium": 2008 Obama campaign speech in Florida as quoted in *A Sea in Flames*, p. 46. "I don't agree . . . very advanced": "Remarks by the President in the Discussion on Jobs and the Economy in Charlotte, North Carolina," April, 2, 2010: http://www.whitehouse.gov/the-press-office/remarks-president-a-discussion-jobs-and-economy-charlotte-north-carolina.
11. "Gulf of Mexico Regional Oil Spill Response Plan," ExxonMobil developed by The Response Group, filed with M.M.S.
12. Hearing on "Drilling Down on America's Energy Future: Safety, Security, and Clean Energy," Subcommittee on Energy and Environment, June 15, 2010.
13. Testimony of Rex Tillerson, ibid.
14. Chairman Bart Stupak's Opening Statement, ibid.
15. Ibid.
16. "For us . . . cheaper": *Chief Executive*, April 1, 2001.
17. National Commission on the B.P. Deepwater Horizon, pp. 50–51, op. cit.
18. Bower, *Oil*, p. 350.
19. *New York Times*, May 8, 2010.
20. ABC News, May 27, 2010.
21. Safina, *A Sea in Flames*, p. 16.
22. National Commission on the B.P. Deepwater Horizon, op. cit.
23. Testimony of Rex Tillerson, Hearing on "Drilling Down on America's Energy Future" op. cit.
24. Notes from the Center for Strategic and International Studies dinner, provided to the author by a participant.

25. Reuters, January 6, 2011.
26. "Drilling Down on America's Energy Future," op. cit.
27. Ibid.
28. Stern, "Stern Review: The Economics of Climate Change," 2007 (final).
29. http://www.climatecentral.org/blogs/climate-in-context-bps-energy-outlook-shines-light-on-future-for-carbon/.
30. CNN, July 2, 2011.
31. *Baltimore Sun*, July 1, 2011.
32. Associated Press, July 8, 2011; *Jakarta Post*, August 9, 2011. "nothing to do": E-mail from spokesman David Eglinton to *Upstreamonline.com*.
33. U.S. Geological Survey Newsroom, July 23, 2008.
34. National Snow and Ice Data Center, Arctic Sea Ice News, nside.org.
35. "scary to utter": *New York Times*, August 30, 2011.
36. Ibid., August 19, 2011.
37. Negative $5.7 trillion: figure calculated using "Monthly Receipts, Outlays, and Deficit or Surplus, Fiscal Years 1981–2010," published by the Financial Management Service. Positive $493 billion: ExxonMobil 10-Ks, 2000 to 2010, and 10-Q for the second quarter of 2011. The merger closed November 30, 1999. The corporation's operating cash flow for December 1999 and the period from July 1 to August 5, 2011, is extrapolated from the average monthly rates of 2000 and the first six months of 2011, respectively.

Bibliography

BOOKS

Arbogast, Stephen V. *Resisting Corporate Corruption: Lessons in Practical Ethics from the Enron Wreckage.* Salem, MA: M & M Scrivener Press, 2008.

Auty, Richard. *Sustaining Development in Mineral Economies: The Resource Curse Thesis.* London: Routledge, 1993.

Azaiki, Steve S. *Oil, Gas and Life in Nigeria.* Ibadan, Nigeria: Y-Books, 2007.

Azevedo, Mario J. *The Roots of Violence: A History of War in Chad.* London: Routledge, 1998.

Baker, Peter, and Susan Glasser. *Kremlin Rising: Vladimir Putin's Russia and the End of Revolution.* New York: Scribner, 2005.

Bower, Tom. *Oil: Money, Politics, and Power in the 21st Century.* New York: Grand Central Publishing, 2009.

Browne, John. *Beyond Business: An Inspirational Memoir from a Visionary Leader.* London: Weidenfeld & Nicolson, 2010.

Bushell, Sharon, and Stan Jones. *The Spill: Personal Stories from the* Exxon Valdez *Disaster.* Kenmore, WA: Epicenter Press, 2009.

Chandrasekaran, Rajiv. *Imperial Life in the Emerald City: Inside Iraq's Green Zone.* New York: Knopf, 2006.

Chernow, Ron. *Titan: The Life of John D. Rockefeller, Sr.* New York: Random House, 1998.

Clarke, Duncan. *Crude Continent: The Struggle for Africa's Oil Prize.* London: Profile Books, 2008.

Coll, Steve. *Ghost Wars: The Secret History of the C.I.A., Afghanistan and Bin Laden, from the Soviet Invasion to September 10, 2001.* New York: Penguin Press, 2004.

DonPedro, Ibiba. *Oil in the Water: Crude Power and Militancy in the Niger Delta.* Lagos: Forward Communications, 2006.

Dressler, Andrew, and Edward A. Parson. *The Science and Politics of Global Climate Change: A Guide to the Debate.* Cambridge: Cambridge University Press, 2006.

Feith, Douglas J. *War and Decision: Inside the Pentagon at the Dawn of the War on Terrorism.* New York: HarperCollins, 2008.

Flint, Julie, and Alex de Waal. *Darfur: A New History of a Long War.* London: Zed Books, 2008.

Friedman, Thomas L. *Hot, Flat, and Crowded: Why We Need a Green Revolution—And How It Can Renew America.* New York: Farrar, Straus and Giroux, 2008.

Gallagher, Kelly Sims. *Acting in Time on Energy Policy.* Washington, D.C.: Brookings Institution Press, 2009.

Gellman, Barton. *Angler: The Cheney Vice Presidency.* New York: Penguin Press, 2008.

Ghazvinian, John. *Untapped: The Scramble for Africa's Oil.* New York: Harcourt, 2007.

Goldman, Marshall I. *Petrostate: Putin, Power, and the New Russia.* New York: Oxford University Press, 2008.

Gore, Al. *Earth in the Balance: Ecology and the Human Spirit.* Boston: Houghton Mifflin, 1992.

Hale, William E., Robert H. Davis, and Mike Long. *One Hundred and Twenty-Five Years of History: ExxonMobil.* Irving, TX: ExxonMobil Corp., 2007.

Halperin, Mark, and John F. Harris. *The Way to Win: Taking the White House in 2008.* New York: Random House, 2006.

Hoffman, David E. *The Oligarchs: Wealth and Power in the New Russia.* New York: Public Affairs, 2002.

Karl, Terry Lynn. *The Paradox of Plenty: Oil Booms and Petro-States.* Berkeley: University of California Press, 1997.

Keeble, John. *Out of the Channel: The Exxon Valdez Oil Spill in the Prince William Sound.* Spokane: Eastern Washington University Press, 1999.

Klitgaard, Robert. *Tropical Gangsters: One Man's Experience with Development and Decadence in Deepest Africa.* New York: Basic Books, 1991.

Kolbert, Elizabeth. *Field Notes from a Catastrophe: Man, Nature, and Climate Change.* New York: Bloomsbury, 2006.

LeVine, Steve. *The Oil and the Glory: The Pursuit of Empire and Fortune on the Caspian Sea.* New York: Random House, 2007.

Maass, Peter. *Crude World: The Violent Twilight of Oil.* New York: Knopf, 2009.

Margonelli, Lisa. *Oil on the Brain: Adventures from the Pump to the Pipeline.* New York: Doubleday, 2007.

Mearsheimer, John J. *The Tragedy of Great Power Politics.* New York: Norton, 2001.

M.H.H., Greg. *Fuel on the Fire: Oil and Politics in Occupied Iraq.* London: The Bodley Head, 2011.

Miller, T. Christian. *Blood Money: Wasted Billions, Lost Lives, and Corporate Greed in Iraq.* New York: Little, Brown, 2006.

Mooney, Chris. *The Republican War on Science.* New York: Basic Books, 2005.

Ott, Riki. *Not One Drop: Betrayal and Courage in the Wake of the Exxon Valdez Oil Spill.* White River Junction, VT: Chelsea Green, 2008.

Patton, Patty Sue. *Eternal Threads: A Journey Towards Discovery.* Bloomington, IN: AuthorHouse, 2005.

Reid, Anthony, ed. *Verandah of Violence: The Background to the Aceh Problem.* Singapore: Singapore University Press (in association with the University of Washington Press), 2006.

Ricks, Thomas E. *Fiasco: The American Military Adventure in Iraq.* New York: Penguin Press, 2006.

Roberts, Adam. *The Wonga Coup: Guns, Thugs, and a Ruthless Determination to Create Mayhem in an Oil-Rich Corner of Africa.* New York: Public Affairs, 2006.

Roberts, Paul. *The End of Oil: On the Edge of a Perilous New World.* New York: Mariner Books, 2004.

Roderick, Jack. *Crude Dreams: A Personal History of Oil & Politics in Alaska.* Fairbanks, AK: Epicenter Press, 1997.

Safina, Carl. *A Sea in Flames: The Deepwater Horizon Oil Blowout.* New York: Crown, 2011.

Shaxson, Nicholas. *Poisoned Wells: The Dirty Politics of African Oil.* New York: Palgrave Macmillan, 2007.

Simmons, Matthew R. *Twilight in the Desert: The Coming Saudi Oil Shock and the World Economy.* Hoboken, NJ: Wiley, 2005.

Suskind, Ron. *The Price of Loyalty.* New York: Simon & Schuster, 2004.

Tarbell, Ida M. *The History of the Standard Oil Company.* New York: McClure, Phillips, 1904.

Taylor, Jean Gelman. *Indonesia: Peoples and Histories.* New Haven, CT: Yale University Press, 2004.

Woodward, Bob. *Plan of Attack.* New York: Simon & Schuster, 2004.

Yergin, Daniel. *The Prize: The Epic Quest for Oil, Money & Power.* New York: Free Press, 2008.

————. *The Quest: Energy, Security, and the Remaking of the Modern World.* New York: Penguin Press, 2011.

DOCUMENTS

Alaska Resources Library and Information Services: Transcripts of telephone recordings from the Alyeska Emergency Center, from 4:57 a.m. on March 24, 1989, and on March 26, 1989, and other selected documents.

"Excerpts from the May 31, 2000, ExxonMobil Annual General Meeting." Transcript produced by Campaign ExxonMobil.

ExxonMobil Transcripts: Transcript of Exxon-Mobil merger press conference, December 1, 1998, and conference call with Emil Jacobs and J. Craig Venter, July 14, 2009.

Exxon Valdez Oil Spill Trustee Council: 2009 Status Report and other selected documents.

Fair Disclosure: ExxonMobil Analyst Meeting Transcripts, 2000–2011.

"Hydrogen and Fuel Cells: Opportunities and Challenges." Slides presented in Brussels, Belgium, November 25, 2003, author's files and www.exxonmobil.com/files/PA/Europe/Blewis final_Amcham.pdf.

"Operations Integrity Management System." www.exxonmobil.com, examined and typed, June 24, 2010.

Records of the Permanent Subcommittee on Investigations, Committee on Governmental Affairs, United States Senate, July 15, 2004, investigation of Riggs National Bank.

Records of the Securities and Exchange Commission, EDGAR database. www.sec.gov/edgar .shtml.

Securities and Exchange Commission. "Concept Release on Possible Revisions to the Disclosure Requirements Relating to Oil and Gas Reserves," December 12, 2007.

United Nations Human Development Reports, 1993–2010. http://hdr.undp.org/en/statistics. National Security Archive, Electronic Briefing Books, "Future of Iraq Project." http://www .gwu.edu/~nsarchiv/NSAEBB/NSAEBB198/index.htm.

www.exxonmobil.com, including Exxon, Mobil, and ExxonMobil Annual Reports, 1997–2010, and ExxonMobil's Chad/Cameroon Development Project Updates.

JOURNAL ARTICLES, REPORTS, AND MANUSCRIPTS

"Aceh Under Martial Law: Inside the Secret War." Human Rights Watch, December 2003.

"All the Presidents' Men: The Devastating Story of Oil and Banking in Angola's Privatized War." Global Witness, January 2002.

"Balancing Natural Gas Policy: Fueling the Demands of a Growing Economy." National Petroleum Council, September 2003. http://fossil.energy.gov/programs/oilgas/publications/npc/03gas study/npcgas03_preface.pdf.

"Carbon Dioxide and Climate: A Scientific Assessment." Report of an Ad Hoc Study Group on Carbon Dioxide and Climate, July 23–27, 1979. National Academy of Sciences, Washington, D.C.

"Chad-Cameroon Petroleum Development and Pipeline Project: Overview." World Bank, December 2006.

Cohen, Arile, and Gerald P. O'Driscoll Jr. "The Road to Economic Prosperity for a Post–Saddam Iraq." Heritage Foundation Backgrounder no. 1594, September 24, 2002.

"Crude Awakening: The Role of the Oil and Banking Industries in Angola's Civil War and the Plunder of State Assets, A." Global Witness, 1999.

Davis, Stephen. "The Potential for Peace and Reconciliation in the Niger Delta." Coventry Cathedral, February 2009.

"Deep Water: The Gulf Oil Disaster and the Future of Offshore Drilling." Report to the President, National Commission on the B.P. *Deepwater Horizon* Oil Spill and Offshore Drilling, January 2011.

"Defense Management: Actions Needed to Address Stakeholder Concerns, Improve Interagency Collaboration, and Determine Full Costs Associated with the U.S. Africa Command." General Accounting Office, February 2009.

"Deposition of Philip Cooney." House Committee on Oversight and Government Reform, March 12, 2007.

"Force Structure: Preliminary Observations on the Progress and Challenges Associated with Establishing the U.S. Africa Command." General Accounting Office, July 2008.

Freudenburg, William R. "Seeding Science, Courting Conclusions: Reexamining the Intersection of Science, Corporate Cash, and the Law." *Sociological Forum* 20, no. 1 (March 2005).

Gary, Ian, and Terry Lynn Karl. "Bottom of the Barrel: Africa's Oil Boom and the Poor." Catholic Relief Services, June 2003.

———, and Nikki Reisch. "Chad's Oil: Miracle or Mirage; Following the Money in Africa's Newest Petro-State." Catholic Relief Services, February 2005.

"The Global Energy Market: Comprehensive Strategies to Meet Geopolitical and Financial Risks—The G8, Energy Security, and Global Climate Issues." Baker Institute, Houston, Report no. 37, July 2008.

"Grounding of U.S. Tankship *Exxon Valdez* on Bligh Reef, Prince William Sound Near Valdez, AK, March 24, 1989." National Transportation Safety Board, Report no. MAR-90-04.

Hansen et al., "Global Climate Changes as Forecast by Goddard Institute for Space Studies Three-Dimensional Model." *Journal of Geophysical Research* 93, no. D8 (August 20, 1988) pps. 9341–64.

Hanson, Stephanie. "China, Africa and Oil." Council on Foreign Relations Backgrounder, June 6, 2008.

Hartley, Peter R., and Kenneth B. Medlock. "Climate Policy and Energy Security: Two Sides of the Same Coin?" Baker Institute, Houston, November 19, 2008.

"The Hydrogen Economy: Opportunities, Costs, Barriers and R&D Needs." National Research Council and the National Academy of Engineering, National Academy of Sciences Press, 2004.

"Interagency Support on Conflict Assessment and Mission Performance Planning for Chad." March 20, 2006.

"IPCC Second Assessment: Climate Change 1995." Intergovernmental Panel on Climate Change, and subsequent assessments.

"Keeping Foreign Corruption Out of the United States: Four Case Histories." Majority and Minority Staff Report, United States Senate, Permanent Subcommittee on Investigations, Committee on Homeland Security and Governmental Affairs, February 4, 2010.

Levi, Michael A. "The Canadian Oil Sands: Energy Security vs. Climate Change." Council on Foreign Relations Special Report no. 47, May 2009.

"Liquefied Natural Gas: Understanding the Basic Facts." Department of Energy, August 2005.

Malik, Rajiv. "Three Days in N'djamena," *Foreign Service Journal*, May 2008.

McCright, Aaron M., and Riley E. Dunlap. "Challenging Global Warming as a Social Problem: An Analysis of the Conservative Movement's Counter-Claims." *Social Problems* 47, no. 4 (November 2000), pp. 499–522.

"More Fight—Less Fuel." Report of the Defense Science Board Task Force on Department of Defense Energy Strategy, February 2008.

Morse, Edward. "Low and Behold: Making the Most of Cheap Oil." *Foreign Affairs*, September–October 2009.

"Nigeria: Petroleum, Pollution and Poverty in the Niger Delta." Amnesty International, 2009.

Oreskes, Naomi. "Behind the Ivory Tower: The Scientific Consensus on Climate Change." *Science*, December 2004.

Pacala, Stephen, and Robert H. Socolow. "Stabilization Wedges: Solving the Climate Problem for the Next 50 Years with Current Technologies." *Science*, August 2004.

Patria, Nezar. "Islam and Nationalism in the Free Aceh Movement, 1976–2005." master's thesis, International History Department, London School of Economics, London.

Ploch, Lauren. "Africa Command: U.S. Strategic Interests and the Role of the U.S. Military in Africa." Congressional Research Service Report for Congress, May 16, 2007.

"Political Interference with Climate Change Science Under the Bush Administration." House Committee on Oversight and Government Reform, December 2007.

"Responsible Actions: A Plan for Alberta's Oil Sands." Government of Alberta Treasury Board, February 2009.

Schierow, Linda-Jo, and Margaret Mikyung Lee. "Phthalates in Plastics and Possible Human Health Effects." Congressional Research Service Report for Congress, updated July 29, 2008.

Shurtleff, Len. "A Foreign Service Murder." *Foreign Service Journal*, October 2007.

Skjærseth, Jon Birger. "ExxonMobil: Tiger or Turtle on Social Responsibility?" The Fridtjof Nansen Institute, Lysaker, Norway, July 2003.

"Smoke, Mirrors and Hot Air." Union of Concerned Scientists, 2007.

Socolow, Robert H., and Stephen Pacala. "A Plan to Keep Carbon in Check." *Scientific American*, September 2006.

Stern, Nicholas. "Stern Review: The Economics of Climate Change." Cambridge: Cambridge University Press, 2007.

"Toward an Angola Strategy: Prioritizing U.S.-Angola Relations." Council on Foreign Relations, Report of an Independent Commission Sponsored by the Center for Preventative Action, 2007.

"Undue Diligence: How Banks Do Business with Corrupt Regimes." Global Witness, March 2009.

Venzke, Ben. "Saudi Compound Bombings." IntelCenter, May 16, 2003.

"Well Oiled: Oil and Human Rights in Equatorial Guinea." Human Rights Watch, July 9, 2009.

Zycher, Benjamin. "Some Historical Lessons from the World Oil Market." Briefing prepared for the Under Secretary of Defense for Policy, April 28, 2005.

COURT DOCUMENTS

Baker v. Exxon, No. 04.35182. United States Court of Appeals for the Ninth Circuit, CV. 89-0095-HRH.

Carey Brothers Construction v. Teodoro Nguema Obiang, Los Angeles County Superior Court, West District, SC-075147.

David Fitzpatrick v. Kenneth P. Cohen, 10-CV-00054-GZS, United States District Court, District of Maine.

Dragan Deletic v. Sweetwater Management, Inc., Los Angeles County Superior Court, West District, SC-104745.

Equal Employment Opportunity Commission v. ExxonMobil Corporation, 30-6-cv-1732-K, United States District Court, Northern District of Texas.

Exxon Corp. v. Breezevale Ltd., 05-98-02050-CV, Court of Appeals of Texas, Fifth District, Dallas.

ExxonMobil Corp. v. United States of America, 3-00-cv-00815 M, United States District Court, Northern District of Texas.

ExxonMobil Corp. v. United States of America, 302-cv-0210 D, United States District Court, Northern District of Texas.

Frontera Resources Azerbaijan Corp. v. State Oil Company of the Azerbaijan Republic, 06 Civ. 1125 (RJH), United States District Court, Southern District of New York.

In Re Exxon Valdez *Oil Spill*, United States District Court, District of Alaska, A-89-095 (Civil Consolidated).

Jeff Alban et al. v. ExxonMobil Corp., 03-C-06-010193 2OT, Baltimore County Circuit Court, Maryland.

John Doe I et al. v. ExxonMobil Corporation et al., 01-cv-01357, United States District Court for the District of Columbia.

Larry Bowoto et al. v. Chevron Corp. et al., C 99-02506 SI, United States District Court, Northern District of California.

Lily Panayotti v. Sweetwater Management, Inc., Los Angeles County Superior Court, West District, SC-099588.

Mobil Cerro Negro, Ltd. v. PDVSA Cerro Negro, S.A., 07-CV-11590, United States District Court, Southern District of New York.

Mobil Cerro Negro, Ltd. v. Petróleos de Venezuela, S.A., Claim No. 2008, Folio 61, High Court of Justice, Queen's Bench Division, Commercial Court, Great Britain.

Pacific National Construction, Inc. v. Teodoro Nguema Obiang, Los Angeles County Superior Court, West District, SC-083725.

State of Alaska v. Exxon Corporation, and *United States of America v. Exxon Corporation*, A91-082 CIV and A91-083 CIV.

Suellen Everett Lundquist v. Sweetwater Management, Inc., Los Angeles County Superior Court, West District, SC-097988.

United States of America v. Arthur D. Seale, United States Court of Appeals, Third Circuit, 20 F.3d 1279.

United States of America v. Irene J. Seale, United States Court of Appeals, Third Circuit, 20 F.3d 1279.

United States of America v. Jason Edward Steph, United States District Court, Southern District of Texas, H-07-307.

Veronique Guillem v. Sweetwater Management, Inc., Los Angeles County Superior Court, West District, SC-104747.

Walter B. Meyer et al. v. Teodoro Nguema Obiang, Los Angeles County Superior Court, West District, SC-073545.

Willie J. Garry v. ExxonMobil Corp. et al., 03-0791 I/3, United States District Court, Eastern District of Louisiana.

Index